On Three Levels
Micro-, Meso-, and Macro-Approaches in Physics

NATO ASI Series

Advanced Science Institutes Series

A series presenting the results of activities sponsored by the NATO Science Committee, which aims at the dissemination of advanced scientific and technological knowledge, with a view to strengthening links between scientific communities.

The series is published by an international board of publishers in conjunction with the NATO Scientific Affairs Division

A	Life Sciences	Plenum Publishing Corporation
B	Physics	New York and London
C	Mathematical and Physical Sciences	Kluwer Academic Publishers
D	Behavioral and Social Sciences	Dordrecht, Boston, and London
E	Applied Sciences	
F	Computer and Systems Sciences	Springer-Verlag
G	Ecological Sciences	Berlin, Heidelberg, New York, London,
H	Cell Biology	Paris, Tokyo, Hong Kong, and Barcelona
I	Global Environmental Change	

Recent Volumes in this Series

Volume 317 —Solid State Lasers: New Developments and Applications
   edited by Massimo Inguscio and Richard Wallenstein

Volume 318 —Relativistic and Electron Correlation Effects in Molecules and Solids
   edited by G. L. Malli

Volume 319 —Statics and Dynamics of Alloy Phase Transformations
   edited by Patrice E. A. Turchi and Antonios Gonis

Volume 320 —Singular Limits of Dispersive Waves
   edited by N. M. Ercolani, I. R. Gabitov, C. D. Levermore, and D. Serre

Volume 321 —Topics in Atomic and Nuclear Collisions
   edited by B. Remaud, A. Calboreanu, and V. Zoran

Volume 322 —Techniques and Concepts of High-Energy Physics VII
   edited by Thomas Ferbel

Volume 323 —Soft Order in Physical Systems
   edited by Y. Rabin and R. Bruinsma

Volume 324 —On Three Levels: Micro-, Meso-, and Macro-Approaches in Physics
   edited by Mark Fannes, Christian Maes, and André Verbeure

Series B: Physics

On Three Levels

Micro-, Meso-, and Macro-Approaches in Physics

Edited by

Mark Fannes

Christian Maes

and

André Verbeure

Katholieke Universiteit Leuven
Leuven, Belgium

Plenum Press
New York and London
Published in cooperation with NATO Scientific Affairs Division

Proceedings of a NATO Advanced Research Workshop on
On Three Levels: Micro-, Meso-, and Macro-Approaches in Physics,
held July 19–23, 1993,
in Leuven, Belgium

NATO-PCO-DATA BASE

The electronic index to the NATO ASI Series provides full bibliographical references (with keywords and/or abstracts) to more than 30,000 contributions from international scientists published in all sections of the NATO ASI Series. Access to the NATO-PCO-DATA BASE is possible in two ways:

—via online FILE 128 (NATO-PCO-DATA BASE) hosted by ESRIN, Via Galileo Galilei, I-00044 Frascati, Italy

—via CD-ROM "NATO Science and Technology Disk" with user-friendly retrieval software in English, French, and German (©WTV GmbH and DATAWARE Technologies, Inc. 1989). The CD-ROM also contains the AGARD Aerospace Database.

The CD-ROM can be ordered through any member of the Board of Publishers or through NATO-PCO, Overijse, Belgium.


```
          Library of Congress Cataloging-in-Publication Data
_____

On three levels : micro-, meso-, and macro-approaches in physics /
    edited by Mark Fannes, Christian Maes, and André Verbeure.
         p.   cm. -- (NATO ASI series. Series B, Physics ; v. 324)
      "Published in cooperation with NATO Scientific Affairs Division."
      "Proceedings of a NATO Advanced Research Workshop on On Three
   Levels: Micro-, Meso-, and Macro-Approaches in Physics, held July
   19-23, 1993, in Leuven, Belgium"--T.p. verso.
      Includes bibliographical references and index.
      ISBN 0-306-44704-5
      1. Mathematical physics--Congresses.   I. Fannes, M.  II. Maes,
   Christian.  III. Verbeure, André.  IV. North Atlantic Treaty
   Organization.  Scientific Affairs Division.  V. NATO Advanced
   Research Workshop on On Three Levels: Micro-, Meso-, and Macro
   -Approaches in Physics (1993 : Louvain, Belgium)  VI. Series.
   QC19.2.O5  1994
   530--dc20                                                  94-7389
                                                                  CIP
_____
```


ISBN 0-306-44704-5

©1994 Plenum Press, New York
A Division of Plenum Publishing Corporation
233 Spring Street, New York, N.Y. 10013

All rights reserved

No part of this book may be reproduced, stored in a retrieval system, or transmitted in any form or by any means, electronic, mechanical, photocopying, microfilming, recording, or otherwise, without written permission from the Publisher

Printed in the United States of America

PREFACE

This volume contains the proceedings of a five-day NATO Advanced Research Workshop "On Three Levels, the mathematical physics of micro-, meso-, and macro-phenomena," conducted from July 19 to 23 in Leuven, Belgium.

The main purpose of the workshop was to bring together and to confront where relevant, classical and quantum approaches in the rigorous study of the relation between the various levels of physical description. The reader will find here discussions on a variety of topics involving a broad range of scales.

For the micro-level, contributions are presented on models of reaction-diffusion processes, quantum groups and quantum spin systems. The reports on quantum disorder, the quantum Hall effect, semi-classical approaches of wave mechanics and the random Schrödinger equation can be situated on the meso-level. Discussions on macroscopic quantum effects and large scale fluctuations are dealing with the macroscopic level of description.

These three levels are however not independent and emphasis is put on relating these scales of description. This is especially the case for the contributions on kinetic and hydrodynamical limits, the discussions on large deviations and the strong and weak coupling limits.

The advisory board was composed of J.L. Lebowitz, J.T. Lewis and E.H. Lieb. The organizing committee was formed by Ph.A. Martin, G.L. Sewell, E.R. Speer and A. Verbeure. We thank all of them, together with the many participants for their help in making this workshop successful. The proceedings were prepared with the help of Marino Broidioi, Jacek Miękisz, Bruno Momont, Urban Studer and Koen Vande Velde. Many thanks to all of them and also to Christine Detroye and Anita Raets for their secretarial help.

We also wish to acknowledge the financial support from NATO and the ISF, New York and the logistic support from the KULeuven. Finally, we express our sincere recognition to everybody having contributed to the realization of this work.

The local organizers,

M. Fannes, C. Maes, A. Verbeure

Instituut voor Theoretische Fysica
Celestijnenlaan 200D, B-3001 Leuven, Belgium

February 14, 1994

CONTENTS

Quantum kinetic equations 1
   H. Spohn

Microscopic derivation of hydrodynamics with phase transition in a plasma model 11
   G.L. Sewell

Ferromagnetism in itinerant electron systems: rigorous examples from flat-band Hubbard models 23
   H. Tasaki

Homogeneity in the ground state of the two-dimensional Falicov-Kimball model   35
   T. Kennedy

Diffusive limit of the asymmetric simple exclusion: The Navier-Stokes correction   43
   R. Esposito, R. Marra, H.T. Yau

Weak coupling limit: Feynman diagrams 53
   L.J. Landau

Limit laws for recurrence times in expanding maps of an interval 73
   P. Collet

Stochastic geometric aspects of some quantum spin chains 81
   B. Nachtergaele

The two species totally asymmetric simple exclusion process 91
   E.R. Speer

How to reconstruct a heat bath 103
   B. Kümmerer

Interacting particle systems on non-commutative spaces 115
   T. Matsui

Non-self-averaging effects in sums of random variables, spin glasses, random maps and random walks 125
   B. Derrida

Quantum adiabatic evolution 139
   A. Joye and C.-E. Pfister

Semi-classical inelastic S-matrix for one-dimensional N-states systems 149
  Ph. A. Martin and G. Nenciu

Gibbsian versus non-Gibbsian measures: some results and some questions in
    renormalization group theory and stochastic dynamics 155
  A.C.D. van Enter, R. Fernández and A.D. Sokal

Stabilities and instabilities in classical lattice gas models without periodic
    ground states 161
  J. Miękisz

One-dimensional anomaly of the Fermi surface 165
  G. Gallavotti

Quantum fluctuation limit: Examples from solid state physics 175
  A. Verbeure and V.A. Zagrebnov

Large deviations and the thermodynamic formalism: A new proof of the
    equivalence of ensembles 183
  J.T. Lewis, C.-E. Pfister and W.G. Sullivan

Finitely correlated pure states 193
  R.F. Werner

Stationary states of Hamiltonian systems with noise 203
  J. Fritz

Stochastic regularization of coherent-state path integrals and quantum Hall effect 215
  R. Alicki

An ADE-$\mathcal{O}$ classification of minimal incompressible quantum Hall fluids 225
  J. Fröhlich, U. Studer, E. Thiran

Simple random walks: New developments 233
  S.B. Shlosman

Symmmetry breaking and long range order in Heisenberg antiferromagnets 239
  T. Koma

Integrable $s = 1/2$ quantum spin chains with short-range exchange 245
  V.I. Inozemtsev

Who is afraid of Griffiths' singularities? 253
  A. Klein

Micro spectral properties of crystals and their band structure 259
  P. Kurasov and B. Pavlov

Stochastically forced Burgers equation 265
  L. Bertini, N. Cancrini, G. Jona-Lasino

Glauber evolution for Kac potentials. Analysis of critical fluctuations:
    convergence to a nonlinear stochastic PDE 271
  B. Rüdiger

Local structure of interfaces in a Kawasaki + Glauber particle model 275
  G. Giacomin

Quantum chaos, fractal spectra and atomic stabilization 281
  G. Casati

Statistical properties of random banded matrices: Analytical results 289
  Y.V. Fyodorov and A. Mirlin

On the Wulff construction as a problem of equivalence of statistical ensembles . 295
  S. Miracle-Sole and J. Ruiz

Scaling profiles of a spreading drop from Langevin or Monte-Carlo dynamics ... 303
  F. Dunlop and M. Plapp

Rigorous calculation of collective excitations in a mean field model 309
  B. Momont

The fluid-dynamical limit for the BBGKY hierarchy of a discrete velocity model 315
  V. Gorunovich

Isospectral deformation of discrete random Laplacians 321
  O. Knill

Scattering and the role of operators in Bohmian mechanics 331
  M. Daumer, D. Dürr, S. Goldstein, N. Zanghí

Towards the Euclidean formulation of quantum statistical mechanics 339
  R. Gielerak, L. Jakóbczyk, R. Olkiewicz

Large deviations in the spherical model 347
  A.E. Patrick

Entropy density and the split property 355
  H. Narnhofer

Perturbations of quantum canonical relations and Q-independence 361
  W.A. Majewski and M. Marciniak

The spectrum of the spin-boson model 367
  M. Hübner and H. Spohn

Some results on the projected two-dimensional Ising model 373
  J. Lőrinczi

The second virial coefficient for quantum-mechanical sticky spheres 381
  M.D. Penrose and O. Penrose

Bethe-Ansatz solution of a modified SU(3)-XXZ model 385
  H. Grosse and E. Raschhofer

The quantum mean field state as a limit of canonical states: Maxwell-Boltzmann
    statistics 393
  N. Angelescu

Multifractal properties of discrete stochastic mappings 399
    U. Behn, J.L. van Hemmen, R. Kühn, A. Lange, V.A. Zagrebnov

Gibbs states of the Chern-Simons charged particle system in the mean-field
    type limit 405
    W.I. Skrypnik

Rigorous Bethe Ansatz for the nonlinear Schrödinger model 409
    T.C. Dorlas

Flat nonregular states on Weyl algebras 417
    F. Acerbi

Second virial coefficient for one-dimensional systems 423
    P. Kurasov, V. Kurasov, B. Pavlov

Self-adjointness and the existence of deterministic trajectories in quantum
    theory 429
    K. Berndl, D. Dürr, S. Goldstein, N. Zanghí

An application of the Maes-Shlosman constructive criteria 435
    H. de Jong

Brownian trapping with grouped traps 441
    L.V. Bogachev, A.M. Berezhkovskii, Yu. A. Makhnovskii

Local thermodynamic equilibrium and continuous media: A programme 445
    H. Roos and R.N. Sen

Stochastic model of a quantum diffusion-reaction process 451
    V.P. Belavkin

Crystalisation of itinerant electrons 457
    A. Messager

Molecular chain: dynamical variables, quantization and statistical mechanics ... 467
    R.F. Alvarez-Estrada

Quasiparticle's spin and fractional statistics in the fractional quantum Hall effect   471
    D. Li

Participants 477

Index 481

QUANTUM KINETIC EQUATIONS

H. Spohn

Theoretische Physik
Ludwig-Maximilians-Universität München
Theresienstr. 37, D-80333 München, Germany

INTRODUCTION

In a very practical sense transport equations are at the heart of understanding the dynamical properties of condensed and fluid matter. Only in rare exceptions one can deal with the full microscopic evolution of particles and one is thus forced to use transport equations as an indispensable intermediary towards experimentally observed phenomena. In such a situation it is clear that there is a considerable amount of physical modelling on the level of transport equations and it is difficult, if not impossible, to establish their precise link to the microscopic world. Given the unquestionable empirical success of the whole procedure, there is little room for doubt. Nevertheless one would like to have, at least in a few prototypical cases, a *complete* understanding of how the microscopic level and the level of transport equations are connected.

For interacting particles governed by the laws of classical mechanics there has been some modest progress which is explained in [1]. Very often I was asked how these results transcribe to the quantum world. The Three Levels Conference at Leuven is certainly the appropriate occasion to collect some answers. However, I should warn the reader from the outset: I have no breathtaking results to report. Rather I will outline the general framework and embed therein existing work. Even this seems to be a worthwhile undertaking. Much of the theoretical literature on the derivation of transport equations is so much engaged in developing general formalisms that it is sometimes difficult to detect a clear question, even less the answer.

I decided to discuss a few kinetic equations for interacting quantum systems. Some related models will be mentioned in the Outlook. The notion "kinetic" is rather loose in general, but intended to have a sharp meaning here. Kinetic means that, after a suitable limiting procedure, the state in a spatial region of a size such that it contains only a finite number of particles is that of an ideal gas (either Fermi, Bose, or classical). A static illustration is provided by the Thomas-Fermi theory of neutral atoms. There the nucleus is fixed and the number, N, of electrons tends to infinity. If one focuses attention on a small region containing a roughly constant number of electrons (of linear dimension $N^{-2/3}$ in atomic units, cf. [2,3]), then as $N \to \infty$ one obtains a portion of an infinitely extended ideal Fermi gas with constant density. Of course, this fact is not in contradiction with having on a large scale (of linear dimension $N^{-1/3}$ in atomic units) a spatially varying electron density as given by the solution of the Thomas-Fermi variational problem. In the same spirit for a kinetic limit we expect a space-time variation on a scale large compared to interparticle distances and the goal is to obtain the equation which governs its behavior. Kinetic limits are to be contrasted with hydrodynamic limits where the local state is a thermal equilibrium (Gibbs) state with non zero interactions.

Let me introduce the standard model of particles interacting through a short range, smooth, central potential V. Since their mass, m, plays no particular role, we set $m = 1$. Equally $\hbar = 1$. It is convenient to keep track of the spatial extent by restricting the particles to the volume $\Lambda \subset \mathsf{R}^3$. (Eventually we will set $\Lambda = \mathsf{R}^3$ for simplicity and require the density to decrease sufficiently rapidly towards infinity.) We allow for a variable number of particles. The appropriate Hilbert space is then the Fock space $\mathcal{F}$ with single particle space $L^2(\Lambda, d^3x)$ and the Hamiltonian is given by

$$H = \frac{1}{2}\int_\Lambda d^3x \nabla a^+(\mathbf{x}) \cdot \nabla a(\mathbf{x}) + \frac{1}{2}\int_\Lambda d^3x \int_\Lambda d^3y\, a^+(\mathbf{x})a^+(\mathbf{y})V(\mathbf{x}-\mathbf{y})a(\mathbf{y})a(\mathbf{x}). \quad (1)$$

Here $a(\mathbf{x})$, $a^+(\mathbf{x})$ are the annihilation and creation operators of a Bose, or Fermi, field with commutation relations $[a(\mathbf{x}), a^+(\mathbf{x}')] = \delta(\mathbf{x}-\mathbf{x}')$, or anticommutation relations $\{a(\mathbf{x}), a^+(\mathbf{x}')\} = \delta(\mathbf{x}-\mathbf{x}')$. In the kinetic limit the local state will be noninteracting. Therefore we might as well impose such a state initially. Thus at $t = 0$ the state, ρ, of the system is quasi-free with

$$\mathrm{tr}\left[\rho\, a^+(\mathbf{x}_m)\ldots a^+(\mathbf{x}_1)\,a(\mathbf{y}_1)\ldots a(\mathbf{y}_n)\right] = \delta_{nm}\,\mathrm{per}^{\det}\left\{<\mathbf{x}_i |R| \mathbf{y}_j> \; i,j = 1,\ldots,m\right\}. \quad (2)$$

R is a self-adjoint operator on the one-particle space with $R \geq 0$ and for fermions also $R \leq 1$. In addition we require $\mathrm{tr}\,R < \infty$, i.e. $\mathrm{tr}[\rho N] < \infty$.

Clearly, a kinetic equation can be valid only under particular conditions. To discuss them is the task of the following Section.

SCALES

The applicability of a kinetic equation relies on a good separation between the scale of the typical interparticle distance and the "macroscopic" scale, i.e. the scale on which the solution to the macroscopic equation is varying. It is convenient to introduce a parameter, which I like to call ε, which measures the ratio of microscale to macroscale. For a concrete physical system ε has a given small value. The validity of a kinetic equation will be proved in the idealized limit $\varepsilon \to 0$, i.e. limit of infinite scale separation.

For a classical system position and momentum can be rescaled separately. This is no longer so quantum mechanically and classical systems equivalent under scale transformation become distinct upon naive quantization. Physically the correct procedure is to start with the system in microscopic units and from thereon to determine the space-time scale of macroscopic variation. We will consider three distinct limits: low density (LD), weak coupling (WC1 and WC2), and mean field (MF). I discuss them in parallel so to be able to point out similarities and differences.

Since the number of particles will tend to infinity, we scale the spatial volume as
$$\varepsilon^{-1}\Lambda = \{\mathbf{x} \mid \varepsilon \mathbf{x} \in \Lambda\}. \qquad \text{(LD, WC, MF)} \quad (3)$$
For weak coupling and mean field the local density is bounded. Thus the number of particles
$$N \cong \varepsilon^{-3}. \qquad \text{(WC, MF)} \quad (4)$$
On the other hand for low density we increase N less slowly,
$$N \cong \varepsilon^{-2} \qquad \text{(LD)} \quad (5)$$
which implies that the density is of order ε. (At this stage the power -2 looks arbitrary. It is determined by requiring a space-time variation on the scale ε, cf. below.) The typical interparticle distance increases then as $\varepsilon^{-1/3}$ [micro units]. This signals already an important difference. For weak coupling and mean field the local state should be an ideal Bose, or Fermi gas, whereas at low density the statistics plays no role and the local state is classical.

Together with the enlarged volume $\varepsilon^{-1}\Lambda$ we impose an initially slow spatial variation of order ε. Both information can be encoded directly into the initial state. For simplicity let $\Lambda = \mathbb{R}^3$. We prescribe a smooth classical density $f(\mathbf{r}, \mathbf{v})$ on $\mathbb{R}^3 \times \mathbb{R}^3$, $f \geq 0$, which vanishes rapidly as $|\mathbf{r}|, |\mathbf{v}| \to \infty$. In particular $\int d^3r \int d^3v\, f(\mathbf{r}, \mathbf{v}) < \infty$. For each ε we choose a trace class operator R^ε with position space kernel $< \mathbf{x}|R^\varepsilon|\mathbf{y} >$ which via (2) defines the initial state ρ^ε. We require that

$$\lim_{\varepsilon \to 0} < \varepsilon^{-1}\mathbf{r} + \frac{1}{2}\eta |R^\varepsilon|\varepsilon^{-1}\mathbf{r} - \frac{1}{2}\eta > = \int d^3v\, e^{i\eta \cdot \mathbf{v}} f(\mathbf{r}, \mathbf{v}), \qquad \text{(WC, MF)} \quad (6')$$

$$\lim_{\varepsilon \to 0} < \varepsilon^{-1}\mathbf{r} + \frac{1}{2}\eta |\varepsilon^{-1}R^\varepsilon|\varepsilon^{-1}\mathbf{r} - \frac{1}{2}\eta > = \int d^3v\, e^{i\eta \cdot \mathbf{v}} f(\mathbf{r}, \mathbf{v}). \qquad \text{(LD)} \quad (6'')$$

This means that the rescaled Wigner distribution of the one particle density matrix converges to $f(\mathbf{r}, \mathbf{v})$. For fermions we need $R^\varepsilon \leq 1$ which implies $\sup f \leq 1$ for WC, MF and which always holds for LD provided ε is small enough.

Let us pause for a moment to discuss a few properties of the initial state ρ^ε. They are verified by applying the definitions (2) and (6). Let $\tau_\mathbf{r}$ be the spatial shift by $\mathbf{r}$. For a local observable A we define the spatial average

$$\varepsilon^3 \int d^3r\, \chi(\varepsilon \mathbf{r})\, \tau_\mathbf{r} A = n^\varepsilon(A, \chi) \qquad \text{(WC, MF)} \quad (7)$$

with χ a smooth averaging function. Let us also denote by $< \cdot > (h(\mathbf{k}))$ the translation invariant quasi-free state of an infinitely extended ideal Bose or Fermi gas with density $\int d^3k\, h(\mathbf{k})$ and momentum distribution $h(\mathbf{k})/\int d^3k\, h(\mathbf{k})$. This state is again defined by (2) with the operator R now given through multiplication by $h(\mathbf{k})$ in momentum space ($h(\mathbf{k}) \geq 0$ and for fermions $h(\mathbf{k}) \leq 1$). We have then

$$\lim_{\varepsilon \to 0} \operatorname{tr}[\rho^\varepsilon n^\varepsilon(A, \chi)] = \int d^3r\, \chi(\mathbf{r})\, <A>(h_\mathbf{r}(\mathbf{k})) \qquad \text{(WC, MF)} \quad (8)$$

with $h_\mathbf{r}(\mathbf{k}) = f(\mathbf{r}, \mathbf{k})$ and

$$\lim_{\varepsilon \to 0} \operatorname{tr}[\rho^\varepsilon n^\varepsilon(A, \chi)^2] = \left[\int d^3r\, \chi(\mathbf{r})\, <A>(h_\mathbf{r}(\mathbf{k}))\right]^2. \qquad \text{(WC, MF)} \quad (9)$$

Thus spatially averaged observables become deterministic. This is a *law of large numbers*. In (8) we are allowed to replace χ by a delta-function which tells us that in the large volume $\varepsilon^{-1}\Lambda$ close to the location $\varepsilon^{-1}\mathbf{r}$ the state converges as $\varepsilon \to 0$ to the one of an ideal Bose, or Fermi, gas with constant density $\int d^3v\, f(\mathbf{r},\mathbf{v})$ and momentum distribution $f(\mathbf{r},\mathbf{k})/\int d^3v\, f(\mathbf{r},\mathbf{v})$.

For low density we have to modify our definitions. Since the number of particles in a region of linear size ε^{-1} increases as ε^{-2}, the spatially averaged observable is now

$$\varepsilon^2 \int d^3r\, \chi(\varepsilon\mathbf{r})\, \tau_\mathbf{r} A = \overline{n}^\varepsilon(A,\chi). \qquad \text{(LD)} \quad (10)$$

It again satisfies a law of large numbers in the sense that fluctuations vanish as $\varepsilon \to 0$. To discuss the local state we recall that the typical interparticle distance is $\varepsilon^{-1/3}$. To have observables with a well defined classical limit we use the Wigner operator in the form

$$W^\varepsilon(g) = (2\pi)^{-3} \int d^3\xi\, d^3\eta\, \varepsilon^{-1}\, \widehat{h}\left(\varepsilon^{-1/3}\xi,\eta\right) e^{i(\xi\cdot\mathbf{x}+\eta\cdot\mathbf{p})} \qquad (11)$$

with $\widehat{}$ denoting Fourier transform and $\mathbf{x}$, $\mathbf{p}$ the position and momentum operator in one-particle space. Let $A(g)$ be the corresponding second quantized observable,

$$A(g) = \int d^3x\, a^+(\mathbf{x})\, W^\varepsilon(g)\, a(\mathbf{x}). \qquad (12)$$

Then

$$\lim_{\varepsilon \to 0} \operatorname{tr}[\rho^\varepsilon\, \tau_{\varepsilon^{-1}\mathbf{r}}\, A(g)] = \int d^3v\, f(\mathbf{r},\mathbf{v}) \int d^3r'\, g(\mathbf{r}',\mathbf{v}). \qquad (13)$$

Similarly for a second quantized two-particle observable

$$A(g_1, g_2) = \int d^3x_1\, d^3x_2\, a^+(\mathbf{x}_1)\, a^+(\mathbf{x}_2)\, W^\varepsilon(g_1)\, W^\varepsilon(g_2)\, a(\mathbf{x}_2)\, a(\mathbf{x}_1) \qquad (14)$$

we have

$$\lim_{\varepsilon \to 0} \operatorname{tr}[\rho^\varepsilon\, \tau_{\varepsilon^{-1}\mathbf{r}}\, A(g_1,g_2)] = \int d^3v f(\mathbf{r},\mathbf{v}) \int d^3v' f(\mathbf{r},\mathbf{v}') \int d^3r'\, g_1(\mathbf{r}',\mathbf{v}) \int d^3r''\, g_2(\mathbf{r}'',\mathbf{v}')$$
$$(15)$$

and correspondingly for n-particle observables. Thus we conclude that the local state at $\varepsilon^{-1}\mathbf{r}$ converges to a classical Poisson distribution with uniform density $\int d^3v\, f(\mathbf{r},\mathbf{v})$ and velocity distribution $f(\mathbf{r},\mathbf{v})/\int d^3v\, f(\mathbf{r},\mathbf{v})$.

So far we have just discussed the initial state. Now we have to go on stage and take up the real issue, namely dynamics. The kinetic equations to be derived all have a part corresponding to force free motion. Since the spatial variation is of order ε, to have the free motion scale invariant we must consider

$$\text{time} \cong \varepsilon^{-1}\, t. \qquad (16)$$

For low density we retain the short range potential

$$\text{(LD)} \quad V(\mathbf{x}). \qquad (17)$$

Then the mean free path and therefore also the mean free time are of the order of the inverse density, i.e. of order ε^{-1}. Thus on the time scale considered a given particle suffers only a finite number of collisions. Free motion and collisions are of the same order of magnitude. If we rescale positions to the macroscopic volume Λ, then the number of particles in Λ increases as ε^{-2} and the interaction potential scales as $V(\mathbf{x}/\varepsilon)$. In the classical context, this is precisely the Boltzmann-Grad limit which yields the Boltzmann equation as kinetic equation [10].

For weak coupling we want to make the effect in a single collision of the order ε so that on the time scale $\varepsilon^{-1}t$ the global effect is of order one. To achieve this, one possibility is to scale down the interaction strength as

$$\text{(WC1)} \quad \sqrt{\varepsilon}\, V(\mathbf{x}). \tag{18}$$

Recall that in the Born approximation the scattering amplitude equals $\sqrt{\varepsilon}\,\widehat{V}(\mathbf{k})$. A second possibility is to scale down the range of the potential as

$$\text{(WC2)} \quad V(\mathbf{x}/\sqrt{\varepsilon}). \tag{19}$$

Then the collision cross section is of order ε. Note that on the macroscopic scale WC2 corresponds to a potential $V(\mathbf{x}/\varepsilon^{3/2})$ and the number of particles in Λ increasing as ε^{-3}. Thus for classical particles LD and WC2 are isomorphic. Quantum mechanically this is no longer so and distinct kinetic equations result.

For mean field the potential is weak and of long range,

$$\text{(MF)} \quad \varepsilon^3\, V(\varepsilon\mathbf{x}), \tag{20}$$

with total strength of order one. The average force acting on a given particle is of order ε and again on the time scale $\varepsilon^{-1}t$ the global effect is of order one.

The central claim of the kinetic limit is that the structure which was imposed initially will be preserved in the course of time. *Thus it is conjectured that at time $\varepsilon^{-1}t$ the limits (8), (9) (for low density with scaling (10)) still hold provided $f(\mathbf{r},\mathbf{v})$ is replaced by the $f(\mathbf{r},\mathbf{v},t)$, where $f(\mathbf{r},\mathbf{v},t)$ is the solution to the appropriate kinetic equation with initial condition $f(\mathbf{r},\mathbf{v})$.* We will discuss the kinetic equations in the following Section. Here I should still emphasize that the time-evolved state $\rho^\varepsilon(\varepsilon^{-1}t)$ must have complicated statistical correlations induced by the dynamics and cannot be of the simple form (2). Only as $\varepsilon \to 0$ and in the sense of expectations of spatially averaged local observables the state at time $\varepsilon^{-1}t$ resembles the structure of the initial state.

If true, the existence of a kinetic limit represents a deep dynamical fact. Certainly for times of order 1 the limits (8), (9) hold. But then $f(\mathbf{r},\mathbf{v})$ has not changed yet. The main point is that over long times the particles organize themselves in such a way as to follow the solution of the kinetic equation.

KINETIC EQUATIONS

We discuss each one of the kinetic limits separately.

(i) *low density.* We remarked already that at low density particles are typically far apart and the statistics is of no relevance, of course, with the exception of collisions. Thus the appropriate kinetic equation should be the Boltzmann equation with the collision cross section computed quantum mechanically,

$$\frac{\partial}{\partial t} f(\mathbf{r},\mathbf{v},t) = -\mathbf{v} \cdot \nabla_r f(\mathbf{r},\mathbf{v},t) + \int d^3v_1\, d^3v'\, d^3v'_1\, W(\mathbf{v},\mathbf{v}_1 \mid \mathbf{v}',\mathbf{v}'_1)$$
$$[f(\mathbf{r},\mathbf{v}',t)\, f(\mathbf{r},\mathbf{v}'_1,t) - f(\mathbf{r},\mathbf{v},t)\, f(\mathbf{r},\mathbf{v}_1,t)]. \qquad (21)$$

W is the rate of collision from the incoming pair $(\mathbf{v},\mathbf{v}_1)$ to the outgoing pair $(\mathbf{v}',\mathbf{v}'_1)$ and we used already the time reversal invariance $W(\mathbf{v},\mathbf{v}_1 \mid \mathbf{v}',\mathbf{v}'_1) = W(-\mathbf{v}',-\mathbf{v}'_1 \mid -\mathbf{v},-\mathbf{v}_1)$. Let $f(\mathbf{k},\mathbf{k}')$ denote the scattering amplitude for the single particle hamiltonian $-\Delta + V$ (this is the hamiltonian for the relative motion of two particles). Then

$$W(\mathbf{v},\mathbf{v}_1 \mid \mathbf{v}',\mathbf{v}'_1) = |f(\mathbf{v}-\mathbf{v}_1,\mathbf{v}'-\mathbf{v}'_1) \pm f(\mathbf{v}_1-\mathbf{v},\mathbf{v}'-\mathbf{v}'_1)|^2$$
$$\delta(\mathbf{v}+\mathbf{v}_1-\mathbf{v}'-\mathbf{v}'_1)\, \delta\!\left((v^2+v_1^2-v'^2-v'^2_1)/2\right). \qquad (22)$$

The $+$ signs stands for bosons, the $-$ sign for fermions. We note that compared to classical scattering quantum mechanics enforces the additional symmetry $W(\mathbf{v},\mathbf{v}_1 \mid \mathbf{v}',\mathbf{v}'_1) = W(\mathbf{v}_1,\mathbf{v} \mid \mathbf{v}',\mathbf{v}'_1)$. Waldmann [4] in his Handbuch article explains under what conditions quantum effects become visible in transport properties of low density noble gases. A reading teaches one also that the determination of the effective scattering amplitude for atoms (which are composite particles) is an undertaking of its own.

(ii) *weak coupling*. Since the statistics is retained locally, we expect the solution to the kinetic equation to approach in the long time limit the equilibrium Bose, or Fermi, distribution. This suggest a kinetic equation of the form

$$\frac{\partial}{\partial t} f(\mathbf{r},\mathbf{v},t) = -\mathbf{v} \cdot \nabla_r f(\mathbf{r},\mathbf{v},t) + \int d^3v_1\, d^3v'\, d^3v'_1\, W(\mathbf{v},\mathbf{v}_1 \mid \mathbf{v}',\mathbf{v}'_1)$$
$$[f(\mathbf{r},\mathbf{v}',t)(1 \pm f(\mathbf{r},\mathbf{v}',t))\, f(\mathbf{r},\mathbf{v}'_1,t)(1 \pm f(\mathbf{r},\mathbf{v}'_1,t))$$
$$-f(\mathbf{r},\mathbf{v},t)(1 \pm f(\mathbf{r},\mathbf{v},t))\, f(\mathbf{r},\mathbf{v}_1,t)(1 \pm f(\mathbf{r},\mathbf{v}_1,t))], \qquad (23)$$

usually called the Uehling-Uhlenbeck equation [5]. If the strength of the potential becomes weak as in WC1, then the scattering amplitude is taken only in the Born approximation and the collision rate becomes

$$W(\mathbf{v},\mathbf{v}_1 \mid \mathbf{v}',\mathbf{v}'_1) = \left|\widehat{V}(\mathbf{v}'-\mathbf{v}) \pm \widehat{V}(\mathbf{v}'-\mathbf{v}_1)\right|^2$$
$$\delta(\mathbf{v}+\mathbf{v}_1-\mathbf{v}'-\mathbf{v}'_1)\, \delta\!\left((v^2+v_1^2-v'^2-v'^2_1)/2\right) \qquad (24)$$

with $\widehat{V}(\mathbf{k}) = 4\pi \int d^3x\, e^{i\mathbf{k}\cdot\mathbf{x}}\, V(\mathbf{x})$. As before $+$ stands for bosons and $-$ for fermions. On the other hand if the collision cross section is scaled down as in WC2, then the full two-particle quantum scattering must be used and the collision rate is as in (22).

For a spatially uniform ideal quantum gas with momentum distribution $f(\mathbf{k})$ the entropy per unit volume is easy to compute. Since we consider here slow spatial variation, the total entropy is approximately the sum of the entropies of the local pieces. This consideration suggests the H-functional

$$H(f) = \int d^3r \left[\int d^3k\, \{f(\mathbf{r},\mathbf{k}) \log f(\mathbf{r},\mathbf{k}) \mp (1 \pm f(\mathbf{r},\mathbf{k})) \log(1 \pm f(\mathbf{r},\mathbf{k}))\} \right]. \qquad (25)$$

Indeed, under the Uehling-Uhlenbeck equation there is an H-Theorem in the sense that $dH(f(t))/dt \leq 0$. The only spatially homogeneous solutions with $dH/dt = 0$ are equilibrium distributions

$$f(\mathbf{v}) = \left(\exp\left[\beta(\frac{1}{2}(\mathbf{v}-\mathbf{v}_0)^2 - \mu)\right] \mp 1\right)^{-1}. \tag{26}$$

Here $\mathbf{v}_0$ is the mean velocity, β the inverse temperature, and μ the chemical potential. It is of interest to note that the Uehling-Uhlenbeck equation tells us that in the case of bosons we have missed some important physics. For simplicity let us consider the spatially homogeneous case and let us choose the initial condition $f(\mathbf{v})$ for Equation (23). We require mean velocity zero, $\int d^3 v f(\mathbf{v})\mathbf{v} = 0$, and bounded kinetic energy, $\int d^3 v(\mathbf{v}^2/2)f(\mathbf{v})/\int d^3 v f(\mathbf{v}) < \infty$. Both properties are preserved in time. Now suppose that

$$\int d^3 v \, f(\mathbf{v}) > \rho_c = \int d^3 v \, (\exp[\beta \mathbf{v}^2/2] - 1)^{-1}. \tag{27}$$

Since, at least formally, the total mass is conserved in time, our initial condition has no equilibrium solution to converge to. Of course, physically the Bose gas tries to form a superfluid condensate – an event not forseen when writing down the kinetic equation (23). Thus we must either require that $\int d^3 v \, f(\mathbf{r},\mathbf{v},t) < \rho_c$ for all $\mathbf{r}$ and t or we have to use a coupled set of kinetic equations, one for the normal fluid and one for the Bose condensate. In the latter case the gauge invariance imposed in (2) is no longer valid. We would have a two-fluid model on the kinetic rather than on the more standard hydrodynamic level [6]. [In parantheses we remark that for coagulation models it is known that formally mass conservation holds whereas the actual solution loses mass at infinity [7]. Possibly the bosonic Uehling-Uhlenbeck equation admits a similar mathematical mechanism.]

The Uehling-Uhlenbeck equation is used widely in solid state physics [8] even under conditions when the interaction is not weak. I think the reason must be that in a comparatively simple way it incorporates both dissipation and the quantum equilibrium.

(iii) *mean field*. Particles at some given location experience the average force due to the other particles. This should yield the Vlasov equation

$$\frac{\partial}{\partial t} f(\mathbf{r},\mathbf{v},t) = -\mathbf{v}\cdot\nabla_r \, f(\mathbf{r},\mathbf{v},t) - \int d^3 r' \, d^3 v' \, f(\mathbf{r}',\mathbf{v}',t) \, \nabla V(\mathbf{r}'-\mathbf{r}) \cdot \nabla_v \, f(\mathbf{r},\mathbf{v},t). \tag{28}$$

The Vlasov equation is reversible, in particular the Boltzmann H-function does not change in time.

SOME MATHEMATICAL EVIDENCE

Perhaps it is useful to recall the sort of minimal assertion one would like to prove. Let $\rho_1^\varepsilon(t)$ be the one particle density matrix defined by

$$< \mathbf{x} \, |\rho_1^\varepsilon(t)| \, \mathbf{y} > = \text{tr}[e^{-iH^\varepsilon t} \rho^\varepsilon e^{iH^\varepsilon t} \, a^+(\mathbf{x}) \, a(\mathbf{y})], \tag{29}$$

where a superscript ε is added to H because the interaction potential depends on ε. We consider its Wigner distribution

$$f^\varepsilon(\mathbf{r},\mathbf{v},t) = (2\pi)^{-3} \int d^3\eta\, e^{-i\eta\cdot\mathbf{v}} <\varepsilon^{-1}\mathbf{r} + \frac{1}{2}\eta|\rho_1^\varepsilon(\varepsilon^{-1}t)|\varepsilon^{-1}\mathbf{r} - \frac{1}{2}\eta> \qquad (30)$$

on the space scale $\varepsilon^{-1}\mathbf{r}$ and the time scale $\varepsilon^{-1}t$. For low density a prefactor ε^{-1} should be inserted. Then the following limit should hold pointwise

$$\lim_{\varepsilon\to 0} f^\varepsilon(\mathbf{r},\mathbf{v},t) = f(\mathbf{r},\mathbf{v},t), \qquad (31)$$

where $f(\mathbf{r},\mathbf{v},t)$ is the solution to the corresponding kinetic equation with initial condition $f(\mathbf{r},\mathbf{v})$. For the law of large numbers one would have to prove that in addition the two-particle Wigner distribution converges to $f(\mathbf{r}_1,\mathbf{v}_1,t)\,f(\mathbf{r}_2,\mathbf{v}_2,t)$. Here one could encounter exceptional subsets on which the convergence fails.

Narnhofer and Sewell [9] prove the validity of the Vlasov equation provided the potential V is such that the force is bounded with bounded Fourier transform. By a compactness argument they establish that the solution of the Wigner form of the quantum BBGKY hierarchy converges to the solution of the Vlasov hierarchy, which by initial factorization implies the validity of the Vlasov equation [10]. It remains open to prove that the local state is quasi-free with parameters determined by the solution of the Vlasov equation.

Hugenholtz [11] investigates the weak coupling limit with scaling (18) for an infinitely extended, spatially uniform Fermi gas. In second order perturbation he identifies the collision term in the Uehling-Uhlenbeck equation with rate (24). From the higher orders he obtains a power series solution to the spatially homogeneous equation (23). For a complete proof several gaps remain. The most obvious one is that the perturbation expansion is established to converge only for times of order $\varepsilon^{-1/2}$ and not, as needed, of order ε^{-1}.

For the derivation of the Boltzmann equation there are various formal schemes — none of them produces the desired result with ease. Since the work of Lanford [12] the derivation of the classical Boltzmann equation is so much better understood, that it is tempting to reconsider the quantum case. This has been done by Wittwer [13] in his diploma thesis. I just want to make a few comments for those familiar with the work of Lanford and King [14]. The first step is to rewrite the quantum BBGKY hierarchy in terms of Wigner distributions. In this form the collision term describes only infinitesimal collisions, whereas in the Boltzmann collision term completed collisions appear. On the classical level this difficulty is circumvented by an appropriate redefinition of correlation functions exploiting that the range of the interaction potential is finite. When mimicking such a procedure, the obstacle arises that quantum mechanical collision cannot be strictly localized. Even if this could be handled by some suitable approximation scheme, one is still left with a difficult N-body, $N = 1,2,\ldots$, scattering problem. Moreover: classically one has a bound on the n-th time-dependent correlation function as $c(t)^n$ yielding a finite radius of convergence for the BBGKY perturbation series. Quantum mechanically one only has a bound as $c(t)^n n!$, cf. (2), implying naively a zero radius of convergence.

To conclude: Although the derivation of the Boltzmann and Uehling-Uhlenbeck equations are perturbational in spirit, present day mathematical techniques are simply not powerful enough to prove a limit as (31).

FLUCTUATIONS

If one cannot do the first step it is unwise to try the second one. Still I simply have to point out that in a very natural way a further structure is added on. From probability theory one knows that in many circumstances the small fluctuations away from the mean are governed by a central limit theorem. The kinetic limit is a law of large numbers for the density field on the one-particle phase space and therefore the fluctuations form a Gaussian field. It is an infinite-dimensional Ornstein-Uhlenbeck process governed by a Langevin equation with the linearized kinetic equation as drift operator and noise added on according to the extended local equilibrium hypothesis [1]. Since the generation of noise is necessarily linked to dissipation, for the Vlasov equation the random noise term is in fact absent.

Quantum mechanically we expect a similar structure. The fluctuation field for the local observable A is defined by

$$\xi^\varepsilon(A, \chi, t) = \varepsilon^{3/2} \int d^3r \, \chi(\varepsilon \mathbf{r}) \left(\tau_{\mathbf{r}} \, A(\varepsilon^{-1}t) - \text{tr}\, [\rho^\varepsilon \, \tau_{\mathbf{r}} \, A(\varepsilon^{-1}t)] \right), \qquad (32)$$

compare with (7). Here $A(t) = \exp[iH^\varepsilon t] \, A \exp[-iH^\varepsilon t]$. (For low density $\varepsilon^{3/2}$ has to be replaced by ε.) Note that the commutator for fluctuation fields is of order one. Therefore the quantum mechanical fluctuations should have the properties of an ideal Bose field. This structure is known from the quantum central limit theorem [15,16] and can be proved for a certain type of mean field dynamics [16] and for short range dynamics on a time scale of order one [17]. For the irreversible Boltzmann and Uehling-Uhlenbeck equations the fluctuations should be dissipative and governed by a quasi-free dynamical semigroup. I do not know how to identify its generator.

Fluctuations are linked to the behavior of time-correlations. E.g. we could consider the density-density time correlation for concreteness in thermal equilibrium. The problem is then to determine its behavior in the kinetic limit on the space-time scale $\varepsilon^{-1}\mathbf{r}$, $\varepsilon^{-1}t$. If in (32) we insert for A the local density $a^+(\mathbf{x}) \, a(\mathbf{x})$ and form second moments in equilibrium, then this just gives the density time correlation. Thus its kinetic limit would identify the two-point structure of the fluctuation field.

OUTLOOK

My brief tour makes it clear that we have not even scratched the surface. On the more optimistic side let me mention that for tracer particle models there has been some progress since the last listing [18]: a quantum particle in a random potential at weak coupling [19,20] and a nailed down N-level atom immersed in a low density gas [21]. This latter work has been extended considerably through the link with quantum stochastic processes [22,23].

Acknowledgments

I am grateful to Oliver Penrose for a critical reading of a preliminary version of this manuscript. He explained to me the second version of the weak coupling limit and the kinetic two-fluid model. I thank J.L.Lebowitz and D. Ruelle for inviting me to the Institut des Hautes Etudes Scientifiques, where most of the manuscript was prepared.

REFERENCES

1. H. Spohn, "Large Scale Dynamics of Interacting Particles," Springer, Heidelberg, 1991.
2. W. Thirring, "Quantum Mechanics of Large Systems," Springer, Wien, 1980.
3. H. Narnhofer and W. Thirring, *Ann. Phys.* **134**, 128 (1981).
4. L. Waldmann, in "Handbuch der Physik," Vol. XII, ed. S. Flügge. Springer, Berlin, (1958).
5. E.A. Uehling and G.E. Uhlenbeck, *Phys. Rev.* **43**, 552 (1933).
6. O. Penrose and L. Onsager, *Phys. Rev.* **104**, 576 (1956).
7. M. Aizenman and P. Bak, *Comm. Math. Phys.* **65**, 203 (1979).
8. N.W. Ashcroft and N.D. Mermin, "Solid State Physics," Holt, Rinehart and Winston, New York, (1976).
9. H. Narnhofer and G.L. Sewell, *Comm. Math. Phys.* **79**, 9 (1981).
10. H. Spohn, *Math. Meth. Appl. Sciences* **3**, 445 (1981).
11. N.M. Hugenholtz, *J. Stat. Phys.* **32**, 239 (1983).
12. O.E. Lanford, in "Time Evolution of Large Classical Systems," ed. J. Moser, Lecture Notes in Physics Vol. 38. Springer, Heidelberg, (1975).
13. P. Wittwer, Zur Quantenmechanik der Boltzmanngleichung. Diplomarbeit, ETH Zürich, 1980, unpublished.
14. F. King, "BBGKY Hierarchy for Positive Potentials," Ph. D. Thesis, Dept. of Mathematics, UCB, 1975, unpublished.
15. D. Goderis and P. Vets, *Comm. Math. Phys.* **122**, 249 (1989).
16. D. Goderis, A. Verbeure and P. Vets, *Comm. Math. Phys.* **128**, 533 (1990).
17. D. Goderis, A. Verbeure and P. Vets, *J. Math. Phys.* **29**, 2581 (1988).
18. H. Spohn, *Rev. Mod. Phys.* **52**, 569 (1980).
19. T.G. Ho, L.J. Landau, and A.J. Wilkins, *Rev. Math. Phys.* **5**, 209 (1993).
20. L.J. Landau, Observation of quantum particles on a large space-time scale, preprint, 1993.
21. R. Dümcke, *Comm. Math. Phys.* **97**, 331 (1985).
22. S. Rudnicki, R. Alicki and S. Sadenski, *J. Math. Phys.* **33**, 2607 (1992).
23. L. Accardi and Y.G. Lu, *J. Phys.* $A24$, 3483 (1991).

MICROSCOPIC DERIVATION OF HYDRODYNAMICS WITH PHASE TRANSITION IN A PLASMA MODEL*

G.L. Sewell**

Department of Physics
Queen Mary and Westfield College
Mile End Road, London E1 4NS

1. INTRODUCTION

We present a microscopic derivation of the hydrodynamics of a classical plasma model, and show that this undergoes a transition from a deterministic (Eulerian) to a stochastic flow, that is governed by its initial charge density and drift velocity profiles. Thus, the model exhibits a phase transition far from equilibrium.

In recent years, there have been developments in the rigorous statistical mechanical theory of continuum mechanics, based on classical *stochastic* models of many-particle systems [1, 2]. These have led to caricatures of physically interesting phenomena, including phase transitions far from equilibrium, such as those involved in the formation of shock waves [2, 3, 4].

On the other hand, the progress made in this area on the basis of the Hamiltonian dynamics of conservative many-particle systems is rather limited. In fact, to the best of our knowledge, the only rigorous treatments of the passage from microscopic Hamiltonian dynamics to continuum mechanics are the quantum mechanical derivations by Davies [5] of Fourier's law of heat conduction for a certain model of interacting atoms, and by the present author [6, 7] of the hydrodynamics of a plasma model. In the last of these works [7], it was shown that the plasma exhibited a phase transition, far from equilibrium, from a deterministic (Eulerian) to a stochastic flow.

The present article is devoted to a relatively simple microscopic derivation of the hydrodynamics of the *classical* version of this plasma model, and of its transition from deterministic to stochastic flow.

---

* Lecture given at the Leuven Conference "On Three Levels", July 19-23, 1993
** Partially supported by European Capital and Mobility Contract No. CHRX-CT92-0007

We present our treatment as follows. In §2, we formulate the $N-$ particle Hamiltonian model of the plasma on both macroscopic and microscopic length scales. We also formulate a modification of this model, in which the singular Coulomb potential is 'smoothed out' by means of a short distance cut-off.

In §3, we present our treatment of the modified plasma model. Thus, we first note that the model satisfies the conditions of Braun and Hepp [8] for a Vlasov dynamics, and that this corresponds to the Liouville equation for a certain *Lagrangian* hydrodynamics, governing the evolution of the time (t)-dependent position, $X_t(x)$, of a 'fluid particle' located initially at the point x. Furthermore, we show that this reduces to a deterministic Eulerian hydrodynamics if and only if the function X_t is invertible: otherwise it corresponds to a stochastic flow, i.e. one whose local density and drift velocity have macroscopic dispersions. We then provide an explicit example of a class of initial conditions, which lead to a phase transition, far from equilibrium, from a deterministic to a stochastic flow.

In §4, we sketch the corresponding theory, given by the classical analogue of [7], for the (unmodified) plasma model, where again we have a hydrodynamic phase transition from a deterministic to a stochastic flow.

We conclude, in §5, with some brief further comments on the results obtained here and on related open problems; and with a simple explanation of why the classical and quantum versions of the models concerned lead to the same hydrodynamic properties.

2. THE MODEL

Let $\Sigma^{(N,L)}$ be a classical plasma consisting of N 'electrons', i.e., particles of charge $-e(<0)$ and mass m, in a cube, $K^{(L)}$, of side L, with uniform neutralising positive charge background. We assume periodic boundary conditions, thereby rendering $K^{(L)}$ a torus. Our objective will be to derive the hydrodynamics of $\Sigma^{(N,L)}$ from its microscopic equations of motion, in a limit where N and L tend to infinity and the mean particle density,

$$\bar{n} = N/L^3, \tag{2.1}$$

remains fixed and finite. Thus, we shall be considering a sequence of systems $\{\Sigma^{(N,L)}\}$.

Denoting the position vectors and momenta of the particles of $\Sigma^{(N,L)}$ by $X_1, \ldots, X_N$ and $P_1, \ldots P_N$,* the Hamiltonian of the model takes the form

$$H^{(N,L)} = \sum_{j=1}^{N} P_j^2/2m + e^2 \sum_{j,k(>j)=1}^{N} U^{(L)}(X_j - X_k), \tag{2.2}$$

where $U^{(L)}(X)$ is the difference between $|X|^{-1}$, periodicised w.r.t. $K^{(L)}$, and its space average over that cube, i.e.

$$U^{(L)}(X) = \frac{4\pi}{L^3} \sum^{(L)} \frac{\exp(iQ.X)}{Q^2}, \tag{2.3}$$

the superscript (L) over Σ signifying that summation is taken over the non-zero vectors $Q = (2\pi/L)(n_1, n_2, n_3)$, with the n's integers. The Newtonian equations of

---

* Since $K^{(L)}$ is a torus, $X_1, \ldots, X_N; P_1, \ldots, P_N$ correspond to local coordinates in the cotangent bundle over $(K^{(L)})^N$.

motion for the system, with T the time, are then

$$\frac{dX_j(T)}{dT} = P_j(T); \quad \frac{dP_j(T)}{dT} = -e^2 \sum_{k \neq j} \nabla^{(L)} U(X_j(T) - X_k(T)), \tag{2.4}$$

where $\nabla^{(L)}$ is the gradient operator for $K^{(L)}$. For $B > 0$, let $NF^{(N)}(B)$ be the number of particles of $\Sigma^{(N,L)}$ with momenta of magnitude greater than B at $t = 0$. We assume that

$$F^{(N)}(B) \to 0, \; uniformly \; w.r.t. \; N, \; as \; B \to \infty. \tag{2.5}$$

We base our macroscopic description of the model on scales of length, time and particle momentum given by L, ω^{-1} and $mL\omega$, where ω is the classical plasma frequency, i.e.

$$\omega = (4\pi \bar{n} e^2 / m)^{1/2}. \tag{2.6}$$

Thus, in this description, we employ the rescaled position, velocity and time variables $x = X/L$, $v = P/mL\omega$, $t = \omega T$, respectively. Under this rescaling, $\Sigma^{(N,L)}$ is mapped onto a system $\Sigma^{(N)}$ of particles in a unit cube, K, with periodic boundaries; and the equations of motion for this system, corresponding to (2.4), are

$$\frac{dx_j(t)}{dt} = v_j(t); \quad \frac{dv_j(t)}{dt} = -N^{-1} \sum_{k \neq j} \nabla U(x_j(t) - x_k(t)), \tag{2.7}$$

where ∇ is the gradient operator for K and

$$U(x) = U_c(x) := \sum_q^{(1)} \exp(iq.x)/q^2, \tag{2.8}$$

the superscript (1) over Σ signifying that summation is taken over the non-zero vectors $2\pi(n_1, n_2, n_3)$, with the n's integers. Our reason for introducing the symbol U_c here is that we want to consider the equations of motion (2.7), not only when U is the Coulomb potential U_c but also when it is a modification thereof, obtained by introducing a short-distance cutoff. Note that the equations (2.7) are those generated by the Hamiltonian

$$H^{(N)} = \frac{1}{2} \sum_{j=1}^{N} p_j^2 + N^{-1} \sum_{j,k(>j)=1}^{N} U(x_j - x_k), \tag{2.9}$$

the particle mass being unity. Thus $H^{(N)}$ is the Hamiltonian for $\Sigma^{(N)}$. We note also that it follows from (2.8) that

$$\Delta U_c(x) = 1 - \delta(x), \tag{2.10}$$

where Δ is the Laplacian and δ the Dirac distribution for K. For $b > 0$, let $Nf^{(N)}(b)$ be the number of particles of $\Sigma^{(N)}$ with velocities of magnitude greater than b at $t = 0$. Then it follows from condition (2.5), equation (2.1) and the above definition of v_j that

$$\lim_{N \to \infty} f^{(N)}(b) = 0, \; \forall b > 0. \tag{2.11}$$

Let $\mu_0^{(N)}$ denote the probability measure on $K \times \mathbf{R}^3$ corresponding to the initial distribution of the particles, i.e.,

$$\int \phi(x,v) d\mu_0^{(N)} = N^{-1} \sum_{j=1}^{N} \phi(x_j(0), v_j(0)), \; \forall \phi \in C_0(K \times \mathbf{R}^3), \tag{2.12}$$

C_0 denoting the continuous functions with compact support. We shall assume that $\mu_0^{(N)}$ converges, in the vague topology, to a measure μ_0, as $N\to\infty$, i.e.,

$$\lim_{N\to\infty} \int \phi d\mu_0^{(N)} = \int \phi d\mu_0 \quad \forall \phi \in C_0(K\times \mathbf{R}^3). \tag{2.13}$$

Then, in view of condition (2.11), it follows from (2.12) and (2.13) that

$$\int \phi(x,v)d\mu_0 = \int \phi(x,0)d\mu_0 \quad \forall \phi \in C_0(K\times \mathbf{R}^3),$$

which implies that μ_0 must be of the form

$$\mu_0(dxdv) = \nu_0(dx)\delta_0(dv), \tag{2.14}$$

where ν_0 is a probability measure on K and δ_0 is the Dirac measure with support at 0. This last factor implies that the model is initially a cold plasma, when viewed on the *macroscopic* scale, though not (cf. (2.5)) on the microscopic one.

We shall assume that the probability ν_0, governing the initial particle configuration in K, is absolutely continuous w.r.t. the Lebesgue measure, i.e., that

$$\nu_0(dx) = \sigma_0(x)dx$$

for some measurable density function σ_0 on K. Hence, (2.14) takes the form

$$\mu_0(dxdv) = \sigma_0(x)dx\delta_0(dv). \tag{2.15}$$

We shall assume that σ_0 is continuously differentiable. We now generalise the condition (2.15) by imposing a space-dependent drift velocity $u_0(x)$,*which converts (2.15) to the form

$$\mu_0(dxdv) = \sigma_0(x)dx\delta(u_0(x)|dv), \tag{2.15}'$$

where $\delta(u|.)$ denotes the Dirac measure on $\mathbf{R}^3$ with support at u.

We formulate the dynamics of the model in terms of the probability measure, $\mu_t^{(N)}$, on $K\times\mathbf{R}^{3N}$, governing the distribution of positions and momenta of the particles at time t and defined by

$$\int \phi(x,v)d\mu_t^{(N)} = N^{-1}\sum_{j=1}^N \phi(x_j(t), v_j(t)), \quad \forall \phi \in C_0(K\times \mathbf{R}^3). \tag{2.16}$$

In order to avoid difficulties stemming from the singularity in the Coulomb potential U_c, we shall now modify the model by 'smoothing out' that potential.

The Modified Plasma

We define $\Sigma_g^{(N)}$ to be the model obtained by replacing the Coulomb potential U_c, by a suitably regular one, U_g, given by

$$U_g(x) = \int_K dy g(x-y)U_c(y); \text{ with } g(x)\geq 0 \ \forall x\in K \text{ and } \int dx g(x) = 1, \tag{2.17}$$

---

* This corresponds to the imposition of a drift velocity profile $Lwu_0(X/L)$ on $\Sigma^{(N,L)}$.

where the 'smoothing function' g, and hence U_g, is twice continuously differentiable. Thus, the dynamics of the modified plasma $\Sigma_g^{(N)}$ is still governed by equations (2.7), (2.9), (2.13), (2.15)' and (2.16), but with $U = U_g$, instead of U_c.

One sees now from the formula (2.9) for the Hamiltonian of $\Sigma_g^{(N)}$ (with $U = U_g$) that, in view of the above specifications of U_g and its scaling by N^{-1}, the model is of the mean field theoretic type that exhibits a Vlasov dynamics in the limit $N \to \infty$ (cf. [8]). Furthermore, the same is true for the Coulomb system $\Sigma^{(N)}$, subject to certain regularity assumptions on the dynamics (cf. §4).

3. HYDRODYNAMICS OF THE MODIFIED PLASMA

As we have just noted, this model enjoys a Vlasov dynamics, which is specified by the following Proposition.

Proposition 3.1. [8] *Under the above conditions on $\Sigma_g^{(N)}$, $\mu_t^{(N)}$ converges, in the vague topology, to a probability measure μ_t on $K \times \mathbf{R}^3$, for all $t \in \mathbf{R}$; and this measure evolves according to the weak form of the Vlasov equation, i.e.,*

$$\frac{d}{dt}\int \phi(x,v) d\mu_t(x,v) = \int v \cdot \frac{\partial}{\partial x}\phi(x,v) d\mu_t(x,v)$$

$$-\int \nabla U_g(x-y) \cdot \frac{\partial}{\partial v}\phi(x,v) d\mu_t(x,v) d\mu_t(y,w) \quad \forall \phi \in C_0^{(1)}(K \times \mathbf{R}^3). \qquad (3.1)$$

Furthermore, this equation has a unique solution, which may be expressed in terms of that of the Newtonian mean field theoretic problem

$$\frac{d\chi_t(x,v)}{dt} = \mathcal{V}_t(x,v); \quad \frac{d\mathcal{V}_t(x,v)}{dt} = -\int d\mu_0(y,w) \nabla U_g(\chi_t(x,v) - \chi_t(y,w)), \qquad (3.2)$$

with

$$\chi_0(x,v) = x; \quad \mathcal{V}_0(x,v) = v, \qquad (3.3)$$

by the formula

$$\int d\mu_t(x,v)\phi(x,v) = \int d\mu_0(x,v)\phi(\chi_t(x,v), \mathcal{V}_t(x,v)) \forall \phi \in C_0(K \times \mathbf{R}^3). \qquad (3.4)$$

Corollary 3.2. *Under the initial condition (2.15)', the solution (3.2)-(3.4) of the Vlasov equation (3.1) may be expressed as follows. Let*

$$X_t(x) := \chi_t(x, u_0(x)); \quad V_t(x) := \mathcal{V}_t(x, u_0(x)). \qquad (3.5)$$

Then

$$\frac{dX_t(x)}{dt} = V_t(x); \quad \frac{dV_t(x)}{dt} = -\int dy\sigma_0(y) \nabla U(X_t(x) - X_t(y)); \quad X_0(x) = x; \quad V_0(x) = u_0(x) \qquad (3.6)$$

and

$$\int d\mu_t(x,v)\phi(x,v) = \int dx\sigma_0(x)\phi(X_t(x), V_t(x)). \qquad (3.7)$$

Comment. The Newtonian mean field theory, given by (3.2)- (3.3), corresponds to a *Lagrangian* hydrodynamics, in which $X_t(x)$ and $V_t(x)$ are the position and velocity, respectively, of a 'fluid particle'; and the Vlasov equation (3.1) is just the Liouville equation representing its probabilistic description.

Our aim now is to investigate the conditions under which the Vlasov dynamics reduces to a deterministic Eulerian hydrodynamics. In fact, we shall show that it does so, provided that X_t is an invertible function of position; and that otherwise it is stochastic. Note here that the invertibility of the canonical transformation $(x,v) \to (\chi_t(x,v), \mathcal{V}_t(x,v))$ does *not* imply that of the mapping $x \to X_t(x,v) \equiv \chi_t(x, u_0(x))$.

Case (a): X_t Invertible. In this case, the Jacobean

$$J_t(x) = \frac{\partial(X_{1,t}, X_{2,t}, X_{3,t})}{\partial(x_1, x_2, x_3)} \qquad (3.8)$$

with x_j (resp. $X_{j,t}$) the j'th component x (resp. X_t), is strictly positive, and so we can define

$$\sigma_t(x) = \sigma_0(X_t^{-1}(x))/J_t(x) \qquad (3.9)$$

and

$$u_t(x) = V_t(X_t^{-1}(x)). \qquad (3.10)$$

Thus, since, by (3.7),(3.9) and (3.10),

$$\int d\mu_t(x,v)\phi(x,v) = \int dx \sigma_t(x)\phi(x, u_t(x)) \quad \forall \phi \in C_0(K \times \mathbf{R}^3), \qquad (3.11)$$

i.e., formally,

$$d\mu_t(x,v) = \sigma_t(x)\delta(v - u_t(x))dxdv,$$

it follows that $u_t(x)$ and $\sigma_t(x)$ are the drift velocity and normalised particle density, respectively, at position x and time t.

It follows now from (2.8), (2.17), (3.6), (3.9) and (3.10) that u_t, σ_t evolve according to the following hydrodynamical equations:

$$\frac{\partial \sigma_t}{\partial t} + \nabla.(\sigma_t u_t) = 0, \qquad (3.12)$$

$$\frac{\partial u_t}{\partial t} + (u_t.\nabla)u_t = E_t, \qquad (3.13)$$

where,

$$\nabla.E_t = (\sigma_t^{(g)} - 1) \qquad (3.14)$$

and

$$\sigma_t^{(g)}(x) = \int dy g(x-y)\sigma_t(y). \qquad (3.15)$$

Thus, we have a modified Euler-cum-Maxwell hydrodynamics, the modification being from the replacement of σ_t by $\sigma_t^{(g)}$ in equation (3.14).

Case (b): X_t Non-Invertible. In this case, we cannot employ the formulae (3.9) and (3.10) to define the time-dependent density and drift velocity. Instead, we have to consider the situation where the equation

$$X_t(y) = x \qquad (3.16)$$

has several solutions, labelled by an index set J, for y as a function of x and t, i.e.,

$$y = \{Y_t^{(j)}(x) | j \in J\}. \tag{3.17}$$

In this case, equation (3.7) implies that

$$\int d\mu_t(x,v)\phi(x,v) = \sum_{j \in J} \int dx \sigma_t^{(j)}(x)\phi(x, u_t^{(j)}(x)), \tag{3.18}$$

where

$$\sigma_t^{(j)}(x) = \sigma_0(Y_t^{(j)}(x))|K_t^{(j)}(x)|; \quad u_t^{(j)}(x) = V_t(Y_t^{(j)}(x)) \tag{3.19}$$

and

$$K_t^{(j)}(x) = \frac{\partial(Y_{1,t}^{(j)}, Y_{2,t}^{(j)}), Y_{3,t}^{(j)})}{\partial(x_1, x_2, x_3)}. \tag{3.20}$$

Thus, (3.18) signifies that, formally,

$$d\mu_t(x,v) = \sum_{j \in J} \sigma_t^{(j)} \delta(v - u_t^{(j)}(x)) dx dv,$$

i.e., the state of the system at time t corresponds to a statistical mixture of different streams, the j'th of which has density $\sigma_t^{(j)}$ and drift velocity $u_t^{(j)}$. This implies that the local density and drift velocity are now *stochastic* variables of a hydrodynamics still governed by the Vlasov equation.

We may summarise the above observations in the following form.

Proposition 3.3. *If X_t is invertible, then the macroscopic dynamics of the system corresponds to a hydrodynamics given by the modified Euler-cum-Maxwell equations (3.12)-(3.15) Otherwise, it corresponds to the flow of a mixture of streams, and its evolution is of a stochastic type, governed by the Vlasov equation (3.1).*

Comment. Unless the domain of non-invertibility of X_t is confined to a surface, the resultant mixture of streams does not correspond to a shock wave. On this basis, it will be seen that the example of Eulerian hydrodynamic breakdown provided below is not that of a shock front.

Example of a Hydrodynamic Phase Transition

We shall now provide an example of initial conditions, which lead to a transition from a deterministic to a stochastic flow. These are conditions where both σ_0 and u_0 are functions of just a single coordinate, say x_1, and u_0 is directed along Ox_1. In this case, the flow becomes effectively one-dimensional, for the following reasons. If

$$X_t^{(b)}(x) := X_t(x+b) - b$$

for arbitrary vectors b in the plane Ox_2x_3, then, in view of the periodicity of K, if X_t is a solution of the Newtonian problem (3.6), so too is $X_t^{(b)}$. Hence, by the uniqueness of its solution, $X_t \equiv X_t^{(b)}$, which implies that the component $X_{1,t}$ of X_t depends on the coordinate x_1 only, and that $X_{2,t}(x), X_{3,t}(x)$ reduce to $x_2 + \xi_2(t)$, $x_3 + \xi_3(t)$, where the ξ's are functions of t only. Furthermore, it follows from the x_2- and x_3-components of (3.6) that these functions are both zero. In other words, the component $X_{1,t}$ of

X_t depends only on x_1, while $X_{2,t}, X_{3,t}$ remain fixed at x_2, x_3, respectively. Hence, the dynamics reduces to a one-dimensional flow. For notational convenience, we shall henceforth drop the suffix 1 from $X_{1,t}$ and x_1.

Thus, by (3.6),

$$X_t(x) = x + u_0(x)t + \int_0^t ds(t-s) \int_0^1 dy \sigma_0(y) F(X_s(x) - X_s(y)), \qquad (3.21)$$

where

$$F(x) = -\int_0^1 dx_2 \int_0^1 dx_3 \frac{\partial U_g}{\partial x}(x, x_2, x_3). \qquad (3.22)$$

Let

$$J_t(x) = \frac{\partial X_t(x)}{\partial x}. \qquad (3.23)$$

Then, by Prop. 3.3 and the implicit function theorem, the neccessary and sufficient condition for deterministic hydrodynamics is that J_t has no zeroes. We note also that the definition (3.23) permits us to re-express (3.21) in the form

$$X_t(x) = x + u_0(x)t + \int_0^t ds(t-s) \int_0^1 dy \sigma_0(y) F(\int_y^x dz J_s(z))). \qquad (3.21)'$$

Since F is a continuously differentiable function, it follows from (3.21) and (3.23) that

$$J_t(x)(1 - \int_0^t ds(t-s) \int_0^1 dy \sigma_0(y) F'(x-y)) = 1 + u_0'(x)t, \qquad (3.24)$$

where the primes denote differentiation w.r.t. x. Thus, defining

$$\|F'\| = sup\{|F'(x)|\ |x \in [0,1]\},$$

and t_0, t_1 to be the times given by

$$t_0 = 2/\|F'\|^{1/2} \qquad (3.25)$$

and

$$t_1 = min\{t(>0)|\ (1 + u_0'(x)t) = 0\ for\ some\ x \in [0,1]\}, \qquad (3.26)$$

it follows immediately from (3.24)-(3.26) that J_t is strictly positive, and hence that X_t is invertible, if $0 \le t < min(t_0, t_1)$. Therefore, by Prop. 3.3, the system evolves, in this regime, according to a deterministic hydrodynamics, given by the modified Euler-cum-Maxwell equations (3.12)-(3.15).

On the other hand, if the initial conditions are such that $t_1 < t_0$, then it follows from (3.24)-(3.26) that J_t changes sign during the interval $t \in (t_1, t_0)$ over some spatial domain $D(\subset [0,1])$. Hence, by Prop. 3.3, the hydrodynamics of the model becomes stochastic, and is (still) governed by the Vlasov equation (3.1).

We may summarise these results as follows.

Proposition 3.4. *The model exhibits a deterministic hydrodynamics, given by the modified Euler-cum-Maxwell equations (3.12)-(3.15), over the time interval $0 \le t <$*

$min(t_0, t_1)$. However, if the initial velocity profile is such that $t_1 < t_0$, then the flow undergoes a transition to stochasticity at time t_1.

Comment. It will be seen from the derivations of this result that, in the stochastic phase, the domain of non-invertibility of X_t is, in general, not confined to a single value of x, i.e., to a surface in K. Thus, in view of the comment following Prop. 3.3, the hydrodynamic phase transition described here does not correspond to the formation of a shock wave.

4. NOTE ON THE PLASMA HYDRODYNAMICS

The theory of the previous Section has been generalised [7] to the quantum version of the plasma model $\Sigma^{(N)}$, by means of a treatment of the BBGKY hierarchy of equations of motion. This treatment is based on the assumption of certain regularity conditions on the dynamics, which essentially demand that the interelectronic repulsion keeps the electrons apart in such a way as to tame the singularity in the Coulomb potential. These assumptions are implicitly conditions on the initial state of the system, since, in the limit $N \to \infty$, certain initial states can lead to a subsequent collapse and thus to an unbounded internal electric field. Although the theory of [7] was quantum mechanical, it may be easily recast into the present classical framework. We shall now summarise, within the terms of this latter framework, the results obtained on the basis of our regularity conditions [7], which we shall not specify here.

The key point is that those conditions render the Coulomb singularity harmless, and so the system, like $\Sigma_g^{(N)}$, exhibits a Vlasov dynamics, given by equations (3.1)-(3.7), but with U_g replaced by U_c. Again, the resultant hydrodynamics is deterministic if X_t is invertible, and otherwise it is stochastic. In the former case, the flow is given by the Maxwell-cum-Euler equations

$$\frac{\partial \sigma_t}{\partial t} + \nabla.(\sigma_t u_t) = 0; \quad \frac{\partial u_t}{\partial t} + (u_t.\nabla)u_t = E_t; \quad \nabla.E_t = (\sigma_t - 1). \tag{4.1}$$

Example of a Hydrodynamic Phase Transition

Again, we consider the situation where σ_0 and u_0 are functions of a single coordinate x_1 only. This leads again to a one-dimensional flow, satisfying equations (3.21)-(3.22), with $x_1 = x$, but with U_g replaced by U_c. On differentiating the first of these equations w.r.t. x and using (2.8), we obtain the following equation for $J_t := \partial X_t / \partial x$.

$$J_t(x) = 1 + u_0'(x)t + \int_0^t ds(t-s)J_s(x)(-1 + \int_0^1 dy \sigma_0(y)\delta(X_s(x) - X_s(y))), \tag{4.2}$$

where δ is the Dirac distribution on $[0,1]$, subject to periodic boundary conditions.

Let us first examine the consequences of assuming that X_t is invertible, i.e. that J_t is strictly positive, over a time interval $0 \le t < \tau_0$, for some positive τ_0. In this case,

$$J_s(x)\delta(X_s(x) - X_s(y)) \equiv \delta(x-y) \quad \forall s \in [0, \tau_0)$$

and therefore (4.2) reduces to

$$J_t(x) = 1 + u_0'(x)t + \int_0^t ds(t-s)(\sigma_0(x) - J_s(x)), \tag{4.3}$$

19

i.e.,
$$(\frac{d^2}{dt^2}+1)J_t(x) = \sigma_0(x); \text{ with } J_0(x) = 1; \; \dot{J}_0(x) = u'_0(x), \tag{4.4}$$

where $\dot{J}_t = dJ_t/dt$. Hence,
$$J_t(x) = \sigma_0(x) + (1-\sigma_0(x))\cos(t) + u'_0(x)\sin(t). \tag{4.5}$$

In view of the non-negativity of σ_0, this equation implies that J_t is strictly positive for all $t \geq 0$ if and only if
$$(\sigma_0(x)-1)^2 + (u'_0(x))^2 < (\sigma_0(x))^2 \;\; \forall x \in [0,1]. \tag{4.6}$$

Otherwise, it is positive only when $t < \tau$, defined as the least positive value of t for which
$$\sigma_0(x) + (1-\sigma_0(x))\cos(t) + u'_0(x)\sin(t) = 0 \; \text{for some } x \in [0,1]. \tag{4.7}$$

Hence, if the condition (4.6) is violated, the assumption that X_t is invertible, i.e. that the hydrodynamics is deterministic, is untenable for $t > \tau$. Hence, by a reductio ad absurdum argument, the flow must be stochastic for $t > \tau$ if (4.6) is violated. The proof that the flow is deterministic for $t < \tau$ if (4.6) is violated, and for all positive t if that condition is fulfilled follows from the assumptions underlying the Vlasov dynamics in [7]. Thus, we have the following result.

Proposition 4.1. *Under the assumptions of [7], the hydrodynamics of the Coulomb model is given by the Euler-cum-Maxwell equations (4.1) at all times, provided that the initial density and velocity profiles satisfy the condition (4.6). Otherwise there is a transition to a stochastic flow at a certain time τ, given by the least positive value of t satisfying (4.7).*

Comment. Again, the domain of non-invertibility of X_t in the stochastic phase is not confined to a single value of x, i.e. to a surface in K; and thus the hydrodynamical phase transition does not correspond to the formation of a shock wave.

5. CONCLUDING REMARKS

We have obtained a microscopic derivation of the hydrodynamics of the models $\Sigma^{(N)}$ and $\Sigma_g^{(N)}$, and shown that both undergo transitions from deterministic to stochastic flow.

On the other hand, we have not obtained any explicit properties of the stochastic phase, beyond the point made, in the comments following Props. 3.4 and 4.1, that they do not correspond to shock waves. It would obviously be interesting to know whether they carry sufficient stochasticity to be candidates for turbulent flow. In this connection we remark that, since the Vlasov dynamics, which is at the core of our treatment, is simply the Liouville probabilistic version of Lagrangian hydrodynamics, there is some parallel between the scheme presented here and that designed by Foias [9] for the treatment of turbulence on the basis of a Liouville equation governing the stochastics of Navier-Stokes flows.

We note here that the hydrodynamics obtained here is completely inviscid. The reason for this, as in [6] (cf. discussion there at the end of §1), is that our macroscopic description is effected on the largest available length scale, L, and that, consequently, the viscous forces are 'scaled away'. Thus, the hydrodynamic picture we have obtained should be regarded as no more than a skeletal version of that of a real plasma.

Finally, we remark that the results presented here also ensue from a quantum mechanical treatment of the models concerned [7]. The essential reason for this is that the transformation $(X, P) \to (x, p) \equiv (X/L, P/mL\omega)$ from the microscopic to the macroscopic description leads to a rescaling of the commutator of position and momentum, and hence of the Planck constant, by a factor $(mL^2\omega)^{-1}$. In other words, the effective Planck constant governing the dynamics of the models $\Sigma^{(N)}$ and $\Sigma_g^{(N)}$ is not the usual $\hbar$, but $\hbar/mL^2\omega$, which tends to zero as $N \to \infty$, by (2.1). As a consequence, the dynamics of these models becomes classical in this limit.

REFERENCES

1. A. De Masi, N. Ianiro, A. Pellegrinotti and E. Presutti, *in:* "Studies in Statistical Mechanics", **XI**, 123, W Montroll and J.L. Lebowitz Eds. (1984).
2. H. Spohn, "Large-Scale Dynamics of Interacting Particles", Springer (1991).
3. D. Wick, *J. Stat. Phys.*, **38**, 1015 (1985).
4. J. Gärtner and E. Presutti, *Ann. Inst. H. Poincaré. Phys. Theor.*, **53**, 1 (1990).
5. E.B. Davies, *J. Stat. Phys.*, **18**, 161 (1978).
6. G.L. Sewell, *J. Math. Phys.*, **26**, 2324 (1985).
7. G.L. Sewell, Hydrodynamical phase transition in a quantum plasma model, Preprint.
8. W. Braun and K. Hepp, *Commun. Math. Phys.*, **56**, 101 (1977).
9. C. Foias, *Russian Math. Surveys*, **29**, 293 (1974).

FERROMAGNETISM IN ITINERANT ELECTRON SYSTEMS: RIGOROUS EXAMPLES FROM FLAT-BAND HUBBARD MODELS

H. Tasaki

Gakushuin University
Toshima-ku, Tokyo 171, Japan

INTRODUCTION

The Hubbard model has been attracting considerable interest in physics community. In short, the Hubbard model is an "over simplified" model of electrons in a solid, which takes into account the itinerant nature of electrons and the short range Coulomb interaction between them. Like all the other "good models" in physics, the simplicity of the definition does not imply the model is easy to solve or understand. Even on heuristic levels, it is not yet clear which phenomena observed in the real itinerant electron systems can be reproduced within the Hubbard model[1]. Rigorous results directed to such physical questions should be welcome, but are still very rare. In the present note, we will describe our recent attempts to understand mechanisms underlying the well-known phenomena of ferromagnetism[2]. The main topic is the recent first rigorous example of three dimensional itinerant electron ferromagnetism which is stable under fluctuation of the electron density [27, 20].

Let us start by describing some historical backgrounds. In some solids, electronic spins spontaneously align with each other to form strong ferromagnetic ordering. A familiar example is Fe, which maintains long range magnetic order up to the Curie temperature, 1043K. Given the fact that interactions between electrons in a solid are almost spin-independent, the existence of such a strong order may sound as a mystery. As we shall describe below, this has indeed been an interesting open problem in theoretical physics for quite a long time.

In 1928, Heisenberg [5] pointed out that the spin-independent Coulomb interaction between electrons, when combined with the Pauli exclusion principle, can generate effective interaction between electron spins. Heisenberg's picture of ferromagnetism

---

[1] A recent notorious problem (which, of course, is not yet solved) has been to explain the high-T_c superconductivity from a version of the Hubbard model. A very preliminary mathematical approach to such a problem can be found, for example, in [28].

[2] The present note is based on a joint work with Andreas Mielke.

was that the relevant electrons are mostly localized at atomic sites, and their spin degrees of freedom interact with each other via "exchange interaction". It has been realized, however, that his exchange interaction usually has the sign which leads to antiferromagnetic interaction rather than ferromagnetic one. (See [6] for a review.) Nevertheless, Heisenberg's idea still plays a fundamental role in modern theories of ferromagnetism.

A somewhat different approach to ferromagnetism, which was originated by Bloch [2], is to look for a mechanism of ferromagnetism in which the itinerant nature of electrons, as well as the Coulomb interaction and Pauli principle, play fundamental roles. This project, combined with sophisticated band-theoretic techniques, has led to many approximate theories [7]. A common feature of all these theories is that they are based on the Hartree-Fock approximation (*i.e.*, mean field theory) and its perturbative corrections. Although such approximations can lead to reasonable conclusions in some situations, they have serious disadvantage from a theoretical point of view. The basic strategy of the approximations is to treat electrons with up and down spins as different species of particles, and then introduce some self-consistency conditions. By doing this, one severely destroys the original $SU(2)$ invariance of the model and gets Z_2 invariant self-consistent equations. The existence of ferromagnetism then reduces to a problem of spontaneous breakdown of the discrete Z_2 symmetry, which is essentially different from the original problem, a spontaneous breakdown of the continuous $SU(2)$ symmetry. As a consequence, the approximate theories give only two ferromagnetic states with the net magnetization pointing up or down, instead of expected infinitely many states with an arbitrary direction of magnetization. Since a continuous symmetry breaking is a very subtle phenomenon in general, results based on the Hartree-Fock approximation and its improvements are not conclusive enough to answer the fundamental question whether spin-independent Coulomb interaction alone can be the origin of a realistic ferromagnetism in an itinerant electron system. We stress that such a critical point of view has been held by many physicists. See, for example, the review of Herring [7].

Given the subtlety of the problem, it is desirable to have idealized models in which one can develop concrete scenarios for the itinerant electron ferromagnetism. The so called Hubbard model [8, 9] is a simple but nontrivial model suitable for developing such scenarios. There has been a considerable amount of heuristic works (mostly based on the Hartree-Fock approximation and its improvements) devoted to theories of ferromagnetism in the Hubbard model. Since the literature is too large to catalogue here, we only refer to the pioneering work of Kanamori [9] and a review by Herring [7]. The Hartree-Fock approximation applied to the Hubbard model yields the so called Stoner criterion for ferromagnetism [7, 8, 9]. It says that ferromagnetism occurs if $D_F U > 1$, where D_F is the single particle-density of states at the fermi level and U is the strength of the Coulomb repulsion. Although this is certainly not true in general, large values of D_F and/or U determine the region in parameter space where one may find ferromagnetism.

In 1965, the first rigorous example of ferromagnetism in the Hubbard model was given by Nagaoka [21], and independently by Thouless [29, 13]. It was proved that certain Hubbard models have ground states with saturated magnetization when there is exactly one hole and the Coulomb repulsion is infinite. Recently, it was pointed out that the theorem extends to a general class of models which satisfy a certain connectivity condition [26]. Whether the Nagaoka-Thouless ferromagnetism survives in the models with finite density of holes and/or finite Coulomb repulsion is a very interesting but totally unsolved problem [3, 22, 24, 25, 30].

In 1989, Lieb proved an important theorem on general properties of the ground states of the Hubbard model [14]. As a consequence of the theorem, he showed that a class of Hubbard models on asymmetric bipartite lattices with finite Coulomb interaction have ferromagnetic ground states at half filling. It is sometimes argued that Lieb's examples represent ferrimagnetism (*i.e.*, antiferromagnetism on a bipartite lattice in which the number of the sites in two sublattices are different) rather than ferromagnetism. Although this might be true when the Coulomb repulsion is sufficiently large and the dimension is high enough, we believe mechanisms underlying his examples are much richer in general situations.

A more recent example is due to Mielke, who studied Hubbard models on general line graphs [16, 17]. A special feature of his models is that the single-electron Schrödinger equation corresponding to the Hubbard model has highly degenerate ground states. He proved that these models with nonvanishing Coulomb interaction have ground states with saturated magnetization when the electron number is exactly equal to the dimension of degeneracy of the single-electron ground states. Mielke also extended his results to a finite range of electron filling factor in certain two dimensional models [18]. See [19] for the recent results for a more general class of models.

The latest example is due to Tasaki [27], who proved the existence of ferromagnetism in a class of Hubbard models with nonvanishing Coulomb interaction on decorated lattices. As in the Mielke's models, the class of models have highly degenerate ground states in the corresponding single-electron Schrödinger equation. It was proved that the ground states of the Hubbard model exhibit ferromagnetism in a finite range of electron filling factor. This work covered models in three dimensions as well. These results were further refined by Mielke and Tasaki [20]. We will discuss this example in the present note.

A common feature in the examples of Lieb, Mielke, and Tasaki is that each model has a completely degenerate band in the corresponding single-electron spectrum, and the ferromagnetism (in interacting many electron problem) is proved when the degenerate band is exactly or nearly half-filled. Such ferromagnetism may be called "flat band ferromagnetism". This situation, where D_F is infinite, is, in view of the Stoner criterion, in some sense dual to the aforementioned theorem of Nagaoka where U is infinite.

Possibility of "flat band ferromagnetism" has been discussed for a long time since the pioneering work of Slater, Statz and Koster [23] in 1953. But general consensus on whether ferromagnetism appears or not has been lacking. See Section VIII.2 of [7].

HUBBARD MODELS WITH DEGENERATE SINGLE-ELECTRON GROUND STATES

As a warmup, we discuss a class of Hubbard models, in which one can easily construct exact ground states which are ferromagnetic. We shall also give some general definitions.

We take a finite lattice Λ with $|\Lambda|$ sites and consider a Hubbard model on Λ. Throughout the present note, we denote by $|S|$ the number of elements in a set S. The Hamiltonian is

$$H = H_{\text{hop}} + H_{\text{int}}, \tag{1}$$

where
$$H_{\text{hop}} = \sum_{x,y \in \Lambda, \sigma=\uparrow,\downarrow} t_{xy} c_{x\sigma}^* c_{y\sigma}, \tag{2}$$

and
$$H_{\text{int}} = \sum_{x \in \Lambda} U_x n_{x\uparrow} n_{x\downarrow}. \tag{3}$$

As usual $c_{x\sigma}^*$ and $c_{x\sigma}$ are the creation and the annihilation operators, respectively, of an electron at site $x \in \Lambda$ with spin $\sigma = \uparrow, \downarrow$. They satisfy the anticommutation relations

$$\{c_{x\sigma}, c_{y\tau}^*\} = \delta_{xy}\delta_{\sigma\tau}, \quad \{c_{x\sigma}, c_{y\tau}\} = \{c_{x\sigma}^*, c_{y\tau}^*\} = 0, \tag{4}$$

for any $x, y \in \Lambda$ and $\sigma, \tau = \uparrow, \downarrow$, where $\{A, B\} = AB + BA$. The number operator is defined as $n_{x\sigma} = c_{x\sigma}^* c_{x\sigma}$. The hopping matrix (t_{xy}) is real symmetric, and the on-site Coulomb repulsion U_x is strictly positive. The total electron number operator is $\hat{N}_e = \sum_{x \in \Lambda}(n_{x\uparrow} + n_{x\downarrow})$, and we denote by N_e its eigenvalue. A standard prescription is to consider the eigenspace of $\hat{N}_e$ with a given eigenvalue N_e, or to consider certain grand canonical ensemble with the expectation value of $\hat{N}_e$ fixed. The quantity $N_e/(2|\Lambda|)$ is called the electron filling factor.

We also define the spin operators by

$$S_x^{(\alpha)} = \sum_{\sigma,\tau=\uparrow,\downarrow} c_{x\sigma}^* p_{\sigma\tau}^{(\alpha)} c_{x\tau}/2, \tag{5}$$

where $p_{\sigma\tau}^{(\alpha)}$ with $\alpha = 1, 2, 3$ are the Pauli matrices,

$$p^{(1)} = \begin{pmatrix} 0 & 1 \\ 1 & 0 \end{pmatrix}, \quad p^{(2)} = \begin{pmatrix} 0 & -i \\ i & 0 \end{pmatrix}, \quad p^{(3)} = \begin{pmatrix} 1 & 0 \\ 0 & -1 \end{pmatrix}. \tag{6}$$

We denote by $S_{\text{tot}}(S_{\text{tot}} + 1)$ the eigenvalue of

$$(\mathbf{S}_{\text{tot}})^2 = \sum_{x,y \in \Lambda} \sum_{\alpha=1,2,3} S_x^{(\alpha)} S_y^{(\alpha)}, \tag{7}$$

which is the square of the total spin operator. We say that a state exhibits ferromagnetism if it has S_{tot} proportional to the system size $|\Lambda|$.

The single-electron Schrödinger equation corresponding to the hopping Hamiltonian (2) is

$$\sum_{y \in \Lambda} t_{xy} \varphi_y = \varepsilon \varphi_x, \tag{8}$$

where $\varphi_x \in \mathbf{C}$, and ε is the single-electron energy. Suppose that our hopping matrix (t_{xy}) has a special property that the (single-electron) ground states of the Schrödinger equation (8) are N_d-fold degenerate. We denote the ground state energy by ε_0, and the space of degenerate ground states by $\mathbf{H}_0$. Let $\{\varphi^{(u)}\}_{u \in V}$ be a complete linear independent basis for the space $\mathbf{H}_0$, where V (with $|V| = N_d$) is the set of indices. The wave function and the creation operator corresponding to a basis state $\varphi^{(u)}$ are denoted as $\{\varphi_x^{(u)}\}_{x \in \Lambda}$ and

$$a_{u\sigma}^* = \sum_{x \in \Lambda} \varphi_x^{(u)} c_{x\sigma}^*, \tag{9}$$

respectively.

Consider a ferromagnetic state

$$\Phi_{A\uparrow} = \prod_{u \in A} a^*_{u\uparrow} \Phi_0, \qquad (10)$$

where A is an arbitrary subset of the index set V, and Φ_0 is the vacuum state, *i.e.*, the state with no electrons. Since the basis $\{\varphi^{(u)}\}_{u \in V}$ is linear independent, the state $\Phi_{A\uparrow}$ is nonvanishing. The electron number of the state $\Phi_{A\uparrow}$ is given by $N_e = |A|$. It is not difficult to see that $H_{\text{hop}} \Phi_{A\uparrow} = N_e \varepsilon_0 \Phi_{A\uparrow}$, where $N_e \varepsilon_0$ is the lowest possible eigenvalue of H_{hop} in the subspace with the electron number is fixed to N_e. On the other hand, we already know from the construction that $H_{\text{int}} \Phi_{A\uparrow} = 0$, where 0 is the minimum possible eigenvalue of H_{int}. Therefore we have found the following.

Theorem 1 *Consider a Hubbard model with the Hamiltonian described by (1), (2) and (3). In the subspace with the electron number fixed to $N_e \leq N_d$, the ground state energy is $N_e \varepsilon_0$, and the ferromagnetic state (10) with an arbitrary subset $A \subset V$ with $|A| = N_e$ is a ground state.*

Such a construction of ferromagnetic ground states may be standard, but is not sufficient to draw any meaningful conclusion about the magnetism of the system. A really important (and much more delicate) problem is whether these ferromagnetic states are the only ground states, or what are the other ground states, if any. Note that the single-electron density of states D_F is infinite for $N_e \leq N_d$. Therefore the Stoner criterion $D_F U > 1$ predicts the appearance of only the ferromagnetic ground states for any value of $U_x > 0$. But it soon turns out that the situation is not that simple[3].

When $N_e = N_d$, Mielke [19] recently gave a necessary and sufficient condition for the ferromagnetic ground state (10) to be the unique ground state.

DEFINITION OF THE MODEL

We will introduce a class of models in which we can get almost complete information about the ground states. See [20] for other models which can be treated by the same method.

Consider a d-dimensional $L \times \cdots \times L$ hypercubic lattice, where L is an even integer, and denote by V the set of sites[4]. We impose periodic boundary conditions. Let

$$B = \{\{v, w\} \mid v, w \in V, |v - w| = 1\}, \qquad (11)$$

be the set of bonds, where $|v - w|$ denotes the euclidean distance between the sites v and w. For each bond $\{v, w\}$ in B, we denote by $m(v, w)$ the point taken in the middle of the sites v and w. We define

$$M = \{m(v, w) \mid \{v, w\} \in B\}, \qquad (12)$$

and consider the decorated hypercubic lattice $\Lambda = V \cup M$.

---

[3] A trivial counter example is the model with $t_{xy} = 0$ for all x, y, which has the degeneracy $N_d = |\Lambda|$. Any state with no doubly occupied site is a ground state of this model, so there can be no magnetic ordering. We have paramagnetism, in contrast to the conclusion based on the Stoner criterion.

[4] In the previous section, we denoted by V the index set for a basis of $\mathbf{H}_0$. We use the same symbol here since one can label a basis state by a site in V. See [27, 20] for details.

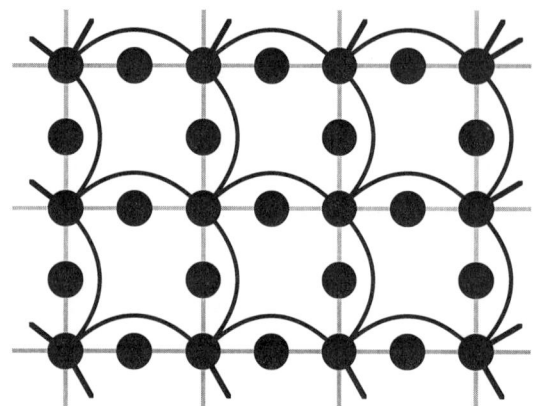

Figure 1. The decorated square lattice. The hopping matrix elements are given by $t_{xy} = t$ for a black line, $t_{xy} = \lambda t$ for a gray line, $t_{xx} = 4t$ for a site x of the square lattice, and $t_{xx} = \lambda^2 t$ for a site x at the middle of a bond, where $t, \lambda > 0$. The on-site Coulomb repulsion is nonvanishing for any site. It is proved that the ground states exhibit ferromagnetism when the electron filling factor ρ is not more than and sufficiently close to $\rho_0 = 1/6$, and paramagnetism when ρ is sufficiently small.

We will study the Hubbard model on the lattice Λ. We again write our Hamiltonian as $H = H_{\text{hop}} + H_{\text{int}}$ where

$$H_{\text{hop}} = \sum_{\sigma=\uparrow,\downarrow} \sum_{\{v,w\}\in B} t(c^*_{v\sigma} + c^*_{w\sigma} + \lambda c^*_{m(v,w)\sigma})(c_{v\sigma} + c_{w\sigma} + \lambda c_{m(v,w)\sigma}), \quad (13)$$

and

$$H_{\text{int}} = U \sum_{u \in V} n_{u\uparrow} n_{u\downarrow} + U' \sum_{x \in M} n_{x\uparrow} n_{x\downarrow}, \quad (14)$$

with $t > 0$, $\lambda > 0$, $U > 0$, and $U' > 0$. Note that the above hopping Hamiltonian and interaction Hamiltonian can be written in the form of (2) and (3) by suitably choosing the hopping matrix (t_{xy}) and the interaction U_x. See Figure 1.

The special form of the hopping Hamiltonian (13) makes the present model fall into the class considered in the previous section. To see this, we should study the eigenstates of the single-electron Schrödinger equation (8) corresponding to (13). A standard way is to use the Fourier transformation to directly solve the eigenvalue problem. One easily finds that the eigenstates can be classified into $(d+1)$ bands, whose dispersion relations are given by

$$\varepsilon_i(k) = \begin{cases} 0 & \text{for } i = 1, \\ \lambda^2 t & \text{for } i = 2, 3, \cdots, d, \\ \lambda^2 t + 2t \sum_{j=1}^{d}(1 + \cos k_j) & \text{for } i = d+1, \end{cases} \quad (15)$$

where $i = 1, 2, \cdots, d+1$ is the index for the bands, and $k = (k_1, \cdots, k_d)$ is the wave vector with $-\pi < k_j \leq \pi$. Note that the present model has a singular band structure, in which most of the bands are dispersion-less.

We have thus found that[5], in the single-electron Schrödinger equation (8) corresponding to the hopping Hamiltonian (13), the ground states have $\varepsilon = 0$, and are

---

[5]In our paper [20], the degeneracy of the ground states are determined not by using the direct Fourier transformation calculation.

$N_d = L^d$-fold degenerate. From Theorem 1, we get the following preliminary result about ferromagnetism.

Corollary 2 *In the subspace with the electron number fixed to $N_e \leq N_d(= L^d)$, the ground state energy of the full Hubbard Hamiltonian H (defined by (13) and (14)) is 0. Among the ground states, there are the ferromagnetic states defined as (10).*

FERROMAGNETISM FOR A SPECIAL ELECTRON NUMBER

Our first nontrivial result about ferromagnetism deals with the model with a special electron number.

Theorem 3 *In the subspace with the electron number fixed to $N_e = N_d(= L^d)$, the ground states of H have $S_{tot} = N_e/2$ and are nondegenerate apart from the $(2S_{tot}+1)$-fold spin degeneracy.*

The theorem establishes that the ground states exhibit the maximum possible ferromagnetism. Mielke [16, 17] proved similar result for a general class of Hubbard models on line graphs and on some decorated graphs similar to our model but with additional hopping matrix elements between the sites $m(v, w)$. Recently Mielke [19] extended his results to a general class of Hubbard models with a degenerate single-electron ground states. The result of [19] includes that of both [16, 17] and the above Theorem 3. Note that, in the band-theoretic language, the degenerate single-electron ground state band (indexed as $i = 1$ in (15)) is exactly half-filled when $N_e = N_d$.

Note that the above theorem applies to the model with $d = 1$ as well. This does not contradict with the general result of Lieb and Mattis [15, 1], which inhibits ferromagnetic order in one dimension, since our model contains non-nearest-neighbor hoppings.

We can show [20] that the coherence length in the ground state is equal to λ^{-1}. This is extremely short if $\lambda \gg 1$. In this limit, our model resembles that of nearly localized electrons (as in the Heisenberg's work [5]), and the origin of the ferromagnetism may be interpreted as a "super exchange interaction" via the nonmagnetic atom on the site between two magnetic atomic sites. We also expect that the model with $N_e = N_d$ describes a kind of Mott insulator, at least when $\lambda \gg 1$ and $U \gg 1$.

When $\lambda \ll 1$, on the other hand, the coherence length λ^{-1} becomes large, and it is no longer possible to regard the present model as that of localized electrons. It seems that even the simplest model with $N_e = N_d$ contains many interesting physics, which remain to be understood.

FERROMAGNETISM IN A FINITE RANGE OF ELECTRON NUMBERS

Let us investigate whether the ferromagnetism established for the special electron number is stable when the electron number is changed. We stress that any physically realistic model of ferromagnetism should possess such stability.

The ground states of H are highly degenerate for $N_e < N_d$. To get a physically meaningful results, we have to consider the average over the degenerate ground states.

Instead of fixing the electron number explicitly, we will employ the grand canonical formalism, and control the expectation value of the electron number by choosing appropriate chemical potential. The reason for using the grand canonical formalism here is mainly technical.

For an arbitrary operator O, we define the grand canonical-like average by

$$\langle O \rangle_\mu = \frac{\text{Tr}[O \exp(\mu \hat{N}_e) P_0]}{\text{Tr}[\exp(\mu \hat{N}_e) P_0]}, \tag{16}$$

where P_0 is the orthogonal projection operator onto the eigenspace of H with the eigenvalue 0. It is expected that, by choosing a suitable (dimensionless) chemical potential μ in (16), we recover zero-temperature properties of the system with a desired electron filling factor. If the electron filling factor had a pathological behavior as a function of μ, the use of the grand canonical formalism could not be justified. The following theorem guarantees that this is not the case.

Theorem 4 *For arbitrary values of μ, we have the upper and lower bounds*

$$(1 + \frac{e^{-\mu}}{2})^{-1} \geq \frac{\langle \hat{N}_e \rangle_\mu}{N_d} \geq 3^{-d}(1 + \frac{e^{-\mu}}{2})^{-1}. \tag{17}$$

In the dimensions $d \geq 2$, there are positive finite constants μ_1, c_1 which depend only on the dimension d, and for any $\mu \geq \mu_1$, we have the lower bound

$$\frac{\langle \hat{N}_e \rangle_\mu}{N_d} \geq 1 - c_1 e^{-\mu}. \tag{18}$$

The above bounds determine the behavior of the electron filling factor for extreme values of the chemical potential. When μ is negative and its absolute value is large (compared to 1), (17) implies

$$\frac{\langle \hat{N}_e \rangle_\mu}{2|\Lambda|} \approx e^\mu. \tag{19}$$

When μ is positive and large, (17) and (18) imply

$$\rho_0 - \frac{\langle \hat{N}_e \rangle_\mu}{2|\Lambda|} \approx e^{-\mu}, \tag{20}$$

where the maximum value of the electron filling factor is defined as $\rho_0 = N_d/(2|\Lambda|) = (2d+2)^{-1}$. The relation $\approx$ means that the both sides behave equally apart from multiplication by a uniformly bounded function of μ.

Now we can state our main theorem.

Theorem 5 *In the dimensions $d \geq 2$, there are finite constants c_2, c_3, μ_1, μ_2 (with $c_2, c_3, \mu_1 > 0$ and $\mu_2 < 0$) which depend only on the dimension d and not on the size of the lattice. For any $\mu \geq \mu_1$, we have*

$$S_{\max}(S_{\max} + 1) \geq \langle (\mathbf{S}_{\text{tot}})^2 \rangle_\mu \geq S_{\max}(S_{\max} + 1)(1 - c_2 e^{-\mu}), \tag{21}$$

where $S_{\max} = N_d/2$. For any $\mu \leq \mu_2$, we have

$$\frac{3}{4}\langle \hat{N}_e \rangle_\mu \leq \langle (\mathbf{S}_{\text{tot}})^2 \rangle_\mu \leq \frac{3}{4}\langle \hat{N}_e \rangle_\mu + c_3|V|e^{2\mu}. \tag{22}$$

Note that, when the bounds (21) hold, the total spin of the model is proportional to the number of sites $|\Lambda|$. When the bounds (22) hold, on the other hand, the total spin is proportional to the square root of $|\Lambda|$. Therefore Theorem 5 establishes that the ground states of our Hubbard model exhibit ferromagnetism when the filling factor is not more than and sufficiently close to ρ_0, and paramagnetism when the filling factor is sufficiently small. In the band-theoretic language, ferromagnetism appears when the degenerate ground state band is nearly half-filled.

The requirement that the dimension is not less than two in Theorem 5 is essential in controlling ferromagnetic region by using a kind of Peierls argument [20]. The paramagnetic part (the bounds (22)) of Theorem 5 easily extends to the case with $d = 1$. We believe that the one-dimensional model exhibits paramagnetism for all the values of $\mu < \infty$.

DISCUSSIONS

The main physical significance of the results [27, 20] we have briefly described is that they provided the first example of three dimensional itinerant electron system which was proved to exhibit ferromagnetism in a finite range of electron filling factor. Our results and their proof demonstrate that there is a mechanism generated by the Coulomb interaction which selects ferromagnetic states as ground states. The selection mechanism works most effectively when the degenerate single-electron band is nearly half filled (in the sense that the electron number is nearly equal to the dimension of degeneracy in the single- electron ground states), but becomes ineffective when the electron density is too small.

Although the models treated in the present note are still artificial, we hope that such a selection mechanism generally takes place in a Hubbard model with a large density of states at the bottom of the single-electron energy band. Such a Hubbard model should exhibit ferromagnetism for suitable electron filling factors when the Coulomb interaction is sufficiently large. In other words, we hope the present examples to provide the simplest models in a "universality class" of Hubbard models which exhibit realistic and robust ferromagnetism. The recent numerical and heuristic works of Kusakabe and Aoki [12] for closely related models (which were proposed in [17, 20]) indeed suggest that the ferromagnetism in the present models is robust and not pathological. It is a challenging problem to extend our results to more general situations.

It is also important to investigate whether ferromagnetic order in the three dimensional model is present at finite temperatures. Recall that, in one and two dimensions, ferromagnetic order in any Hubbard model is destroyed by thermal fluctuation at finite temperatures [4, 10].

It is encouraging that ferromagnetism observed in transient metals, like Ni, might have similar properties as our ferromagnetism. The 3d band of Ni has large (single-electron) density of states at the top of the band, and the filling factor of the band is close to one [9]. After the electron-hole transformation, the situation becomes quite similar to that in the models treated in the present note. When the number of electrons is equal to the degeneracy of the single-electron ground states, our model might also resemble certain ferromagnetic ionic crystals.

A reader might feel that the Hubbard models treated in the present work have "nonstandard" hopping matrix elements (including the next nearest neighbor hopping) when compared with the "standard" models with uniform nearest neighbor hoppings.

(See eq.(13) and Figure 1.) It should be remarked that, unlike in the lattice field theories, for example, the hopping matrix in the Hubbard model need not be the naive discretization of the Laplacian. It is determined from overlap integrals between electron orbits, and can be quite complicated in general. Another important reason for studying a "nonstandard" model is the remarkable "nonuniversality" of the Hubbard model. These days it is suspected that the simplest Hubbard model with uniform hopping does not exhibit neither ferromagnetism nor superconductivity for reasonable parameter values. However the chance is big that a particular version of the Hubbard model with specific hopping matrix (*i.e.*, band structure) and filling factor shows such interesting properties [28]. This is reminiscent of the rich nonuniversal behavior one observes in the actual itinerant electron systems in nature.

Acknowledgments

I wish to thank Andreas Mielke for enjoyable collaboration, Tohru Koma for stimulating discussions and valuable comments, Arisato Kawabata, Kenn Kubo and Koichi Kusakabe for useful discussions. I also thank the organizers for inviting me to give a talk.

REFERENCES

1. Aizenman, M. and E.H. Lieb, *Phys. Rev. Lett.*, **65**, 1470 (1990).
2. Bloch, F., *Z. Phys.*, **57**, 545 (1929).
3. Douçot, B. and X. G. Wen, *Phys. Rev.*, **B40**, 2719 (1989).
4. Ghosh, D.K., Phys. *Rev. Lett.*, **27**, 1584 (1971).
5. Heisenberg, W.J., *Z. Phys.*, **49**, 619 (1928).
6. Herring, C., in "Magnetism" IIB, G. T. Rado and H. Suhl eds., Academic Press, New York, London, (1966).
7. Herring, C., in Magnetism IV, G. T. Rado and H. Suhl eds., Academic Press, New York, London, (1966).
8. Hubbard, J., *Proc. Roy. Soc. A London*, **276**, 238 (1963).
9. Kanamori, J., *Prog. Theor. Phys.*, **30**, 275 (1963).
10. Koma, T. and H. Tasaki, *Phys. Rev. Lett.*, **68**, 3248 (1992).
11. Kusakabe, K. and H. Aoki, *J. Phys. Soc. Jpn.*, **61**, 1165 (1992).
12. Kusakabe, K. and H. Aoki, Robustness of the Ferromagnetism in Flat Bands, preprint.
13. Lieb, E.H., in "Phase Transitions, Proceedings of the Fourteenth Solvay Conference," Wiley, Interscience, New York, (1971), p.45.
14. Lieb, E.H., *Phys. Rev. Lett.*, **62**, 1201 (1989).
15. Lieb, E.H. and D. Mattis, *Phys. Rev.*, **125**, 164 (1962).
16. Mielke, A., *J. Phys. A : Math. Gen.*, **24**, L73 (1991).
17. Mielke, A., *J. Phys. A: Math. Gen.*, **24**, 3311 (1991).
18. Mielke, A., *J. Phys. A: Math. Gen.*, **25**, 4335 (1992).
19. Mielke, A., *Phys. Lett. A.*, **174**, 443 (1993).
20. Mielke, A. and H. Tasaki, *Comm. Math. Phys.*, to appear.
21. Nagaoka, Y., *Phys. Rev.*, **147**, 392 (1966).
22. Shastry, B.S., H.R. Krishnamurthy and P.W. Anderson, *Phys. Rev.*, **B41**, 2375 (1990).
23. Slater, J.C., H. Statz and G.F. Koster, *Phys. Rev.*, **91**, 1323 (1953).
24. Sütő, A., *Phys. Rev.*, **B43**, 8779 (1991).
25. Sütő, A., *Comm. Math. Phys.*, **140**, 43 (1991).

26. Tasaki, H., *Phys. Rev.*, **B40**, 9192 (1989).
27. Tasaki, H., *Phys. Rev. Lett.*, **69**, 1608 (1992).
28. Tasaki, H., *Phys. Rev. Lett.*, **70**, 3303 (1993).
29. Thouless, D.J., *Proc. Phys. Soc. London*, **86**, 893 (1965).
30. Tóth, B., *Lett. Math. Phys.*, **22**, 321, (1991).

HOMOGENEITY IN THE GROUND STATE OF THE TWO-DIMENSIONAL FALICOV-KIMBALL MODEL

T. Kennedy

Department of Mathematics
University of Arizona
Tucson, AZ 85721, USA

INTRODUCTION

The Falicov-Kimball model is one of the simplest models of itinerant fermions with an on site interaction. It consists of spinless fermions on a lattice which may hop between nearest neighbor sites and which interact with another species of particles which are not allowed to hop. There are several physical interpretations of this model. We will use the interpretation in which one thinks of the hopping fermions as electrons and the static particles as nuclei. The Hamiltonian is

$$H = \sum_{x,y:|x-y|=1} c_x^\dagger c_y - U \sum_x c_x^\dagger c_x W(x)$$

where $c_x^\dagger$ and c_x are creation and annihilation operators for the electrons, $W(x) = 1$ if there is a nuclei at x and $W(x) = 0$ if there is not.

The electrons do not interact with each other, so H is the second quantized form of the single electron Hamiltonian consisting of a hopping term and the potential $W(x)$. The model is not trivial because we consider all possible $W(x)$. We will only consider the ground state problem for a fixed density of nuclei and a fixed density of electrons. This means that for each W with the desired density of nuclei, we compute the ground state energy of H with the desired density of electrons. We then look for the W which minimizes this energy. We will refer to such a W as a ground state, although it is only a partial description of the ground state since it does not include the electronic wave function.

Rigorous and non-rigorous work has suggested two general principles. The first principle concerns the neutral case in which the density of electrons is equal to the density of nuclei. It has been proved that when these densities both equal 1/2, then for all nonzero U the ground state is obtained by putting the nuclei all on one of the two sublattices (a checkerboard configuration)[1,2,3]. Thus the nuclei like to spread out as much as possible. For one dimension Lemberger[4] give a precise definition of the phrase "most homogeneous" and proved that for large U the ground state is the most homogeneous configuration. (Some results in this direction had been obtained previously [5,6]. A stronger result has been proved for low densities [7,8].) The appropriate definition of most homogeneous in more than one dimension is not clear. In two dimensions for densities $1/3, 1/4, 1/5$, the ground state has been determined nonrigorously by computing the perturbation theory in $1/U$ for the first two nontrivial orders[9]. The ground states they found appear to be good candidates for the most homogeneous configurations with those densities. (See figure 1.)

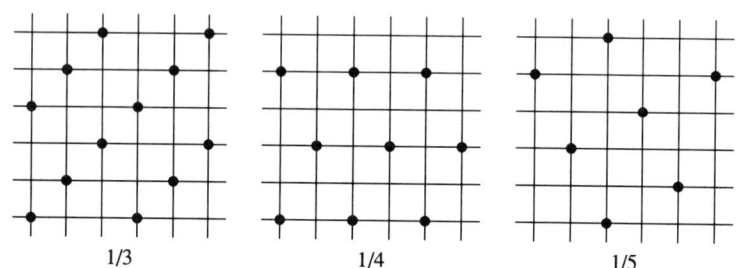

Figure 1. The ground state configuration in the neutral case for densities 1/3, 1/4 and 1/5.

The second general property is known as the segregation principle [10]. It says that in the nonneutral case if U is large enough then in the ground state the nuclei cluster together. In one dimension Lemberger[4] proved that if the model is not neutral then for large enough U the ground state is obtained by putting all the nuclei next to each other. It should be emphasized that as the electron and nuclear densities approach each other one must take U larger and larger for the segregation to occur.

In this note we will state and sketch the proofs of some theorems related to these two properties. Complete proofs may be found in Ref. 11. *All our results are for a square lattice in two dimensions with U large and positive.* Two different hole-particle symmetries imply that the model with interaction strength U, nuclear density ρ_n and electron density ρ_e is equivalent to the models with parameters $(-U, 1-\rho_n, \rho_e)$, $(-U, \rho_n, 1-\rho_e)$ and $(U, 1-\rho_n, 1-\rho_e)$. Thus statements about the attractive model in the neutral case $(\rho_n = \rho_e)$ may be translated into statements about the repulsive model in the half-filled case $(\rho_n + \rho_e = 1)$ and about the attractive model in the neutral case with the density ρ replaced by $1-\rho$. We state our results only for the attractive model with the nuclear density in $[0, 1/2]$, and leave the symmetries to the reader.

Precise statements of our theorems will follow. Here we will give somewhat imprecise statements to help explain what the theorems say. For the neutral model with a rational density between $1/4$ and $1/2$ and with some mild assumptions on the nature of the ground state, there is a slope m such that every line of lattice sites with slope m is either completely occupied or completely empty. This result does not determine the ground state, but it does rule out various configurations as ground states. As we will argue, for certain densities this property implies that the configurations that one might naturally consider to be the "most homogeneous" are not ground states.

For the neutral model with density $1/3, 1/4$ or $1/5$ we will prove that for large U the ground states are indeed those given in Ref. 9. Moreover, if $\rho_0 = 1/2, 1/3, 1/4$ or $1/5$ and two densities are both close to ρ_0 but not necessarily equal, then the fraction of the lattice in which the ground state equals some ground state of the neutral model with density ρ_0 must be close to 1. In particular the ground state is not segregated. Thus there are open regions in the nuclear density - electron density plane where the ground state is far from being segregated. (As $U \to \infty$ these regions presumably shrink. Thus this result complements the segregation principle rather than contradicting it.)

THE NEUTRAL MODEL

In the following the phrase "W is constant along lines with slope m" will mean that along each line of lattice sites with slope m, W is constant. Since W only takes on the values 0 and 1, this means that each such line of sites is either completely occupied by nuclei or completely free of nuclei.

Theorem 1: Let ρ be a rational density in $[1/4, 1/2]$. We consider the neutral model with density ρ on a square with periodic boundary conditions and sides of length L where L is such that $L\rho$ is an integer. Suppose there exists U_0 and a two-dimensional nuclear configuration W such that for all $U \geq U_0$, W is a ground state. Suppose also that W is periodic. Then there is a slope m such that W is constant along lines with slope m. If the density $\rho \in [1/3, 1/2]$ then m is either 1 or -1. If the density $\rho \in [1/4, 1/3]$ then m is either $1/2, -1/2, 2$ or -2. Moreover, in the case of $\rho \in [1/3, 1/2]$ each occupied line with slope m must be followed by either one or two empty lines with slope m before the next occupied line. In the case of $\rho \in [1/4, 1/3]$ each occupied line with slope m must be followed by either two or three empty lines with slope m before the next occupied line.

This theorem implies that for some densities the ground state must be somewhat inhomogeneous. For example, consider a density that is less than but very close to $1/2$. A natural candidate for the ground state for large U would be to take the checkerboard configuration and remove nuclei in a periodic way so that the removed nuclei are as widely separated as possible. The theorem says that this cannot be the ground state. The defects with respect to the checkerboard configuration must be arranged in diagonal lines.

Sketch of the proof: Let $H(W)$ be the ground state energy for the configuration W with the number of electrons equal to the number of nuclei. The proof is based on an expansion of $H(W)$ in powers of $1/U$. Gruber, Jedrzejewski and Lemberger[9] developed such an expansion and computed the first two nontrivial terms in the expansion, the U^{-1} and U^{-3} terms. We will refer to the sum of these terms as the second order Hamiltonian. At these orders the terms are supported on either a four site plaquette or two sites separated by a distance $1, \sqrt{2}$ or 2. We can rewrite the sum of these terms as a sum over three by three blocks. For a three by three block b we define H_b to be the sum of the U^{-1} and U^{-3} terms which depend only on sites in the block b and weighted so that $\sum_b H_b$ equals the second order Hamiltonian. (For example, a nearest neighbor bond is contained in 6 different blocks, so the coeffecient of such a term in H_b is taken to be $1/6$ of its coeffecient in the second order Hamiltonian.)

It is usually not possible to find a configuration which minimizes each H_b individually. However, it may be possible to find another decomposition of the second order Hamiltonian so that this can be done. Let K_b be another Hamiltonian that depends only on the spins in block b and such that $\sum_b K_b = constant$. Then $\sum_b [H_b + K_b]$ is equivalent to $\sum_b H_b$. One can then try to find K_b such that there is a configuration which minimizes each $H_b + K_b$ individually. This method was used[9] to show that certain configurations were the ground state of the second order Hamiltonian for densities 1/3 and 1/5. We will always be interested in the ground state for a fixed nuclear density. Thus one of the terms that we may include in K_b is a density term.

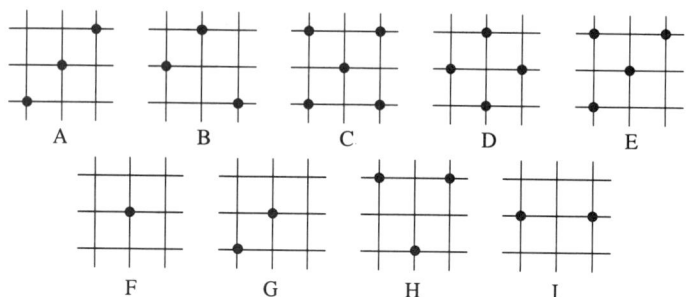

Figure 2. Under the hypothesis of theorem 1 for the neutral model, all three by three blocks in the ground state must be equal (up to lattice symmetries) to one of A, B, C, D or E if the density is between 1/3 and 1/2, and to one of A, B, F, G, H or I if the density is between 1/4 and 1/3.

It is possible to choose K_b so that $H_b + K_b \geq 0$ with equality if and only if b is (up to symmetries of the lattice) one of the blocks A,B,C,D or E in figure 2. All the other blocks have $H_b + K_b \geq gU^{-3}$ for some constant $g > 0$. Likewise it is also possible to chose K_b so that the same statement is true but with blocks A,B,C,D,E replaced by A,B,F,G,H,I. (The details are tedious[11].)

Next we show that for rational $\rho \in [1/3, 1/2]$ one can construct a configuration with nuclear density ρ such that every three by three block is (up to symmetries) one of A,B,C,D,E. Choose a horizontal line of 0's and 1's such that each pair of consecutive 1's is separated by one or two 0's and with the density of 1's equal to ρ. (This is possible for $1/3 \leq \rho \leq 1/2$.) Now fill out the lattice so that it is constant along lines with slope 1. (We could also use slope -1). It is straightforward to check that the resulting configuration has the desired property. For $\rho \in [1/4, 1/3]$ we choose a horizontal line of 1's and 0's in which consecutive 1's are separated by two or three 0's. We then extend it so that it is constant along lines with slope 1/2. The resulting configuration contains only the three by three blocks A,B,F,G,H,I (up to symmetries). (We could also use slope $\frac{-1}{2}$. To use slope 2 or -2 we must arrange the original sequence of 0's and 1's along a vertical line rather than a horizontal line.)

The preceding shows that it is possible to find a configuration in which $H_b + K_b$ attains its minimum for every block b. The periodicity of the ground state implies that one may consider a finite volume with periodic boundary conditions. Then for large enough U the energies in the higher order terms are too small to make it worthwhile to have a single block in W for which $H_b + K_b$ does not attain its minimum. So in the ground state W, $H_b + K_b$ must equal zero for every block b.

Thus in the ground state each three by three block is (up to symmetries) one of A,B,C,D,E in figure 2 for $\rho \in [1/3, 1/2]$ and one of A,B,F,G,H,I for $\rho \in [1/3, 1/4]$. If we know the configuration in a particular block, then the possibilities for the blocks which share six sites with the known block are rather limited. By pursuing this idea and checking lots of possible cases one can prove that if every block in a periodic configuration is one of A,B,C,D,E, or their symmetries then it must either be constant along lines with slope 1 or along lines with slope -1. If it only contains blocks A,B,F,G,H,I and their symmetries then it must be constant along lines with slope m where m is one of $2, -2, 1/2$ or $-1/2$. (The details are quite involved[11].) ∎

Theorem 2: If the nuclear density is $1/5, 1/4$ or $1/3$ and U is sufficiently large then the ground states for a square with periodic boundary conditions are given (up to symmetries of the lattice) by figure 1. (We assume that the size of the square is commensurate with the configuration in figure 1.)

We will only make a few comments about the proof. It begins in the same way as the preceding theorem. However, this theorem does not assume that the ground state is periodic and independent of U for large U. Thus to show that the ground state for the second order Hamiltonian is the ground state for the full Hamiltonian requires a much more involved argument than the cheap argument used above. We make use of the fact that although the full Hamiltonian contains terms with arbitrarily long range, the size of such terms decays exponentially with the size of their range.

For each of the three densities there is a choice of the K_b such that $H_b + K_b \geq 0$ and equality is attained only when b is a block that appears in one of the putative ground states for that density. One then shows that a configuration which contains only such blocks must in fact be one of the putative ground states.

THE NON-NEUTRAL MODEL

If we fix some ρ_n and ρ_e with $\rho_n \neq \rho_e$ then as $U \to \infty$ the ground state is expected to eventually be a segregated configuration. Suppose that instead we fix a large value of U and let $\rho_e \to \rho_n$. For rational $\rho_e = \rho_n$ we expect a periodic ground state, and for densities $1/2, 1/3, 1/4, 1/5$ this has been proven. We will show that if ρ_n and ρ_e are both close to one of the values $1/2, 1/3, 1/4, 1/5$ then the ground state is close to the neutral ground state.

Theorem 3: Let ρ_0 be $1/2, 1/3, 1/4$ or $1/5$. Consider a square with periodic boundary conditions and sides of length L where L is a multiple of $1/\rho_0$. Let $f(W)$ be the fraction of three by three blocks for which the configuration W is not equal to one of the blocks that may be found in a ground state for the neutral model with density ρ_0 (figure 1). There are positive constants c and U_0 such that if $U \geq U_0$ and W_g is a ground state for nuclear density ρ_n and electron density ρ_e then

$$f(W_g) \leq c[|\rho_n - \rho_0| + |\rho_e - \rho_0|]U^p$$

where $p = 2$ for $\rho_0 = 1/2$ and $p = 4$ in the other three cases.

Remark: In a connected region in which all the three by three blocks are equal to a block that may be found in one of the ρ_0 ground states, one can show that the configuration in the region agrees with a single ρ_0 ground state. Thus the density

$f(W_g)$ is some measure of how close the configuration W_g is to the ground state for the neutral case with density ρ_0. However, even if $f(W_g)$ is tiny, W_g can still contain a domain wall with respect to a ρ_0 ground state. Thus W_g might agree with a ρ_0 ground state in one half of the lattice and with a translation or rotation of this ground state in the other half. So $f(W_g)$ being small does not imply the strict long range order that is present in the neutral model with density ρ_0. But small $f(W_g)$ certainly implies that the ground state is far from being segregated.

Proof: We will let $E(W, \rho_e)$ denote the ground state energy for the configuration W with the electron density equal to ρ_e. Then $E(W_g, \rho_e) \leq E(W, \rho_e)$ for any configuration W with nuclear density ρ_n.

Each eigenvalue of the single particle operator associated with H has absolute value no bigger than $U + 4$. Hence

$$E(W_g, \rho_e) \geq E(W_g, \rho_0) - (U+4)|\rho_e - \rho_0|L^2$$

Let W' be a configuration which is obtained from W_g by changing W_g at $|\rho_n - \rho_0|L^2$ sites in such a way that W' has nuclear density ρ_0. Then the change in H when we replace W_g by W' has norm at most $U|\rho_n - \rho_0|L^2$, so we have

$$E(W_g, \rho_0) \geq E(W', \rho_0) - U|\rho_n - \rho_0|L^2$$

Let W_0 be the ground state for nuclear density ρ_0 and electron density ρ_0 (figure 1). Theorem 2 says that $E(W', \rho_0) \geq E(W_0, \rho_0)$, but the proof actually shows much more[11]. There is a constant $\gamma > 0$ such that

$$E(W', \rho_0) \geq E(W_0, \rho_0) + \gamma f(W')L^2 U^{-p+1}$$

By the construction of W', $f(W') \geq f(W_g) - 9|\rho_n - \rho_0|$.

Let $\tilde{W}$ be a configuration which is obtained from W_0 by changing $|\rho_0 - \rho_n|$ sites so that $\tilde{W}$ has density ρ_n. Then by the same arguments we used above,

$$E(W_0, \rho_0) \geq E(\tilde{W}, \rho_e) - (U+4)|\rho_e - \rho_0|L^2 - U|\rho_n - \rho_0|L^2$$

But we must have $E(\tilde{W}, \rho_e) \geq E(W_g, \rho_e)$. Combining all our inequalities we find

$$\gamma f(W_g) U^{-p+1} \leq 9\gamma |\rho_n - \rho_0| U^{-p+1} + 2U|\rho_n - \rho_0| + 2(U+4)|\rho_e - \rho_0|$$

The inequality in the theorem follows. ∎

The above theorem says that as both densities approach ρ_0, the ground state configuration must approach that for the neutral model with density ρ_0. A natural question to ask is how fast is this convergence, i.e., is there a lower bound on $f(W_g)$. For $\rho_0 = 1/2$ we have proved[11] that $f(W_g)$ is bounded below by an amount proportional to $|\rho_n - 1/2| + |\rho_e - 1/2|$.

We conclude with some open problems. Theorem 1 says nothing about densities below $1/4$. It would be interesting to find analogous results for these densities. This will probably require looking at higher order terms in the perturbation theory. Our results are all two dimensional. Are there counterparts in higher dimensions?

Theorem 1 says that the ground state is determined by what it does along a horizontal or vertical line. Thus we can think of the theorem as reducing the search for the ground state from a two-dimensional problem to a one-dimensional one. Is the arrangement along this line the "most homogeneous" one?

All of these results are for the ground state. There is some hope that a low temperature version of theorem 2 can be proved by a Peierls argument. Theorem 1 is a different story. It proof depends crucially on the fact that we are only considering the ground state. Moreover, the present theorem must assume certain properties of the ground state, e.g., periodicity. The first step should probably be to prove theorem 1 without these assumptions on the ground state. As for theorem 3, it might be possible to use the ideas of its proof to show that configurations with a large $f(W)$ have large energy compared to the ground state energy. Combined with an energy-entropy argument this might yield a bound on the thermal average of $f(W)$ for low temperatures.

Finally, we should note that very little has been done for the nonneutral model in dimensions greater than one. Even the existence of regions in the density-density plane where segregation occurs has not been proven.

Acknowledgments

It is a pleasure to thank Pirmin Lemberger and Elliott Lieb for useful correspondence. This work was supported in part by NSF grants DMS-9103621 and DMS-9303051.

REFERENCES

1. U. Brandt and R. Schmidt, Exact results for the distribution of the f-level ground state occupation in the spinless Falicov-Kimball model, *Z. Phys. B*, **63**, 45 (1986).
2. T. Kennedy and E. H. Lieb, An itinerant electron model with crystalline or magnetic long range order, *Physica*, **138A**, 320 (1986).
3. C. Gruber, J. Iwanski, J. Jedrzejewski and P. Lemberger, Ground states of the spinless Falicov-Kimball model, *Phys. Rev. B*, **41**, 2198 (1990).
4. P. Lemberger, Segregation in the Falicov-Kimball model, *J. Phys. A*, **25**, 715 (1992).
5. U. Brandt, Phase separation in the spinless Falicov-Kimball model, *J. Low Temp. Phys.*, **84**, 477 (1991).
6. C. Gruber, Spinless Fermi gas on one-dimensional lattice: rigorous results, *Helv. Phys. Acta*, **64**, 668 (1991).
7. C. Gruber, J. L. Lebowitz and N. Macris, Ground-state energy and low-temperature behavior of the one-dimensional Falicov-Kimball model, *Europhys. Lett.*, **21**, 389 (1993).
8. C. Gruber, J. L. Lebowitz and N. Macris, Ground state configurations of the one dimensional Falicov-Kimball Model, to appear in *Phys. Rev. B* (1993).
9. C. Gruber, J. Jedrzejewski and P. Lemberger, Ground states of the spinless Falicov-Kimball model. II, *J. Stat. Phys.*, **66**, 913 (1992).
10. F.K. Freericks and L.M. Falicov, Two-state one-dimensional spinless Fermi gas, *Phys. Rev. B*, **41**, 2163 (1990).
11. T. Kennedy, Some rigorous results on the ground states of the Falicov-Kimball model, submitted to *Rev. Math. Phys.*

DIFFUSIVE LIMIT OF THE ASYMMETRIC SIMPLE EXCLUSION: THE NAVIER-STOKES CORRECTION

R. Esposito[1], R. Marra[2] and H.T. Yau[3]

1. Dipartimento di Matematica, Università di Roma Tor Vergata
   Via della Ricerca Scientifica 00133 Roma, Italy

2. Dipartimento di Fisica, Università di Roma Tor Vergata
   Via della Ricerca Scientifica 00133 Roma, Italy

3. Courant Institute of Mathematical Sciences
   New York University, NY, NY, 10012

PROBLEM AND RESULTS

We consider the simple exclusion process (SEP) on a cubic sublattice, $\Lambda_L \subset \mathcal{Z}^d$, of size $2L+1$, with periodic boundary conditions. The dynamics is as follows. Let e denote one of the $2d$ possible directions in $\mathcal{Z}^d$. A particle in the site x, independently of the others, waits for an exponential time and jumps, with probability proportional to $p_e \geq 0$, to the site $x + e$, if it is empty; otherwise a particle stays in x and the process starts again. We denote by $\eta_x(\tau) = 0, 1$ the number of particles in the site x at time τ and by $\mathcal{L}$ the generator of the process: $\mathcal{L}f = \sum_b \mathcal{L}_b f$, the sum running on the set of all oriented bonds $b = (x, y)$ in $\mathcal{Z}^d$ such that $y - x = e$ and

$$\mathcal{L}_b f = p_e \eta_x [f(\eta^b) - f(\eta)] \tag{1}$$

with

$$(\eta^b)_z = (\eta^{x,y})_z = \begin{cases} \eta_y, & \text{if } z = x \\ \eta_x, & \text{if } z = y \\ \eta_z, & \text{otherwise.} \end{cases}$$

It is convenient to choose the normalization $p_{-e} + p_e = 2$ for all e.

In the case $p_e = p_{-e}$ for all e (symmetric simple exclusion process) the hydrodynamical limit is non trivial only in the diffusive scaling, $z = \varepsilon x$, $t = \varepsilon^2 \tau$, where x and τ are the microscopic space and time, while z and t are the macroscopic ones, and $\varepsilon = L^{-1}$. Under this scaling, as $\varepsilon \to 0$, the limiting equation reduces to the diffusion equation with diffusion coefficient 1. The case $p_e \neq p_{-e}$, called the asymmetric simple exclusion process (ASEP), is much more difficult.

Define the product measure in Λ_L, with chemical potential β, as

$$d\mu_{\beta,L} = Z_\beta^{-1} \exp[\beta \sum_{x \in \Lambda_L} \eta_x], \quad \eta_x \in \{0,1\}, \tag{2}$$

Z_β being the normalization. $\mu_{\beta,L}$ is invariant for the dynamics because of the periodic b.c., but the dynamics is not reversible with respect to it. The behavior of the system is non trivial already in the hyperbolic (Euler) scaling ($\tau = \varepsilon^{-1}t$) and is given by the non viscous Burgers equation (see ref. [1])

$$\partial_t \rho + \delta \cdot \nabla_z[\rho(1-\rho)] = 0, \tag{3}$$

$\delta_e = p_e - p_{-e}$. Since the system moves already on the Euler scale, it is not clear the meaning of the diffusive limit for this system. For weak asymmetries, namely, for the case δ_e of order ε, in the diffusive scaling the process converges to a diffusion with a non linear drift [2,3]. However, this case is a small perturbation of the symmetric one and does not manifest the asymmetric nature of the problem. In fact, the diffusive constant obtained in this way is still 1.

We consider[4] the ASEP with δ of order 1, but in a restricted class of initial data for which there is a different mechanism to get a well defined diffusive limiting behavior. This is suggested by the analogy with the study of the incompressible regime for a fluid[5]. Suppose that the initial density $\rho_0(\varepsilon x)$ is of the form

$$\rho_0(\varepsilon x) = \theta - \varepsilon u_0(\varepsilon x)$$

with $\theta \in (0,1)$. Then, assuming that this form of the density persists at time t, the macroscopic current $J = \delta[\rho(1-\rho)]$ becomes

$$J = \delta\theta(1-\theta) - \varepsilon v u - \varepsilon^2 \delta u^2, \quad v = (1-2\theta)\delta.$$

The equation for the field u is expected to be

$$\partial_t u + \varepsilon^{-1} v \cdot \nabla_z u + \delta \cdot \nabla_z u^2 = \sum_{i,j=1}^{d} D_{i,j} \frac{\partial^2 u}{\partial z_i \partial z_j}, \tag{4}$$

where $D_{i,j}$ is a diffusion matrix. A change of reference frame

$$m(z,t) = u(z + \varepsilon^{-1}vt, t)$$

allows to remove the diverging term and the equation for $m(z,t)$ is the viscous Burgers equation:

$$\partial_t m + \delta \cdot \nabla_z m^2 = \sum_{i,j=1}^{d} D_{i,j} \frac{\partial^2 m}{\partial z_i \partial z_j}. \tag{5}$$

This suggests us to assume an initial measure with density w.r.t $\mu_{\beta,L}$ given by

$$\psi_0 = Z_\lambda^{-1} \exp\{\varepsilon \sum_x \lambda(\varepsilon x)\eta_x\} \tag{6}$$

for some smooth function $\lambda(z)$, Z_λ being the normalization. Since the product measure with constant chemical potential is invariant, it is natural to think of $\psi_0 d\mu_{\beta,L}$ as a *local equilibrium* for the system.

The average occupation number in the initial local equilibrium state is, up to terms of higher order in ε, $\theta - \varepsilon m_0(\varepsilon x)$, with $\theta = e^\beta/(1+e^\beta)$ and $m(z) = -\theta(1-\theta)\lambda(z)$.

By above heuristic arguments one expects that, starting with this initial datum, the occupation number at time t, in the moving frame, differs from θ by $\varepsilon m(z,t)$ with $m(z,t)$ solving (5) with initial datum $m_0(z)$. To state our result we introduce the empirical measure

$$\nu_\varepsilon(z,t) = \varepsilon^{d-1} \sum_x \delta(z - \varepsilon x)(\theta - \eta_x(\varepsilon^{-2}t)).$$

and $\hat{\nu}_\varepsilon(z,t) = \nu_\varepsilon(z - \varepsilon^{-1}vt, t)$. Next theorem states the weak convergence of $\hat{\nu}_\varepsilon(z,t)$ to $m(z,t)$, if $d \geq 3$.

THEOREM 1.1 *For $d \geq 3$, $\hat{\nu}_\varepsilon(z,t)$ converges weakly in probability to $m(z,t)$ solving the equation (5). The diffusion matrix satisfies $1 < D < \infty$ (as a matrix) and it is formally given by the Green-Kubo formula.*

Remarks

1. The diffusion matrix is expected to be infinite in dimension 1 and 2 [1]. Actually our method relies on a multiscale analysis which fails in dimension less than 3.

2. The fact that $D > 1$ is, in our opinion, related to the presence of a Navier-Stokes correction in this model. In fact, we interpret the non-reversibility of the generator $\mathcal{L}$ as due to a *deterministic* motion added to the stochastic motion due to the symmetric part of the generator. The last one produces a diffusion process with diffusion coefficient 1. Hence, there is a non vanishing contribution to the diffusion due to the antisymmetric part of the generator.

3. The Green-Kubo formula does not make really sense. Hence, we give a variational formula for D which is a way to give a rigorous sense to the Green-Kubo formula.

4. We cannot expect a isotropic diffusion because of the presence of the cubic lattice which interferes in some complicated way with the driving field δ.

Theorem 1.1 is an immediate consequence of the fact that we can construct a local equilibrium measure which approximates, in a suitable sense, up to order ε^2, the true distribution density.

Let f_t be the distribution density at time t with respect to $\mu_{\beta,L}$. Then f_t satisfies the rescaled forward equation

$$\partial_t f_t = \varepsilon^{-2} \mathcal{L}^* f_t$$

Here $\mathcal{L}^*$, defined by

$$\mathcal{L}^* f = \sum_b \mathcal{L}_b^* f, \quad \mathcal{L}_b^* f = p_{-\varepsilon} \eta_x [f(\eta^b) - f(\eta)],$$

is the adjoint of $\mathcal{L}$ in $L_2(d\mu_{\beta,L})$ with periodic b.c. The distribution density is initially $f_0 = \psi_0$.

Recall that the relative entropy of the densities f and g (w.r.t. $d\mu_\beta$) is defined by

$$s(f|g) = \varepsilon^d \int f \log \frac{f}{g} d\mu_\beta \tag{7}$$

and put $s(f) \equiv s(f|1)$.

Note that $s(f_0) = O(\varepsilon^2)$, so that $s(f_t) = O(\varepsilon^2)$ too because $s(f_t)$ is constant in time. A better approximation is provided by a local equilibrium ψ_t with the right choice of the chemical potential. In fact, let

$$\psi_t = \frac{1}{Z(t)} \exp\{\varepsilon \sum_x \lambda(\varepsilon x - \varepsilon^{-1}vt, t)\eta_x\}, \tag{8}$$

where $Z(t)$ is the normalization (w.r.t. $d\mu_\beta$) and λ satisfies the equation

$$\frac{\partial}{\partial t}\lambda(z,t) - \theta(1-\theta)\sum_{e>0}\delta_e(\partial_e\lambda^2)(z,t) = \sum_{e>0}\sum_{e'>0}D_{ee'}(\partial_e\partial_{e'}\lambda)(z,t). \tag{9}$$

Here $\partial_e\lambda(z) = e \cdot \nabla_z\lambda$ and $D_{e,e'} = e \cdot De'$. We prove the following

THEOREM 1.2. *Let f_t be the density at time t relative to $d\mu_\beta$ with initial datum ψ_0. Then, if $d \geq 3$,*

$$\lim_{\varepsilon \to 0}\varepsilon^{-2}s(f_t|\psi_t) = 0. \tag{10}$$

For any positive a and any smooth function $J(z)$ we denote by A the event

$$\{\eta \in \{0,1\}^{\mathbb{Z}^d} \text{ s.t.} |\int J(z)\hat{v}_\varepsilon(dz,t) - \int dz J(z)m(z,t)| > a\}.$$

Since one can check the following estimate for the probability of the event A in the local equilibrium:

$$P^{\psi_t}(A) \leq \text{const.}\exp[-\text{const.}\varepsilon^{2-d}\},$$

Theorem 1.1 follows from Theorem 1.2 and the entropy inequality.

OUTLINE OF THE PROOF OF THEOREM 1.2

The proof of our result is based on the entropy method introduced in refs. [6, 7]. The general idea is to compare the non-equilibrium measure with the local equilibrium modelled according to the behavior of the system on the hydrodynamic space-time scales. However, in this case, the dynamics is non-gradient and non-reversible. Since we look at times of order ε^2, we expect that the local equilibrium ψ_t, which works fine on the Euler scale, is not sufficient for our purposes and we need to add a correction which is not in a product form. In fact, it is enough to take a correction of order ε^2 which does not affect the relative entropy up to order ε^2. The specific form of the correction is dictated by the method of ref. [8] for the non-gradient systems, which, in its original form [8,9], applies to reversible systems with sensible changes only on the diffusive scale. The extension to non-reversible systems given in ref. [10], where the asymmetric simple exclusion process with a mean zero condition is considered, cannot be applied in our case, because the mean zero condition prevents the system from moving on the Euler scale. In our case the restriction to the Euler contributions is only in the average, while for typical configurations big displacements are possible on the Euler scale. Our method relies strongly on a multiscale analysis and the logarithmic Sobolev inequality for the corresponding symmetric process in the canonical ensemble proved in ref. [11]. It is more general than the method in ref. [10] but is confined to $d \geq 3$.

The local equilibrium ψ_t is replaced by the density $\tilde{\psi}_t$ defined as follows:

$$\tilde{\psi}_t = \tilde{Z}_t^{-1}\exp\left\{\varepsilon\sum_x(\tilde{\omega}*\lambda(\varepsilon y - \varepsilon^{-1}vt,t))(\varepsilon x,t)\eta_x + \varepsilon^2\Phi(\eta)\right\}. \tag{11}$$

Here the convolution is on the variable y and $\Phi(\eta) = -\sum_{x,e>0}(\partial_e\alpha)(\varepsilon x,t)(\tilde{\omega}*\tau_y F_e)(x)$ is the correction of order ε^2 discussed above. F_e are local functions such that $\hat{F}_e(\theta) = 0 = \partial\hat{F}_e/\partial y|_{y=\theta}$, $\hat{F}_e(u) = E^{\mu_h}[F_e]$, u and h being dual variables in the sense that $u = P'(h)$ and $P(h) = \log[e^h + 1]$ is the pressure.

In the definition of $\tilde{\psi}_t$ we have also smeared $\lambda(z)$ convolving it with $\tilde{\omega}$, which is the normalized characteristic function of a cubic region $\tilde{\Lambda}_k$ of size $2k+1$, with some corridors removed. To define it, let us introduce $\bar{\ell} = \ell^{d+2}$ and the partition of the cube Λ_k of size $2k+1$ into blocks $\Lambda_{\bar{\ell},\sigma}$ of size $2\bar{\ell}+1$ with centers $\sigma \in (2\bar{\ell}+1)\mathcal{Z}^d$. Now remove from Λ_k corridors of width ℓ^{1/d^2} around each block $\Lambda_{\bar{\ell},\sigma}$. $\tilde{\Lambda}_k$ is the union of all these regions. We do not enter in the rather technical explanation of the choice of $\bar{\ell}$ and of the need of removing above corridors. It is a typical procedure in multiscale analysis and in this way we can manage more easily some boundary terms.

The first step in the proof is to get an integral inequality for the relative entropy. The time derivative of the relative entropy can be bounded in terms of two contributions of different origin: a large deviation term and a non-gradient term. The first one is a standard entropy method contribution like in [6,7]. It involves a quadratic variation of η similar to the large deviation functional in ref. [7]. Therefore, one substitutes η_x with its average in a large block Λ_k. Due to diffusive scaling, we have to dominate an extra factor ε^{-1}, so we use the large deviation estimate in a very big block Λ_k with $k = \varepsilon^{-2/d}$.

Such a choice of k makes more difficult to deal with the non-gradient contribution. The non-gradient term arises in the computation of $\mathcal{L}\eta_x$, which is the divergence of the current $w_{x,e} = \nabla_e \eta_x + \delta_e[\eta_x \eta_{x+e} - (\eta_x + \eta_{x+e})/2]$ with $\nabla_e \eta_x = \eta_{x+e} - \eta_x$. To use large deviations we have to replace $w_{x,e}$ with $E^{\mu_\beta}[w_{x,e}|\bar{\eta}_k]$. Would k be independent on ε one could use the usual one-block estimate to control the error due to this replacement. The dependence of k on ε forces us to use a multiscale arguments and log Sobolev inequality to reduce the estimate to a block of fixed size $\bar{\ell}$.

The estimate of the rate of production of relative entropy is given by the following

Theorem 2.1 *Let $d \geq 3$ and F_e any local function such that $E^\mu[F_e] = 0 = \sum_x E^\mu[\eta_x F_e]$. Then*

$$\lim_{\ell \to \infty} \lim_{\varepsilon \to 0} \varepsilon^{-2} s(f_T|\tilde{\psi}_T) \leq C \lim_{\ell \to \infty} \lim_{\varepsilon \to 0} \int_0^T \varepsilon^{-2} s(f_t|\tilde{\psi}_t) dt + CT \sum_{e>0} V(\tilde{q}_e, \beta) \quad (12)$$

Here C is a positive constant and

$$\tilde{q}_e = \delta_e(\eta_0 - \theta)(\eta_e - \theta) - \sum_{e'} \tilde{D}_{ee'} \nabla_{e'} \eta_0 - \mathcal{L}^* F_e \quad (13)$$

with $\tilde{D}_{ee'} = \delta_{ee'} - D_{ee'}$.

The "variance" $V(G, \beta)$ is defined as follows: let

$$V_\ell(G, y) = (2\ell_1 + 1)^{-d} \left\langle \left[\sum_{|x| \leq \ell_1} (\tau_x G - \alpha_\ell(G)) \right] (-\mathcal{L}_{s,\ell})^{-1} \left[\sum_{|x| \leq \ell_1} (\tau_x G - \alpha_\ell(G)) \right] \right\rangle_{\mu_{\ell,y}} \quad (14)$$

where $\ell_1 = \ell - \ell^{1/d^2}$ and $\mu_{\ell,y}$ is the canonical Gibbs state in Λ_ℓ with density y, namely, $\mu_{\ell,y} = Q_{\ell,y}^{-1} \delta(\bar{\eta}_\ell - y) \mu_\ell$. Here μ_ℓ denotes the counting measure on the configurations in Λ_ℓ and $Q_{\ell,y}$ is the normalization. The generator $\mathcal{L}_{s,\ell}$ is defined by $\mathcal{L}_{s,\ell} = 1/2(\mathcal{L}_\ell + \mathcal{L}_\ell^*)$ with $\mathcal{L}_\ell = \sum_{b \in \Lambda_\ell} \mathcal{L}_b$ and $\mathcal{L}_\ell^*$ is defined similarly. Moreover, $\alpha_\ell(G) = E^{\mu_\beta}[G|\bar{\eta}_\ell]$. Finally, $V(G, \beta)$ is defined by

$$V(G, \beta) = \limsup_{\ell \to \infty} E^{\mu_\beta}[V_\ell(G, \bar{\eta}_\ell)]. \quad (15)$$

The proof of Theorem 1.2 is achieved once we can find local functions F_e and a matrix D such that $V(\tilde{q}_e, \beta) = 0$.

The problem now is to show that we can replace the current with a term proportional to $\nabla_e \eta$ plus another one which is given by the action of the generator on some suitable function. To do this we use a geometric argument, namely, we construct a Hilbert space of local functions and show that it is spanned by functions of the form $\alpha \cdot \nabla \eta = \sum_{e>0} \alpha_e \nabla_e \eta$ and $\mathcal{L}G$. Here $e > 0$ means e with positive components.

Let $\mathcal{G}$ be the following linear space:

$$\mathcal{G} = \{g | g \text{ is a local function s.t. } \hat{g}(\theta) = 0 = \frac{\partial \hat{g}}{\partial y}|_{y=\theta}.\}.$$

Here $\hat{g}$ is defined by $\hat{g}(y) = E^{\mu\beta}[g]$, $y = P'(\beta)$. Previous conditions are equivalent to $E^{\mu\beta}[g] = 0 = \sum_x E^{\mu\beta}[g; \eta_x]$ and $E^{\mu\beta}[A; B] = E^{\mu\beta}[AB] - E^{\mu\beta}[A]E^{\mu\beta}[B]$.

We use the variance $V(G, \beta)$ to construct the inner product. Its definition (15) is not easy to manage, but next theorem provides an alternative variational characterization.

Theorem 2.2 *Let $d \geq 3$. The variance $V(g)$ has the following variational representation:*

$$\frac{1}{2}V(g,\beta) = \sup_{\alpha_e \in \mathcal{R}, G \in \mathcal{G}} \left\{ \sum_{e>0} \alpha_e t_e(g) + \langle g, G \rangle_0 - \frac{1}{4} \sum_{e>0} E^\mu [\alpha_e \nabla_e \eta + \nabla_e \sum_x \tau_x G]^2 \right\}. \quad (16)$$

Here

$$\langle g, G \rangle_0 = E^\mu [g \sum_x \tau_x G], \quad t_e(g) = \lim_{\ell \to \infty} E^\mu [g \sum_{x \in \Lambda_\ell} (e \cdot x) \eta_x]. \quad (17)$$

The inner product is defined by polarizing the variance V:

$$V(g, h) = \frac{1}{4}[V((g+h), \beta) - V((g-h), \beta)]. \quad (18)$$

We denote the above inner product by $\ll \cdot, \cdot \gg$. Theorem 2.2 is a way to give a rigorous sense to the formal expression $\langle f, \mathcal{L}_s^{-1} h \rangle_0$. In fact, by the conservation of the number of particles, the inverse of $\mathcal{L}_s$ is well defined only on a space with fixed number of particles (canonical ensemble) and one is forced to work at finite volume and take the limit afterwards. In Theorem 2.2, instead we have an infinite volume formulation, which makes possible the geometric argument given below. A fundamental ingredient in its proof is again the multiscale analysis, which restricts it validity to dimensions larger than 2.

A consequence of Theorem 2.2 is that $\mathcal{G}$ is in the closures of the direct sum of the orthogonal subspaces: $\mathcal{G}^{(0)} = \{\sum_{e>0} \alpha_e \nabla_e \eta\}$ and $\{\mathcal{L}_s \mathcal{G}\}$. Moreover, for $g, G \in \mathcal{G}$, we have:

$$V(\mathcal{L}_s g) = \frac{1}{2} \sum_{e>0} E^\mu [(\nabla_e \sum_x \tau_x g)^2], \quad (19)$$

$$\langle \nabla_e \eta, G \rangle_0 = 0, \quad (20)$$

$$t_{e'}(\nabla_e \eta) = \delta_{e,e'} V(\nabla_e \eta) = E^\mu[(\nabla_e \eta) \eta_e] = \frac{1}{2} E^\mu[(\nabla_e \eta)^2] \quad (21)$$

and for $\xi \in \mathcal{G}$,

$$\ll \mathcal{L}_s g, \xi \gg = -\langle g, \xi \rangle_0. \quad (22)$$

We now have a decompositon of the Hilbert space but this is not the one we need, because we have to deal with the non symmetric generator $\mathcal{L}$.

The next step is to show that the Hilbert space can also be decomposed using functions of the form $\mathcal{L}G$. In fact, we can prove that

$$\overline{\mathcal{L}\mathcal{G} + \mathcal{G}^{(0)}} = \overline{\mathcal{G}} = \overline{\mathcal{L}^*\mathcal{G} + \mathcal{G}^{(0)}}. \tag{23}$$

Recall the definition of the currents $w_{x,e} = \nabla_e \eta_x + \delta_e[\eta_x \eta_{x+e} - (\eta_x + \eta_{x+e})/2]$ and $w^*_{x,e} = \nabla_e \eta_x - \delta_e[\eta_x \eta_{x+e} - (\eta_x + \eta_{x+e})/2]$, whose divergences give $\mathcal{L}\eta_x$ and $\mathcal{L}^*\eta_x$ respectively. Note that $w_e \notin \mathcal{G}$. We introduce

$$\sigma_e = (\nabla_e \eta) + \delta_e(\eta_0 - \theta)(\eta_e - \theta) \tag{24}$$

so that $\sigma_e \in \mathcal{G}$ and $\langle g, w_e \rangle_0 = \langle g, \sigma_e \rangle_0$ for $g \in \mathcal{G}$. Similarly, one can define $\sigma^*_e = (\nabla_e \eta) - \delta_e(\eta_0 - \theta)(\eta_e - \theta)$ so that $\langle g, w^*_e \rangle_0 = \langle g, \sigma^*_e \rangle_0$. Let $\mathcal{G}_w = \{\sum_{e>0} \alpha_e \sigma_e\}$. Then we have also

$$\overline{\mathcal{G}_w + \mathcal{L}\mathcal{G}} = \overline{\mathcal{G}} = \overline{\mathcal{G}_{w^*} + \mathcal{L}^*\mathcal{G}}.$$

The above relations justify (up to some approximation procedure) the replacement of σ^*_e with $\sum_{e'>0} D_{e,e'} \nabla_{e'} \eta + \mathcal{L}^* G_e$ for some matrix D and some $G_e \in \mathcal{G}$, or, equivalently $\nabla_e \eta = \sum_{e'>0}(D^{-1})_{e,e'} \sigma^*_{e'} + \mathcal{L}^* \tilde{G}_e$.

The above decomposition of the Hilbert space is not orthogonal and we need some orthogonality relation to determine the diffusion coefficient D. Heuristically, we can write

$$0 = \langle \nabla_e \eta, G \rangle_0 = \langle \mathcal{L}^{-1} \nabla_e \eta, \mathcal{L}^* G \rangle_0 = - \ll \mathcal{L}_s \mathcal{L}^{-1} \nabla_e \eta, \mathcal{L}^* G \gg. \tag{25}$$

We don't know how to give sense to $\mathcal{L}^{-1}$ but we can give a sense to a map T, formally equivalent to $\mathcal{L}_s \mathcal{L}^{-1}$, in such a way that the (25) becomes the desired orthogonality condition between $T\nabla \eta$ and $\mathcal{L}^* \mathcal{G}$.

Definition of T: given $\xi \in \mathcal{G}$ of the form $\xi = \sum_{e>0} \alpha_e \sigma_e + \mathcal{L}g$, $T\xi$ is given by

$$T(\sum_{e>0} \alpha_e \sigma_e + \mathcal{L}g) = \sum_{e>0} \alpha_e \nabla_e \eta + \mathcal{L}_s g. \tag{26}$$

Theorem 2.3 *T is bounded above by 1. Moreover,*

$$T\nabla_e \eta \perp \mathcal{L}^* \mathcal{G}, \quad \ll T\nabla_e \eta, \sigma^*_{e'} \gg = \delta_{ee'} V(\nabla_e \eta) \equiv 2\chi \delta_{ee'}.$$

Let Q be the matrix: $Q_{ee'} = \ll \nabla_e \eta, T\nabla_{e'} \eta \gg$. Then the diffusion matrix D is given by

$$D = V(\nabla_e \eta) Q^{-1} > 1. \tag{27}$$

Let R be the linear transformation from $\mathcal{G}$ to $\mathcal{G}$ s.t.

$$R(\sum_{e>0} \alpha_e \nabla_e \eta + \mathcal{L}_s g) = \sum_{e>0} \alpha_e \sigma_e + \mathcal{L}g. \tag{28}$$

Then R is a bounded linear transformation, $R = T^{-1}$ and in consequence D is bounded above.

The expression (27) is formally equivalent to a Green-Kubo formula. In fact denote by $\bar{G}$ the value of G such that $\sigma - D\nabla \eta + \mathcal{L}\bar{G} = 0$. Using the short notation $\alpha \cdot \sigma$ for $\sum_e \alpha_e \sigma_e$ we get from (27):

$$\alpha \cdot (2\chi D)^{-1} \alpha = \ll \alpha \cdot \nabla \eta, T\alpha \cdot \nabla \eta \gg.$$

Hence, by the representation of $\nabla \eta$ and using the definition of the map T, we have

$$\alpha \cdot 2\chi D\alpha = \ll \alpha \cdot (\sigma - \mathcal{L}\bar{G}), \alpha \cdot (\nabla \eta - \mathcal{L}_s \bar{G}) \gg =$$

$$\ll \alpha \cdot \sigma, \alpha \cdot \nabla \eta \gg + \ll \mathcal{L}\alpha \cdot \bar{G}, \mathcal{L}_s \alpha \cdot \bar{G} \gg + \ll \mathcal{L}\alpha \cdot \bar{G}, \alpha \cdot \nabla \eta \gg + \ll \alpha \cdot \sigma, \mathcal{L}_s \alpha \cdot \bar{G} \gg.$$

The first term is just $2\chi\alpha \cdot \alpha$. Moreover, since $\sigma + \sigma^* = \nabla \eta$,

$$\ll \mathcal{L}\alpha \cdot \bar{G}, \alpha \cdot \nabla \eta \gg + \ll \alpha \cdot \sigma, \mathcal{L}_s \alpha \cdot \bar{G} \gg = - <\alpha \cdot \sigma^*, \alpha \cdot \bar{G}>_0 - <\alpha \cdot \sigma, \alpha \cdot \bar{G}>_0 = 0.$$

Therefore, we get the following expression for D:

$$\alpha \cdot D\alpha = \alpha \cdot \alpha + (2\chi)^{-1} \ll \mathcal{L}_s \alpha \cdot \bar{G}, \mathcal{L}_s \alpha \cdot \bar{G} \gg.$$

We compare now this expression with the heuristic Green-Kubo formula for D given in ref. [12]:

$$D_{ee'} = \delta_{ee'} + (2\chi)^{-1} \int_0^\infty dt \sum_x [< \sigma_e \partial_{0e} e^{\mathcal{L}t} \sigma_{x,x+e'} >$$

where $\partial_{0e} f(\eta) = f(\eta^{0e})$. Since $\partial_{0e} \sigma = -\sigma^*$, we have

$$\alpha \cdot D\alpha = \alpha \cdot \alpha + (2\chi)^{-1} \langle \alpha \cdot \sigma^*, \mathcal{L}^{-1} \alpha \cdot \sigma \rangle_0$$

Since the inner product $\langle , \rangle_0$ is degenerate, only the non-gradient component of the current contributes. Therefore,

$$\alpha \cdot D\alpha = \alpha \cdot \alpha + (2\chi)^{-1} < \mathcal{L}^* \alpha \cdot \bar{G}, \mathcal{L}^{-1} \mathcal{L}\alpha \cdot \bar{G} > = \alpha \cdot \alpha + (2\chi)^{-1} < \alpha \cdot \bar{G}, \mathcal{L}_s \alpha \cdot \bar{G} >_0$$

From the definition of the inner product $\ll, \gg$ it follows that our representation of the diffusion coefficient is a way to give a sense to the above Green-Kubo formula.

Acknowledgments

Work partially supported by US-NSF Grant n. DMS 9101196, A.P.Sloan Foundation Fellowship and David and Lucile Packard Foundation Fellowship

REFERENCES

1. H. Spohn, " Large Scale Dynamics of Interacting Particles", Springer-Verlag, New-York (1991).
2. A. De Masi, E. Presutti and E. Scacciatelli, The weakly asymmetric simple exclusion, *Ann. Instit. H. Poincaré A*, **25**, 1 (1989).
3. C. Kipnis, S. Olla and R. S. R. Varadhan, Hydrodynamics and large deviations for simple exclusion process, *Commun. Pure Appl. Math*, **42**, 115 (1989).
4. R. Esposito, R. Marra and H. T. Yau, Diffusive limit of asymmetric simple exclusion, to appear in *Rev. Math. Phys.* (1993).
5. A. De Masi, R. Esposito and J. L. Lebowitz, Incompressible Navier-Stokes and Euler limits of the Boltzmann equation, *Commun. Pure Appl. Math.*, **42**, 1189 (1989).
6. H. T. Yau, Relative entropy and hydrodynamics of Ginsburg-Landau models, *Letters Math. Phys.*, **22**, 63 (1991).
7. S. Olla, S. R. S. Varadhan and H. T. Yau, Hydrodynamical limit for a Hamiltonian system with weak noise, (1991).
8. S. R. S. Varadhan, Nonlinear diffusion limit for a system with nearest neighbor interactions II, *in:* "Proc. Taniguchi Symp.", Kyoto (1990).

9. J. Quastel, Diffusion of color in the simple exclusion process, *Commun. Pure Appl. Math.*, **40**, 623 (1992).
10. L. Xu, Diffusion limit for the lattice gas with short range interactions, Thesis, New York University (1993).
11. S. Lu and H.T. Yau, Logarithmic Sobolev inequality for the Kawasaki dynamics, in preparation.
12. S.Katz, J.L. Lebowitz and H.Spohn, Nonequilibrium steady states of stochastic lattice gas models of fast ionic conductors, *J. Stat. Phys*, **34**, 497 (1984).

WEAK COUPLING LIMIT: FEYNMAN DIAGRAMS

L. J. Landau

Mathematics Department, King's College London

1. INTRODUCTION

An elementary discussion is given of the derivation of irreversible behavior in the weak coupling limit for a Fermi gas in a random potential based on the Feynman-diagrammatic description of the individual terms in the Dyson perturbative expansion, which is one aspect of the study[1], joint work with T.G.Ho and A.J.Wilkins. (See reference[1] for technical details concerning self-adjointness of the Hamiltonian, and control of the perturbative expansion.) The model is a quantum version of the Lorentz gas[2], proposed by Lorentz in 1905 as a model for electron conduction in metals, and describes a gas of non-interacting particles in the presence of randomly distributed static impurities on which the gas particles scatter elastically. The diagrammatic analysis is presented in the same spirit as Hugenholtz's discussion[3] of the Boltzmann equation for a self-interacting Fermi gas. The difficulties[1] with Hugenholtz's treatment are not present in the simpler model considered here. The impurities are represented as a random potential. The weak coupling limit for a single quantum particle in a random potential has been considered by Martin and Emch[4], Spohn[5], and Dell'Antonio[6]. Some discussion of their results is given in reference[1]. The Fermi gas is described by the canonical anticommutation relations generated by the creation operators $a^*(x)$ and the destruction operators $a(x)$ satisfying $\{a(x_1), a^*(x_2)\} = \delta(x_1 - x_2)$ or their Fourier transforms $a^*(p)$ and $a(p)$ satisfying $\{a(p_1), a^*(p_2)\} = \delta(p_1 - p_2)$. The free Hamiltonian is $H_0 = \int dp\, \varepsilon(p) a^*(p) a(p)$, where for the non-relativistic gas considered here $\varepsilon(p) = p^2/2m$. The random potential is

$$V = \int dx\, v(x) a^*(x) a(x) = \int dp\, dq\, \tilde{v}(q) a^*(p-q) a(p)$$

where $v(x)$ is a random field. The free evolution of a creation or destruction operator $a^\#(k)$ is denoted $a_t^\#(k)$ and is given by

$$a_t^\#(k) = e^{itH_0} a^\#(k) e^{-itH_0} = e^{\pm it\varepsilon(k)} a^\#(k)$$

The total Hamiltonian $H_\lambda = H_0 + \lambda V$. The weak coupling limit is defined by $t \to \infty, \lambda \to 0, \lambda^2 t = \tau$.

1.1 Goal

Let A be a product of creation and destruction operators: $A = a^{\#}(f_1) \cdots a^{\#}(f_N)$, where $a^{\#}(f) = \int dp \, f(p) a^{\#}(p)$ are regularized (bounded) operators. Set $A^{\lambda}(\tau) = e^{itH_{\lambda}} A e^{-itH_{\lambda}}$ where $t = \tau/\lambda^2$. $(a^{\#}(f))^{\lambda}(\tau) = a^{\#}(e^{\pm itH_{\lambda}}f)$ since H generates a one-particle evolution.) Since H_{λ} depends on the random field $v(x)$, so does $A^{\lambda}(\tau)$. Given an initial state S of the Fermi gas, the time evolved state S_{τ}^{λ} is defined by

$$S_{\tau}^{\lambda}(A) \doteq S\left(A^{\lambda}(\tau)\right) \tag{1}$$

We study the behavior of S_{τ}^{λ} in the weak coupling limit. Our present goal is to show, to each order in perturbation theory, that S_{τ}^{λ} converges to an asymptotic state S_{τ} which is a gauge invariant quasi-free state (i.e. only the gauge invariant 2-point truncated function is different from zero). If the initial 2-point function is $S(a^*(f)a(g)) = \int dp \, f(p) \Gamma(p) g(p)$, the 2-point function of S_{τ} is

$$\int dp \, f(p) \{e^{-\tau \mathcal{L}} \Gamma\}(p) g(p) \tag{2}$$

where $\mathcal{L}$ is a certain operator on functions of p.

To show this, expand the expectation value $S(A^{\lambda}(\tau))$ as a product of truncated functions

$$S^T\left(A_1^{\lambda}(\tau)\right) \cdots S^T\left(A_N^{\lambda}(\tau)\right) \tag{3}$$

and show

1. Any higher order truncated function or non-gauge invariant 2-point function tends to zero as $\lambda \to 0$ in the sense that

$$\left\langle S^T\left(A^{\lambda \, *}(\tau)\right) S^T\left(A^{\lambda}(\tau)\right)\right\rangle \longrightarrow 0 \tag{4}$$

as $\lambda \to 0$. (Note that $S^T(A^*) = \overline{S^T(A)}$. The convergence (4) is L^2-convergence as functions of the random field. The boundedness of the truncated functions then implies convergence in all L^p, $1 \leq p < \infty$.) This follows in perturbation theory by showing that each Feynman diagram containing such a truncated function S^T goes to zero in the weak coupling limit.

2. (a) The average over the random field $\langle S_{\tau}^{\lambda}(a^*(f)a(g)) \rangle$ converges to (2). This follows in perturbation theory by showing that the sum of the nth-order Feynman diagrams containing only the gauge-invariant 2-point function converges to

$$1/N! \int dp \, f(p) \{(-\tau \mathcal{L})^N \Gamma\}(p) g(p)$$

where $n = 2N$.

   (b)

$$\left\langle S_{\tau}^{\lambda}(a^*(\bar{g})a(f)) S_{\tau}^{\lambda}(a^*(f)a(g))\right\rangle - \left\langle S_{\tau}^{\lambda}((a^*(\bar{g})a(f)))\right\rangle \left\langle S_{\tau}^{\lambda}((a^*(f)a(g)))\right\rangle \longrightarrow 0$$

as $\lambda \to 0$. This follows in perturbation theory by showing that any Feynman diagram which does not factorize tends to zero as $\lambda \to 0$.

2. DYSON PERTURBATION SERIES

The time evolution of $a^*(k)$ with respect to the total Hamiltonian is expressed in the form $W(t)a_t^*(k)W(t)^*$ where $W(t) = e^{itH_\lambda}e^{-itH_0}$. Using $dW(t)/dt = i\lambda W(t)V(t)$, where

$$V(t) = e^{itH_0}Ve^{-itH_0} = \int dp\, dq\, \tilde{v}(q)a_t^*(q-p)a_t(p)$$

leads to

$$e^{itH_\lambda}a^*(k)e^{-itH_\lambda} = a_t^*(k) + i\lambda \int_0^t ds\, W(s)\,[V(s), a_t^*(k)]\, W(s)^*$$

Iterating this gives the Dyson series

$$e^{itH_\lambda}a^*(k)e^{-itH_\lambda}$$
$$= a_t^*(k) + \sum_{n=1}^\infty (i\lambda)^n \int_0^t dt_1 \int_0^{t_1} dt_2 \cdots \int_0^{t_{n-1}} dt_n\, [V(t_n),\ldots,[V(t_1),a_t^*(k)]\ldots]$$

Computing $[V(t'), a_t^*(k)]$ gives

$$\int dq\, \tilde{v}(q) e^{i(t-t')\varepsilon(k)} a_{t'}^*(q-k)$$

and iterating this leads to

$$(i\lambda)^n \int_0^t dt_1 \cdots \int_0^{t_{n-1}} dt_n\, [V(t_n),\cdots,V(t_1),a_t^*(k)]$$
$$= (i\lambda)^n \int_0^t dt_1 \cdots \int_0^{t_{n-1}} dt_n \int dq_1 \cdots dq_n\, \tilde{v}(q_1)\cdots\tilde{v}(q_n)$$
$$\times\ \exp\left[i\varepsilon(k-q_1)(t_1-t_2)\right]\exp\left[i\varepsilon(k-q_1-q_2)(t_2-t_3)\right]\cdots$$
$$\times\ \exp\left[i\varepsilon(k-q_1-\cdots-q_{n-1})(t_{n-1}-t_n)\right]\exp\left[i\varepsilon(k-q_1-\cdots-q_n)t_n\right]$$
$$\times\ a^*(k-q_1-\cdots-q_n)$$

We represent this expression by diagram 1. The solid line will be referred to as a **creation-line** and the dotted lines as **potential-lines**. There is a factor $i\lambda$ for each vertex on this creation-line, a factor $e^{i\varepsilon(p)(t_{j-1}-t_j)}$ for a line from the t_{j-1}-vertex to the t_j-vertex carrying momentum p, a factor $a^*(p)e^{i\varepsilon(p)t_n}$ for an arrow starting at t_n carrying momentum p, and a factor $\tilde{v}(q)$ for a potential-line carrying momentum q. All momenta $q_1,\ldots,q_n$ and times $0 \leq t_n \leq \cdots \leq t_1 \leq t$ are integrated over.

By taking adjoints we obtain a similar expression for the nth-order term associated with the destruction operator. Using $\overline{\tilde{v}(q)} = \tilde{v}(-q)$ we represent the expression by diagram 2. The solid line will be referred to as a **destruction-line**. On a destruction-line each vertex gives a factor $-i\lambda$, and an open arrow ending at t_n carrying momentum p gives a factor $a(p)e^{-i\varepsilon(p)t_n}$. The remaining factors are the same as for a creation-line. In particular, the line from the t_2-vertex to the t_1-vertex carrying momentum $k+q_1$ gives the factor $e^{i\varepsilon(k+q_1)(t_2-t_1)}$. Creation-lines and destruction-lines may be referred to as **Fermi-lines**.

3. FEYNMAN DIAGRAMS

Associated with the translation invariant initial state S of the Fermi gas are the truncated functions, which have a single δ-function representing momentum conservation.

$$S^T\left(a^{\#}(p_0),\ldots,a^{\#}(p_n)\right) = \delta(\epsilon_0 p_0 + \cdots + \epsilon_n p_n)\Gamma(p_1,\ldots,p_n) \tag{5}$$

where $\epsilon_j = 1$ for a creation operator and -1 for a destruction operator. (We have suppressed the dependence of Γ on $\epsilon_0,\ldots,\epsilon_n$. The distinguished momentum p_0 eliminated in Γ need not necessarily be the first argument.) The function $\Gamma(p_1,\ldots,p_n)$ will be supposed "smooth". The expression (5) is represented by diagram 3, where the arrow points toward the bubble if $a^{\#}$ is a creation operator and away from the bubble if it is a destruction operator. We shall use the convention that the distinguished momentum eliminated in (5) is drawn above the bubble while the remaining momenta are drawn below the bubble. Such a solid bubble will be called a **Fermi-bubble**. We shall also use the convention that in a gauge invariant 2-point function the distinguished momentum shall be that of the destruction operator, so a gauge invariant 2-point function is represented as in diagram 4. A translation invariant random field has truncated functions

$$\langle \tilde{v}(q_0),\ldots,\tilde{v}(q_n)\rangle^T = \delta(q_0 + \cdots + q_n)\gamma(q_1,\ldots,q_n) \tag{6}$$

where again $\gamma(q_1,\ldots,q_n)$ is assumed smooth. (The 1-point function $\langle v(x)\rangle = c$, a constant. Shifting $v(x)$ by the constant c does not change the time-evolution of observables and consequently we shall suppose the 1-point function is zero.) The expression (6) is represented by diagram 5. Such an open bubble will be called a **potential-bubble**.

As explained in §1.1 the basic object of interest is

$$\langle S^T\left(A_1^\lambda(\tau)\right)\cdots S^T\left(A_N^\lambda(\tau)\right)\rangle \tag{7}$$

where each A_j is a product of creation and destruction operators. (In fact a product of only two truncated functions S^T is required, but we consider here the more general expression.) Applying the perturbative expansion of §2 to each creation and destruction operator and expanding the average of products of $\tilde{v}(q)$ into truncated functions of the random field, we arrive at a typical perturbative contribution to (7), which may be represented as a Feynman diagram. Illustrated in diagram 6 is an 8th-order term consisting of one 6-point truncated function and two 2-point truncated functions of the Fermi gas and one 4-point truncated function and two 2-point truncated functions of the random field.

Ordering of Vertices

The vertices in a diagram will be ordered from lower left to upper right: A Fermi-bubble to the left of another Fermi-bubble comes before the other bubble, and all Fermi-lines associated with the first bubble come before the Fermi-lines of the other bubble. The upper Fermi-line of a bubble comes after all the lower Fermi-lines, and the lower lines are ordered from left to right. Vertices are ordered according to the Fermi-line on which they are located and a vertex below another vertex on the same Fermi-line comes before the other vertex.

3.1 Feynman Rules

The rules for obtaining the perturbative contribution corresponding to a diagram follow from the previous discussion.

1. Each k-variable k_j at the bottom of the diagram ($k_1 - k_7$ in diagram 6) is an independent integration variable which is controlled by the smooth function $f_j(k_j)$.

2. Each Fermi-bubble produces one dependent k-variable (say, k': at the top of a bubble) which is a linear combination of the other k-variables and q-variables associated with the lines of that bubble. There will thus arise a factor $f'(k')$. In obtaining upper bounds such functions are expressed in the form

$$f'(k') = \int dx\, \tilde{f}'(x) e^{ix \cdot k'}$$

and the integral over x is estimated using $\|\tilde{f}'\|_1$ after the momentum integrations are estimated.

3. Each Fermi-bubble gives a factor of the form $\Gamma(k'_1, \ldots, k'_n)$ where k'_j is a linear combination of the independent k- and q-variables. In diagram 6 the truncated 6-point function gives the factor $\Gamma(k_1, k_2 - q_1, k_3 - q_1, k_4, k_5 + q_2)$. As in point 2., such factors are expressed in the form

$$\Gamma(k'_1, \ldots, k'_n) = \int dx_1 \cdots dx_n\, \tilde{\Gamma}(x_1, \ldots, x_n) e^{i(x_1 \cdot k'_1 + \cdots + x_n \cdot k'_n)}$$

and are estimated using $\|\tilde{\Gamma}\|_1$.

4. Each potential-bubble produces a dependent q-variable which equals the negative of the sum of the remaining q-variables entering the bubble. It is convenient to reverse the direction of this potential-line and set its momentum equal to the sum of the remaining q-variables. (The direction of a potential line has no significance other than to keep track of momentum conservation.) Among the q-variables of a potential-bubble, the dependent q-variable is taken to be the last one (the one associated with the last vertex according to the ordering of vertices).

5. The integral over the independent q-variables is controlled by the truncated functions $\gamma(q_1, \ldots, q_n)$.

3.2 Matrix Diagrams

In points 2. and 3. above, imaginary exponentials linear in the momentum variables were introduced. An essential role is played by imaginary exponentials quadratic in the momentum variables. These arise in the non-relativistic case where $\varepsilon(p) = p^2/2m$ from the factors $\exp[i\varepsilon(p)(t' - t'')]$ discussed in §2, where p is a linear combination of momentum variables. It is a straightforward matter to compute the quadratic form in the independent momentum variables associated with a diagram by substituting for the dependent momenta in the energy expressions. This quadratic form (or associated matrix) is used to bound the contribution of a diagram, as discussed in §4. It is thus convenient to associate to the diagram the corresponding matrix rather than the entire perturbative contribution, and in this context the diagram will be referred to as a matrix diagram and the rules to compute the matrix as matrix rules. The matrix diagram may then be manipulated according to matrix relations. For example, eliminating certain lines in the matrix diagram corresponds to taking a submatrix of the associated matrix.

Matrix Rules

We discuss here some properties of the matrix associated with a diagram.

1. The quadratic form involving only the k-variables is a sum of terms, one from each Fermi-bubble. The various exponents in a creation-line add up to $i\varepsilon(k')t$ and in a destruction-line to $-i\varepsilon(k'')t$. The quadratic form associated with a Fermi-bubble then has the form

$$2t \sum_{j=0}^{n} \epsilon_j \varepsilon(k_j) = t[\sum_{j=1}^{n} \epsilon_j k_j^2 + \epsilon_0(\sum_{j=1}^{n} \epsilon_j k_j)^2]/m$$

Hence the matrix is

$$\mathcal{K}_{ij} = t[\epsilon_i \delta_{ij} + \epsilon_0 \epsilon_i \epsilon_j]/m$$

2. We consider now the quadratic form involving only the q-variables. In relation to the ordering of vertices discussed earlier let k', q', and t' be the momenta and time at the *first* vertex, and take q' as one of the independent q-momenta. The only t'-dependence of the quadratic form in the q-variables is then $\pm 2t'\varepsilon(k' \pm q')$ which contributes $\pm t'q'^2/m$. Other t_j-variables may also appear in the complete coefficient of q'^2 and in cross-terms of q' with other q-variables. But the essential point is that t' occurs linearly in the diagonal element associated with q', and it occurs *nowhere else*.

3. Suppose the first vertex lies in a gauge invariant 2-point function and that the potential line from this vertex connects, via a 2-point potential-bubble, another gauge invariant 2-point function, as in diagram 7. (There may be other vertices and Fermi-bubbles not shown in the diagram.) The two potential-lines joined by a 2-point potential-bubble will be said to form a **potential-arc**. Then

$$Q_{11} = [(t'-t) + (t''-t)]/m$$

4. Let a diagram consist of only one Fermi-bubble, that of a gauge invariant 2-point function, and furthermore containing only 2-point potential-bubbles, as in diagram 8. The quadratic form in the q-variables has diagonal coefficients given by $(t_{lower} - t_{upper})/m$. For example, in diagram 8 the coefficient of q_2^2 is $(t_b - t_c)/m$. The coefficient of the cross-terms are given by the intersection of the segments $[t'_{lower}, t'_{upper}] \cap [t''_{lower}, t''_{upper}]$. For example, in diagram 8 the coefficient of $q_1 q_2$ is $[t_b - t_c]/m$ (the first potential-arc graph-shadows the second potential-arc), the coefficient of $q_1 q_3$ is $[t_d - t_e]/m$ (the potential-arcs intersect), the coefficient of $q_1 q_4$ is 0 (the potential-arcs are disjoint).

4. DETERMINANT INEQUALITIES

To obtain an upper bound for a general perturbative term, we must, according to the discussion in §3, estimate quantities such as

$$\mathcal{I} = \int dp_1 \cdots dp_r \, g(p_1, \ldots, p_r) e^{i/2(p, Qp)} e^{i(L, p)} \tag{8}$$

where the integrand contains a well-behaved function g and quadratic and linear imaginary exponentials. The p-variables correspond to k- or q-variables and the t-dependence is contained in Q and L. The standard estimate[5,6] for the integral (8) is

$$|\mathcal{I}| \leq \|\hat{g}\|_1 |\det Q|^{-1/2} \tag{9}$$

where $\|\hat{g}\|_1$ is the L^1-norm of the Fourier transform of g, and $\det Q$ is the determinant of the matrix Q. The upper bound (9) is derived by writing $\mathcal{I} = \lim_{\epsilon \to 0} \mathcal{I}_\epsilon$ where

$$\mathcal{I}_\epsilon = \int dp_1 \cdots dp_r \, g(p_1, \ldots, p_r) e^{-1/2(p,\epsilon[I-iQ]p)} e^{i(L,p)}$$

substituting

$$g(p)e^{i(L,p)} = (2\pi)^{-r/2} \int dx_1 \cdots dx_r \, \hat{g}(x_1 + L_1, \ldots, x_r + L_r) e^{-i(x,p)}$$

doing the complex Gaussian p-integral before the x-integral and then taking the limit $\epsilon \to 0$. A stronger standard estimate is obtained by applying (9) to the integral over a subset $p'_1, \ldots, p'_l$ of the variables $p_1, \ldots, p_r$. Doing the integral over the p'-variables first and applying (9) yields

$$|\mathcal{I}| \leq \|\tilde{g}\|_1 |\det Q'|^{-1/2} \tag{10}$$

where $\tilde{g}$ is the Fourier transform of g with respect to the p'-variables and Q' is the corresponding *submatrix* of Q. (If $\{\alpha_1, \ldots, \alpha_l\}$ is a subset of $\{1, \ldots, r\}$ then $Q'_{jk} = Q_{\alpha_j \alpha_k}$. We write $Q' \prec Q$.) Let $\|g\|$ denote the maximum of $\|\tilde{g}\|_1$ over all subsets of the p-variables, and let

$$\max Q = \sup_{Q' \prec Q} |\det Q'|$$

The empty submatrix is included, setting $\det Q' = 1$ in this case. (This corresponds to estimating (8) by the L^1-norm of g.) The estimate (9) is thus strengthened to[5,6]

$$|\mathcal{I}| \leq \|g\| \, [\max Q]^{-1/2} \tag{11}$$

Now observe the vector nature of the momentum variables. Thus $\varepsilon(p) = 1/(2m) \sum_{j=1}^{\nu} p_j^2$ where ν is the number of space dimensions. This has the effect of raising $\det Q$ to the νth power and thus the basic estimate (11) actually takes the form

$$|\mathcal{I}| \leq \|g\|[\max Q]^{-\nu/2} \tag{12}$$

Consequently if the number of space dimensions is three or more (which we shall always suppose), the exponent of $[\max Q]^{-1}$ is greater than 1. In order to effectively apply inequality (12) it is convenient to relate $\max Q$ to $\max \widehat{Q}$, where $\widehat{Q}$ is the submatrix of Q obtained by eliminating the first row and column of Q. Expressing Q in the form

$$Q = \begin{pmatrix} Q_{11} & \bar{v} \\ u & \widehat{Q} \end{pmatrix} \tag{13}$$

Then

$$\det Q = \left[Q_{11} - (v, \widehat{Q}^{-1} u) \right] \det \widehat{Q} \tag{14}$$

Equality (14) follows easily from Cramer's rule. To derive it, first transform to a new orthonormal basis such that the first basis vector is unchanged and the new second basis vector is parallel to v. With respect to this new basis, expression (13) takes the form

$$Q' = \begin{pmatrix} Q_{11} & v' \\ u' & A' \end{pmatrix}$$

where $v'_j = \|v\|\delta_{j1}$. Then

$$\det Q' = Q_{11} \det A' - \|v\| \det A_1$$

where A_1 is obtained by replacing the first column of A' with the vector u'. By Cramer's rule $\det A_1 / \det A'$ equals the first component of $A'^{-1} u'$. Hence

$$\det Q' = [\, Q_{11} - (v', A'^{-1} u') \,] \det A'$$

Equality (14) now follows by the invariance of the determinant and scalar product under a change of orthonormal basis. Equality (14) is applied to inequality (11) in the following way. Let Q_1 be the submatrix of $\widehat{Q}$ satisfying $\max \widehat{Q} = |\det Q_1|$. Extend Q_1 to the submatrix Q_2 of Q obtained by adding 1 to the subset of $\{1, \ldots, r\}$ defining Q_1. From (14),

$$\det Q_2 = [\, Q_{11} - (v, Q_1^{-1} u) \,] \det Q_1$$

where v and u are obtained by taking the appropriate subset of elements from the first row and first column of Q. Then, since $\max Q \geq |\det Q_2|$, we have

$$\max Q \geq |Q_{11} - f(Q)| \max \widehat{Q}$$

where $f(Q)$ does not depend on Q_{11}. Since $\max Q \geq \max \widehat{Q}$ we have

$$\max Q \geq H\left(Q_{11} - f(Q)\right) \max \widehat{Q} \qquad (15)$$

where $H(x) = \max(|x|, 1)$. Inequality (15) serves as the basic tool in controlling the weak coupling limit of perturbative terms. It is applied to the integral over q-variables.

One further determinant identity is useful in estimating the integral over k-variables, and shows that the contribution of higher order and non-gauge invariant 2-point truncated functions of the Fermi gas tends to zero in the weak coupling limit. The identity is

$$\det(A + |u\rangle\langle v|) = [1 + (v, A^{-1} u)] \det A \qquad (16)$$

where $(|u\rangle\langle v|)_{ij} = u_i \bar{v}_j$. To prove this write

$$\det(A + |u\rangle\langle v|) = \det A \det(I + |A^{-1} u\rangle\langle v|)$$

and evaluate $\det(I + |w\rangle\langle v|)$ by transforming to a new orthonormal basis such that the first basis vector is parallel to v. Then $\det(I + |w\rangle\langle v|) = 1 + \|v\| w_1 = 1 + (v, w)$.

4.1 Applications

A. Free Evolution

The 0th-order term in the perturbative expansion is the free evolution. According to §3.1 the matrix $\mathcal{K}$ associated with an rth-order truncated function of the Fermi gas is

$$\mathcal{K} = t/m[\, I_\epsilon + |u\rangle\langle v| \,]$$

where $[\, I_\epsilon \,]_{ij} = \epsilon_i \delta_{ij}$ and $u_i = \epsilon_0 \epsilon_i$, $v_i = \epsilon_i$, $i,j = 1,\ldots,r-1$. Then according to equality (16)

$$\det \mathcal{K} = (t/m)^{r-1} \sum_{i=0}^{r-1} \epsilon_i \prod_{j=0}^{r-1} \epsilon_j$$

and

$$|\det \mathcal{K}| = (t/m)^{r-1} |\sum_{i=0}^{r-1} \epsilon_i| \qquad (17)$$

Similarly

$$|\det \widehat{\mathcal{K}}| = (t/m)^{r-2} |\sum_{i \neq 1} \epsilon_i| \qquad (18)$$

Now $\max \mathcal{K} \geq \max(|\det \mathcal{K}|, |\det \widehat{\mathcal{K}}|)$. If the truncated function is not gauge invariant $\sum_{i=0}^{r-1} \epsilon_i \neq 0$ and hence the truncated function is bounded by $Ct^{-(r-1)\nu/2}$. If the truncated function is gauge invariant $\sum_{i=0}^{r-1} \epsilon_i = 0$ but $\sum_{i \neq 1} \epsilon_i \neq 0$. Then the truncated function is bounded by $Ct^{-(r-2)\nu/2}$. Hence higher order and non-gauge invariant 2-point functions tend to zero in the long-time limit[7] of the free evolution.

B. Time Integral Bound

To bound the time integral

$$\int_{t_a}^{t_b} dt' \, [\max Q]^{-\alpha}$$

where $\alpha > 1$ and the t'-vertex is the first vertex according to the ordering of vertices in §3, substitute inequality (15) and extend the t'-integral to $[-\infty, \infty]$. Using $Q_{11} = \pm t'/m + T$ where T does not depend on t' and the fact that t' occurs *only* in Q_{11} yields the bound

$$\int_{t_a}^{t_b} dt' \, [\max Q]^{-\alpha} \leq h_\alpha [\max \widehat{Q}]^{-\alpha} \qquad (19)$$

where $h_\alpha = m \|H^{-\alpha}\|_1 = 2m\alpha/(\alpha - 1)$.

C. Double Integral Bounds

Inequality (15) combined with $\max Q \geq |Q_{11}|$ yields

$$[\max Q]^{-\alpha} \leq |Q_{11}|^{-\alpha\epsilon} H(Q_{11} - f(Q))^{-\alpha(1-\epsilon)} [\max \widehat{Q}]^{-\alpha(1-\epsilon)} \qquad (20)$$

where ϵ is chosen so that $\alpha\epsilon < 1 < \alpha(1-\epsilon)$.

C1

In relation to matrix rule 3 in §3, let $Q_{11} = [(t' - t) + (t'' - t)]/m$ where t' occurs only in Q_{11} and t'' does not occur in $\widehat{Q}$. Then

$$\int_{t_a}^{t_b} dt' \int_{t_c}^{t_d} dt'' \, [\max Q]^{-\alpha}$$

$$\leq \int_0^t dt'' [(t - t'')/m]^{-\alpha\epsilon} \int_{-\infty}^{\infty} dt' \, H(t'/m + T)^{-\alpha(1-\epsilon)} \left[\max \widehat{Q}\right]^{-\alpha(1-\epsilon)}$$

$$\leq m^{\alpha\epsilon} t^{(1-\alpha\epsilon)}/(1 - \alpha\epsilon) h_{\alpha(1-\epsilon)} \left[\max \widehat{Q}\right]^{-\alpha(1-\epsilon)}$$

C2

In relation to matrix rule 4 in §3, let $Q_{11} = (t' - t'')/m$ where t' occurs only in Q_{11} and t'' does not occur in $\widehat{Q}$. Then for any c, $t_a \leq c \leq t_b$,

$$\int_{t_a}^{c} dt'' \int_{c}^{t_b} dt' [\max Q]^{-\alpha}$$
$$\leq \int_{0}^{c} dt'' [(c-t'')/m]^{\alpha \epsilon} \int_{-\infty}^{\infty} dt' H(t'/m + T)^{-\alpha(1-\epsilon)} [\max \widehat{Q}]^{-\alpha(1-\epsilon)}$$
$$\leq t^{(1-\alpha\epsilon)}/(1-\alpha\epsilon) m^{-\alpha\epsilon} h_{\alpha(1-\epsilon)} [\max \widehat{Q}]^{-\alpha(1-\epsilon)}$$

The same inequality holds in the case $t_a \leq t' \leq c \leq t'' \leq t_b$.

5. WEAK COUPLING LIMIT

The contribution from a Feynman diagram has, according to the Feynman Rules of §3.1, the general form of equation (8), integrating over the independent k- and q-variables. The basic upper bound (12) is applied, the matrix Q being computed according to the Matrix Rules of §3.2. Consider first the quadratic form in the independent q-variables only. The independent q-variables are chosen and ordered according to the vertex ordering of §3. If the order of the diagram is n and there are r independent q-variables, then r t-integrals and the corresponding potential lines in the matrix diagram may be eliminated by successively applying the Time Integral Bound (19) to each q-variable in turn, giving an upper bound

$$C\lambda^n h_{\nu/2}^r$$

where we have made explicit the factor of λ associated with each vertex. The remaining $n - r$ t-integrals give a final upper bound

$$C h_{\nu/2}^r \lambda^n t^{n-r} = C' \tau^{n/2} t^{n/2-r}$$

Now if the diagram contains a potential bubble of order greater than two, then $r > n/2$ and the bound tends to zero in the weak coupling limit. We have thus shown

A. *If a diagram contains a potential bubble of order higher than second order, then its contribution tends to zero in the weak coupling limit. Thus only the second order truncated function of the random field contributes in the weak coupling limit.*

Consider now the effect of the k-variables. Applying the upper bound (12) to the submatrix Q associated with the independent q-variables and then to the submatrix $\mathcal{K}$ associated with the independent k-variables gives

$$|\mathcal{I}| \leq C [\max \mathcal{K}]^{-\epsilon\nu/2} [\max Q]^{(1-\epsilon)\nu/2}$$

where ϵ is chosen so that $\epsilon\nu/2 < 1 < (1-\epsilon)\nu/2$. The matrix $\mathcal{K}$ depends only on t and not on the t-variables at the vertices. Proceeding as in the previous discussion yields the upper bound

$$[\max \mathcal{K}]^{-\epsilon\nu/2} C h_{(1-\epsilon)\nu/2}^r \lambda^n t^{n-r} = [\max \mathcal{K}]^{-\epsilon\nu/2} C' \tau^{n/2} t^{n/2-r}$$

Now $r \geq n/2$ so $t^{n/2-r}$ remains bounded as $t \to \infty$. If the diagram contains a Fermi bubble of order greater than second or a non-gauge invariant 2nd order bubble, $\max \mathcal{K} \to 0$

as $t \to \infty$, according to the result in §4.1A. We have thus shown

B. *If a diagram contains a Fermi bubble of order higher than second or a non-gauge invariant 2nd order bubble, then its contribution tends to zero in the weak coupling limit.* Thus only the gauge invariant 2-point truncated function of the initial state of the Fermi gas contributes in the weak coupling limit.

According to results A. and B. we shall consider diagrams containing only 2-point potential bubbles and gauge invariant 2-point Fermi bubbles. If the diagram contains a potential arc connecting two Fermi bubbles, eliminate potential arcs from the Matrix Diagram one-by-one, as discussed above, until the said potential arc is associated with the first q-variable. According to Matrix Rule 3 of §3.2, $Q_{11} = [(t'-t) + (t''-t)]/m$. Now apply the Double Integral Bound of §4.C1, eliminating the t-integrals associated with *both* endpoints of the potential line, and thereby introducing a factor $ct^{1-\epsilon\nu/2}$. Eliminating the remaining potential arcs as above yields, in place of the bound $C'\tau^{n/2}t^{n/2-r} = C'\tau^{n/2}$ since $r = n/2$, the bound

$$C''\tau^{n/2}t^{n/2-r-\epsilon\nu/2} = C''\tau^{n/2}t^{-\epsilon\nu/2}$$

which tends to zero as $t \to \infty$. We have thus shown

C. *If a diagram contains a potential arc connecting two different Fermi bubbles, then its contribution tends to zero in the weak coupling limit.* Hence the expectation over the random field of products of 2-point truncated functions of the Fermi gas factorizes in the weak coupling limit — to each order in perturbation theory. (See 2(b) of §1.1.)

According to results A., B., and C. we shall consider diagrams containing only 2-point potential bubbles and gauge invariant 2-point Fermi bubbles, with potential arcs which do not connect different Fermi bubbles. The contribution from such a diagram factorizes as a product of terms, one for each Fermi bubble. Hence we need only consider diagrams with a single Fermi bubble.

Totally Ordering the Time Variables

Observe that the vertex-times are totally ordered on the creation-line of the Fermi bubble and totally ordered on the destruction-line of the Fermi bubble, but that there is no order relation between the times on the creation-line and the times on the destruction-line. It is convenient to decompose the time-integrals into the $\binom{l_1 + l_2}{l_1}$ totally ordered sectors (where there are l_1 vertices on the creation-line and l_2 vertices on the destruction-line). We may therefore consider that *the vertex-times of the diagram are totally ordered.* (Do not confuse this time-ordering with the vertex ordering of §3.)

If the diagram contains a potential arc connecting vertices with *non-consecutive* times, eliminate the potential arcs from the Matrix Diagram one-by-one, until the said potential arc is associated with the first q-variable. According to Matrix Rule 4 of §3.2, $Q_{11} = (t'-t'')/m$. As t', t'' are not consecutive, there is a vertex-time $\hat{t}$ satisfying $t' \leq \hat{t} \leq t''$ (or $t'' \leq \hat{t} \leq t'$). Now (holding $\hat{t}$ fixed) apply the Double Integral Bound of §4.C2, eliminating the t-integrals associated with *both* endpoints of the potential line and, as in the previous discussion of point C., obtaining in place of the bound $C'\tau^{n/2}$ the bound

$$C''\tau^{n/2}t^{-\epsilon\nu/2}$$

which tends to zero as $t \to 0$. We have thus shown

D. *If a diagram contains a potential arc connecting vertices with non-consecutive times, then its contribution tends to zero in the weak coupling limit.*

Remark. The careful reader may have considered the possibility that the $\hat{t}$-vertex comes *before* the t'-vertex with respect to the vertex ordering and hence may have been eliminated before the Double Integral Bound is applied. However, we may suppose the potential line under consideration is the *first* such non-consecutive potential line and thus if the $\hat{t}$-vertex comes before the t' vertex then the $\hat{t}'$-vertex which is paired with $\hat{t}$ is consecutive with $\hat{t}$ and thus also satisfies $t' \leq \hat{t}' \leq t''$. Hence if $\hat{t}$ were eliminated we may use $\hat{t}'$ in its place.

6. BOLTZMANN EQUATION

The generic term Boltzmann equation will be used to denote the irreversible equation describing the behavior of a reversible system in an appropriate limit. According to the results of §5 we shall consider Feynman diagrams which contain only (1)a single gauge invariant 2-point Fermi-bubble, (2)2-point potential-bubbles, and (3)totally ordered vertex-times which are consecutively paired by the potential arcs. Let $0 \leq t_n \leq \cdots \leq t_1 \leq t$ be the totally ordered vertex times. The order $n = 2N$ is necessarily even. The arrangement of vertices in the diagram is conveniently parametrized by the $2N$ parameters $\sigma_1, \cdots, \sigma_N, \epsilon_1, \cdots, \epsilon_N$ as follows. If t_{2j-1} and t_{2j} lie on the same Fermi-line then $\sigma_j = 0$; if one lies on the creation-line and the other on the destruction-line then $\sigma_j = 1$. If in the vertex ordering t_{2j-1} comes before (lies below) t_{2j} then $\epsilon_j = 1$; otherwise $\epsilon_j = -1$. (See diagram 9.)

A typical Feynman diagram is illustrated in diagram 10 where the Fermi-bubble contributes the factor $\Gamma(k - q_1)$. In general the Fermi-bubble contributes the factor $\Gamma(k - \sum_{j=1}^{N} \sigma_j q_j)$. The exponential associated with diagram 10 is

$$\exp i \left[\varepsilon(k)(t - t_2) + \varepsilon(k - q_1)(t_2 - t_3) + \varepsilon(k - q_1 - q_2)(t_3 - t_4) + \varepsilon(k - q_1)t_4 \right.$$
$$\left. + \varepsilon(k - q_1)(-t_5) + \varepsilon(k - q_1 - q_3)(t_5 - t_6) + \varepsilon(k - q_1)(t_6 - t_1) + \varepsilon(k)(t_1 - t)\right]$$

Substituting $\varepsilon(p) = p^2/2m$ gives an exponent

$$i/(2m) \left[(t_2 - t_1)\{q_1^2 - 2k \cdot q_1\} + (t_3 - t_4)\{q_2^2 - 2k \cdot q_2\} + (t_5 - t_6)\{q_3^2 - 2k \cdot q_3\} \right.$$
$$\left. + 2(t_3 - t_4)q_1 \cdot q_2 + 2(t_5 - t_6)q_1 \cdot q_3\right]$$
$$= i/(2m) \left[\sum_{j=1}^{N} \epsilon_j(t_{2j-1} - t_{2j})(q_j^2 - 2k \cdot q_j) + 2\sum_{k<j} \sigma_k \epsilon_j(t_{2j-1} - t_{2j})q_j \cdot q_k\right]$$

where in the last line we have written the general formula, as can easily be verified.

The contribution from all such diagrams of order $2N$ is then

$$\sum_{\sigma_1 \cdots \sigma_N} \sum_{\epsilon_1 \cdots \epsilon_N} (i\lambda)^{2N}(-1)^{\sum_{j=1}^{N} \sigma_j} \int_0^t dt_1 \cdots \int_0^{t_{2N-1}} dt_{2N} \int dk\, f(k)g(k)$$

$$\int dq_1 \cdots dq_N \gamma(q_1) \cdots \gamma(q_N) \times \Gamma\left(k - \sum_{j=1}^{N} \sigma_j q_j\right)$$

$$\exp i/(2m)\left[\sum_{j=1}^{N} \epsilon_j(t_{2j-1} - t_{2j})(q_j^2 - 2k \cdot q_j) + 2\sum_{k>j} \sigma_k \epsilon_j(t_{2j-1} - t_{2j})q_j \cdot q_k\right] \quad (21)$$

The weak coupling limit is facilitated by introducing the variables $\tau_j = \lambda^2 t_{2j-1}$, $s_j = t_{2j-1} - t_{2j}$. Then $0 \leq \tau_N \leq \cdots \leq \tau_1 \leq \tau$ and $0 \leq s_j \leq t_{2j-1} - t_{2j+1} = (\tau_j - \tau_{j+1})/\lambda^2$, where we set $\tau_{N+1} = 0$. Noting that the integrand of the time-integrals, $\mathcal{F}(\epsilon_1(t_1 - $

$t_2), \ldots, \epsilon_N(t_{2N-1} - t_{2N}))$, depends on the time variables only in the form $\epsilon_j(t_{2j-1} - t_{2j})$, the expression (21) is transformed to

$$(-1)^N \sum_{\sigma_1 \cdots \sigma_N} (-1)^{\sum_{j=1}^N \sigma_j} \sum_{\epsilon_1 \cdots \epsilon_N} \int_0^\tau d\tau_1 \cdots \int_0^{\tau_{N-1}} d\tau_N$$

$$\times \int_0^{(\tau_1 - \tau_2)/\lambda^2} ds_1 \cdots \int_0^{\tau_N/\lambda^2} ds_N \mathcal{F}(\epsilon_1 s_1, \ldots, \epsilon_N s_N)$$

As $\lambda \to 0$ this converges to

$$(-1)^N \sum_{\sigma_1 \cdots \sigma_N} (-1)^{\sum_{j=1}^N \sigma_j} \sum_{\epsilon_1 \cdots \epsilon_N} \tau^N/N! \int_0^\infty ds_1 \cdots \int_0^\infty ds_N \mathcal{F}(\epsilon_1 s_1, \ldots, \epsilon_N s_N)$$

$$= (-\tau)^N/N! \sum_{\sigma_1 \cdots \sigma_N} (-1)^{\sum_{j=1}^N \sigma_j} \int_{-\infty}^\infty ds_1 \cdots \int_{-\infty}^\infty ds_N \mathcal{F}(s_1, \cdots, s_N)$$

Substituting $\int_{-\infty}^\infty ds\, e^{isy/(2m)} = 4\pi m \delta(y)$ gives

$$(-\tau)^N/N! \sum_{\sigma_1 \cdots \sigma_N} (-1)^{\sum_{j=1}^N \sigma_j} \int dk\, f(k)g(k) \int dq_1 \cdots dq_N \gamma(q_1) \cdots \gamma(q_N) \times$$

$$\prod_{j=1}^N \left\{ 4\pi m \delta \left(q_j^2 - 2k \cdot q_j + 2\sum_{k>j} \sigma_k q_j \cdot q_k \right) \right\} \Gamma \left(k - \sum_{j=1}^N \sigma_j q_j \right) \qquad (22)$$

Introducing the operator $\mathcal{L}$ by

$$(\mathcal{L}h)(k) = 4\pi m \int dp\, \gamma(k-p) [h(k) - h(p)] \delta(p^2 - k^2)$$

$$= 4\pi m \int dq\, \gamma(q) [h(k) - h(k-q)] \delta(q^2 - 2k \cdot q)$$

$$= \sum_{\sigma=1,0} (-1)^\sigma 4\pi m \int dq\, \gamma(q) h(k - \sigma q) \delta(q^2 - 2k.q)$$

the expression (22) is

$$1/N! \int dk\, f(k) g(k) \left[(-\tau \mathcal{L})^N \Gamma \right](k)$$

Summing over N gives

$$\int dk\, f(k) \left(e^{-\tau \mathcal{L}} \Gamma \right)(k) g(k)$$

Thus in the weak coupling limit the 2-point function of the Fermi gas is given by

$$\Gamma_\tau = e^{-\tau \mathcal{L}} \Gamma$$

and satisfies the irreversible Boltzmann equation[5,6]

$$d/d\tau\, \Gamma_\tau = -\mathcal{L} \Gamma_\tau$$

The equation is linear because there is no back reaction of the Fermi gas on the random field.

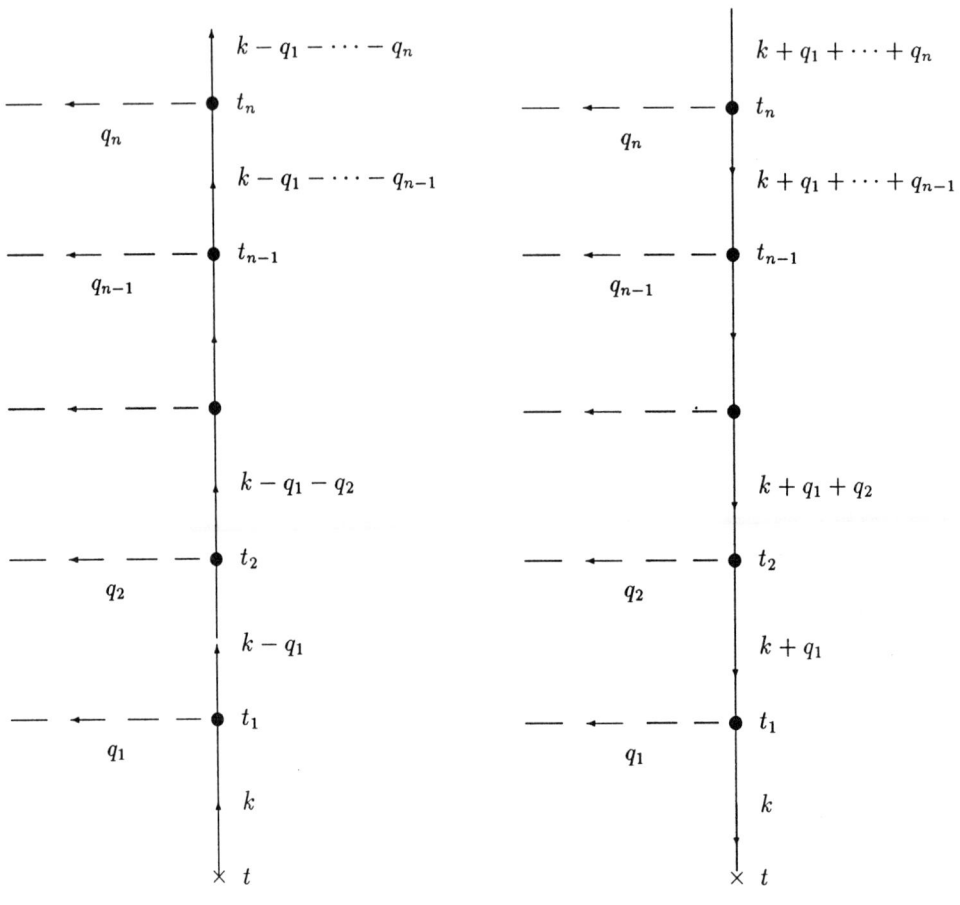

Diagram 1
Creation-line with potential-
lines and vertices

Diagram 2
Destruction-line with potential-
lines and vertices

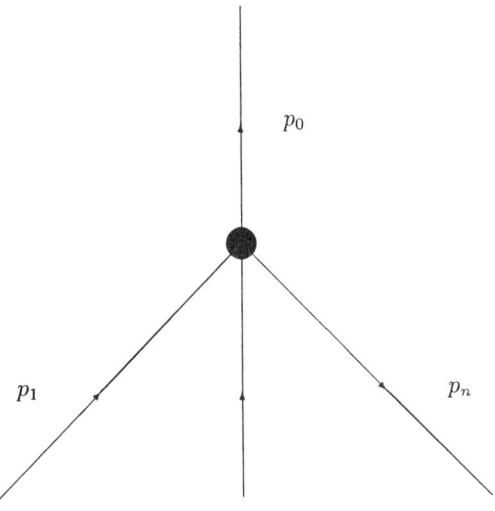

Diagram 3

Fermi-bubble

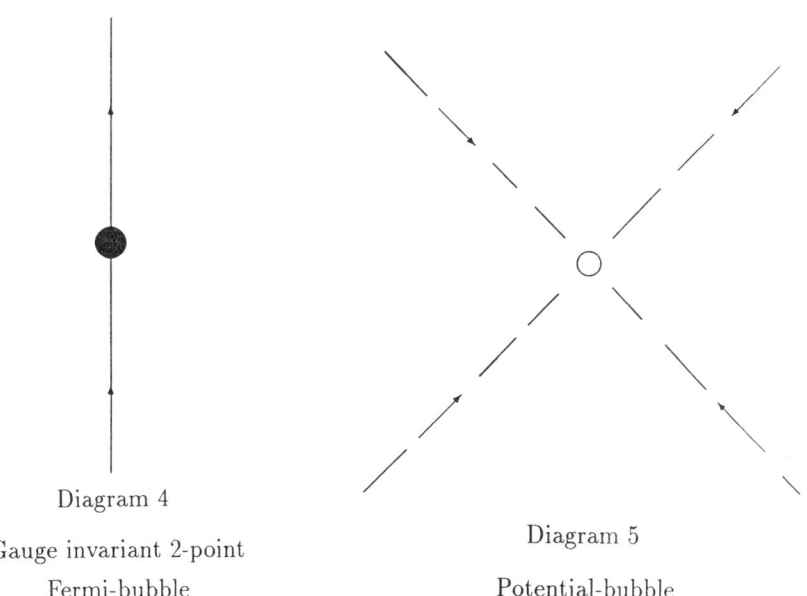

Diagram 4

Gauge invariant 2-point
Fermi-bubble

Diagram 5

Potential-bubble

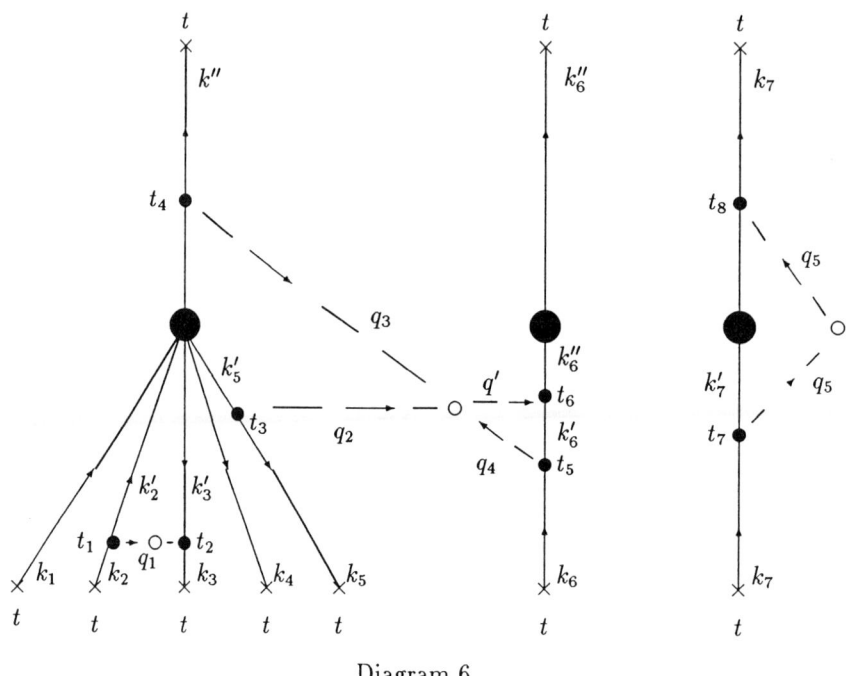

Diagram 6

An 8th-order Feynman diagram

Conservation of momentum holds. For example, $k'' = k_1 + k_2 + k_3 - k_4 - k_5 - q_2 - q_3$.

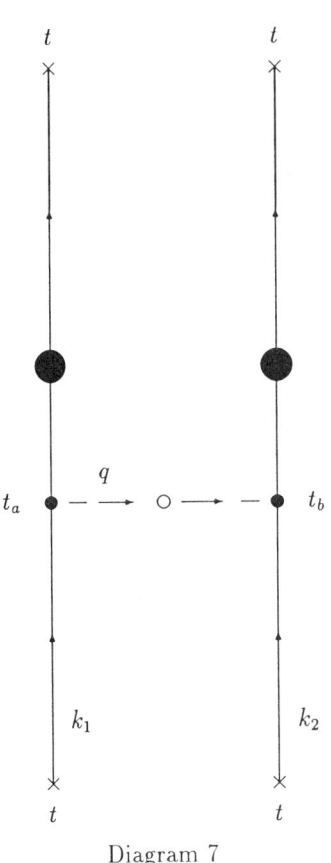

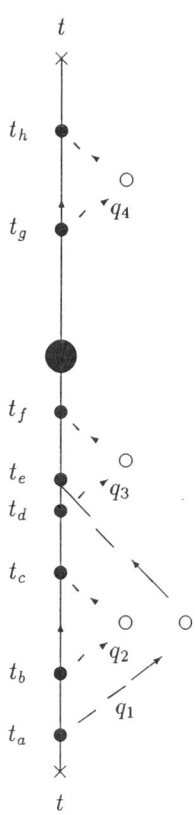

Diagram 7

Diagram 8

Two Fermi-bubbles joined by a potential arc

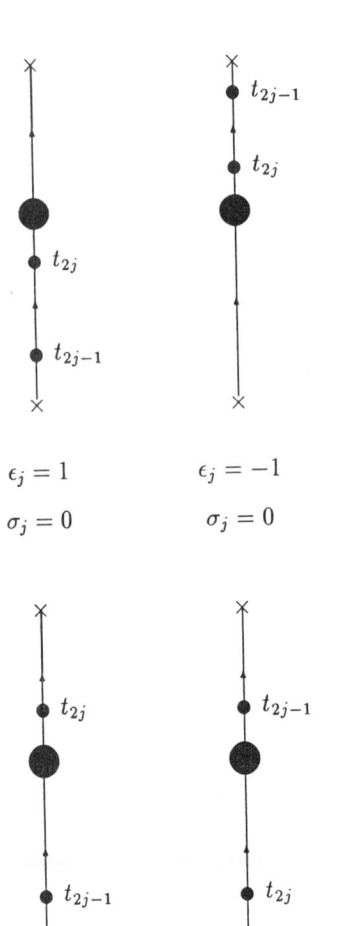

Diagram 9
Parametrization

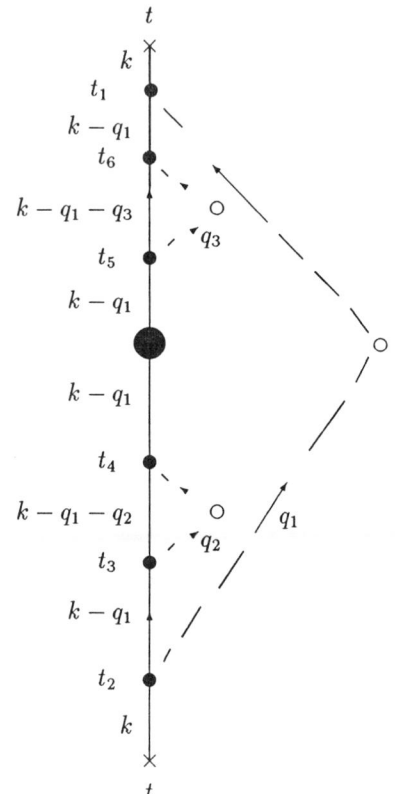

Diagram 10

REFERENCES

1. T.G. Ho, L.J. Landau and A.J. Wilkins, On the weak coupling limit for a Fermi gas in a random potential, *Rev. Math. Phys.*, (to appear).
2. H. Spohn, Kinetic equations from hamiltonian dynamics: markovian limits, *Rev. Mod. Phys.*, **53**, 569 (1980).
3. N.M. Hugenholtz, Derivation of the Boltzmann equation for a Fermi gas, *J. Stat. Phys.*, **32**, 231 (1983).
4. P. Martin and G.G. Emch, A rigorous model sustaining van Hove's phenomenon, *Helv. Phys. Acta.*, **48**, 59 (1975).
5. H. Spohn, Derivation of the transport equation for electrons moving through random impurities, *J. Stat. Phys.*, **17**, 385 (1977).
6. G.F. Dell'Antonio, Large time, small coupling behaviour of a quantum particle in a random potential, *Ann. Inst. Henri Poincaré A*, **39**, 339 (1983).
7. O.E. Lanford and D.W. Robinson, Approach to equilibrium of free quantum systems, *Commun. Math. Phys.*, **24**, 193 (1972).

LIMIT LAWS FOR RECURRENCE TIMES IN EXPANDING MAPS OF AN INTERVAL

P. Collet

Centre de Physique Théorique, CNRS UPR 14
Ecole Polytechnique
F-91128 Palaiseau Cedex (France)

I. INTRODUCTION

The problem of entrance time dates back to the first developments of ergodic theory. A well known historical example being the raging discussions at the beginning of the century about the period of Poincaré recurrence. As for many questions in ergodic theory, the main step came much later with Birkhoff's ergodic theorem. We recall that if we have a space X equipped with a σ-algebra $\mathcal{A}$ and a measure μ on this σ-algebra which is invariant and ergodic for a measurable transformation f of X, then for any measurable set A of positive measure, and for μ almost every initial condition x, the orbit of x spends (asymptotically) in A a fraction of time equal to $\mu(A)$. In other words, for μ almost every x, we have

$$\lim_{n \to \infty} \frac{1}{n} \sum_{j=0}^{n-1} \chi_A(f^j(x)) = \mu(A) ,$$

where as usual χ_A denotes the characteristic function of the set A, and f^j is the jth iterate of f. This means that roughly the orbit of x recurs to A with a period equal to $1/\mu(A)$, and this is a natural time scale associated to any measurable set A with positive measure. At this level of generality one cannot say much more about the law of recurrence time to A [1], but by analogy with the techniques of probability theory, one can hope to get interesting asymptotic results when the measure of A tends to zero. To be more precise we define the first entrance time to A as the random variable given by

$$\tau_A(x) = \min\{n \geq 0 \,;\, f^n(x) \in A\} .$$

Note that it follows immediately from Birkhoff's ergodic theorem that τ_A is almost surely finite.

---

[1] There is however an interesting result by M.Kac about the average of the return time to a set A when one starts in A. We thank L.Landau for pointing out this reference[1] to us.

Similarly, we define recursively the process of successive visits to A by setting $\tau_A^1 = \tau_A$, and for $p > 1$

$$\tau_A^p(x) = \min\{n > \tau_A^{p-1}(x);\ f^n(x) \in A\}\ .$$

As before, all these random variables are μ almost surely finite. Our primary goal will be to obtain an asymptotic law for $\tau_A \mu(A)$ when $\mu(A)$ tends to zero.

We will now make some further hypothesis on the dynamical system. Instead of formulating an abstract theory, we will consider a concrete family of dynamical systems: the piecewise regular expanding maps of an interval. Our phase space will be the interval $[0, 1]$ on which there is given a finite partition in sub-intervals with boundary points $0 = a_0 < a_1 < \cdots < a_q < a_{q+1} = 1$. The mapping f is regular on each open interval of the partition, and can be extended to a regular map on each closed interval $[a_j, a_{j+1}]$ (a simple example is the map $f(x) = 2x \pmod{1}$). Moreover, we assume that there is a number $\rho > 1$ such that for each $0 \leq j \leq q$

$$|f'_{[a_j, a_{j+1}]}| > \rho\ .$$

It is in fact enough to assume that some iterate of f satisfies this condition but for simplicity we will not work with this (easy) generalization.

It is well known[2] that these maps have always an absolutely continuous invariant measure. Moreover, eventually taking some iterate and after a finite decomposition into invariant sub-intervals, we can assume that the dynamical system is ergodic and mixing, which implies that it is exponentially mixing on functions of bounded variation[2]. We will denote below by h the density of μ. We also mention that all the results below are also true using a probability equivalent to the invariant measure.

We recall that any finite state Markov chain can be represented by such a dynamical system with a map f piecewise affine (some branches may have slope of modulus 1). This case was treated previously by Pitskel using different techniques[3]. Also most random number generators are of this form or based on related systems[4] We also mention that the results explained below are also valid for Gibbs measures on subshifts of finite type with essentially the same proofs (in fact slightly simpler because of the topological Markov property). This case was treated independently by Hirata[5] using ζ functions.

II. RESULTS ABOUT ENTRANCE TIMES

Theorem 1[6]**.** *Let (K_j) be a sequence of intervals such that $\mu(K_j) > 0$, and $|K_j|$ tends to 0 if $j \to \infty$. Then there is a diverging sequence of positive numbers (T_j) such that for any $t > 0$ we have*

$$\lim_{j \to \infty} \mu(\{\tau_{K_j} > t T_j\}) = e^{-t}\ .$$

In other words, we have convergence in law to an exponential distribution. Note that at this level of generality we do not have a good control over the normalizations T_j. It can be proven however that $\liminf_{j \to \infty} T_j \mu(K_j) > 0$. The hypothesis that the sets K_j are intervals (or finite union of intervals) is crucial in the proof. This is probably due to the fact that there are many other invariant measures, singular with respect to the Lebesgue measure. We also remark that there is no condition on the convergence of the sequence (K_j). This means that the above result is about small intervals although it is convenient to formulate it with sequences.

In order to get more precise information we will now make some more hypothesis on the sequence of intervals.

Theorem 2[6]. Let (K_j) be a sequence as in Theorem 1. Assume there is a diverging sequence of integers (m_j) such that for any integer $0 < l < m_j$

$$f^l(K_j) \cap K_j = \emptyset,$$

then

$$\lim_{j \to \infty} T_j \mu(K_j) = 1.$$

Moreover, if $\liminf_{j \to \infty} \inf_{x \in K_j} h(x) > 0$, then the renormalized process of return times $(\mu(K_j)\tau^n_{K_j})$ converges in law to a Poisson point process of density one.

We remark that the first condition says that the sequence of intervals (K_j) do not accumulate too fast to a periodic orbit. The periodic orbits are dense in the phase space but we do not impose any condition on the divergence rate of the sequence (m_j), it can be quite slow and a simple upper bound obtained from the expansiveness is given by $\mathcal{O}(-\log \mu(K_j))$.

The second condition in the above Theorem about the density is convenient for proving the Poisson property since we have to control a conditional expectation. Technically it is used to control the division of two vanishingly small quantities.

We also mention that the conditions of the Theorem are μ generic. Indeed, for μ almost every point x one can find a sequence (K_j) converging to x and satisfying the hypothesis of Theorem 2. This is almost obvious since the periodic points form only a countable set, and the set where the density vanishes is of measure zero.

An equivalent result can be formulated in terms of the number of visits to a set up to a given time.

Theorem 3[6]. For t a fixed positive number, define the number $N_{K_j}(t, x)$ by

$$N_{K_j}(t, x) = \sum_{0 \leq j \leq t/\mu(K_j)} \chi_{K_j}(f^j(x)).$$

Then, under the hypothesis of Theorem 2, for any integer k, we have

$$\lim_{j \to \infty} \mu(\{N_{K_j} = k\}) = \frac{e^{-t} t^k}{k!}.$$

Note however the difference with Birkhoff's ergodic theorem. Here we consider a suitable diagonal limit and we get at the end a random variable with non trivial distribution.

III. OTHER RESULTS

The ideas presented above can probably be developed in many directions. We will just mention below two related results.

The first result concerns the occurrence time of sensitive dependence to initial conditions (s.d.i.c. for short). It is well known that since the maps f we are considering have a slope larger than one in modulus, they have this property[7]. We recall that intuitively this means that if one starts with two nearby initial conditions, the trajectories will separate exponentially fast up to a time where the distance between the two is of the order of the size of the (bounded) phase space. Then the non-linearities of the map fold the phase space and after sufficiently many foldings, the two trajectories may come close together again. For a fixed positive number ϵ (small), and two initial conditions x and y, we define the time of first ϵ-close encounter as the smallest integer n such that

$$|f^n(x) - f^n(y)| < \epsilon.$$

For the product measure $\mu \times \mu$ this is again an almost surely finite random variable, and we can define as before the process of successive ϵ-close encounters. We have the following theorem.

Theorem 4[8]. *Assume the density h of the invariant measure is bounded below away from zero. Then the process of ϵ-close encounters renormalized by ϵ^{-1} converges in law when ϵ tends to 0 to an integerly marked Poisson process. The density of the Poisson process and the marking measure on the integers can be computed in terms of f and h.*

An intuitive way of describing this result is as follows. On the (discrete) time line one puts a dot when the two orbits are ϵ close. Then one changes the scale by ϵ and take the limit. However, when two orbits get ϵ close they may come at a distance much smaller than ϵ. S.d.i.c. will eventually separate them at a distance larger than ϵ, but this may take several iterates, and we get on the time line a cluster of neighboring points. When we rescale by ϵ, such clusters collapse to a unique point, and the marking by integers is a remnant of this effect in the limit.

Another extension concerns maps which are Markov, expanding except at an indifferent fixed point. These mappings appear in some model of the time intermittence phenomenon[9]. We consider for simplicity a map with two branches defined on the intervals $[0, 1/2]$ and $[1/2, 1]$. On the second interval we take the map to be $f(x) = 2x - 1$, and on the first interval we take an increasing regular function with slope larger than 1 except at 0 where it is one and such that $f(0) = 0$ and $f(1/2) = 1$. If an orbit falls at some time near zero, it will remain in a neighborhood of 0 for a long time since the slope of the map is near one. This is called a laminar phase, while in the interval $]1/2, 1]$ where one has s.d.i.c. one speaks of a turbulent phase. A typical signal is composed of laminar phases separated by turbulent bursts. We also observe that entering a long laminar phase is equivalent to entering a small neighborhood of zero.

The problem of the initial distribution is here more involved. A typical trajectory spends most of the time near 0, and it is known that the SRB measure for this system is the Dirac measure at the origin. There is also an absolutely continuous invariant measure which is however not normalizable, the density has a singularity $1/x$ at the origin. This measure describes in some sense the frequency of visit to closed intervals which do not contain the origin, but instead of having a number of visits proportional the time (here the number of iterates n), one gets numbers proportional to $n/\log n$[10]. In the following we use for initial distribution the Lebesgue measure on a closed interval not containing the origin. We can now state our main theorem, refering to the original[11] paper for the technical assumptions on the map.

Theorem 5[11]. *There is a positive function β_a diverging if a tends to zero such that the first entrance time in the interval $]0, a]$ renormalized by β_a converges in law to an exponential random variable, and the process of successive visits converges in law to a Poisson point process.*

When a tends to zero, the function β_a diverges like $a^{-1} \log a^{-1}$. Similar results were also obtained by Campanino and Isola[12]. The connection with the previous results is by an induction. One considers the decreasing sequence of points $1/2 = x_0 > x_1 > x_2 > \cdots$ such that $f(x_j) = x_{j-1}$ for $j \geq 1$. One now defines a new map F by setting

$$F(x) = \begin{cases} f(x) & \text{if } x \in]1/2, 1] \\ f^{j+1}(x) & \text{if } x \in]x_j, x_{j-1}] \text{ for } j \geq 1. \end{cases}$$

The map F has the effect of anihilating the laminar phases. The map F is piecewise expanding but with an infinite number of branches, however similar results as those of section 2 can be proven and lead to the above Theorem.

IV. PROOFS

For the details of the proofs we refer to the original papers. We will simply give here an idea of the proof of the basic Theorem 1. Suppose we consider a set K, then the event $\{\tau_K > n\}$ is simply the fact that the orbit has stayed in the complement K^c of K at least up to time n. Therefore, in terms of characteristic functions of sets we have

$$\chi_{\{\tau_K > n\}} = \prod_{j=0}^{n} \chi_{K^c} \circ f^j .$$

We are interested in the probability of the event $\{\tau_K > n\}$, which is the integral of the characteristic function with respect to the measure $d\mu = h\,dx$. At this point it is important to introduce the transfer operator P associated to the map f. This operator which acts on functions (which are densities of absolutely continuous measures) is the adjoint of the Koopman operator with respect to the (non invariant) Lebesgue measure. It is explicitly given by the expression

$$Pg(x) = \sum_{y,\, f(y)=x} \frac{g(y)}{|f'(y)|} .$$

It is easy to verify that $P(g_1 \circ f\, g_2) = g_1 P(g_2)$, and if g_1 and g_2 are bounded we have (by definition)

$$\int_0^1 g_1 \circ f\, g_2\, dx = \int_0^1 g_1 P g_2\, dx .$$

Using these two elementary properties, we get at once

$$\mathbb{P}(\{\tau_K > n\}) = \int_0^1 (P\chi_{K^c})^{n+1} h\, dx ,$$

where by $P\chi_{K^c}$ we mean multiplication by the characteristic function of the set K^c and then applying P.

The interesting fact about the operator P is that in the present setting the spectrum is well known[2] in the set BV of functions of bounded variation. It consists for a mixing system of the simple isolated eigenvalue 1, the rest of the spectrum being inside a disk centered at the origin of radius strictly smaller than one. We recall that BV can be considered as a Banach space with norm

$$\|u\| = \vee u + \|u\|_{L^1} .$$

The eigenvector corresponding to the eigenvalue 1 is the function h, and the corresponding eigenform is the integration with respect to the Lebesgue measure.

One can try to use these results to prove Theorem 1. Indeed, since $\chi_{K^c} = 1 - \chi_K$, we have

$$P\chi_{K^c} = P - P\chi_K ,$$

and if K is small, one may expect that the second term is small in which case it would be possible to use perturbation theory. This would lead to an isolated dominant eigenvalue for the operator $P\chi_{K^c}$ and an exponential decay in n for $\mathbb{P}(\{\tau_K > n\})$ which is essentially the result of Theorem 1. However this does not work as such because of the nature of the Banach space where the spectral result holds. In particular the variation of the characteristic function of an interval is always equal to 2, and as a consequence the operator $P\chi_K$ is not small. Note however that it is here that the hypothesis that K is an interval is crucial, for general measurable sets, the operator $P\chi_K$ is not even defined on the set of functions of bounded variations.

The proof is based on two main steps. First we will control some fixed but unknown constants by considering a power of the operator. We have

$$(P\chi_{K^c})^q = (P - P\chi_K)^q = P^q + R_q ,$$

where R_q contains a priori $2^q - 1$ terms which by the above argument should have each a norm of order unity. The strategy is however as follows. The number q will be chosen large enough but fixed (later on). Then, the interval K will be of length as small as we like. There is then for any fixed q, and the interval K small enough, a resummation of the terms in R_q which can be written as a sum of at most $\mathcal{O}(q^2)$ terms of the form $P^j \chi_{\tilde{K}} P^{q-j}$ where the sets $\tilde{K}$ are intervals (possibly different form K).

The second step starts with an estimate of the norm of R_q. We first recall the basic estimate of Lasota and Yorke for the operator $P^{(2)}$. There is a positive number $\alpha < 1$ and a positive number Γ such that if u is a function of bounded variation, for any integer n we have the inequality

$$\vee P^n u \leq \alpha^n \vee u + \Gamma \|u\|_{L^1} .$$

Combining this inequality with the re-summation of the first step we obtain

$$\vee R_q u \leq \mathcal{O}(1) q^2 \alpha^q \vee u + \mathcal{O}(1) q^2 \|u\|_{L^1}$$

and

$$\|R_q u\|_{L^1} \leq \mathcal{O}(1) |K| q^2 (\vee u + \|u\|_{L^1}) .$$

As explained before, we will first fix q large enough to control the constants, and then take $|K|$ small enough. If we do this, we can arrange that all the coefficients are small except the last one in the first inequality. However this term is in some sense "off diagonal" and we can try to better "condition" the two inequalities by choosing a more adequate equivalent norm. This norm depends on a positive number θ and is given by

$$\|u\|_\theta = \theta \vee u + \|u\|_{L^1} .$$

The number θ is chosen exponentially small in q such that in the new norm the operator R_q is small. Note that any exponential decay in q of θ, no matter how slow, will compensate (for large q) the q^2 term, and it is here that the re-summation of the first part is crucial. There is however another condition to be satisfied, namely we have

$$P^q = P_0 + \Delta_q$$

where P_0 is a rank one operator, and Δ_q has a norm deceasing to zero exponentially fast in q. This estimate holds however only in the initial norm, and gets multiplied by θ^{-1} in the new norm. In order to get a control of this quantity we have to take θ to be slowly exponentially converging to zero. Finally, as explained before, once the exponential rate for θ has been chosen, we take q large enough to control the constants and such that for $|K|$ small enough, we are in the domain of perturbation theory. We refer to the original paper[6] for the complete details.

As already mentioned, Theorems 2 and 3 follow by similar techniques if one can control a conditional expectation. Theorem 5 is also proven in a similar way. Theorem 4 is based on direct estimates of factorial moments using a tree decay estimate.

REFERENCES

1. M. Kac, *Bul. Amer. Math. Soc.*, **53**, 1002 (1947).
2. F. Hofbauer and G. Keller, *Math. Zeit.*, **180**, 119 (1982).
3. B. Pitskel, *Ergod. Th. & Dynam. Sys.*, **11**, 501 (1991).
4. P. L'Ecuyer, *Commun. of the ACM*, **33**, 85 (1990).
5. M. Hirata, Preprint, University of Tokyo 1991.
6. P. Collet and A. Galves, Preprint 1991.
7. P. Collet and J.-P. Eckmann, "Iterated Maps on the Interval as Dynamical Systems", Birkhäuser, Basel Boston Stuttgart (1980).
8. Z. Coelho and P. Collet, Preprint 1992.
9. Y. Pomeau and P. Manneville, *Commun. Math. Phys.*, **74**, 189 (1980).
10. P. Collet and P. Ferrero, *Ann. Inst. H. Poincaré*, **52**, 283 (1990).
11. P. Collet and A. Galves, *Journ. Stat. Phys*, to appear
12. M. Campanino and S. Isola, Preprint 1993.

STOCHASTIC GEOMETRIC ASPECTS OF SOME QUANTUM SPIN CHAINS

Bruno Nachtergaele

Department of Physics
Princeton University, Jadwin Hall
Princeton NJ 08544-0708, USA
E-mail: bxn@math.princeton.edu

Work supported in part by NSF Grant PHY-9214654

ABSTRACT

We discuss some rigorous results that were recently obtained using a new Poisson integral representation for the Gibbs kernel of a class of antiferromagnetic quantum spin models.

1. INTRODUCTION

The ground states of quantum spin models have structural features which may have an interesting dependence on one or more parameters in the Hamiltonian, i.e. the ground state can be in one of several possible phases characterized by symmetry breaking, decay of correlations, and the nature of the low-lying energy spectrum. In this contribution we show how these structural features can be given a stochastic geometric interpretation using a Poisson process representation of the Gibbs kernel $e^{-\beta H}$. Although such a representation can be derived for a quite general class of Hamiltonians, including both ferro- and antiferromagnetic models, we restrict ourselves here to the study of a particular family of one-dimensional models introduced by Affleck[1]:

$$H_L = -(2S+1) \sum_{x=-L}^{L} J_x P_{x,x+1}^{(0)} \tag{1}$$

where $P_{x,y}^{(0)}$ is the orthogonal projection onto the singlet state of two quantum spins of magnitude S, and $J_x > 0$. We are mainly interested in the case with translation invariant (all $J_x = J$) or staggered coupling constants (two different values for x even and x odd). The singlet state is determined by the vector φ given by:

$$\varphi = \sum_{\alpha=-S}^{S} (-1)^{\alpha} |\alpha\rangle \otimes |-\alpha\rangle \tag{2}$$

where $\{|\alpha\rangle \mid \alpha = -S, -S+1, \ldots, S\}$ is an orthonormal basis of eigenstates of the third component of the spin S^3 with eigenvalues α. The interaction can of course also be expressed as a polynomial in the Heisenberg interaction $\mathbf{S}_x \cdot \mathbf{S}_{x+1}$. For $S = 1/2$ and $S = 1$ one obtains

$$P_{x,y}^{(0)} = \begin{cases} \frac{1}{4} - \mathbf{S}_x \cdot \mathbf{S}_{x+1} & \text{for } S = 1/2 \\ \frac{1}{3}(\mathbf{S}_x \cdot \mathbf{S}_{x+1})^2 - \frac{1}{3} & \text{for } S = 1 \end{cases} \tag{3}$$

See Ref. 2 for the analogous expressions for general S.

We will now review our main results for the ground states of (1) which can be derived by means of the stochastic geometric representation explained in Section 3 of the present paper. Only a brief hint of the arguments will be given in Section 4. For proofs we refer the reader to Ref. 3.

2. THE ISSUES AND MAIN RESULTS

One is mainly interested in properties of the ground states of (1) in the thermodynamic limit ($L \to \infty$). The existence of some specific thermodynamic limits is guaranteed by Theorem 0 of our study.

Theorem 0. *For each $L \geq 0$ the Hamiltonian H_L has a unique ground state ψ_L and for all local observables A (finite algebraic combinations of the spin matrices $S_x^i, i = 1, 2, 3$ and $x \in \mathbb{Z}$), the following two limits exist:*

$$\langle A \rangle_{\text{even}} = \lim_{\substack{L \to \infty \\ L \text{ even}}} \frac{\langle \psi_L \mid A\psi_L \rangle}{\langle \psi_L \mid \psi_L \rangle} \quad , \quad \langle A \rangle_{\text{odd}} = \lim_{\substack{L \to \infty \\ L \text{ odd}}} \frac{\langle \psi_L \mid A\psi_L \rangle}{\langle \psi_L \mid \psi_L \rangle} \tag{4}$$

A priori it is not obvious whether $\langle \cdot \rangle_{\text{even}}$ and $\langle \cdot \rangle_{\text{odd}}$ represent pure phases or not, and also these two limits are not necessarily distinct.

Our main results are the following:

a) Long range order

The interaction $-P^{(0)}$ is obviously antiferromagnetic in nature: it favours states where nearest neighbour spins are antiparallel. It is therefore natural to ask whether the ground states posses long-range antiferromagnetic order in the sense that

$$\langle S_0^3 S_x^3 \rangle \sim (-1)^x m^2 \quad \text{for } x \text{ large} \tag{5}$$

with $m \neq 0$. This kind of behaviour is called Néel order. Ignoring the quantum fluctuations such a state could be pictorially represented by

$$\cdots \uparrow \downarrow \uparrow \downarrow \uparrow \downarrow \uparrow \downarrow \uparrow \downarrow \uparrow \downarrow \uparrow \downarrow \uparrow \downarrow \uparrow \downarrow \uparrow \downarrow \cdots$$

It turns out that Néel order does not occur in the models with Hamiltonian (1). Absence of Néel order is in fact expected to hold for any one-dimensional quantum antiferromagnet. So far, a general proof of this has not been found but for the family of models under consideration here we have the following theorem.

Theorem 1. *The infinite volume ground states described in (4) do not have Néel order:*

$$\lim_{x \to \infty} \langle S_0^3 S_x^3 \rangle_{\text{even(odd)}} = 0$$

The interaction $-P^{(0)}$ could however give rise to another type of long-range order. Although for a chain of length ≥ 3 there is no state in which all nearest neighbour pairs of spins are exactly in the singlet state, there are two states in which half of the nearest neighbour pairs do form a singlet: all the spins at an even site could form a singlet with their nearest neighbour on the right or, alternatively, with their nearest neighbour on the left. This leads to two states of periodicity 2 with respect to lattice translations. Pictorially these two states would look as follows:

This kind of long-range order is called dimerization for obvious reasons. Again we are ignoring the quantum fluctuations and one can actually not expect that these are the true ground states. Also in cases where there is only a preference of the spins at even sites to form singlet states with their left or right neighbours, we call the state (partially) dimerized.

The stochastic geometric representation permits us to prove the following result related to the Affleck-Lieb dichotomy[4].

Theorem 2. *For the ground states of the translation invariant model (1) ($J_x \equiv J$), one of the following holds:*

- *either the translation symmetry is spontaneously broken in the infinite volume ground states*

- *or the spin-spin correlation function decays slowly (non-exponential) with*

$$\sum_x |x \langle S_0^3 S_x^3 \rangle| = +\infty \qquad (6)$$

In the first case, the symmetry breaking is manifested in the non-invariance of the pair correlation:

$$\langle \mathbf{S}_0 \cdot \mathbf{S}_1 \rangle_{\text{even}} \neq \langle \mathbf{S}_1 \cdot \mathbf{S}_2 \rangle_{\text{even}} = \langle \mathbf{S}_0 \cdot \mathbf{S}_1 \rangle_{\text{odd}} \quad . \qquad (7)$$

The stochastic geometric representation in fact gives us a detailed picture of the correlations in the dimerized phase. We find the following behaviour: in the state where the spins on the even sites are more correlated with their neighbours to the right one finds that with probability 1 some spins on the left half-infinite chain $(-\infty, x]$ form a singlet with some spins on the right half-infinite chain $[x+1, +\infty)$, for each even x. In the same state this probability is < 1 for odd x.

b) Decay of correlations

Apart from the condition (6) Theorem 1 does not specify how fast $\langle S_0^3 S_x^3 \rangle$ converges to zero as $x \to \infty$. The decay rate by itself is interesting information and in particular one would like to show that the decay is exponentially fast (existence of a finite correlation length) in the cases where the translation symmetry is broken. More generally one can consider the truncated correlation function of any two local observables A and B. For any local observable C we denote by C_x the observable obtained by translating C over x and let $\mathrm{Supp}(C)$ be the smallest interval $[a, b]$ in the chain such that C is localized in $[a+1, b-1]$. We have an estimate for the truncated correlation function

$$|\langle A; B_x \rangle| = \langle AB_x \rangle - \langle A \rangle \langle B_x \rangle$$

in terms of the truncated two-point function $\tau(x; y)$ of an associated two-dimensional Potts model. For the translation invariant case the Potts model is at its self-dual point and for the staggered model it can be taken to be in the high temperature phase. The precise definition of this Potts model is given in Section 4. The comparison theorem reads then as follows.

Theorem 3. *Let $\langle \cdot \rangle$ denote the expectation in the ground state of a finite chain containing an even number of sites, or in one of the limiting states $\langle \cdot \rangle_{\mathrm{even}}$ or $\langle \cdot \rangle_{\mathrm{odd}}$. Then, for any pair of local observables A and B of the quantum spin chain there are constants C_A and C_B such that*

$$|\langle A; B_x \rangle| \leq C_A C_B \sum_{\substack{y \in \mathrm{Supp}(A) \\ z \in \mathrm{Supp}(B)_x}} \tau(y; z) \tag{8}$$

c) The spectral gap

One says that the system has a spectral gap of magnitude at least $\gamma > 0$ in an infinite volume ground state $\langle \cdot \rangle$ if for any local observable A the following inequality holds

$$\lim_L \langle A^*[H_L, A] \rangle \geq \gamma (\langle A^* A \rangle - |\langle A \rangle|^2)$$

This is equivalent to the GNS Hamiltonian having a spectral gap separating the ground state from the rest of the spectrum. It is therefore obvious that the existence of a spectral gap would also follow from an estimate of the type

$$|\langle A e^{-tH} B \rangle| \leq C_A C_B e^{-\gamma t} \tag{9}$$

One can actually prove the analogue of Theorem 3 for the observable B being translated in imaginary time instead of in space and this implies the following theorem.

Theorem 4. *Whenever the two-dimensional Potts model associated with the quantum spin chain has an exponentially decaying truncated two-point function there is a spectral gap in the ground states $\langle \cdot \rangle_{\mathrm{even}}$ and $\langle \cdot \rangle_{\mathrm{odd}}$.*

We now turn to a brief discussion of the stochastic geometric representation that allows one to prove Theorems 0-4.

3. THE STOCHASTIC GEOMETRIC REPRESENTATION

It is convenient to absorb some trivial constants in the definition of the Hamiltonian:
$$H_L = \sum_{x=-L}^{L} J_x \{1 - (2S+1) P^{(0)}_{x,x+1}\} \tag{10}$$

In this section we are considering finite chains of spins of magnitude S.

We now describe a Poisson integral representation for the Gibbs state at inverse temperature β of the models. The ground states can be recovered by taking the limit $\beta \to \infty$. The following formula is fairly simple to derive[3]:

$$\langle A \rangle_{L,\beta} \equiv \frac{\text{Tr } A e^{-\beta H_L}}{\text{Tr } e^{-\beta H_L}} = \int \mu^J_{L,\beta}(d\omega) \, E_\omega(A) \tag{11}$$

where $\mu^J_{L,\beta}(d\omega)$ and $E_\omega(A)$ are defined as follows.

- $\mu^J_{L,\beta}(d\omega) = \mathcal{Z}^{-\infty}_{\mathcal{L},\mathcal{J},\beta} \rho^{\mathcal{J}}_{\mathcal{L},\beta}(d\omega)(\in\mathcal{S}+\infty)^{l(\omega)}$, where $\rho^J_{L,\beta}(d\omega)$ is the probability measure of independent Poisson processes, one for each bond $\{x, x+1\}$ in the chain, running over the time interval $[0,\beta]$, and with rates J_x. It is useful to draw the configurations ω for this process as in Figure 1, i.e. the "beeps" of the Poisson process at the bond $\{x, x+1\}$ are recorded by drawing a double line connecting x and $x+1$ at the appropriate time. $l(\omega)$ is then the number of loops one counts in ω assuming periodic boundary conditions in the vertical (imaginary time) direction. $\mathcal{Z}_{\mathcal{L},\mathcal{J},\beta}$ is the normalization factor which makes $\mu^J_{L,\beta}(d\omega)$ a probability measure.

- For any observable A we define the random variables $E_\omega(A)$ by:

$$E_\omega(A) = \frac{\text{Tr } A K(\omega)}{\text{Tr } K(\omega)}$$

where $K(\omega)$ is a product of operators $(2S+1)P^{(0)}_{x,x+1}$, one for each bond occurring in ω and ordered according to the times at which they occur. If $K(\omega)$ is a positive operator, then, considered as a functional of A, $E_\omega(A)$ is a state. But this is typically not the case and therefore we call the functionals E_ω quasi-states.

The expression (11) can now be interpreted as a decomposition of the quantum state into the quasi-states E_ω in which, in some sense, correlations are always 0-1, i.e. two spins are either independent or rigidly correlated. This can be seen by analyzing first the structure of the quasi-states restricted to the observables generated by the S^3_x alone (or any component of the spin for that matter).

We denote by $\sigma = \{\sigma_x\}$ a configuration of joint eigenvalues values for the commuting family of observables $\{S^{(3)}_x\}$: $\sigma_x(t) \in \{-S, -S+1, \ldots, S\}$. These configurations form a natural parametrization for an orthonormal basis of the Hilbert space of the system. We then consider time dependent configurations $\sigma(t) = \{\sigma_x(t)\}$, with the time t ranging over the interval $[0,\beta]$, and $\sigma(0) = \sigma(\beta)$. σ is called consistent with a configuration ω if $(-1)^x \sigma_x(t)$ is constant on each loop in ω. It is obvious that the number of spin configurations consistent with ω is given by $(2S+1)^{l(\omega)}$. One can then see that for any function f of $\{S^3_x \mid -L \leq x \leq +1\}$:

$$E_\omega(f) = \frac{1}{(2S+1)^{l(\omega)}} \sum_{\sigma \text{ consistent with } \omega} f(\sigma(t=0)) \tag{12}$$

This means that as far as the third component of the spin is concerned, E_ω is a state in which the spins are rigidly correlated (in a staggered, antiferromagnetic way) within the loops of ω and the clusters of spins on different loops are completely independent. It is worth remarking that nevertheless the functionals E_ω are invariant under spin rotations. For a cluster of more than two spins no quantum state exhibits rigid antiferromagnetic correlations and rotation invariance at the same time. E_ω can therefore not be a state in general and the quantum fluctuations they lack are restored in the state $\langle \cdot \rangle_{L,\beta}$ after integration over all random loop configurations ω with the probability measure $\mu^J_{L,\beta}(d\omega)$.

It is now straightforward to derive an expression for the spin-spin correlation function of the system:

$$E_\omega(S^3_x S^3_y) = (-1)^{|x-y|} C(S) \mathbf{I}[\,(x,0) \text{ and } (y,0) \text{ are on the same loop}\,] \tag{13}$$

with $C(S) = \frac{1}{2S+1} \sum_{m=-S}^{S} m^2 = \frac{1}{3}S(S+1)$ and where $\mathbf{I}[\cdot]$ denotes the indicator function of the event described between the brackets.

Hence, the spin-spin correlation is proportional to the probability, with respect to the probability measure $\mu^J_{L,\beta}(d\omega)$, that two sites are on the same loop of ω:

$$\langle S^3_x S^3_y \rangle = (-1)^{|x-y|} C(S) \text{Prob}_\mu(\,(x,0) \text{ and } (y,0) \text{ are on the same loop}) \tag{14}$$

The functionals E_ω can be extended to the full algebra of observables in order to obtain expressions for the expectation values of general observables (see Ref. 3 for more details). The quasi-state decomposition described above can be derived in a much more general context. The models need not be one-dimensional and there is a much wider class of interactions (both ferro- and antiferromagnetic) for which the weights are non-negative, although the treatment of antiferromagnetic interactions is restricted to models with a bipartite structure.

4. HINT TO THE ARGUMENTS AND RELATION WITH THE POTTS MODEL

Formula (14) and similar relations make it possible to translate properties of the quantum spin chain into statements about the probability measure $\mu^J_{L,\beta}(d\omega)$. Typically one has to compute or estimate the probability of an event with a more or less simple geometric interpretation. Before we give some hints about the essential ideas involved in the proof of theorems stated in Section 2, we briefly explain the connection of the quantum spin models (1) with certain two-dimensional Potts models.

a) Relation with the Potts model

By construction the loops in ω are non-intersecting and hence are the boundaries of the elements of a partition of $[-L, L+1] \times [0, \beta]$ considered as a subset of the plane. It is useful to color the plane in two colors A and B as in Figure 1. We thus obtain two collections of connected clusters, the A-clusters and the B-clusters. A configuration of A-clusters consists of pieces of A-strips, and is obtained by considering the bonds of ω occurring in the A-strips as disconnecting, and the bonds in the B-strips as connecting. In this way $\mu^J_{L,\beta}(d\omega)$ becomes the probability measure of a random cluster model. The Gibbs probability measures of two-dimensional q-state Potts models also lead to random cluster models by the so-called Fortuin-Kasteleyn (FK) representation[6]. The random cluster measures obtained from the spin chains here are in fact Potts measures in the

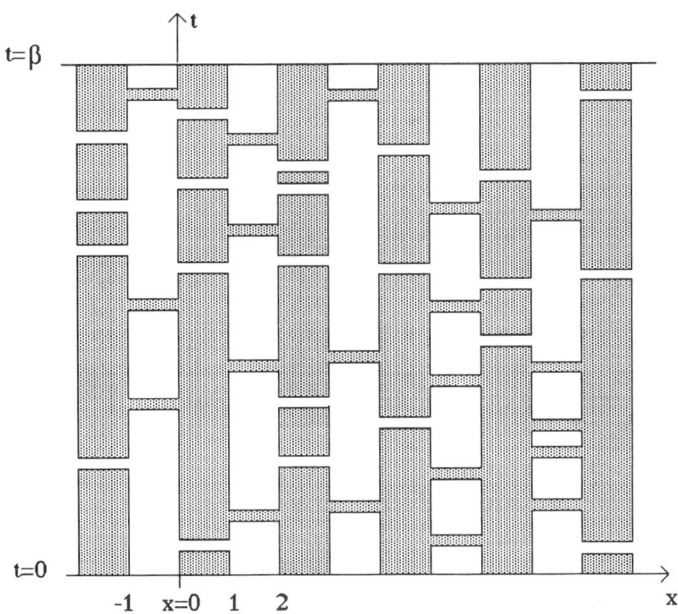

Figure 1. *A space-time configuration ω for the $-P^{(0)}$ quantum spin chain, at an inverse temperature β.* The spins are correlated within loops drawn by following the lines in the space-time. A special feature of this interaction is that the loops can be viewed as the boundaries of the connected clusters of two random cluster models, dual to each other. The shaded areas are the connected A-clusters, the connected B-clusters are left blank.

FK-representation[3]. An algebraic relation between the spin chains and Potts models was realized before but in a rather different framework[1,2,5]. There are in fact two Potts models, an A- and a B-model, dual to each other, and for which the clusters are given by the A- and B-clusters respectively. I restrict myself here to giving the definition of the required Potts models as well as the relations between the parameters of the spin chains with the ones of the Potts models.

The Potts models are defined over a 1+1 dimensional "lattice", $\mathbb{Z} \times \mathbb{R}$, in which one of the directions (corresponding to the "time" of Section 2) is continuous. The Potts variables of the A-model reside on the vertical lines bisecting the A-strips and the variables of the B-model are situated on the lines bisecting the B-strips.

For a volume $[0,T] \times [L_-, L_+]$, the Potts configurations of the A-model are functions $\xi(x,t)$, $x = \text{even} + \frac{1}{2} \in [L_-, L_+]$, which are piecewise constant in time and take values in $\{1,\ldots,q\}$. In the time direction we always take the periodic boundary conditions. Other than that, there are two natural boundary conditions for a Potts model, which are exchanged under the standard duality map: the *free* and the *wired*, the latter corresponding to adding an extra strip to the left and to the right of the volume where the spins are required to assume a common value (we assume the convention that the partition function includes also the sum over this common value). It turns out that the relevant boundary conditions here depend on the label of the strip along the boundary. If it is A then the A-model gets the *free* b.c. and the B-model the *wired* one, and otherwise it is the other way around.

The partition function of the A-model (with the relevant boundary terms) is given by

$$Z_A^{\text{Potts}} = \int \rho_A^{J^V}(d\omega) \sum_\xi{}^\omega \exp\{ \sum_{\substack{\S=\text{even}+\frac{1}{2} \\ L_- - \infty \leq \S \leq L_+}} \int_I^T d\sqcup \mathcal{J}_\S^\mathcal{H}(\delta_{\xi(\S,\sqcup),\xi(\S+\in,\sqcup)} - \infty)\} \tag{15}$$

where $J^V = \{J_x^V\}$ and $\{J_x^H\}$ are sets of positive constants (these are the ferromagnetic coupling constants in the vertical and horizontal direction respectively and in which we have absorbed the inverse temperature of the Potts model), $\rho_A^{J^V}$ is a product of independent Poisson point processes on the lines $\{x = 2n+\frac{1}{2}\} \times [0,T]$ with intensity J_x^V, and $\sum^\omega$ denotes the sum over all configurations ξ for which the discontinuities happen only at points (x,t) in the configuration ω of the Poisson process. The sum in (15) should be interpreted as incorporating the boundary condition convention explained in the previous paragraph. The B-model is defined analogously.

The parameters of the A- and B-Potts models associated with the quantum spin chain with interaction $-P^{(0)}$ on the interval $[L_-, L_+]$, are determined by:

$$J_x^H = (2S+1)J_{x+\frac{1}{2}}, \quad J_x^V = (2S+1)^{-1}J_{x-\frac{1}{2}} \tag{16}$$

$$q = (2S+1)^2, \quad \beta = T \tag{17}$$

b) A hint of the arguments

For the proof of Theorem 0, the existence of the thermodynamic limits, we make use of the FKG-structure of the random cluster measures[7]. Theorem 1, the absence of Néel order, follows from the fact that simultaneous percolation of both the A- and B-clusters does not occur in the kind of models at hand[8]. For the A-L dichotomy, Theorem 2, we have a topological argument: if the loops are small, i.e., by (13), if the spin-spin correlations decay fast enough, then the number of loops encircling any point is finite and one can define the total winding number of a random loop configuration w.r.t. a given site of the chain. It is then easy to see that the total winding number must alternate between even and odd under translations of the reference site, indicating breaking of the translation invariance. Some additional arguments then prove (7). For Theorem 3 and 4 one has to exploit the full power of the stochastic geometric representation, as they are statements about arbitrary local observables. Also there the FKG-structure plays an essential rôle.

Acknowledgments

The results described in this paper were obtained in collaboration with Michael Aizenman.

REFERENCES

1. I. Affleck: Exact results on the dimerization transition in SU(n) antiferromagnetic chains, *J. Phys. C: Cond. Matter* **2** (1990), 405–415
2. A. Klümper, The spectra of q-state vertex models and related antiferromagnetic quantum spin chains, *J.Phys. A: Math. Gen.* **23** (1990), 809–823
3. M. Aizenman and B. Nachtergaele, Geometric Aspects of Quantum Spin States, preprint and in preparation
4. I. Affleck and E.H. Lieb, A proof of part of Haldane's conjecture on quantum spin chains, *Lett. Math. Phys.* **12** (1986), 57–69
5. M.T. Batchelor and M. Barber, Spin-s quantum chains and Temperley-Lieb algebras, *J. Phys. A: Math. Gen.* **23** (1990), L15–L21
6. C.M. Fortuin and P.W. Kasteleyn, On the random cluster model I, *Physica* **57** (1972), 536–564
7. C.M. Fortuin, P.W. Kasteleyn, and L. Ginibre, Correlation inequalities on some partially ordered sets, *Commun. Math. Phys.* **22** (1971), 89–103
8. A. Gandolfi, M. Keane, and L. Russo, On the uniqueness of the infinite occupied cluster in dependent two-dimensional site percolation, *Ann. Prob.* **16** (1988), 1147–1157

THE TWO SPECIES TOTALLY ASYMMETRIC SIMPLE EXCLUSION PROCESS

Eugene R. Speer

Department of Mathematics
Rutgers University
New Brunswick, New Jersey 08903 USA

The *two species totally asymmetric simple exclusion process*[1,2], or *two species TASEP*, is an interacting particle system[3] in which two types of particles, *first class* and *second class*, live on the sites of a one dimensional lattice, hopping at random times to the adjacent site to their right; the model is called totally asymmetric because particles can jump only to the right, while more general ASEP models permit jumps in both directions, with a preference for one or the other. Here we study steady states—time invariant measures on the space of all configurations—for this model, and in particular discuss a family of explicitly computable translation invariant steady states on the infinite lattice; we show that these are precisely the extremal members of the set of all translation invariant steady states for the system. The family is parameterized by the densities ρ_1 and ρ_2 of the two species, and includes states with all densities lying within the triangle $\rho_1, \rho_2 \geq 0$, $\rho_1 + \rho_2 \leq 1$. On the boundary of this triangle the model reduces to the the standard one species TASEP[3] and the states constructed here reduce to the product states invariant for this simpler model.

To carry out the construction we first consider the model on a finite lattice with periodic boundary conditions; for this case the unique invariant states are constructed explicitly in terms of traces of products of linear operators on an infinite dimensional Hilbert space (a method originating in Ref. 4). We then obtain the infinite volume states as a limit of these finite volume ones. The construction is sufficiently explicit to permit the derivation of some interesting consequences. For example, we can show that the states we construct are not Gibbs states for any absolutely summable interaction; moreover, we can derive from our knowledge of the two species model an explicit description of microscopic density profiles near a macroscopic shock in the standard one species TASEP.

Many of the results discussed here were first developed in Ref. 1 (see also Ref. 2). We review briefly some additional results of this reference, as well as applications of the methods to related models, at the end of this article.

1. THE TWO SPECIES TASEP AND ITS EXACT SOLUTION FOR PERIODIC BOUNDARY CONDITIONS

In the two species TASEP, sites of a one dimensional lattice may be occupied by a single particle of one of the two species or may be empty; empty sites are called *holes* and will be treated notationally as particles of zeroth class. A typical configuration will thus be written as $\tau = (\tau(i))$, where i runs over the lattice and $\tau(i) = 0, 1,$ or 2. We will consider both periodic lattices $\mathbb{Z}_N = \{-N, \ldots, 0, \ldots, N\}$ of $2N+1$ sites and the infinite lattice $\mathbb{Z}$; the corresponding configuration spaces for the model are $Y_N \equiv \{0,1,2\}^{\mathbb{Z}_N}$ and $Y \equiv \{0,1,2\}^{\mathbb{Z}}$. The dynamics are governed by independent alarm clocks associated with the sites, which ring at random times, Poisson-distributed with rate one. When the alarm at site i sounds, the particle at that site, if there is one, attempts to jump to the adjacent site $i+1$. The jump succeeds if site $i+1$ is empty, in which case site i is empty after the jump, or if the particle at site i is a first class particle and that at site $i+1$ is a second class particle, in which case the two sites exchange particles; that is, the possible transitions are

$$1\,0 \to 0\,1, \qquad 1\,2 \to 2\,1, \qquad \text{and} \qquad 2\,0 \to 0\,2. \tag{1}$$

Note that the number of particles of each species is conserved. As mentioned in the introduction, if the system contains no first class particles, no second class particles, or no empty sites, then the model reduces to the standard one species TASEP; in the last case the second class particles play the role of holes.

Consider now the periodic system on $\mathbb{Z}_N$. For each site i of the lattice let η_i^0, η_i^1, and η_i^2 be random variables taking value 1 if site i is occupied by the corresponding species (a hole, a first class particle, or a second class particle respectively) and value 0 otherwise. For any measure ν on Y_N we write $\nu(f)$ for the expectation of the random variable f, so that $\nu\left(\prod_{i=-N}^{N} \eta_i^{\tau(i)}\right)$ is the probability of the configuration τ. If there are no second class particles in the system—that is, in the case of a one species TASEP—the invariant measure for the dynamics assigns equal probability to all configurations with the same number of particles. The two species TASEP with one or more second class particles present, however, also has a remarkably simple exact solution.

Theorem 1: *Suppose that X_0 and X_1 are bounded operators on some Hilbert space which satisfy two conditions:*

(C1) $X_1 X_0 = X_1 + X_0$;

(C2) *The operator $X_2 = [X_1, X_0] = X_1 X_0 - X_0 X_1$ has finite positive trace.*

Then the two species TASEP satisfying periodic boundary conditions on $2N+1$ sites and containing K_1 first class particles, $K_2 > 0$ second class particles, and $K_0 = 2N+1-K_1-K_2$ holes, has (unique) invariant measure $\nu \equiv \nu^{(K)}$ given by

$$\nu\left(\eta_{-N}^{\tau(-N)} \cdots \eta_N^{\tau(N)}\right) = Z^{-1}\,\mathrm{trace}(X_{\tau(-N)} \cdots X_{\tau(N)}), \tag{2}$$

where Z is a normalizing constant.

Proof: We will indicate the main idea of the proof. Condition **(C1)** implies that $X_2^2 = X_2$, i.e., that X_2 is a projection, and that

$$X_1 X_2 = X_2 \qquad \text{and} \qquad X_2 X_0 = X_2. \tag{3}$$

Condition **(C2)**, with the assumption $K_2 > 0$, implies that the trace in (2) is finite. **(C1)**, **(C2)**, and (3), with a little calculation, imply that the trace in (2) is positive and hence in particular that Z is nonzero (for example, $\mathrm{trace}(X_0^{K_0} X_1^{K_1} X_2^{K_2}) = \mathrm{trace}\, X_2$).

Note that the property trace AB = trace BA implies that (2) respects the periodic boundary conditions.

Now consider the time evolution of some arbitrary measure $\tilde{\nu}_t$. If a configuration $\tau = \tau(-N)\cdots\tau(N)$ is divided into blocks of identical species, then it follows from (1) that the rate of change, at time t, of the probability $\tilde{\nu}_t(\prod \eta_i^{\tau(i)})$ of τ is the sum of one contribution from each block boundary: a boundary of the form 10, 12, or 20 represents an opportunity to leave the configuration τ and hence contributes $-\tilde{\nu}_t(\prod \eta_i^{\tau(i)})$, while a boundary 01, 21, or 02 represents a possible entry into the configuration and hence contributes $\tilde{\nu}_t(\prod \eta_i^{\tau'(i)})$, where τ' is the configuration obtained from τ by interchanging the two spins at this block boundary. Thus we may write

$$d\tilde{\nu}_t(\prod \eta_i^{\tau(i)})/dt = -\sum_a \tilde{\nu}_t(\prod \eta_i^{\tau(i)}) + \sum_b \tilde{\nu}_t(\prod \eta_i^{\tau^b(i)}), \tag{4}$$

where a indexes block boundaries of the first type and b those of the second type, and τ^b is obtained from τ by interchanging the spins at the block boundary b. Suppose now that $\tilde{\nu}_t = \nu$, and insert (2) on the right side of (4). In each resulting term the distinguished block boundary will contribute a factor $X_1 X_0$, $X_1 X_2$, or $X_2 X_0$ within the trace; if these factors are replaced, according to **(C1)** and (3), by $X_1 + X_0$, X_2, and X_2 respectively, then all terms cancel—more specifically, each block of 0's or 1's gives rise to two terms, one from the boundary at each end of the block, which cancel, while no additional terms arise corresponding to blocks of 2's. Thus ν is invariant. ∎

Operators X_0 and X_1 satisfying **(C1)** and **(C2)** are easily found by writing $X_0 = I + x_0$ and $X_1 = I + x_1$. Then **(C1)** becomes $x_1 x_0 = I$; if the Hilbert space $\mathcal{H}$ on which X_0 and X_1 act were finite dimensional this would imply that $x_1 = x_2^{-1}$, so that $X_2 = [X_1, X_0] = 0$, contradicting **(C2)**. (Choosing $x_1 = x_2^{-1}$ on a finite dimensional space—even $x_1 = x_2 = 1$ on a one-dimensional space—does lead to the correct invariant measure for the system with no second class particles.) We may, however, take $\mathcal{H} = \ell^2$ with x_0 the right shift and x_1 the left shift operators; in the Dirac notation, with $|1\rangle, |2\rangle,\ldots$ the standard basis, $x_0|k\rangle = |k+1\rangle$ and $x_1 = x_0^*$. Then $X_2 = |1\rangle\langle 1|$ is orthogonal projection on the subspace spanned by $|1\rangle$, so that **(C2)** is satisfied. We will adopt this choice of X_0, X_1, and X_2 in the sequel. Equation (2) can then be written, if for convenience we take $\tau(0) = 2$, as

$$\nu(\eta_{-N}^{\tau(N)}\cdots\eta_{-1}^{\tau(-1)}\eta_0^2\eta_1^{\tau(1)}\cdots\eta_N^{\tau(N)}) = Z^{-1}\langle 1|X_{\tau(1)}\cdots X_{\tau(N)}X_{\tau(-N)}\cdots X_{\tau(-1)}|1\rangle. \tag{5}$$

From $X_2 = |1\rangle\langle 1|$ it is clear that second class particles have a decoupling effect: if we condition on the presence of second class particles at sites i and j then the configuration in the interval (i, j) is statistically independent of the configuration outside this interval.

It is in fact convenient to work with another class of invariant states: *grand canonical ensembles*, which are superpositions of the measures $\nu^{(K)}$ for different numbers of particles. For these measures to be well defined we must consider only ensembles with at least one second class particle, and it is computationally convenient to condition on the presence of such a particle at a fixed site, which as in (5) we take to be 0. Introducing fugacity-like parameters z_0 and z_1 associated with holes and first class particles, respectively, and denoting the conditioning at the origin by a subscript 0, we define the grand canonical measure $\mu_0^{(N)} \equiv \mu_0^{(z,N)}$ by

$$\mu_0^{(N)}\left(\prod_{i=-N}^{N} \eta_i^{\tau(i)}\right) = \delta_{\tau(0)2}\Xi_N^{-1}z_0^{k_0(\tau)}z_1^{k_1(\tau)}\langle 1|X_{\tau(1)}\cdots X_{\tau(N)}X_{\tau(-N)}\cdots X_{\tau(-1)}|1\rangle, \tag{6}$$

where $k_0(\tau)$ and $k_1(\tau)$ are the number of occurrences of 0 and 1, respectively, among $\tau(-N)\cdots\tau(N)$. With $G = z_0 X_0 + z_1 X_1 + X_2$ the partition function may be written as

$$\Xi_N = \sum_{\tau(i)=0,1,2} \delta_{\tau(0)2}z_0^{k_0(\tau)}z_1^{k_1(\tau)}\langle 1|X_{\tau(1)}\cdots X_{\tau(N)}X_{\tau(-N)}\cdots X_{\tau(-1)}|1\rangle = \langle 1|G^{2N}|1\rangle. \tag{7}$$

2. THE INFINITE VOLUME LIMIT

We now define the (weak) infinite volume limit $\mu_0 = \mu_0^{(z)}$ of the conditioned grand canonical ensemble (6) by

$$\mu_0(\eta_{i_1}^{\tau(1)}\cdots\eta_{i_m}^{\tau(m)}) = \lim_{N\to\infty} \mu_0^{(N)}(\eta_{i_1}^{\tau(1)}\cdots\eta_{i_m}^{\tau(m)})$$
$$= \lim_{N\to\infty} \Xi_N^{-1}\langle 1|G^{i_1-1}z_{\tau(1)}X_{\tau(1)}G^{i_2-i_1-1}\cdots z_{\tau(i_p)}X_{\tau(i_p)} \times$$
$$\times G^{2N-i_p+i_{p+1}}z_{\tau(i_{p+1})}X_{\tau(i_{p+1})}\cdots z_{\tau(m)}X_{\tau(m)}G^{-i_m-1}|1\rangle, \quad (8)$$

where we have assumed in the second line that $i_{p+1} < \cdots < i_m < 0 < i_1 < \cdots < i_p$. The existence of the limit in (8) may be established by direct calculation; rather than prove a general theorem, however, we will illustrate the method, and illuminate some interesting properties of μ_0, by treating several special cases in Theorems 2 and 4 and Remark 5 below. Before doing so we make three preliminary comments.

First, we discuss the significance of the fugacity parameters z. It can be shown[1] that for large N,

$$\Xi_N = \langle 1|G^{2N}|1\rangle \simeq (1 - z_0 z_1)\gamma^{2N}, \qquad \text{where} \qquad \gamma = (1+z_0)(1+z_1). \quad (9)$$

Now the mean particle densities $\rho_0^{(N)}$, $\rho_1^{(N)}$, and $\rho_2^{(N)}$ in the finite volume ensemble $\mu_0^{(N)}$ are given by $\rho_\sigma^{(N)} = (2N)^{-1}z_\sigma \partial \log \Xi_N/\partial z_\sigma$. The fluctuations in these densities vanish in the $N \to \infty$ limit[1], so that (9) (or, more properly, an asymptotic analysis of $\partial \Xi_N/\partial z_\sigma$ similar to that which led to (9)) yields the limiting densities

$$\rho_0 = \frac{z_0}{1+z_0}, \qquad \rho_1 = \frac{z_1}{1+z_1}, \qquad \rho_2 = 1 - \rho_0 - \rho_1 = \frac{1-z_0 z_1}{(1+z_0)(1+z_1)}. \quad (10)$$

Clearly there exist nonnegative fugacities z, with $z_0 z_1 \leq 1$, which yield any densities in the triangle $\rho_1, \rho_2 \geq 0$, $\rho_1 + \rho_2 \leq 1$, with the boundaries of the triangle achieved by $z_0 = 0$, $z_1 = 0$, or $z_0 z_1 = 1$. For example, if we set $z_0 = 0$ in (10) we eliminate holes from the system and obtain the usual density-fugacity relations $\rho_2 = 1/(1+z_1)$, $\rho_1 = z_1/(1+z_1)$ for a product measure in a one species system (where now the second class particles play the role of empty sites). The situation on the other boundaries is similar—see also Section 5. In the remainder of this section (except for the final sentence of Remark 5) we will consider only densities lying in the interior of the triangle.

Second, we note that the measure μ_0 may also be obtained[1] from the canonical measures (2) by conditioning on a second class particle at the origin and then taking the infinite volume limit with $K_\sigma/N \to \rho_\sigma$. That is, equivalence of ensembles holds.

Finally, we observe that μ_0 may be obtained by conditioning a translation invariant measure $\mu = \mu^{(\rho)} = \mu^{(\rho_0,\rho_1,\rho_2)}$ (labeled for later convenience by the densities rather than the fugacities) on the presence of a second class particle at the origin. The measure μ arises as the infinite volume limit of translation invariant measures for the periodic system or may be written directly written in terms of μ_0: since the probability of a second class particle at any given site is ρ_2, conditioning on the location of the first second class particle at or to the left of the origin yields

$$\mu^{(\rho)}(\eta_{i_1}^{\tau(1)}\cdots\eta_{i_m}^{\tau(m)}) = \rho_2 \sum_{k=0}^{\infty} \mu_0\left(\prod_{j=0}^{k-1}(1-\eta_j^2)\,\eta_{i_1+k}^{\tau(1)}\cdots\eta_{i_m+k}^{\tau(m)}\right). \quad (11)$$

In Section 4 it will be important to note that μ_0 can also be obtained from μ in an alternate way: if the two species TASEP is in the state μ, then the process derived from it by viewing the system from a selected second class particle, that is, the process in which jumps of this particle are replaced by the reverse jumps of the rest of the system, is in the invariant state μ_0.

We next give a characterization of the measure μ_0 originally observed by Ferrari, Fontes, and Kohayakawa[5] and verified by Derrida[6]; the idea is that, because of the decoupling effect of second class particles, it suffices to give the probability $\pi(\theta)$ that each possible sequence θ of holes and first class particles appears between successive second class particles. Thus let $\mathcal{F}$ denote the set of all finite sequences of binary digits, including the empty sequence $\emptyset$. For $\theta \in \mathcal{F}$ let $k_0(\theta)$, $k_1(\theta)$, and $n(\theta) = k_0(\theta) + k_1(\theta)$ denote the number of 0's, the number of 1's, and the length of θ, respectively, and define

$$\pi(\theta) = R(\theta)\rho_1^{k_1(\theta)}(1-\rho_1)^{k_0(\theta)+1}\rho_0^{k_0(\theta)}(1-\rho_0)^{k_1(\theta)+1}. \tag{12}$$

Here $R(\theta)$ is a combinatorial factor which counts the number of elements of $\mathcal{F}$ which may be obtained from θ by moving 1's to the right. For example, if 0^p denotes a sequence of p 0's, etc., then $R(\emptyset) = 1$, $R(0^p 1^q) = 1$ and $R(1^p 0^q) = \binom{p+q}{p}$ for all $p, q \geq 0$, and $R(1010) = 5$. It is a consequence of the proof of Theorem 2 below, or may be shown by direct calculation, that π extends to a probability measure on $\mathcal{F}$. Finally, let $\hat{\mathcal{F}}$ be the set of doubly infinite sequences $\Theta = (\theta_i)_{i=-\infty}^{\infty}$ of elements of $\mathcal{F}$, and to each $\Theta \in \hat{\mathcal{F}}$ associate the infinite particle configuration $U(\Theta) \equiv \cdots 2\theta_{-1} 2\theta_0 2\theta_1 2 \cdots$, where the second class particle immediately preceding θ_0 is at the origin.

Theorem 2: *The measure μ_0 is the image $U\hat{\pi} \equiv \hat{\pi} \circ U^{-1}$ of the product measure $\hat{\pi} = \bigotimes \pi$ on $\hat{\mathcal{F}}$.*

Proof: For $\theta \in \mathcal{F}$ let $Q(\theta) \equiv \langle 1|X_{\theta(1)} \cdots X_{\theta(n(\theta))}|1\rangle$; we first show by induction on $n = n(\theta)$ that $R(\theta) = Q(\theta)$. For if $\theta = 0^p$ then $R(\theta) = Q(\theta) = 1$ by (3), while if $\theta(n) = 1$ then with θ' obtained by omitting this last 1 from θ we find that $R(\theta) = R(\theta') = Q(\theta') = Q(\theta)$, by the induction assumption and (3). If neither of these cases occurs then θ must contain a 1 and end with 0; let θ' be obtained by omitting the last 1 from θ, and θ'' be obtained by omitting the 0 succeeding this 1. Then $R(\theta')$ counts the number of possible right shifts of 1's in θ with the last 1 shifted to the end, and $R(\theta'')$ the number of other shifts, so that $R(\theta) = R(\theta') + R(\theta'')$. Since $Q(\theta) = Q(\theta') + Q(\theta'')$ by **(C1)**, the equality is verified.

Now for $\theta \in \mathcal{F}$ we compute from (8) the probability that the sequence of holes and first class particles specified by θ appears immediately to the right of the origin, followed by a second class particle:

$$\begin{aligned}
\mu_0\left(\prod_{i=1}^{n(\theta)} \eta_i^{\theta(i)} \eta_{n(\theta)+1}^2\right) &= \gamma^{-n(\theta)-1} z_0^{k_0(\theta)} z_1^{k_1(\theta)} \langle 1|X_{\theta(1)} \cdots X_{\theta(n(\theta))}|1\rangle \\
&\quad \times \lim_{N\to\infty} \Xi_N^{-1} \gamma^{n(\theta)+1} \langle 1|G^{2N-n(\theta)-1}|1\rangle \\
&= \gamma^{-n(\theta)-1} z_0^{k_0(\theta)} z_1^{k_1(\theta)} \langle 1|X_{\theta(1)} \cdots X_{\theta(n(\theta))}|1\rangle \\
&= \pi(\theta). \tag{13}
\end{aligned}$$

Since $\rho_2 > 0$, there will almost surely be a second class particle somewhere to the right of the origin, which implies that π is a probability measure on $\mathcal{F}$. The decoupling effect of the second class particles noted in the previous section, which persists in the infinite volume limit, now implies the general characterization of μ_0 given in the theorem. ∎

Combining this theorem with (11) gives a relatively simple and explicit form for the translation invariant measure μ, which in turn may be used to show that μ is not an equilibrium measure for any well-behaved interaction (see e.g. Ref. 7 for definitions). This result was obtained independently by K. Vande Velde[8].

Theorem 3: *The measure μ is not Gibbsian for any absolutely summable Hamiltonian.*
Proof: If μ were Gibbsian then there would exist well defined probabilities $\mu(\eta_1^\sigma|\tau)$ for the occupation of site 1 conditioned on a configuration τ in the complement of this site, continuous as functions of τ. In particular, suppose that for $p, q \geq 0$, τ^{pq} has $\tau_0^{pq} = 2$, $\tau^{pq}(i) = 1$ for $2 \leq i \leq p+1$, $\tau^{pq}(i) = 0$ for $p+2 \leq i \leq p+q+1$, and $\tau^{pq}(p+q+2) = 2$; for our model, the decoupling effect of second class particles implies that $\mu(\eta_1^\sigma|\tau^{pq})$ is independent of any further specification of τ^{pq} and can be computed from (11) and Theorem 2. Thus, writing $f_\sigma^{pq} = \eta_0^2 \eta_1^\sigma \prod_{i=2}^{p+1} \eta_i^1 \prod_{i=p+2}^{p+q+1} \eta_i^0 \eta_{p+q+2}^2$, we have

$$\mu(f_0^{pq}) = \rho_2 R(01^p 0^q) \rho_1^p (1-\rho_1)^{q+2} \rho_0^{q+1}(1-\rho_0)^{p+1}, \qquad (14)$$
$$\mu(f_1^{pq}) = \rho_2 R(1^{p+1} 0^q) \rho_1^{p+1}(1-\rho_1)^{q+1} \rho_0^q (1-\rho_0)^{p+2}, \qquad (15)$$
$$\mu(f_2^{pq}) = \rho_2 R(\emptyset) R(1^p 0^q) \rho_1^p (1-\rho_1)^{q+2} \rho_0^q (1-\rho_0)^{p+2}, \qquad (16)$$

so that, since $R(\emptyset) = 1$, $R(1^p 0^q) = R(01^p 0^q) = \binom{p+q}{p}$, and $R(1^{p+1} 0^q) = \binom{p+q+1}{p+1}$,

$$\mu(\eta_1^1|\tau^{pq}) = \frac{\mu(f_1^{pq})}{\mu(f_0^{pq}) + \mu(f_1^{pq}) + \mu(f_2^{pq})}$$
$$= \frac{(p+q+1)\rho_1(1-\rho_0)}{(p+1)(1-\rho_1)\rho_0 + (p+q+1)\rho_1(1-\rho_0) + (p+1)(1-\rho_1)(1-\rho_0)}. \qquad (17)$$

But continuity of the conditional probabilities would imply that

$$\lim_{p \to \infty} \sup_{q,q' \geq 0} |\mu(\eta_0^1|\tau^{pq}) - \mu(\eta_0^1|\tau^{pq'})| = 0, \qquad (18)$$

which fails for (17), since the supremum has value $(1-\rho_1)/(1-\rho_0\rho_1)$ for all p. ∎

The next property of μ_0 which we discuss is simple to state, but we lack an equally simple proof.

Theorem 4: *The distribution under μ_0 of first class particles to the right of the origin is Bernoulli with density ρ_1, that is, if $0 < i_1 < \cdots < i_m$ then*

$$\mu_0(\eta_{i_1}^1 \cdots \eta_{i_m}^1) = \rho_1^m. \qquad (19)$$

Similarly, holes to the left of the origin have Bernoulli distribution with density ρ_0.
Proof: We verify (19) by induction on m; the $m = 0$ case is trivial. Let $H = [G, z_1 X_1] = X_2(G - (1+z_0)(1+z_1)I)$. We note that (i) if $f(X)$ is any polynomial in X_0, X_1, X_2 then from $X_2 = |1\rangle\langle 1|$ it follows that $\langle 1|f(X)H = c_f\langle 1|H$ and $Hf(X)H = d_f H$ for some constants c_f, d_f, and (ii) if $k \geq 0$, then $H(z_1 X_1)^k = H g_k(G)$ for some polynomial g_k, where the $k = 1$ result is an explicit calculation—$Hz_1 X_1 = H(G - z_0 I)$—and the general case follows by induction, using the definition of H and (i). Now from (9), $\lim_{N \to \infty} \Xi_N^{-1}\langle 1|HG^{2N-k}|1\rangle = 0$ for any $k \in \mathbb{Z}$, and with (i) and (ii) this implies that

$$\lim_{N \to \infty} \Xi_N^{-1}\langle 1|f(X)Hg(z_1 X_1, G)G^{2N-k}|1\rangle = 0 \qquad (20)$$

for any polynomials f and g. Thus by repeatedly commuting factors X_1 to the left, using (20) to eliminate commutator terms, we find that

$$\mu_0(\eta_{i_1}^1 \cdots \eta_{i_m}^1) = \lim_{N \to \infty} \Xi_N^{-1}\langle 1|G^{i_1-1} z_1 X_1 \cdots G^{i_m-i_{m-1}-1} z_1 X_1 G^{2N-i_m}|1\rangle$$
$$= \lim_{N \to \infty} \Xi_N^{-1}\langle 1|(z_1 X_1)^m G^{2N-m}|1\rangle$$
$$= \lim_{N \to \infty} \Xi_N^{-1}\langle 1|(G - X_2 - z_0 X_0)(z_1 X_1)^{m-1} G^{2N-m}|1\rangle$$
$$= (1 - (1+z_0)/\gamma)\rho_1^{m-1} = \rho_1^m. \qquad \blacksquare \qquad (21)$$

Remark 5: As a last special case of (8) we consider the single site densities $\mu_0(\eta_i^0)$, $\mu_0(\eta_i^1)$, and $\mu_0(\eta_i^2)$. Note that, by Theorem 4, $\mu_0(\eta_i^1) = \rho_1$ for $i > 0$ and $\mu_0(\eta_i^0) = \rho_0$ for $i < 0$; since $\mu_0(\eta_i^0) + \mu_0(\eta_i^1) + \mu_0(\eta_i^2) = 1$, all the single site densities will be determined by $\mu_0(\eta_i^2)$. Further, Theorem 2 implies that $\mu_0(\eta_i^2) = \mu_0(\eta_{-i}^2)$ (a related symmetry may be established[1] by a proof independent of any details of μ_0). Consider then $i > 0$; in Ref. 1 it is shown that $\mu_0(\eta_i^2) > \rho_2$ for all such i, that $\lim_{i \to \infty} \mu_0(\eta_i^2) = \rho_2$, and that asymptotically

$$\mu_0(\eta_i^2) - \mu_0(\eta_{i+1}^2) \simeq Ci^{-3/2}\xi^i \quad \text{for} \quad \xi = \left(\sqrt{\rho_1(1-\rho_0)} + \sqrt{\rho_0(1-\rho_1)}\right)^2. \quad (22)$$

Thus the single site density decays exponentially to its asymptotic value. The characteristic length $-\log \xi$ of this decay diverges as ρ_2^{-2} in the limit $\rho_2 \searrow 0$, but there remain an infinite number of second class particles in the system (see the discussion of bound states in Section 5) and the limiting decay is algebraic: $\mu_0(\eta_i^2) \simeq Ci^{-1/2}$ for $\rho_2 = 0$.

3. EXTREMAL INVARIANT MEASURES FOR THE TWO SPECIES TASEP

In this section we establish that the measures $\mu^{(\rho)}$ defined in (11) are the extremal translation invariant steady states for the two species TASEP, and that together with certain blocking measures they comprise all extremal steady states. Our proofs are based closely on results of Liggett for the one species ASEP.

We let $\mathcal{I}$ and $\mathcal{S}$ denote the sets of measures on $Y = \{0,1,2\}^{\mathbb{Z}}$ which are respectively invariant for the two species TASEP dynamics and translation invariant, and for any closed, convex set $\mathcal{P}$ of probability measures we write $\mathcal{P}_e$ for the extreme elements of $\mathcal{P}$. The blocking measure $\hat{\mu}^{(a,b)}$, defined for $a,b \in \mathbb{Z} \cup \{-\infty, \infty\}$ with $a \leq b$ and at least one of a,b finite, is the point mass supported on the configuration τ with $\tau(i) = 0$ for $i < a$, $\tau(i) = 2$ for $a \leq i < b$, and $\tau(i) = 1$ for $b \leq i$; note that these measures permit only finitely many second class particles unless either first class particles or holes are absent. With this notation we can state the main result of this section.

Theorem 6: (a) $(\mathcal{I} \cap \mathcal{S})_e = \{\mu^{(\rho)} \mid \rho_0, \rho_1, \rho_2 \geq 0, \rho_0 + \rho_1 + \rho_2 = 1\}$;
(b) $\mathcal{I}_e = (\mathcal{I} \cap \mathcal{S})_e \cup \{\hat{\mu}^{(a,b)} \mid a,b \in \mathbb{Z} \cup \{-\infty, \infty\}, a \leq b, \text{ and } a \text{ or } b \text{ finite}\}$.

As a preliminary to the proof we discuss some results of Liggett[3,9], beginning with the construction of an interacting particle system representing coupled copies of the one species ASEP. The configuration space for a single copy is $X = X^1 = \{0,1\}^{\mathbb{Z}}$ and that for a system of m copies is $X^m = \{(\zeta_1, \ldots, \zeta_m) \mid \zeta_k \in X\}$. The dynamics is such that particles move only within their own copy of X, but move jointly whenever possible; specifically, when a Poisson alarm clock associated with site i rings, a particle at site i in any copy jumps to its right if site $i+1$ in that copy is vacant. We will let $T : X^m \to X^m$ be translation to the right—$T\zeta(i) = \zeta(i-1)$—and let T act also on random variables defined on X^m via $Tf \equiv f \circ T^{-1}$ and on measures on X^m via $(\lambda T)(f) \equiv \lambda(Tf)$. Finally, for $\zeta \in X^m$ and $1 \leq k, l \leq m$ we write $\zeta_k \leq \zeta_l$ if $\zeta_k(i) \leq \zeta_l(i)$ for all $i \in \mathbb{Z}$.

Now let $\mathcal{I}^{(m)}$ and $\mathcal{S}^{(m)}$ denote respectively the invariant and translation invariant measures on X^m. The proof of the next lemma follows exactly that of Lemma 2.15, in Chapter 2 of Ref. 3.

Lemma 7: Let $\mathcal{Q}^{(m)}$ stand for $\mathcal{I}^{(m)}$, $\mathcal{I}_e^{(m)}$, $\mathcal{I}^{(m)} \cap \mathcal{S}^{(m)}$, or $(\mathcal{I}^{(m)} \cap \mathcal{S}^{(m)})_e$. If $\lambda_1 \in \mathcal{Q}^{(m_1)}$ and $\lambda_2 \in \mathcal{Q}^{(m_2)}$ then there exists $\lambda \in \mathcal{Q}^{(m_1+m_2)}$ such that λ has marginals λ_1 and λ_2 under the natural product structure $X^{m_1+m_2} = X^{m_1} \times X^{m_2}$.

The next lemma is Corollary 5.3 of Ref. 9.

Lemma 8: If $\lambda \in \mathcal{I}_e^{(2)}$ satisfies

$$\sup_{m \leq n} \left| \sum_{i=m}^n [\lambda(\{\zeta \mid \zeta_1(i) = 1\}) - \lambda(\{\zeta \mid \zeta_2(i) = 1\})] \right| \leq \infty \qquad (23)$$

then precisely one of the conditions $\lambda(\{\zeta \mid \zeta_1 = \zeta_2\}) = 1$, $\lambda(\{\zeta \mid \zeta_1 \leq \zeta_2, \zeta_1 \neq \zeta_2\}) = 1$, and $\lambda(\{\zeta \mid \zeta_1 \geq \zeta_2, \zeta_1 \neq \zeta_2\}) = 1$, holds.

Finally, we need the result for the one-species ASEP corresponding to Theorem 6. For $0 \leq \alpha \leq 1$ let $\nu^{(\alpha)}$ be the Bernoulli measure on X with density α, and for $a \in \mathbb{Z}$ let $\hat{\nu}^{(a)}$ be the point mass at $\zeta \in X$, where $\zeta(i) = 0$ for $i < a$ and $\zeta(i) = 1$ for $i \geq a$.

Lemma 9: (a) $(\mathcal{I}^{(1)} \cap \mathcal{S}^{(1)})_e = \{\nu^{(\alpha)} \mid 0 \leq \alpha \leq 1\}$;
(b) $\mathcal{I}_e^{(1)} = \{\nu^{(\alpha)} \mid 0 \leq \alpha \leq 1\} \cap \{\hat{\nu}^{(a)} \mid a \in \mathbb{Z}\}$.

To apply these results to the two species TASEP we observe that this model may be viewed as two coupled one species TASEP's by identifying Y with the subset $\{(\zeta_1, \zeta_2) \mid \zeta_2 \leq \zeta_1\}$ of X^2, through the correspondence $0 \leftrightarrow (0,0)$, $1 \leftrightarrow (1,1)$, and $2 \leftrightarrow (1,0)$ (this is the form in which second class particles were defined in Ref. 10). In coupling two two species TASEP's we will thus be led to consider measures λ on $Y \times Y \subset X^4$ with prescribed marginals μ_1 and μ_2 on the two copies of Y; in this case we will write λ_+ and λ_- for the marginals of λ on the (ζ_1, ζ_3) and (ζ_2, ζ_4) subsystems.

Proof of Theorem 6: (a) We first show that $\mu^{(\rho)} \in (\mathcal{I} \cap \mathcal{S})_e$ for all ρ. If $\rho_0 = 0$, $\rho_1 = 0$, or $\rho_2 = 0$ then the two species TASEP reduces to a one species TASEP, $\mu^{(\rho)}$ reduces to a Bernoulli measure, and the result follows from Lemma 9, so that we may assume that $\rho_\sigma > 0$ for all σ. We know that $\mu^{(\rho)} \in \mathcal{I} \cap \mathcal{S}$; to complete the proof it suffices[3] to establish the exponential decay of correlations in $\mu^{(\rho)}$, for example, to show that if f and g are random variables such that f depends only on $(\tau(i))_{i \leq 0}$ and g only on $(\tau(i))_{i > 0}$, then $e_j \equiv \mu^{(\rho)}(fT^j g) - \mu^{(\rho)}(f)\mu^{(\rho)}(g)$ converges exponentially to zero as $j \to \infty$. By conditioning on the occurrence or nonoccurrence of the event E_j that a second class particle appears in the interval $[1, j]$ and using the decoupling effect of second class particles, we see that $|e_j| \leq c(f,g)\mu^{(\rho)}(E_j)$ for some constant $c(f,g)$. But from (11) and (13),

$$\mu^{(\rho)}(E_j) = \rho_2 \sum_{k \geq j} \mu_0^{(\rho)}(E_k) = \rho_2 \sum_{k \geq j} \sum_{\substack{\theta \in \mathcal{F} \\ n(\theta) \geq k}} \pi(\theta), \qquad (24)$$

and it follows from estimates in Ref. 1 that this sum over k is exponentially convergent.

Now suppose conversely that $\mu \in (\mathcal{I} \cap \mathcal{S})_e$. Because μ is translation invariant it gives rise to well defined densities ρ_0, ρ_1, and ρ_2 via $\rho_\sigma = \mu(\eta_i^\sigma)$, and by Lemma 7 there exists a measure λ on $X^4 = X^2 \times X^2$ having marginals μ and $\mu^{(\rho)}$ (and therefore supported on $Y \times Y$) and satisfying $\lambda \in (\mathcal{I}^{(4)} \cap \mathcal{S}^{(4)})_e$. The marginal $\lambda_+ \in (\mathcal{I}^{(2)} \cap \mathcal{S}^{(2)})_e$ then has as its marginals two Bernoulli measures with the same density $\rho_1 + \rho_2$, so that Lemma 8 implies that $\lambda(\{\zeta \mid \zeta_1 = \zeta_3\}) = 1$; The same argument implies that $\lambda(\{\zeta \mid \zeta_4 = \zeta_2\}) = 1$. Thus λ is supported on the diagonal of $Y \times Y$, and $\mu = \mu^{(\rho)}$.

(b) It is easy to verify that $\hat{\mu}^{(a,b)} \in \mathcal{I}$ for any a and b, and certainly $\hat{\mu}^{(a,b)}$ is extremal since it is a point mass. Conversely, we suppose that $\mu \in \mathcal{I}_e$ but μ is not translation invariant, and show that $\mu = \hat{\mu}^{(a,b)}$ for some a, b. The marginals of ν_1 and ν_2 of μ belong to $\mathcal{I}_e^{(1)}$ so that, from Lemma 9 and the fact that μ is supported on Y, there are four cases to consider: (i) ν_1 and ν_2 are both Bernoulli, (ii) $\nu_1 = \hat{\nu}^{(a)}$ and $\nu_2 = \hat{\nu}^{(b)}$ with $a, b \in \mathbb{Z}$ and $a \leq b$, (iii) $\nu_1 = \hat{\nu}^{(a)}$ and $\nu_2 = \nu^{(0)}$, and (iv) $\nu_1 = \nu^{(1)}$ and $\nu_2 = \hat{\nu}^{(b)}$. If case (i) holds then by Lemma 7 there exists a measure λ in $\mathcal{I}_e^{(4)}$ with marginals μ and μT. The marginal $\lambda_+ \in (\mathcal{I}^{(2)} \cap \mathcal{S}^{(2)})_e$ then has as its marginals two Bernoulli measures with the same density, so that Lemma 8 implies that $\lambda(\{\zeta \mid \zeta_1 = \zeta_3\}) = 1$; similarly, $\lambda(\{\zeta \mid \zeta_2 = \zeta_4\}) = 1$. Thus $\mu = \mu T$, contradicting the assumption that $\mu \notin \mathcal{S}$. But case (ii) is precisely $\mu = \mu^{(a,b)}$, case (iii), $\mu = \mu^{(a,\infty)}$, and case (iv), $\mu = \mu^{(-\infty,b)}$. This completes the proof, since it is clear that no $\mu^{(\rho)}$ could be a convex combination of the $\hat{\mu}^{(a,b)}$ and other $\mu^{(\rho)}$. ∎

4. SHOCK PROFILES IN THE SINGLE SPECIES TASEP

A central question in statistical mechanics is the relation between microscopic and macroscopic—that is, atomistic and continuous—descriptions of matter[11-13]. The one species TASEP is a familiar arena in which to study this relation (in fact, most of what we say applies equally to the more general partially asymmetric process, in which particles can jump in either direction); for later convenience we will refer to the (first class) particles in such a one species TASEP as *atoms*. For this model there exists a continuum limit as the lattice spacing goes to zero, in which the limiting atomic density $u(x,t)$ satisfies[14-16] the Burgers' equation

$$u_t + (u[1-u])_x = 0. \tag{25}$$

Among the physically interesting solutions of (25) are those in which the density field has discontinuities, or *shocks*; a simple example is a shock between constant densities:

$$u(x,t) = \begin{cases} u_+, & \text{for } x > x_s(t), \\ u_-, & \text{for } x < x_s(t), \end{cases} \tag{26}$$

where $u_+ > u_-$ and the shock position $x_s(t)$ moves with constant velocity $1-(u_++u_-)$. It is natural to ask how these shocks are reflected in the behavior of the system at the microscopic level. Is there a change in the density on the scale of length appropriate for the microscopic description? Or does the density appear locally constant on the microscopic level and change only on some intermediate, mesoscopic scale? Are there three levels, or only two?

To study the microscopic composition of the simple shock (26) one must first locate—or, more properly, define—the shock position in any particular microscopic realization; this cannot be taken to be $x_s(t)$ because fluctuations in the shock position, although negligible on the macroscopic scale, are unbounded on the microscopic scale[17]. Such a definition can be achieved by the introduction into the system of a single second class particle[10]; because such a particle moves with the group velocity, which has value $1-2\rho$ when the local density of atoms is ρ, it is attracted to the shock, and its position may conveniently be taken to define the shock position. With this definition it has been established[18] that the shock is sharp on the microscopic level—that the density of the atoms, as seen from the second class particle, has a time-invariant distribution that approaches the limiting values $u_\pm$ at $\pm\infty$. Thus the shock structure does not introduce a third, intermediate, scale into the problem, although the fluctuations in its position referred to above may be regarded as doing so.

We now show that the measure μ_0 for the two species TASEP studied in Section 2 determines explicitly the density profile of this shock, and indeed the complete distribution of the atoms as seen from the shock location[1]. To do so we must establish a correspondence between the two species TASEP viewed from a selected second class particle—see the remark following (11)—and the one species TASEP viewed from a single second class particle (Ref. 19 contains a more elaborate version of this idea). Two properties of the two species system make this correspondence possible. First, the model contains one species subsystems in two distinct ways: the first class particles

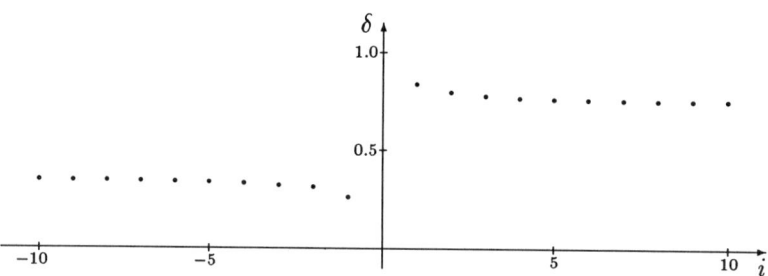

Figure 1. Shock profile δ_i for $\rho_- = 0.35$, $\rho_+ = 0.75$.

alone form a one species TASEP, since from their point of view the second class particles are indistinguishable from holes, and the set of all particles in the system forms a one species TASEP, if we ignore the labels on the particles and the exchanges of first and second class particles. Second, if we select one second class particle, then the second class particles ahead of it are to it indistinguishable from first class particles (here the total asymmetry of the process is crucial). Now if we associate to each configuration $\tau = (\tau(i))_{i=-\infty}^{\infty}$ of the two species TASEP with $\tau(0) = 2$ a configuration $V\tau$ of the one species model with a single second class particle at the origin, so that an atom is present in $V\tau$ at a site $i < 0$ if and only if $\tau(i) = 1$, and at a site $i > 0$ if and only if $\tau(i) = 1$ or $\tau(i) = 2$, then these properties imply that the measure $V\mu_0 \equiv \mu_0 \circ V^{-1}$ is the invariant measure for the one species system, representing a shock with asymptotic atomic densities $\rho_- = \rho_1$ behind the shock and $\rho_+ = \rho_1 + \rho_2$ ahead of the shock (corresponding to u_- and u_+ for the continuum model).

Let us write δ_i for the single site atomic density which defines the shock profile, so that $\delta_i = \mu_0(\eta_i^1)$ for $i < 0$ and $\delta_i = \mu_0(\eta_i^1+\eta_i^2)$ for $i > 0$. Explicit formulas for δ_i are given in Ref. 1, but the essential behavior follows from Remark 5 above: note in particular the behavior $\delta_i - \rho_+ \simeq Ci^{-3/2}\xi^i$ for large i and the symmetry $\delta_i + \delta_{-i} = \rho_- + \rho_+$. Note also the rather unexpected fact that $\delta_i > \rho_+$ for $i > 0$, so that the density profile, rather than being monotonically increasing, has the appearance of Figure 1.

Finally, we may ask what happens to the profile in the limit of a vanishing shock: $\rho_+ \searrow \rho_-$. Physically, this limit yields a uniform system of atoms seen from the point of view of a single second class particle; in terms of the two species model, this corresponds to the $\rho_2 \searrow 0$ limit of Remark 5. The discussion in that remark indicates that the second class particle sees a distorted profile—a "shock" of zero height—in which the atomic density decays to its asymptotic value $\rho_+ = \rho_- = \rho$ as $\delta_i \simeq \rho + C/\sqrt{i}$.

5. BOUND STATES OF SECOND CLASS PARTICLES

In Section 2 we took a particular infinite volume limit of the periodic two species TASEP, characterized by positive densities of holes and of both particle species. Other limits are also of interest. For example[1], we may introduce finite volume ensembles $\bar{\nu}^{(N)} = \bar{\nu}_0^{(z_1,K_2,N)}$ which contain a fixed number $K_2 > 0$ of second class particles, one located at the origin, and in which the distribution of the number of holes and first class particles are controlled by fugacity z_1, by

$$\bar{\nu}_0^{(N)}\left(\prod_{i=-N}^{N} \eta_i^{\tau(i)}\right) = \delta_{\tau(0)2}\delta_{k_2(\tau)K_2}\bar{\Xi}_N^{-1} z_1^{k_1(\tau)}\langle 1|X_{\tau(1)}\cdots X_{\tau(N)}X_{\tau(-N)}\cdots X_{\tau(-1)}|1\rangle, \tag{27}$$

where the notation is as in (6), and then define an infinite volume measure by $\bar{\nu}_0 = \bar{\nu}_0^{(z_1,K_2)} = \lim_{N\to\infty}\bar{\nu}_0^{(N)}$. The state $\bar{\nu}_0$ has densities of holes and first class particles

$$\rho_0 = 1/(1+z_1), \qquad \rho_1 = z_1/(1+z_1) \tag{28}$$

(obtained directly or by setting $z_0 = z_1^{-1}$ in (10)). Note that the condition that a second class particle lie at the origin is necessary to avoid a trivial limit describing a state with only first class particles and holes, and that $\bar{\nu}_0$ is not obtained by conditioning a translation invariant measure on the presence of a second class particle at the origin.

When $K_2 = 1$ the measure $\bar{\nu}_0$ describes the state seen by a single second class particle in a uniform system of first class particles and holes, and thus recovers the $\rho_+ \searrow \rho_-$ limit discussed in the previous section. The situation for $K_2 > 1$ is more interesting. One might expect that the second class particles not conditioned to lie at the origin would escape to infinity during the limiting process, but this is not the case; rather, these particles remain in the system, forming in effect a bound state with the particle at the origin. The binding is quite weak. When $K_2 = 2$, the probability $p(k)$ that the two particles are separated by a distance k is decays as $k^{-3/2}$, so that the expected distance between the particles is infinite. When $K_2 > 2$ the relative distances between successive second class particles are independent, with each distributed according to p.

6. OTHER APPLICATIONS OF THE OPERATOR METHOD

The operator method used in Section 1 to obtain the invariant measure of the two species TASEP with periodic boundary conditions was originally devised[4] to solve the problem of the single species TASEP on a finite lattice with open boundary conditions and a flux of particles through the system. The solution involves only the operators X_0 and X_1, but the solution extends easily to the inclusion of a fixed number of second class particles, trapped in the system, via the introduction of operators X_2 in the natural way. The method may be extended to treat partially asymmetric systems with either one[4] or two[1] species, or to the two species TASEP in which each of the jumps in (1) has a different rate[1], but the resulting computations are much more difficult and have not been worked out. The method may also be extended[20] to compute the diffusion constant of a marked particle in a one species TASEP with periodic boundary conditions.

Acknowledgments

I thank Bernard Derrida, Steve Janowsky, Pablo Ferrari, Shelly Goldstein, and particularly Joel Lebowitz for helpful conversations.

REFERENCES

1. B. Derrida, S. A. Janowsky, J. L. Lebowitz, and E. R. Speer, Exact solution of the totally asymmetric simple exclusion process: shock profiles, to appear in *J. Stat. Phys* **73** (1993).

2. B. Derrida, S. A. Janowsky, J. L. Lebowitz, and E. R. Speer, Microscopic shock profiles: exact solution of a nonequilibrium system, *Europhys. Lett.* **22**, 651–656 (1993).

3. T. M. Liggett, "Interacting Particle Systems," Springer-Verlag, New York (1985), and references therein.

4. B. Derrida, M. R. Evans, V. Hakim, and V. Pasquier, An exact solution of a 1D asymmetric exclusion model using a matrix formulation, *J. Phys. A* **26**, 1493–1517 (1993).

5. P. Ferrari, L. Fontes, and Y. Kohayakawa, personal communication.

6. B. Derrida, personal communication.

7. H.-O. Georgii, "Gibbs Measures and Phase Transitions," Walter de Gruyter, Berlin (1988).

8. K. Vande Velde, personal communication.

9. T. M. Liggett, Coupling the simple exclusion process, *Ann. Prob.* **4**, 339–356 (1976).

10. E. D. Andjel, M. Bramson, and T. M. Liggett, Shocks in the asymmetric exclusion process, *Prob. Th. Rel. Fields* **78**, 231–247 (1988).

11. H. Spohn, "Large-Scale Dynamics of Interacting Particles," Texts and Monographs in Physics, Springer-Verlag, New York (1991).

12. A. De Masi and E. Presutti, "Mathematical Methods for Hydrodynamic Limits," Lecture Notes in Mathematics 1501, Springer-Verlag, New York (1991).

13. J. Lebowitz, E. Presutti, and H. Spohn, Microscopic models of hydrodynamic behavior, *J. Stat. Phys.* **51**, 841–862 (1988).

14. H. Rost, Nonequilibrium behavior of many particle process: density profiles and local equilibria, *Z. Wahrsch. Verw. Gebiete* **58**, 41–53 (1981).

15. A. Benassi and J. P. Fouque, Hydrodynamic limit for the asymmetric simple exclusion process, *Ann. Prob.* **15**, 546–560 (1987).

16. E. D. Andjel and M. E. Vares, Hydrodynamical equations for attractive particle systems on $\mathbb{Z}$, *J. Stat. Phys.* **47**, 265–288 (1987).

17. D. Wick, A dynamical phase transition in an infinite particle system, *J. Stat. Phys.* **38**, 1015–1025 (1985).

18. P. Ferrari, Shock fluctuations in asymmetric simple exclusion, *Prob. Th. Rel. Fields* **91**, 81–101 (1992).

19. P. Ferrari, C. Kipnis, and E. Saada, Microscopic structure of traveling waves in the asymmetric simple exclusion, *Ann. Prob.* **19**, 226–244 (1991).

20. B. Derrida, M. R. Evans, and D. Mukamel, Exact diffusion constant for 1 dimensional asymmetric exclusion models, *J. Phys. A*, to appear.

HOW TO RECONSTRUCT A HEAT BATH

B. Kümmerer

Mathematisches Institut
Universität Tübingen
Auf der Morgenstelle 10
72076 Tübingen
Germany

1 INTRODUCTION

An open quantum system is considered as being a subsystem of a larger closed quantum system. We discuss the problem, how one can decompose the closed system into the open system and a heat bath and, correspondingly, the time evolution of the closed system into a free evolution of system and bath and an interaction part. As an illustration of the general theory we discuss this problem in the simple case of certain open two-level quantum systems.

Problem. Given an open quantum system as a subsystem of a closed system, is it possible to decompose the closed system into the open system and a separated quantum system, the *heat bath*, and, correspondingly, the time evolution of the closed system into a free evolution of the heat bath, a free evolution of the open system, and an interaction part of the evolution. If such a decomposition is possible, then analyse the structure of its components.

Performing such a decomposition means, formally, to decompose the Hamiltonian $I\!H$ of the global evolution as $I\!H = I\!H_{system} + I\!H_{interaction} + I\!H_{bath}$. In the case which is discussed in the following, the open system is small, i.e., finite dimensional, and it is not necessary to separate $I\!H_{interaction}$ from $I\!H_{bath}$, since this can be easily done afterwards.

The above problem is inverse to a standard problem in the theory of open quantum systems, which consists in constructing a heat bath and a coupling between system and bath, and then to compute the reduced evolution of the open system (cf., e.g., [2] or [15], where already the problem of investigating the relation between the reduced evolution and the heat bath is considered).

One reason for studying this problem is the experience, that in the construction of open systems there is frequently a limiting procedure involved (weak coupling limit, singular coupling limit, etc.), which can be controlled on the open subsystem, while, in most cases, the information on the heat bath is lost. The following considerations are an attempt to regain this structure.

Another motivation comes from the fact that the mathematical structure of an open quantum system is identical to the mathematical structure of a quantum stochastic process ([1]). Therefore, the following considerations are also part of the structure theory of these processes: They are a non-commutative counterpart to the reconstruction of a stochastic process from its transition probabilities in classical probability theory (such a reconstruction is not available in the quantum context).

The most satisfactory results on the problem are obtained under certain stationarity conditions of the evolutions (see below). In these notes, some of the general theory in [3], [4], [10], [11], [13], is illustrated for a simple example of an open quantum system.

Notation. For the purpose of the present notes a *quantum system* will be modelled by a pair of the form $(\mathcal{A}, \phi)$, where $\mathcal{A}$ is its von Neumann algebra of observables and ϕ is its state which we assume in this context to be faithful and normal on $\mathcal{A}$. All algebras occuring in the following are assumed to be von Neumann algebras. If $(R_t)_{t \in \mathbb{R}}$ is a pointwise weak* continuous one-parameter group of *-automorphisms of $\mathcal{A}$ which leaves the state ϕ fixed, then $(\mathcal{A}, \phi, (R_t)_t)$ is called a *stationary dynamical system*. The group $(R_t)_t$ may be viewed as the dynamics of the quantum system $(\mathcal{A}, \phi)$.

A *stationary open quantum system* will be modelled by a quadruple $(\hat{\mathcal{A}}, \hat{\phi}, \hat{T}_t; \mathcal{A}_0)$. Here $(\hat{\mathcal{A}}, \hat{\phi}, \hat{T}_t)$ is a stationary dynamical system and $\mathcal{A}_0$ is a distinguished subalgebra of $\hat{\mathcal{A}}$. In addition, we assume the existence of the conditional expectation P_0 : $(\hat{\mathcal{A}}, \hat{\phi}) \to \mathcal{A}_0$, i.e., of a map $P : \hat{\mathcal{A}} \to \hat{\mathcal{A}}$ with $P(\mathbb{1}) = \mathbb{1}$, $P^2 = P$, $P(\hat{\mathcal{A}}) = \mathcal{A}_0$, $\|P\| \leq 1$, and $\hat{\phi} \circ P = \hat{\phi}$. It is well known (cf. [18]) that such a map is automatically completely positive and satisfies the module property $P(axb) = aP(x)b$ for $a, b \in \mathcal{A}_0$, $x \in \hat{\mathcal{A}}$. This justifies its name. If $\hat{\phi}$ is a trace, then the existence of P_0 is automatic, otherwise it may become a rather restrictive condition ([17]). If the conditional expectation exists, it is unique (by the invariance of $\hat{\phi}$).

The subalgebra $\mathcal{A}_0$ is viewed as the algebra of observables of an open quantum system, which is part of the larger quantum system $(\hat{\mathcal{A}}, \hat{\phi})$, equipped with a stationary time evolution. Therefore, the corrrect name of $(\hat{\mathcal{A}}, \hat{\phi}, \hat{T}_t; \mathcal{A}_0)$ would be *closed stationary quantum system with an open subsystem*. We prefer the shorter terminology.

Sometimes, it is convenient to identify $\mathcal{A}_0$ with $j(\mathcal{A})$, where $\mathcal{A}$ is a von Neumann algebra isomorphic to $\mathcal{A}_0$ and $j : \mathcal{A} \to \hat{\mathcal{A}}$ is an injective *-homomorphism with $j(\mathcal{A}) = \mathcal{A}_0$. In this case one puts $P := j^{-1} \circ P_0 : \hat{\mathcal{A}} \to \mathcal{A}$. Associated to a stationary open quantum system is its *reduced evolution* $T_t := P \circ \hat{T}_t \circ j$, which is a pointwise weak* continuous family of completely positive operators on the observables of the open subsystem. Its interpretation is suggested by the name.

As mentioned above, the mathematical structure as outlined so far, can also be interpreted as a stationary quantum stochastic process with values in $\mathcal{A}_0$ ([1]). In particular, if $\hat{\mathcal{A}}$, hence $\mathcal{A}_0$, is commutative, $\mathcal{A}_0 = L^\infty(\Omega, \Sigma, \mu)$, then $(\hat{\mathcal{A}}, \hat{\phi}, \hat{T}_t; \mathcal{A}_0)$ has a natural interpretation as a classical stationary stochastic process with values in Ω. In fact, this stochastic interpretation is mostly adopted.

Throughout the following we will consider stationary open quantum systems of the form $(\hat{\mathcal{A}}, \hat{\phi}, \hat{T}_t; \mathcal{A}_0)$ where $\mathcal{A}_0$ is isomorphic to the algebra M_2 of complex 2×2-matrices. Moreover, we assume that $\hat{T}_t \circ j \begin{pmatrix} 1 & 0 \\ 0 & 0 \end{pmatrix} = j \begin{pmatrix} 1 & 0 \\ 0 & 0 \end{pmatrix}$, equivalently, $T_t \begin{pmatrix} 1 & 0 \\ 0 & 0 \end{pmatrix} = \begin{pmatrix} 1 & 0 \\ 0 & 0 \end{pmatrix}$. For our considerations it is no restriction to assume *minimality*, i.e., we will assume that $\hat{\mathcal{A}}$ is generated, as an algebra, by the subalgebras $\hat{T}_t \circ j(M_2)$, $t \in \mathbb{R}$. Moreover, we may and will assume without restriction that $\hat{\phi}|_{j(M_2)}$ is the normalized trace tr on M_2.

This seems to be the simplest non-trivial example of a stationary open quantum system, where $\mathcal{A}_0$ is non-commutative. In the standard interpretation of M_2 as the algebra of observables of a spin-$\frac{1}{2}$-particle, based on the Pauli matrices σ_x, σ_y, σ_z, the evolutions leave the z-component unaffected, and non-trivial evolutions occur only in the x- and y-components of this system. We will therefore refer to such a system in the following as a *stationary open x-y-system*.

2 A PARADIGMATIC EXAMPLE: A SPIN-$\frac{1}{2}$-PARTICLE IN A STOCHASTIC MAGNETIC FIELD IN Z-DIRECTION

In this section we give a concrete and simple example of a stationary open quantum system ([9]), which should be compared with the general results in the following sections.

Consider standard Brownian motion $(B_t)_{t \geq 0}$ as being represented on the probability space of white noise $(\mathcal{S}'(\mathbb{R}), \mu)$ (cf. [5]): Here, $\mathcal{S}'(\mathbb{R})$ is the dual of the Schwarz space $\mathcal{S}(\mathbb{R})$ of rapidly decreasing functions, which we consider as a subspace of $(L^2(\mathbb{R}), \|\cdot\|)$. By the Bochner-Minlos theorem, there is a unique probability measure μ on $\mathcal{S}'(\mathbb{R})$, the *measure of white noise*, whose characteristic function on $f \in \mathcal{S}(\mathbb{R})$ is given by $\int_{\mathcal{S}'(\mathbb{R})} exp(i < f, f' >) d\mu(f') = exp(-\frac{1}{2} \cdot \|f\|^2)$. For $t > 0$ the function $B_t : \mathcal{S}'(\mathbb{R}) \ni f' \mapsto <f', \chi_{[0,t]}> = \int_0^t f'(s) ds$ is defined for μ-almost all $f' \in \mathcal{S}'(\mathbb{R})$ and the family $(B_t)_{t \geq 0}$ is a realization of standard Brownian motion on $(\mathcal{S}'(\mathbb{R}), \mu)$. In particular, $B_t \in L^2(\mathcal{S}'(\mathbb{R}), \mu)$ with $\|B_t\|^2 = t$.

The left translation on $\mathbb{R}$ induces a group $(\sigma_t)_{t \in \mathbb{R}}$ of measure preserving transformations on $(\mathcal{S}'(\mathbb{R}), \mu)$ with $\sigma_t(\delta_s) = \delta_{s-t}$, called *the flow of Brownian motion* ([5]). The important feature of this representation of Brownian motion is the *additive cocycle identity* $B_{s+t} = B_t + B_s \circ \sigma_t$.

This representation of Brownian motion is cast into algebraic terms as follows: The measure μ induces a faithful normal state ψ_μ on the von Neumann algebra

$L^\infty(S'(\mathbb{R}), \mu)$, the group $(\sigma_t)_t$ gives rise to a stationary automorphism group S_t of $(L^\infty(S'(\mathbb{R}), \mu), \psi_\mu)$. Its extension to a unitary group on $L^2(S'(\mathbb{R}), \mu)$ will still be denoted by $(S_t)_{t \in \mathbb{R}}$ and the cocycle identity reads as

$$B_{s+t} = B_t + S_t(B_s) \qquad (s, t \geq 0) \ . \qquad (*)$$

Define the unitaries $v_t := e^{iB_t}$ in $L^\infty(S'(\mathbb{R}), \mu)$; then the additive cocycle identity $(*)$ is transformed into the *multiplicative cocycle identity*

$$v_{s+t} = v_t \cdot S_t(v_s) \ . \qquad (**)$$

From the properties of Brownian motion it is immediate that $t \mapsto v_t$ is weakly continuous.

There is a unique way to extend the families $(B_t)_{t \geq 0}$ and $(v_t)_{t \geq 0}$ to negative times such that the cocycle identities are preserved, so we will consider these families as being indexed by $\mathbb{R}$.

From these ingredients a stationary open quantum system is obtained as follows: Put

$\hat{A} := M_2 \otimes L^\infty(S'(\mathbb{R}), \mu) \ ,$
$\hat{\phi} := tr \otimes \psi_\mu \ ,$
$A_0 := M_2 \otimes \mathbb{1} \ .$

The algebra $\hat{A}$ will also be realized as $M_2(L^\infty(S'(\mathbb{R}), \mu))$, the algebra of 2×2-matrices with entries in $L^\infty(S'(\mathbb{R}), \mu)$.

Then $u_t := \begin{pmatrix} \mathbb{1} & 0 \\ 0 & v_t \end{pmatrix}$ is a unitary in $\hat{A}$ which induces the inner automorphism $Ad\, u_t : x \mapsto u_t \cdot x \cdot u_t^*$ ($x \in \hat{A}$, $t \in \mathbb{R}$). Finally, put
$\hat{T}_t := Ad\, u_t \circ (Id_{M_2} \otimes S_t) \ .$

It follows from $(**)$ that $(\hat{T}_t)_t$ is a (stationary) group of automorphisms of $(\hat{A}, \hat{\phi})$. It is explicitly given by

$$\hat{T}_t \begin{pmatrix} f_{11} & f_{12} \\ f_{21} & f_{22} \end{pmatrix} = \begin{pmatrix} S_t(f_{11}) & S_t(f_{12}) \cdot v_t^* \\ v_t \cdot S_t(f_{21}) & S_t(f_{22}) \end{pmatrix} \qquad (f_{ij} \in L^\infty(S'(\mathbb{R}), \mu)),$$

in particular,

$$\hat{T}_t \circ j \begin{pmatrix} x_{11} & x_{12} \\ x_{21} & x_{22} \end{pmatrix} = \begin{pmatrix} x_{11} \cdot \mathbb{1} & x_{12} \cdot v_t^* \\ x_{21} \cdot v_t & x_{22} \cdot \mathbb{1} \end{pmatrix} \qquad for \qquad \begin{pmatrix} x_{11} & x_{12} \\ x_{21} & x_{22} \end{pmatrix} \in M_2 \ .$$

From $\psi_\mu(v_t) = \mathbb{E}(e^{iB_t}) = e^{-\frac{1}{2}t}$ one obtains

$$T_t \begin{pmatrix} x_{11} & x_{12} \\ x_{21} & x_{22} \end{pmatrix} = \begin{pmatrix} x_{11} & e^{-\frac{1}{2}t} \cdot x_{12} \\ e^{-\frac{1}{2}t} \cdot x_{21} & x_{22} \end{pmatrix} \ .$$

The constituents of this process are the *free evolution* S_t and the unitary cocycle $v_t = e^{iB_t}$. It is well known that the cocycle is not differentiable. Instead, it satisfies the stochastic differential equation

$$dv_t = v_t \cdot (i\, dB_t - \frac{1}{2}dt)$$

which has to be interpreted in the Itô-sense.

In the following we consider the question, in how far this example reflects the general structure of a stationary open x-y-system.

3 THE GENERAL STRUCTURE OF A STATIONARY OPEN X-Y-SYSTEM

Proposition. *(A) Consider the stationary open x-y-system $(\hat{\mathcal{A}}, \hat{\phi}, \hat{T}_t; \mathcal{A}_0)$, $\mathcal{A}_0 = j(M_2)$.*

(1) There exists a quantum system $(\mathcal{C}, \psi)$ such that $(\hat{\mathcal{A}}, \hat{\phi})$ can be canonically identified with $(M_2 \otimes \mathcal{C}, tr \otimes \psi)$ and $j(x) = x \otimes \mathbb{1}$ for $x \in M_2$. The state ψ is a faithful normal trace on $\mathcal{C}$.

(2) There exists a stationary automorphism group S_t of $(\mathcal{C}, \psi)$ and a continuous unitary right cocycle $(v_t)_{t \in \mathbb{R}}$ of S_t (i.e., $\mathbb{R} \ni t \mapsto v_t \in \mathcal{C}$ is a weakly continuous family of unitaries in $\mathcal{C}$ satisfying $v_{s+t} = v_t \cdot S_t(v_s)$) such that $\hat{T}_t = Ad\, u_t \circ (Id_{M_2} \otimes S_t)$, where u_t is the unitary $u_t := \begin{pmatrix} \mathbb{1} & 0 \\ 0 & v_t \end{pmatrix} \in M_2(\mathcal{C}) = M_2 \otimes \mathcal{C}$. Therefore, we can write

$$(\hat{\mathcal{A}}, \hat{\phi}, \hat{T}_t; \mathcal{A}_0) = (M_2 \otimes \mathcal{C}, tr \otimes \psi, Ad\, u_t \circ (Id_{M_2} \otimes S_t); M_2 \otimes \mathbb{1}) \ .$$

(B) Conversely, if $(\mathcal{C}, \psi, S_t)$ is a stationary dynamical system, where ψ is a trace, and whenever $(v_t)_{t \in \mathbb{R}}$ is a continuous unitary right cocycle of S_t, as above, then $(M_2 \otimes \mathcal{C}, tr \otimes \psi, Ad\, u_t \circ (Id_{M_2} \otimes S_t); M_2 \otimes \mathbb{1})$, with u_t as above, is a stationary open x-y-system.

The assertion in (B) is obvious. For proving the first part of (A) put $\mathcal{C} := \{y \in \hat{\mathcal{A}} : j(x) \cdot y = y \cdot j(x) \text{ for } x \in M_2\}$, the relative commutant of $j(M_2)$ in $\hat{\mathcal{A}}$. The factorization of $\hat{\phi}$ into $tr \otimes \psi$ follows from the assumed existence of the conditional expectation P_0 from $(\hat{\mathcal{A}}, \hat{\phi})$ onto $j(M_2)$: its module property implies that the subalgebra $\mathbb{1} \otimes \mathcal{C}$ is mapped into the center of $\mathcal{A}_0 = j(M_2)$, i.e., into the multiples of $\mathbb{1}$. It thus induces a state ψ on $\mathcal{C}$. Since the modular automorphism group of $\hat{\phi}$ commutes with P_0 and with $\hat{T}_t$, it is trivial on $j(M_2)$ hence on $\hat{\mathcal{A}}$ (by minimality). Therefore, $\hat{\phi}$, hence ψ, is a trace.

For proving the second part one uses an argument which is part of A. Connes' "cocycle trick". For details of the proof we refer to ([9], section 3) and ([8], 5.10).

Without any assumptions on the existence of a stationary state in part (A) one still gets the decompositions $\hat{\mathcal{A}} = j(M_2) \otimes \mathcal{C}$ and $\hat{T}_t = Ad\, u_t \circ (Id_{M_2} \otimes S_t)$.

The proposition establishes a canonical one to one correspondence between stationary open x-y-systems and quadruples of the form $(\mathcal{C}, \psi, S_t; (v_t)_t)$ where $(\mathcal{C}, \psi, S_t)$ is a stationary dynamical system, ψ is a trace, and $(v_t)_{t \in \mathbb{R}}$ is a weakly continuous unitary right cocycle of S_t, and the mathematical investigation of those structures is equivalent. For further reference we call the quadruple $(\mathcal{C}, \psi, S_t; (v_t)_t)$ a *stationary cocycle system*.

The easiest way to obtain examples of stationary cocycle systems is the following: Let $(\mathcal{C}, \psi, S_t)$ be a stationary dynamical system with ψ a trace and let $h \in \mathcal{C}$ be some self-adjoint element. The differential equation

$$\begin{cases} \dot{v}_t = v_t \cdot i\, S_t(h) \\ v_0 = \mathbb{1} \end{cases}$$

has a unique norm continuous global solution $(v_t)_{t \in I\!R}$ in $\mathcal{C}$, and it is elementary to check that v_t is unitary for all $t \in I\!R$ and $(\mathcal{C}, \psi, S_t; (v_t)_t)$ is a stationary cocycle system. It follows that $(\mathcal{C}, \psi, (Ad\, v_t \circ S_t)_t)$ is another stationary dynamical system and the generator of $(Ad\, v_t \circ S_t)_{t \in I\!R}$ is given by $x \mapsto L(x) + i \cdot (hx - xh)$ for x in the domain of L, the generator of $(S_t)_{t \in I\!R}$.

The generator of $\hat{T}_t$ on $M_2 \otimes \mathcal{C}$ is then given as

$$x \mapsto Id_{M_2} \otimes L + i \cdot \left(\begin{pmatrix} 0 & 0 \\ 0 & h \end{pmatrix} x + x \begin{pmatrix} 0 & 0 \\ 0 & h \end{pmatrix} \right)$$

on the domain of $Id_{M_2} \otimes L$.

The physical meaning of the decomposition in part (A) of the above proposition is obvious:

Starting with a stationary open x-y-system, part (1) shows the existence of a heat bath, i.e., the existence of a separated system, such that the whole system is composed of the two-level system and the other system. In part (2) the evolution of the composed system is decomposed into a *free evolution* S_t of the bath and a *coupling* of the two-level-system to the bath, which is represented by the cocycle $(u_t)_t$. The cocycle may be viewed as the time evolution in the *interaction representation*.

The free evolution S_t has a generator which can be used to obtain a physical interpretation for it. In order to give, in a similar way, a physical interpretation also to the unitary cocycle $(u_t)_{t \in I\!R}$ one would try to extract from it the interaction Hamiltonian. This is certainly possible in the above example where the cocycle is differentiable and the interaction Hamiltonian is given by $\begin{pmatrix} 0 & 0 \\ 0 & h \end{pmatrix} \in M_2 \otimes \mathcal{C}$. However, as is shown by the example in the previous section, we can not expect the interaction Hamiltonian to exist in general. Instead, in the following we will look for a stochastic interpretation of the cocycle.

4   THE CASE OF A CLASSICAL HEAT BATH

When the algebra of the heat bath is commutative, it can be viewed as the algebra of observables of a classical system. This case has been studied in detail by M. Epple ([3], [4]).

Consider a stationary cocycle system $(\mathcal{C}, \psi, S_t; (v_t)_t)$, where $\mathcal{C}$ is commutative. From the assumed minimality condition and the continuity of the time evolution it follows that $\mathcal{C}$ has separable predual. Therefore, $(\mathcal{C}, \psi)$ can be represented as $(L^\infty(\Omega, \Sigma, \mu), \psi_\mu)$, where (Ω, Σ, μ) is a Lebesgue probability space and the measure μ induces the state $\psi = \psi_\mu$; moreover, there is a group $(\sigma_t)_{t \in I\!R}$ of measure preserving transformations of (Ω, Σ, μ) which induces S_t by $S_t(f) = f \circ \sigma_t$ ($f \in L^\infty(\Omega, \Sigma, \mu)$, $t \in I\!R$) ([16]).

Theorem. (M. Epple) *The unitary cocycle $(v_t)_{t \in I\!R}$ can be represented as $v_t = e^{iYt + iX_t}$, where $Y : \Omega \to I\!R$ is a real-valued measurable function with $Y \circ \sigma_t = Y$ for all t, and $(X_t)_{t \in I\!R}$ is a family of real-valued measurable functions in $L^2(\Omega, \Sigma, \mu)$ for which $t \mapsto X_t \in L^2(\Omega, \Sigma, \mu)$ is norm continuous and which satisfies the cocycle identity $X_{s+t} = X_t + X_s \circ \sigma_t$.*

Since the unitary cocycle v_t may be viewed as a σ_t-cocycle with values in the torus group $\mathbb{T}$, which is a quotient of the group $\mathbb{R}$, the proof of this theorem in ([3], [4]) solves the cohomological problem of lifting a cocycle from the quotient $\mathbb{T}$ to $\mathbb{R}$. The proof makes essential use of the representation of $(\sigma_t)_{t \in \mathbb{R}}$ as a flow under a function.

If $(\mathcal{C}, \psi, S_t)$ is a commutative stationary dynamical system, $h^* = h \in \mathcal{C}$, put $X_t := \int_0^t S_s(h)ds$, $v_t := e^{iX_t}$ ($t \geq 0$). Then $(X_t)_t$ is an additive cocycle and $(v_t)_t$ satisfies the differential equation in the previous section.

In stochastic terms, Y is a random variable, i.e., a "constant stochastic process", while the cocycle $(X_t)_{t \in \mathbb{R}}$ is a real-valued stochastic process with stationary increments, which has finite variance and is mean square continuous.

In physical terms, such a system $(M_2 \otimes \mathcal{C}, tr \otimes \psi, Ad\, u_t \circ (Id_{M_2} \otimes S_t); M_2 \otimes \mathbb{1})$ allows the following interpretation: The Hamiltonian of a spin-$\frac{1}{2}$-particle in an exterior magnetic field in z-direction can be normalized to be of the form $\begin{pmatrix} 0 & 0 \\ 0 & b \end{pmatrix}$ for some $b \in \mathbb{R}$. It generates the unitary group $\begin{pmatrix} 1 & 0 \\ 0 & e^{itb} \end{pmatrix}$, $t \in \mathbb{R}$. Therefore, this system describes a spin-$\frac{1}{2}$-particle which interacts with a classical magnetic field in z-direction. This field has a component which is constant in time, and the distribution of its possible values is given by the distribution of the random variable Y. Superposed to the constant field is a second magnetic field, whose time behaviour is described by the, possibly generalized, stationary stochastic process, which is obtained as the increment process, or the formal time derivative, of the process $(X_t)_{t \in \mathbb{R}}$. In the example in section 2, $(X_t)_t$ is standard Brownian motion and the increment process is white noise. In general, it is any stationary coloured noise. It is shown in ([3]) that stationary coloured noise can be classified by its spectrum and its characteristic function. This type of model is a standard model in mathematical physics. It is often referred to as *random frequency modulation* (e.g., in ([7], 2.1), where the case of a stationary Gaussian increment process is discussed). Thus the above theorem shows, that every stationary cocycle system with $\mathcal{C}$ commutative can be interpreted as a random frequency modulation.

5 STATIONARY OPEN X-Y-SYSTEMS AS COUPLINGS TO WHITE NOISE

In this section, consider a stationary open x-y-system, given as $(\hat{\mathcal{A}}, \hat{\phi}, \hat{T}_t; \mathcal{A}_0) = (M_2 \otimes \mathcal{C}, tr \otimes \psi, Ad\, u_t \circ (Id_{M_2} \otimes S_t); M_2 \otimes \mathbb{1})$. Here it is no longer assumed that $\mathcal{C}$ is commutative.

The time structure of $(\hat{\mathcal{A}}, \hat{\phi}, \hat{T}_t; \mathcal{A}_0)$ is reflected in the subalgebras $\mathcal{A}_I$ of $\hat{\mathcal{A}}$, generated by $\{\hat{T}_t(x \otimes \mathbb{1}) : x \in M_2, t \in I\}$, where $I \subseteq \mathbb{R}$ are intervals. The conditional expectation from $(\hat{\mathcal{A}}, \hat{\phi})$ onto $\mathcal{A}_I$ exists (since $\hat{\phi}$ is a trace) and is denoted by P_I.

When the interval $I \subseteq \mathbb{R}$ contains 0, define the subalgebra $\mathcal{C}_I \subseteq \mathcal{C}$, which is generated by $\{v_t : t \in I\}$. Obviously, $\mathcal{A}_I = M_2 \otimes \mathcal{C}_I$. For such an interval $I = [-s, t]$, $s, t \geq 0$, the cocycle identity for $(v_t)_t$ implies $S_{s'}(\mathcal{C}_{[-s,t]}) = \mathcal{C}_{[-s+s',t+s']}$ when $0 \leq s' \leq s$, hence the definition $\mathcal{C}_{[s,t]} := S_s(\mathcal{C}_{[0,t-s]})$ for a general interval

$[s, t] \subseteq \mathbb{R}$ makes sense and extends the previous definition (since $(v_t)_t$ is continuous, we need only to consider closed intervals). The *filtrated stationary dynamical system* $(\mathcal{C}, \psi, S_t; (\mathcal{C}_I)_I)$ is an example of an *(algebraic) generalized stationary process* ([11], ch. 4).

Proposition. *([9],[10]) The following conditions are equivalent:*
(a) *The stationary open x-y-system has the Markov property:* $P_{]-\infty,0]}(x) = P_0(x)$ *for all* $x \in \mathcal{A}_{[0,\infty[}$.
(b) *The filtrated stationary dynamical system* $(\mathcal{C}, \psi, S_t; (\mathcal{C}_I)_I)$ *is a generalized white noise:* $\psi(x \cdot y) = \psi(x) \cdot \psi(y)$, *whenever* $x \in \mathcal{C}_{I_1}$, $y \in \mathcal{C}_{I_2}$, *and* I_1 *and* I_2 *are disjoint intervals (or have one point in common).*

The proof follows in this case easily from the definitions ([9], sect. 3).

The contents of this proposition may be summarized as follows: *Markovian stationary open x-y-systems are couplings to generalized white noise.* The system in section 2 is an example of this case.

Condition (a) expresses the Markov condition: $\mathcal{A}_{[0,\infty[}$ and $\mathcal{A}_{]-\infty,0]}$ represent the algebras of observables, which are accessible by observation of the subsystem $M_2 \otimes \mathbb{1}$ in the future and in the past, respectively. By the Markov condition, the best approximation in mean square of an observable in $\mathcal{A}_{[0,\infty[}$ by an observable in $\mathcal{A}_{]-\infty,0]}$ is achieved by an observable at time zero, i.e., in $M_2 \otimes \mathbb{1}$. In this case the reduced evolution on M_2 is a semigroup of the form

$$T_t \begin{pmatrix} x_{11} & x_{12} \\ x_{21} & x_{22} \end{pmatrix} = \begin{pmatrix} x_{11} & x_{12} \cdot e^{-\frac{1}{2}\bar{\lambda}t} \\ x_{21} \cdot e^{-\frac{1}{2}\lambda t} & x_{22} \end{pmatrix}, \quad t \geq 0,$$

where $\operatorname{Re}\lambda \geq 0$ and $e^{-\frac{1}{2}\lambda t} = \psi(v_t)$ (the factor $\frac{1}{2}$ is a convenient normalization).

Condition (b) expresses the stochastic independence of observables belonging to disjoint time intervals. In particular, if $\mathcal{C} = L^\infty(\Omega, \Sigma, \mu)$ is commutative, $\psi = \psi_\mu$, then $\mathcal{C}_{I_k} = L^\infty(\Omega, \Sigma_k, \mu)$ for some σ-subalgebras Σ_k, $k = 1, 2$, which then are independent in the classical sense.

An important feature of the white noise property is the fact that it allows to derive a structure for non-commutative stationary cocycle systems which is similar to the commutative case discussed in the previous section:

The state ψ allows to equip $\mathcal{C}$ with a scalar product $<x,y> := \psi(y^*x)$, $x, y \in \mathcal{C}$. The completion of $\mathcal{C}$ with respect to this scalar product leads to the Hilbert space $L^2(\mathcal{C}, \psi)$, of which the Hilbert spaces $L^2(\mathcal{C}_I, \psi)$ are subspaces in the obvious way. The automorphism group $(S_t)_t$ extends to a weakly continuous unitary group on $L^2(\mathcal{C}, \psi)$ which is still denoted by S_t. For the following discussion it is convenient to normalize the cocycle $(v_t)_t$ such that $\psi(v_t)$, hence λ, is real.

Theorem. *[13] For* $N \in \mathbb{N}$, $t \geq 0$, *put*

$$B_N(t) := \frac{1}{\sqrt{\lambda}} \sum_{k=0}^{N-1} S_{\frac{kt}{N}}(v_{\frac{t}{N}}) - \psi(v_{\frac{t}{N}}) \cdot \mathbb{1} \quad \in \quad \mathcal{C}_{[0,t]} \subseteq L^2(\mathcal{C}_{[0,t]}, \psi).$$

Then the limit $B_t := \lim_{N \to \infty} B_N(t)$ *exists in* $L^2(\mathcal{C}_{[0,t]}, \psi) \subseteq L^2(\mathcal{C}, \psi)$.

The family $(B_t)_{t\geq 0} \subseteq L^2(\mathcal{C}, \psi)$ is continuous in t and satisfies the cocycle identity

$$B_{s+t} = B_t + S_t(B_s) \, .$$

If this limit is performed with the example in section 2, then the cocycle $(B_t)_t$ becomes classical standard Brownian motion, multiplied by the imaginary unit i. In general, $(B_t)_t$ may be viewed as a process with stationary independent increments.

This process can now be used for defining an Itô-type stochastic integral

$$\int_0^t \mu(s) dB_s,$$

where $[0,t] \ni s \mapsto \mu(s) \in L^2(\mathcal{C}_{[0,s]}, \psi)$ is measurable and satisfies the standard L^2-condition $\int_0^t \|\mu(s)\|^2 ds < \infty$ ([13]). This stochastic integral generalizes the classical Itô-integral and some non-commutative stochastic integrals, e.g., in [6] (the definition of the integral can also be extended to the case where the process $\mu(s)$ takes values in some initial Hilbert space). Correspondingly, the theorem of existence and uniqueness of solutions of stochastic differential equations can be formulated and proved in the standard way. This gives sense to the final result:

Theorem. [13] *The unitary cocycle $(v_t)_t$ is obtained as the solution of the stochastic differential equation*

$$dv_t = v_t \cdot (dB_t - \lambda dt) \, .$$

Therefore, also in this case we succeeded in making the unitary cocycle accessible to a physical (stochastic) interpretation. It now only depends on the meaning of the white noise system. This can be classical or Fermionic or free (in the sense of [19], [14]), or still different (the case of Bosonic white noise reduces in this particular case to commutative white noise, since ψ is a trace).

The overlap of this section with the previous section, i.e., when $(\mathcal{C}, \psi, S_t; (\mathcal{C}_I)_I)$ is a commutative white noise, is a special case of the situation considered in [12]: The unitary cocycle $(v_t)_t$ and its logarithm $(X_t)_t$ can be viewed as processes with stationary independent increments, taking values in $\mathbb{T}$ and $\mathbb{R}$, respectively. These processes are completely classified and characterized by the Levy-Chintchin formula: They are always composed of Brownian motions, Poisson processes, and drifts.

Even in this case, stationary open x-y-systems behave very differently from their commutative counterparts, classical stationary Markov processes. In the latter case, all higher correlation functions are determined by the semigroup of transition probabilities (via the Chapman Kolmogorov equations). In our case the role of this semigroup is played by the reduced evolution $(T_t)_t$ which is characterized by the parameter λ. From the higher correlation functions however, one obtains the complete information about the cocycle system $(\mathcal{C}, \psi, S_t; (v_t)_t)$. This is clear on an abstract level by minimality, but it may be instructive to see this more concretely:

The general correlation functions have the form $\hat{\phi}(\hat{T}_{t_1}(x_1) \cdot \ldots \cdot \hat{T}_{t_n}(x_n))$, $x_i \in M_2 \otimes \mathbb{1}$, $t_i \in \mathbb{R}$. It suffices to compute this for the case that x_k is of the form $e_{ij} := f_{ij} \otimes \mathbb{1}$, where f_{ij}, $1 \leq i, j \leq 2$, are the canonical matrix units of M_2. Then it follows from the particular structure of $(\hat{T}_t)_{t\in\mathbb{R}}$ derived in section 3, that the only non-vanishing correlation functions are of the form

$$\hat{\phi}(\hat{T}_{t_1}(e_{12}) \cdot \hat{T}_{t_2}(e_{21}) \cdot \hat{T}_{t_3}(e_{12}) \cdot \ldots \cdot \hat{T}_{t_n}(e_{21})) = \tfrac{1}{2}\psi(v^*_{t_1} \cdot v_{t_2} \cdot v^*_{t_3} \cdot \ldots \cdot v_{t_n})$$

and

$$\hat{\phi}(\hat{T}_{t_1}(e_{21}) \cdot \hat{T}_{t_2}(e_{12}) \cdot \hat{T}_{t_3}(e_{21}) \cdot \ldots \cdot \hat{T}_{t_n}(e_{12})) = \tfrac{1}{2}\psi(v_{t_1} \cdot v^*_{t_2} \cdot v_{t_3} \cdot \ldots \cdot v^*_{t_n})$$

($t_i \in \mathbb{R}$, $1 \le i \le n$) (since $v_0 = \mathbb{1}$ it is enough to consider terms of the first type). Therefore, one observes all correlation functions of the stationary cocycle system $(\mathcal{C}, \psi, S_t; (v_t)_t)$. For example, $\hat{\phi}((e_{12} \cdot \hat{T}_t(e_{21}))^n) = \tfrac{1}{2}\psi(v_t^n)$ gives the higher moments which are known from the Levy-Chintchin formula. They are not determined by the first order moments $\psi(v_t) = e^{-\tfrac{1}{2}\lambda t}$.

Acknowledgements

These notes are based on two talks which were given at the Dublin Institute for Advanced Studies and in Leuven at the workshop "On Three Levels". I would like to thank Prof. J.T Lewis and Professors M. Fannes and A. Verbeure for their kind invitations and their friendly hospitality.

The work under report was part of a research project supported by the Deutsche Forschungsgemeinschaft.

REFERENCES

1. L. Accardi, F. Frigerio and J.T. Lewis, Quantum stochastic processes, *Publ. RIMS*, **18**, 97 (1982).

2. E.B. Davies, "Quantum Theory of Open Systems", Academic Press, New-York (1976).

3. M. Epple, A class of non-commutative stationary processes over the 2×2- matrices, Dissertation, Tübingen (1991).

4. M. Epple, On unitary cocycles of measure-preserving flows, Preprint, Mainz (1993).

5. T. Hida, "Brownian Motion", Springer-Verlag, New-York (1980).

6. R.L. Hudson and J. M. Lindsay, Uses of non-Fock quantum Brownian motion and a quantum martingale representation theorem, *in:* "Quantum Probability and Applications II", Lecture Notes in Mathematics **1136**, 277, Springer-Verlag, Berlin-Heidelberg-New York-Tokyo (1985).

7. R. Kubo, M. Toda and N. Hashitsume, "Statistical Physics II", Springer-Verlag, Berlin (1985).

8. B. Kümmerer, Markov dilations on W*-algebras, *J. Funct. Anal.*, **63**, 139 (1985).

9. B. Kümmerer, Examples of Markov dilations over the 2×2-matrices, *in:* "Quantum Probability and Applications to the Quantum Theory of Irreversible Processes", Lecture Notes in Mathematics **1055**, 228, Springer-Verlag, Berlin-Heidelberg-New York (1984).

10. B. Kümmerer, On the Structure of Markov Dilations on W*-algebras, *in:* "Quantum Probability and Applications II", Lecture Notes in Mathematics **1136**, 318, Springer-Verlag, Berlin-Heidelberg-New York-Tokyo (1985).

11. B. Kümmerer, Stochastic processes with values in M_n as couplings to free evolutions, Preprint, Tübingen (1993).

12. B. Kümmerer and H. Maassen, The essentially commutative dilations of dynamical semigroups on M_n, *Commun. Math. Phys.*, **109**, 1 (1987).

13. B. Kümmerer and J. Prin, Stochastic integration with respect to generalized white noise, In Preparation.

14. B. Kümmerer and R. Speicher, Stochastic integration on the Cuntz algebra O_∞, *J. Funct. Anal.*, **103**, 372 (1992).

15. J.T. Lewis and L.C. Thomas, How to make a heat bath, *in:* "Functional Integration", 97, Oxford University Press (Clarendon), London (1975).

16. G.W. Mackey, Point realizations of transformation groups, *Illinois J. Math.*, **6**, 327 (1962).

17. M. Takesaki, Conditional expectation in von Neumann algebra, *J. Funct. Anal.*, **9**, 306 (1972).

18. M. Takesaki, "Theory of Operator Algebras I", Springer-Verlag, New York (1979).

19. D. Voiculescu, Symmetries of some reduced free product C*-algebras, *in:* "Operator Algebras and their Connections with Topology and Ergodic Theory", Lecture Notes in Mathematics **1132**, 556, Springer-Verlag, Berlin-Heidelberg-New York-Tokyo (1985).

INTERACTING PARTICLE SYSTEMS ON NON-COMMUTATIVE SPACES

Taku Matsui[1,2]

1. Institute of Theoretical Physics, University of Leuven, Celestijnenlaan 200D, 3001 Heverlee, Belgium

2. on leave from Department of Mathematics, Tokyo Metroplitan University, Minami Ohsawa, Hachioji, Japan

INTRODUCTION

In this article, we consider a class of Markov semigroups on non-commutative, infinite dimensional algebras.Throughout this paper, a Markov semigroup on a C^*-algebra always means a(continuous time) semigroup of completely positive maps. We will discuss the construction of Markov semigroups on UHF C^*-algebras, their ergodicity and the nature of invariant states.

In mathematical physics, positivity preserving semigroups on operator algebras appear *naturally* in various contexts. For example, hypercontractive semigroups in the quantum field theory , quantum probability theory which should be relevant to the laser physics, the spectral theory of Dirac operators etc. The following examples of quantum spin models are our starting point.

(i) Correlation inequality for Gibbs state : M.Fannes and A.Verbeure[6] have shown that the translation invariant Gibbs states of quantum spin models on lattices are characterized by certain correlation inequalities. In the proof, the Markov semigroups are used to perturb states and to see change of the energy and the entropy of states. This technique seems still useful for study of quantum states.

(ii) Ground states of quantum spin models : In ref.9, we see a direct connection between classical interacting particle systems[8] and ground states of quantum spin models. In that paper, the quantum spin Hamiltonians are assumed to have the Perron-Frobenious positivity. The ground states are realized by the generators of spin flip processes and reversible measures(Gibbs measures) .(No Trotter-Kato or Feymann-Kac type formulae are used.)

This is an evolution of our previous work[9]. We consider truly quantum Markov semigroups. Their commutative counterpart is the classical interacting particle system[8].

We now describe our systems more precisely. The C^*-algebras we consider are UHF algebras which are either the infinite tensor product of the algebra of N by N matrices or the infinite dimensional Clifford algebras(the algebras of Fermionic creation and annihilation operators). Though the infinite dimensional Clifford algebras and the infinite tensor product of the algebra of 2 by 2 matrices are isomorphic as C^*-algebras, their dynamics will be different due to difference of structures of locality. Here we explain the results for the infinite dimensional Clifford algebras. The case of the infinite tensor product was already discussed in other articles[10][11].

We consider the Fermions on the regular lattice $\mathbf{Z}^d$. Let $a_j^* a_j$ satisfy the CAR (canonical anti-commutation relations) :

$$\{ a_j , a_k \} = \{ a_j^* , a_k^* \} = 0, \{ a_j , a_k^* \} = \delta_{j,k} 1 \quad (1)$$

where $\{ , \}$ is the anti-commutator, $\{A, B\} = AB + BA$, j and k are lattice points in $\mathbf{Z}^d$. Let $\mathcal{A}$ be the C^*-algebra generated by a_j^* and a_j (j in $\mathbf{Z}^d$). We find the following Clifford algebra notation more convenient. Set $X = \mathbf{Z}^d \times \{+,-\}$ and

$$\begin{aligned} C_{(j,+)} &= a_j + a_j^* \\ C_{(j,-)} &= i(a_j - a_j^*) \end{aligned} \quad (2)$$

Then for a and a' in X , C_a and $C_{a'}$ are self-adjoint and

$$\{ C_a , C_{a'} \} = \delta_{a,a'} 1. \quad (3)$$

The lattice translation τ_j (j in $\mathbf{Z}^d$) is an automorphism of $\mathcal{A}$ determined by $\tau_j(C_{(k,.)}) = C_{(j+k,.)}$ ($. = +$ or $-$). We also introduce the Z_2 grading Θ of $\mathcal{A}$ determined by

$$\Theta(C_{a_1} C_{a_2} .... C_{a_n}) = (-1)^n C_{a_1} C_{a_2} .... C_{a_n}$$

The generator of our Markov semigroup is the translation invariant sum of local Lindblad generators.

$$\begin{aligned} L(Q) &= \sum_{k \in \mathbf{Z}^d} L_k(Q), \\ L_k(Q) &= \tau_k L_0(\tau_{-k}(Q)), \\ L_0(Q) &= E_0(Q) - 1/2\{E_0(1), Q\}, \end{aligned} \quad (4)$$

where E_0 is a completely positive map on $\mathcal{A}$.

To make this pregenerator well-defined, it would be natural to assume that E_0 is sufficiently local in the following sense.

$$\lim_{k \to \infty} L_0(\tau_k(Q)) = 0 \quad (5)$$

The locality we assume is specified with the following form of the completely positive map E_0 in (4).

$$E_0(Q) = E_0^{(+)}(Q) + E_0^{(-)}(Q) \quad (6)$$

$$\begin{aligned} E_0^{(+)}(Q) &= \sum_k a_k^* Q a_k \\ E_0^{(-)}(Q) &= \sum_k b_k^* \Theta(Q) b_k \end{aligned} \quad (7)$$

where a_k (resp.b_k) is an even(resp. odd) element of $\mathcal{A}$,

$$\Theta(a_k) = a_k , \Theta(b_k) = -b_k.$$

We will present the following results:

(i) Sufficient condition for generating Markov semigroups:
If a_k and b_k are strictly local or they have exponentially fast decaying tail, it is easy to establish the above L generates a Markov semigroup. We will see that the power law decay condition is sufficient to produce Markov generators. Our method improves the existence of dynamics for the Hamiltonian evolution.(Compare the existing results[2] and ours.) Our proof is a non-commutative translation of that for interacting particle systems[8]. The generator condition for the type II von Neumann algebra are known [1] [3]. A common understanding in operator algebraists is that the C^*-algebras are the sets of continuous functions on non-commutative spaces while the von Neumann algebras are the sets of measurable functions on non-commutative measured spaces. In this context, we may call our semigroups Feller semigroups.

(ii) Ergodicity:
We say a Markov semigroup is ergodic if the invariant state is unique. (This situation is also referred to unique ergodicity.) We present a criterion similar to the $M - \epsilon$ criterion of the commutative case[8].

(iii) Invariant states:
Already in 1970's, many people realized difficulty of constructing non trivial Markov semigroups for which Gibbs states of quantum models are invariant. On the other hand, several rigorous results on ground states have been obtained relatively recently and we may consider the characterization of certain ground states by use of Markov semigroup[7]. We will give one example which complements the previous work of Goderis and Maes[7]. We will also mention the Dobrushin's uniqueness theorem[4] [5] for quantum Gibbs states.

EXISTENCE OF MARKOV SEMIGROUPS

In this section, we consider the Fermion algebra generated by a_j and a_j^*. First we introduce some notations.

Fix an order for elements of $X = \mathbf{Z}^d \times \{+, -\}$. For a finite subset A of X we put

$$C(A) = \prod_{a \in A} C_a, \tag{8}$$

where the product is taken from left to right according to the order of X.
For a in X we define a linear operator δ_a on $\mathcal{A}$ via the formula,

$$\delta_a(Q) = 1/2(Q - C_a\Theta(Q)C_a). \tag{9}$$

Then for a and b in X, we obtain

$$\delta_a(C_a) = C_a \quad \delta_a(C_b) = 0 \ (a \neq b) \tag{10}$$

and

$$\delta_a \delta_b = \delta_b \delta_a \quad \delta_a^2 = \delta_a. \tag{11}$$

Set

$$|||Q||| = \sum_{a \in X} ||\delta_a(Q)|| \tag{12}$$

$$C^1(\mathcal{A}) = \{Q \in \mathcal{A} \mid |||Q||| < \infty\} \tag{13}$$

We assume that a_k and b_k in (6) and (7) have the following expression.

$$a_k = \sum_{|A|:even} g^{(k)}(A)\,C(A) \tag{14}$$

$$b_k = \sum_{|B|:odd} f^{(k)}(B)\,C(B) \tag{15}$$

where $g^{(k)}(A)$ and $f^{(k)}(B)$ are scalars and the sum of (14) (resp. (15)) is taken for all finite subsets of X with even elements (resp. odd).

Theorem 1 *Suppose the following is finite.*

$$\sum_k \sum_A |g^{(k)}(A)|^2 |A|^2 + \sum_k \sum_B |f^{(k)}(B)|^2 |B|^2 < \infty \tag{16}$$

Then the operator L determined by (4), (6) and (7) is well-defined on $C^1(\mathcal{A})$ and the closure of L generates a Markov semigroup S_t on $\mathcal{A}$.
The Markov semigroup S_t leaves $C^1(\mathcal{A})$ invariant and there exists a constant M such that

$$||| S_t(Q) ||| \leq 2e^{tM} ||| Q ||| \tag{17}$$

for any $t > 0$ and Q in $C^1(\mathcal{A})$.

We also obtain the time evolution determined by the following Hamiltonian H.

$$H = \sum_{|A|:even} h(A)C(A) \tag{18}$$

where $h(A)$ is a complex number and the sum is taken over all finite subsets A of X with an even number of elements. We assume the formal self-adjointness of H,

$$h(A) = (-1)^{\frac{1}{2}|A|} \overline{h(A)} \tag{19}$$

and the decay condition,

$$sup_{a \in X} \sum_{a \in A} |h(A)||A| < \infty . \tag{20}$$

Let δ be the derivation defined on $C^1(\mathcal{A})$ determined by $\delta(Q) = i[\,H\,,\,Q\,]$. The closure of δ is the generator of the 1 parameter group of automorphisms α_t of $\mathcal{A}$ satisfying

$$\frac{d}{dt}\alpha_t(Q) = \alpha_t(\delta(Q)) \tag{21}$$

for Q in $C^1(\mathcal{A})$.
The proof of Theorem 1 is similar to that of the classical case[8].

ERGODICITY

Definition 1 *A Markov semigroup is ergodic iff the invariant state is unique.*

Any Markov semigroup has at least one invariant state due to compactness of the state space of $\mathcal{A}$. A simple example of the ergodic semigroup is determined by the generator L_N

$$L_N(Q) = -\sum_{a \in X} \delta_a(Q) \tag{22}$$

In this case, we can compute by hand,

$$e^{tL_N}(C(A)) = e^{-t|A|}C(A). \tag{23}$$

In general, if we have
$$lim_{t \to \infty} |||S_t(Q)||| = 0 \tag{24}$$

for any Q in $C^1(\mathcal{A})$, S_t is ergodic. In the special situation described below, we can prove (24).

Let F_0 be a finite range completely positive unital map which has the realization as in (6) and (7), thus $F_0(1) = 1$ and there exists a finite subset Λ of X such that

$$F_0(C(A)) = C(A) \tag{25}$$

if $A \subset \Lambda^c$.

We also assume there is another $\Lambda_0 \subset \Lambda$ such that

$$\delta_a(F(Q)) = 0 \tag{26}$$

for any a in Λ_0 and Q in $\mathcal{A}$.

Let α_{ab} ($a \in \Lambda_0$, $b \in \Lambda_0{}^c$) be a non negative matrix such that

$$\|\delta_b(F_0(Q)) - F_0(\delta_b(Q))\| \leq \sum_a \alpha_{ab} \|\delta_a(Q)\|. \tag{27}$$

The real number γ is defined by the equation,

$$\sum_{a \in \Lambda_0} \sum_{b \in \Lambda_0{}^c} \alpha_{ab} = \gamma |\Lambda_0|. \tag{28}$$

Let F_1 be another finite range completely positive (not necessarily unital) map satisfying (6), (7). Set

$$E_0^c(Q) = F_0(Q) + c F_1(Q) \tag{29}$$

where c is a positive real number.

Theorem 2 *Suppose that $0 \leq \gamma < 1$. Let $S(c)_t$ be the Markov semigroup generated by (4) with $E_0 = E_0^c$ of (29). There exists a positive constant c_0 and $\epsilon = \epsilon(c)$ such that if $0 \leq c < c_0$, $S(c)_t$ is ergodic, $\epsilon \geq (1-\gamma)|\Lambda_0|$ and*

$$|||S(c)_t(Q)||| \leq 2e^{-t\epsilon}|||Q||| \tag{30}$$

for any Q in $C^1(\mathcal{A})$.

REVERSIBLE STATES

The aim of this section is to show that the ground states we constructed before [9] are characterized by the reversibility of Markov semigroups on UHF algebras.

Definition 2 *A state φ is reversible for the Markov semigroup S_t iff*

$$\varphi(Q_1 S_t(Q_2)) = \varphi(S_t(Q_1) Q_2) \tag{31}$$

for any Q_1 and Q_2 in $\mathcal{A}$.

The condition (31) is also referred to as the detailed balance condition. By use of Schwartz inequality for completely positive maps, we can show that S_t is extendible to a bounded self-adjoint operator on the Hilbert space of the GNS representation for the reversible state φ and the self-adjoint extension of the Markov generator has the negative spectra.

The following observation reveals a connection of the ground states for quantum Hamiltonians and the reversible state of Markov semigroups. Suppose a state φ of $\mathcal{A}$ is characterized by the equation,

$$\varphi(d_k^* d_k) = 0 \quad (k = 1, 2, ...) \tag{32}$$

where d_k is an element of $\mathcal{A}$.
φ is then a ground state for the Hamiltonian H,

$$H = \sum_{k=1,2,3...} d_k^* d_k. \tag{33}$$

The standard definition of the quantum ground state is

$$\varphi(Q^*[H, Q]) \geq 0 \tag{34}$$

for any local element Q provided that $[H, Q]$ is a well-defined derivation.(See ref.2.) Next we consider the Markov generator defined by

$$\begin{aligned} E_k(Q) &= d_k^* Q d_k \\ L(Q) &= \sum_k (E_k(Q) - 1/2\{E_k(1), Q\}) \end{aligned} \tag{35}$$

If L satisfies the assumption of Theorem 1, the state φ is reversible for the associated semigroup. In general, the converse statement may not be valid. However, we can characterize the ground states constructed in our previous work[9] by reversibility of a single semigroup. We next explain the construction of these states and the semigroup.

The system is the spin 1/2 quantum model on the lattice $\mathbf{Z}^d$ and the C*-algebra of observables is the infinite tensor product of 2-by-2 matrices,

$$\mathcal{A} = \overline{\otimes_{\mathbf{Z}^d} M_2(C)}^{C^*}. \tag{36}$$

By $\sigma_\alpha^{(k)}$ ($\alpha = x, y, z \; k \in \mathbf{Z}^d$) we denote the Pauli spin matrix on the site k. For a finite subset A of $\mathbf{Z}^d$, we set

$$\sigma_\alpha(A) = \prod_{k \in A} \sigma_\alpha^{(k)}. \tag{37}$$

The maximal abelian C*-subalgebra $\mathcal{B}$ of $\mathcal{A}$ generated by $\sigma_z^{(k)}$ ($k \in \mathbf{Z}^d$) will be identified with the set of continuous functions on the configuration space $\{1, -1\}^{\mathbf{Z}^d}$ of the classical spin, $\mathcal{B} = C(\{1, -1\}^{\mathbf{Z}^d})$.

The ground states and Hamiltonians are determined by the interaction J_A on $\{1,-1\}^{\mathbf{Z^d}}$. For simplicity, we assume that the interaction is of finite range and translation invariant, so $J_{A+k} = J_A$ (k in $\mathbf{Z^d}$, $A \subset \mathbf{Z^d}$) and $J_A = 0$ if the diameter of A is larger than the range of the interaction. The Hamiltonian of the classical spin is now

$$h = \sum_A J_A \sigma_z(A). \tag{38}$$

The Gibbs state is a measure μ on $\{1,-1\}^{\mathbf{Z^d}}$ satisfying

$$\frac{d\mu(\sigma_k)}{d\mu(\sigma)} = \exp(2 \sum_{k \in A} J_A \sigma_z(A)) \tag{39}$$

where the left-hand side is the Radon Nikodym derivative for the spin flip operation at the site k.
The quantum Hamiltonian we consider is

$$H = \sum_{k \in \mathbf{Z^d}} \left\{ \exp(\sum_{k \in A} J_A \sigma_z(A)) - \sigma_x^{(k)} \right\}. \tag{40}$$

Note that the first and second terms do not commute in the above Hamiltonian. (40) generates the time evolution α_t, $\alpha_t(Q) = e^{itH} Q e^{-itH}$. We constructed the ground state of this Hamiltonian [9]. That ground state is an extension of the Gibbs measure to $\mathcal{A}$. For a Gibbs measure μ on $\{1,-1\}^{\mathbf{Z^d}}$ define the state φ_μ via the equation,

$$\varphi_\mu(\sigma_x(A)\sigma_z(B)) = \mu \left(\sigma_z(B) \exp(\sum_{|A \cap C| \text{ odd}} J_A \sigma_z(C)) \right) \tag{41}$$

We proved the following[9].

Theorem 3 *(i) For any Gibbs measure μ, the state φ_μ is a ground state of the Hamiltonian H (40) satisfying (34).*
(ii) For any translation invariant ground state ψ there exists a Gibbs measure μ such that $\psi = \varphi_\mu$.

The question whether any non translation invariant ground state is described as above or not is still left to solve. We next present the precise form of the generator which yields the Markov semigroup for which the above ground states are reversible. Set

$$d_k = 1 - \exp(-\sum_{k \in A} J_A \sigma_z(A)) \sigma_x^{(k)}. \tag{42}$$

For any Gibbs measure μ we have (32) with $\varphi = \varphi_\mu$. Consider the following generator.

$$\begin{aligned} E_k(Q) &= d_k^* Q d_k + d_k^* \sigma_z^{(k)} Q \sigma_z^{(k)} d_k \\ L(Q) &= \sum_k (E_k(Q) - 1/2\{E_k(1), Q\}) \end{aligned} \tag{43}$$

Theorem 4 *Let S_t be the Markov semigroup generated by L of (43). For any reversible state ψ of S_t, there exists a Gibbs measure μ such that $\psi = \varphi_\mu$.*

Remark 1 *(i) As a corollary of Theorem 4, we see that any reversible state ψ of S_t is a translation invariant ground state of H of (40).*
(ii) If we chose the following $E_k(Q)$ for the generator (43), $E_k(Q) = d_k^* Q d_k$, the statement of Theorem 4 may not be valid. In fact, if we set $J_A = 0$ for any A, both the tracial state and the translation invariant, product state φ satisfying $\varphi(\sigma_x^k) = 1$ are reversible.
(iii) If the interaction $\{J_A\}$ is small (= high temperature), the above semigroup is ergodic and the exponentially fast convergence to the unique invariant state can be shown.

The proof of Theorem 4 is based on ideas of ref.9.

GIBBS STATES

Here we explain briefly our idea for proof of the uniqueness of the quantum Gibbs state by our Markov semigroup technique. In classical spin models, a way of proving uniqueness of Gibbs measures is to reduce the problem to ergodicity of (discrete or continuous time) Markov processes[4][5]. We may ask whether the same idea is valid in the quantum setup. However, talking about use of reversible dynamics, we have a quantum constraint. Namely, it is known that the time evolution of (18) and (23) must commute with any Markov semigroup satisfying (31) for any KMS state (=Gibbs state) φ. This is a reason why the construction of the Glauber dynamics(stochastic Ising model) for quantum spin models has not been successful so far.

We proposed another way of looking at the quantum Dobrushin's uniqueness theorem[11]. As we saw in the previous section, the ground state problem is rather naturally connected with reversible Markov semigroups. So we consider the larger system in which the ground state problem is somehow equivalent to the Gibbs state problem of the smaller system. The strategy is that we replace the commutative algebra B with the quantum observable algebra $\mathcal{A}$ and $\mathcal{A}$ with the tensor product $\mathcal{A} \otimes \mathcal{A}$ in the construction of states of the previous section.

Let φ be a faithful state of the C*-algebra $\mathcal{A}$. (Faithfulness of a state means that $Q = 0$ if $\varphi(Q^*Q) = 0$.) Let $\tilde{\varphi}$ be a state of $\mathcal{A} \otimes \mathcal{A}$ determined by

$$\tilde{\varphi}(A \otimes B) = (\Omega, \pi(A) J \pi(\Gamma(B)) J \Omega). \quad (A, B \in \mathcal{A}) \tag{44}$$

where $\{\pi, \Omega, \mathcal{H}\}$ is the GNS cyclic representation associated with the state φ.(Ω is the cyclic separating vector for the von Neumann algebra $\pi(\mathcal{A})''$ on the Hilbert space $\mathcal{H}$), J is the modular conjugation operator associated with Ω and Γ is a complex conjugation of $\mathcal{A}$. We call $\tilde{\varphi}$ the purification of φ. A non commutative version of (39) is

$$\pi(U)\Omega = J\pi(\alpha_{i\beta/2}(U)) J \Omega \tag{45}$$

where $\alpha_{i\beta/2}$ is the complex analytic extension of the time evolution α_t and U is an arbitrary analytic element of $\mathcal{A}$.
$\tilde{\varphi}$ satisfies the following equation.

$$\tilde{\varphi}(d_U^* d_U) = 0 \tag{46}$$
$$d_U = U \otimes 1 - 1 \otimes \Gamma(\alpha_{i\beta/2}(U)) \tag{47}$$

With suitable choice of a set of U, we may construct reversible Markov semigroups for purified Gibbs states of the given interaction. This is our way of getting rid of the quantum constraint. Finally we have to consider the ergodicity of the Markov semigroup. This scenario works in high temperatures[11].

Acknowledgments

This work was done while the author was a research fellow of Canon Foundation in Europe. The author acknowledges the financial support from Canon Foundation and the warm hospitality of Institute for Theoretical Physics, University of Leuven.

REFERENCES

1. S. Albeverio and R. Hoegh-Krohn, Dirichlet forms and Markov semigroups on C^*-algebras. *Comm. Math. Phys.*, **56**, 173-187 (1977).
2. O. Bratteli and D. Robinson, "Operator Algebras and Quantum Statistical Mechanics.Vol.I and II", Springer-Verlag.
3. E. B. Davies and J. M. Lindsay, Non-commutative symmetric Markov semigroups, *Math. Z.*, **210**, 379-411 (1992).
4. R. L. Dobrushin, *Theory Probab. Its Appl.*, **13**, 197-224 (1968).
5. R. L. Dobrushin and B. Shlosman, Constructive criterion for the uniqueness of Gibbs field *in* : "Statistical Physics and Dynamical Systems. Rigorous Results," A.Jaffe ed., Birkhauser, Basel 1985.
6. M. Fannes and A. Verbeure, Global thermodynamical stability and correlation inequalities. *J. Math. Phys.*, **19**, 558-560 (1978).
7. D. Goderis and C. Maes, Constructing quantum dissipations and their reversible states from classical interacting spin systems. *Ann. Inst. Henri Poincaré*, **55**, 805-829 (1991).
8. T. M. Liggett, "Interacting Particle Systems," Springer Verlag, (1981).
9. T. Matsui, Gibbs measure as quantum ground states.*Comm. Math. Phys.*, **135**, 79-89 (1990).
10. T. Matsui, Markov semigroups on UHF algebras, To appear.
11. T. Matsui, Purification and uniqueness of quantum Gibbs states, Preprint.

NON-SELF-AVERAGING EFFECTS IN SUMS OF RANDOM VARIABLES, SPIN GLASSES, RANDOM MAPS AND RANDOM WALKS

B. Derrida

Service de Physique Théorique, CEN Saclay
F–91191 Gif-sur-Yvette Cedex, France

INTRODUCTION

This is a short review of systems which exhibit non-self-averaging effects: sums of random variables when the distribution has a long tail, mean field spin glasses, random map models and returns of a random walk to the origin. We will see that the non-self-averaging effects are identical in the case of sums of random variables and in the spin glass problem as predicted by the replica approach. Also we will see that for the random map models or for the problem of the returns of a random walk to the origin, the non-self-averaging effects coincide with the results of the replica approach when the number n of replica equals $-1/2$ or -1.

An important outcome [1,2] of the replica theory of spin glasses has been the prediction of non-self-averaging effects: in the low temperature phase, phase space can be thought of as if it was decomposed into infinitely many pure states α, the weights of which remain sample dependent even in the thermodynamic limit (i.e. when the system size becomes infinite). The fluctuations of these weights W_α can be described by considering their moments Y_k defined by

$$Y_k = \sum_\alpha (W_\alpha)^k \tag{1}$$

where in (1) the sum runs over all the pure states α. The fact that these moments Y_k have non-trivial probability distributions, even in the thermodynamic limit appears as a signature of the presence of non-self-averaging effects in the spin glass phase.

The goal of this lecture is to show that similar non-self-averaging effects are present in much simpler systems such as sums of identically distributed random variables or random map models. We will see in particular that the statistical properties of the moments Y_k are identical in the case of sums of random variables and of mean field spin glasses. We will also see that the statistical properties of the moments Y_k for several other systems (random map models, returns of a random walk to its starting point) coincide with what one would obtain from the Parisi scheme for breaking the symmetry between the replica for some unusual limits of the number n of replica ($n = -1/2$ or $n = -1$).

The presentation of this lecture is as follows: the case of sums of positive random variables is discussed in section 2. The replica approach for mean field spin glasses is presented in section 3 and the calculation of the Y_k within the replica scheme is given in section 4. Section 5 shows how the Y_k can be computed for random map models whereas the problem of the returns of a random walk to its starting point is discussed in section 6.

SUMS OF RANDOM VARIABLES

Let us consider the sum S_N of N independent random variables $x_1, \cdots x_N$

$$S_N = \sum_{\alpha=1}^{N} x_\alpha . \tag{2}$$

We assume that these variables are all positive and that they are distributed according to a probability distribution $\rho(x)$ which decays slowly as $x \to \infty$

$$\rho(x) \simeq \frac{A}{x^{1+\mu}} \tag{3}$$

with $0 < \mu < 1$. For large N, it is well known that $S_N/N^{1/\mu}$ has a stable distribution called a Lévy distribution (see [3] and references therein). One can define the weight W_α of the term x_α in the sum as

$$W_\alpha = \frac{x_\alpha}{S_N} . \tag{4}$$

Obviously one has

$$Y_1 = \sum_\alpha W_\alpha = 1 . \tag{5}$$

For large N, one can show that the moments Y_k defined by (1) (for $k > \mu$) have a probability distribution which depends only on μ and k (and becomes independent of N or A). For example, one can calculate $\langle Y_k \rangle$ in the large N limit by using the following identity

$$Y_k = \sum_\alpha \frac{x_\alpha^k}{\left(\sum_\beta x_\beta\right)^k} = \int_0^\infty \frac{t^{k-1} \, dt \, e^{-t\sum_\beta x_\beta}}{\Gamma(k)} \sum_\alpha x_\alpha^k . \tag{6}$$

(This is a direct consequence of the definition of the Γ function: $\Gamma(z) = \int_0^\infty t^{z-1} e^{-t} dt$) When one averages over the x_β, one gets

$$\langle Y_k \rangle = \frac{N}{\Gamma(k)} \int_0^\infty t^{k-1} \, dt \, \langle e^{-tx} \rangle^{N-1} \langle x^k e^{-tx} \rangle . \tag{7}$$

For large N, the integral is dominated by the small t behavior. One can easily see that for small t

$$\langle e^{-tx} \rangle = 1 - \int_0^\infty (1 - e^{-tx}) \rho(x) \, dx \simeq \exp[-t^\mu A(-\Gamma(-\mu))] \tag{8}$$

whereas for $k > \mu$, one has

$$\langle x^k e^{-tx} \rangle = \int_0^\infty x^k e^{-tx} \, \rho(x) \, dx \simeq A \, t^{\mu-k} \, \Gamma(k-\mu) \, . \tag{9}$$

This gives for large N

$$\langle Y_k \rangle \simeq \frac{N \, A \, \Gamma(k-\mu)}{\Gamma(k)} \int_0^\infty t^{\mu-1} \, dt \, \exp\left[-(N-1) \, t^\mu \, A \, (-\Gamma(-\mu))\right] \tag{10}$$

which for $N \to \infty$ leads to

$$\langle Y_k \rangle_{\text{Lévy}} = \frac{\Gamma(k-\mu)}{\Gamma(k) \, \Gamma(1-\mu)} \, . \tag{11}$$

Exact expressions of all the correlation functions between the Y_k can be obtained by similar (but of course longer) calculations. For example, for $k > \mu$ and $k' > \mu$, one can show that

$$\langle Y_k Y_{k'} \rangle_{\text{Lévy}} = \frac{\Gamma(k+k'-\mu)}{\Gamma(1-\mu) \, \Gamma(k+k')} + \mu \, \frac{\Gamma(k-\mu) \, \Gamma(k'-\mu)}{[\Gamma(1-\mu)]^2 \, \Gamma(k+k')} \, . \tag{12}$$

Remark 1: The fact that $\langle Y_k^2 \rangle \neq \langle Y_k \rangle^2$ indicates that the distribution of Y_k remains broad even in the large N limit. Therefore, the moments Y_k and consequently the weights W_α are non-self-averaging quantities.

Remark 2: A consequence of the fact that $\langle Y_k \rangle$ has a non-zero limit as $N \to \infty$ is that the largest term in the sum S_N contributes as a non-zero fraction to the sum S_N. Indeed, if one defines $W_{\max}$ by

$$W_{\max} = \max_\alpha \, W_\alpha \tag{13}$$

one has

$$\langle Y_k \rangle = \left\langle \sum_\alpha W_\alpha^k \right\rangle \leq \left\langle (W_{\max})^{k-1} \right\rangle \, . \tag{14}$$

So the fact (11) that $\langle Y_k \rangle > 0$ implies that $W_{\max} > 0$ with at least a non-zero probability.

Remark 3: There exists an easy way of generating by a Monte Carlo procedure the probability distributions of the Y_k. To do so one can generate an infinite sequence of independent random numbers $z_1, z_2, \cdots$ uniformly distributed between 0 and 1 and one can construct a sequence $\omega_1, \omega_2, \cdots$ by

$$\omega_1 = (-\log z_1)^{-\frac{1}{\mu}} \tag{15}$$

$$\omega_{i+1} = \omega_i \, (1 - \omega_i^\mu \, \log z_{i+1})^{-\frac{1}{\mu}} \, . \tag{16}$$

Then the Y_k computed by

$$Y_k = \sum_{i=1}^\infty \left(\frac{\omega_i}{\sum_{n \geq 1} \omega_n} \right)^k \tag{17}$$

have the same distribution as the Y_k defined by (1) and (4). In fact, one can show that $\omega_1 / \sum_n \omega_n$ has the same distribution as $W_{\max}$, $\omega_2 / \sum_n \omega_n$ has the same distribution as the second largest weight and so on. So a good approximation to the Y_k is to keep only the first few terms in the sums over i and n which appear in (17).

MEAN FIELD SPIN GLASSES

We are going to see now that quantities similar to the Y_k can be introduced also in the spin glass problem. To do that one needs to recall briefly the Parisi replica approach [1, 2, 4] to the Sherrington Kirkpatrick model [5, 6].

The Sherrington Kirkpatrick model is a model of N Ising spins $S_i = \pm 1$ which interact with random long-range interactions. The system has 2^N possible spin configurations and the energy E_α of a configuration $\alpha \equiv \{S_i^\alpha\}$ is given by

$$E_\alpha = - \sum_{1 \leq i < j \leq N} J_{ij} S_i^\alpha S_j^\alpha . \tag{18}$$

In the Sherrington Kirkpatrick model, for each pair ij of spins, there is a random interaction J_{ij} chosen according to

$$\rho(J_{ij}) = \sqrt{\frac{N-1}{2\pi}} \exp\left(-\frac{(N-1) J_{ij}^2}{2}\right) . \tag{19}$$

A given sample corresponds to a random choice of the interactions $\{J_{ij}\}$ and the partition function of each sample is given by

$$Z(\{J_{ij}\}) = \sum_{\alpha=1}^{2^N} \exp\left(-\frac{E_\alpha}{T}\right) \tag{20}$$

where T is the temperature. As usual in the theory of disordered systems, the meaningful quantity to consider is $\langle \log Z \rangle$, the average of $\log Z$ over all the possible realisations of the interactions J_{ij}.

A possible approach to try to calculate $\langle \log Z \rangle$ is to use the replica method. So far, the replica approach is still mathematically poorly understood, but it is very much used in the theory of disordered systems, mostly because in many cases, it is the only available approach. There are, in fact, only a few problems like the random energy models or the mean field theory of directed polymers in a random medium for which there exist alternative approaches to compute $\langle \log Z \rangle$ and for which the replica approach, with a symmetry broken as predicted by the Parisi approach, has been shown to give the right answer [7, 8, 9].

The replica method to calculate $\langle \log Z \rangle$ consists of two steps. First, one computes the integer moments $\langle Z^n \rangle$ of the partition function Z for all integer n. Then, one tries to use the following limiting procedure:

$$\langle \log Z \rangle = \lim_{n \to 0} \frac{\log \langle Z^n \rangle}{n} . \tag{21}$$

In general, it is the second step which is mathematically hard to justify, mostly because the knowledge of $\langle Z^n \rangle$ for integer n is usually not sufficient to determine the value of that quantity for non-integer n.

Let us see how the first step can be done. The calculation of the first moment of Z is trivial and one finds

$$\langle Z \rangle = 2^N \exp \frac{N}{4T^2} . \tag{22}$$

The calculation of the other integer moments of Z requires a little more work. For example the second moment $\langle Z^2 \rangle$ is given by

$$\langle Z^2 \rangle = \sum_{\alpha=1}^{2^N} \sum_{\beta=1}^{2^N} \left\langle \exp\left(-\frac{E_\alpha + E_\beta}{T}\right) \right\rangle . \qquad (23)$$

It can be seen easily from the expressions (18) and (19) that $\langle \exp - (E_\alpha + E_\beta)/T \rangle$ depends only on the overlap $q_{\alpha\beta}$ between the two configurations α and β

$$q_{\alpha\beta} = \frac{1}{N} \sum_{i=1}^{N} S_i^\alpha S_i^\beta . \qquad (24)$$

For large N it can be written as

$$\left\langle \exp\left(-\frac{E_\alpha + E_\beta}{T}\right) \right\rangle \sim \exp[Nf(q_{\alpha\beta})] \qquad (25)$$

with the function $f(q) = (1+q^2)/T^2$. Thus to calculate the large N behavior of $\langle Z^2 \rangle$, it is sufficient to know the number of pairs of configurations α and β having a given overlap $q_{\alpha\beta}$ and to sum over $q_{\alpha\beta}$. The number of pairs of configurations having a given overlap $q_{\alpha\beta}$ is an entropic term which behaves like $\exp[Ns(q_{\alpha\beta})]$ for large N. Therefore, the large N behavior of $\langle Z^2 \rangle$ can be determined by a saddle point method

$$\frac{\log \langle Z^2 \rangle}{N} = \max_{q_{\alpha\beta}} \{f(q_{\alpha\beta}) + s(q_{\alpha\beta})\} . \qquad (26)$$

This reasoning can be easily extended to calculate all higher integer moments and for each n, the large N behavior of $\langle Z^n \rangle$ can be obtained as a saddle point in a space of the $n(n-1)/2$ variables $q_{\alpha\beta}$:

$$\frac{\log \langle Z^n \rangle}{N} = \max_{\{q_{\alpha\beta}\}} g(\{q_{\alpha\beta}\}, n) \qquad (27)$$

where the function $g(\{q_{\alpha\beta}\}, n)$ contains both the energy and the entropy terms.

THE PARISI BROKEN SYMMETRY OF REPLICA

Even when n is an integer, finding the maximum of a complicated function of $n(n-1)/2$ parameters is a difficult task. It turns out that for integer n, the maximum is symmetric [10, 11], i.e. all the $q_{\alpha\beta}$ are equal to some value q. Then the problem of finding the maximum is greatly simplified as one needs to find the maximum of a function of a single variable q. This is what Sherrington and Kirpatrick [5, 6] originally did to calculate the moments $\langle Z^n \rangle$. The n dependence was simple enough that they could obtain an expression of $\langle \log Z \rangle$ via the formula (21). However, they noticed from the very beginning that their expression could not be correct as it predicted a negative entropy at sufficiently low temperature.

Several works followed the one of Sherrington Kirkpatrick to test the assumption that the saddle point was symmetric [12, 13]. In 1979, Parisi proposed an ansatz for which this symmetry was broken [4, 2]. His idea was to assume that the $q_{\alpha\beta}$ depend on the pair $\alpha\beta$ in such a way that the calculation can be done for any value of n, the number of replica. In its simplest version, one groups the n replica into n/μ blocks of

μ replica each. One takes two possible values q_1 and q_2 for the overlap and one looks for a saddle point such that $q_{\alpha\beta} = q_1$ whenever α and β belong to the same group and $q_{\alpha\beta} = q_2$ when α and β belong to different groups. For this saddle point the function $g(\{q_{\alpha\beta}\}, n)$ becomes a function of q_1, q_2 and μ and one tries to find the extremum with respect to these three parameters q_1, q_2 and μ. (In fact what Parisi proposed was a little more complicated as he considered situations where the blocks were themselves decomposed into blocks and so on).

Of course, this kind of saddle point is in principle acceptable only if n is an integer and if it is divisible by μ. However, Parisi proposed to forget the fact that n was an integer, and he suggested that all the calculations be done as if n, n/μ and μ were integers and at the end of the calculation the limit $n \to 0$ be taken, by replacing the constraint that $1 \leq \mu \leq n$, by $0 \leq \mu < 1$. I shall not try to explain why this procedure is believed to lead to the correct expression of $\langle \log Z \rangle$ [1, 2, 4, 14]. Here, I just want to show that within this replica approach the calculation of the moments of the Y_k becomes extremely easy.

Consider a system of n replica (or n objects) grouped into n/μ blocks of μ replica each. If one chooses k different replica among these n replica, the probability that the k replica are in the same group is given by

$$\langle Y_k \rangle_{\text{replica}} = \frac{n(\mu-1)(\mu-2)\cdots(\mu-k+1)}{n(n-1)(n-2)\cdots(n-k+1)} \tag{28}$$

For non-integer n this can be written as

$$\langle Y_k \rangle_{\text{replica}} = \frac{\Gamma(k-\mu)\,\Gamma(1-n)}{\Gamma(1-\mu)\,\Gamma(k-n)} \tag{29}$$

with the convention that when n is an integer, one takes the limit of this expression as n tends to its integer value.

Now let us assume that for a given realisation of the interactions J_{ij}, phase space is decomposed into regions of weights W_α. If one considers k realisations of the same sample (i.e. k copies of the system with the same set of interactions, or k 'typical' spin configurations of the same sample), the probability that the k of them are in the same region of phase space is

$$Y_k = \sum_\alpha (W_\alpha)^k. \tag{30}$$

When one tries to compute this expression within the replica approach [1, 2], each realisation appears as a different replica and on averaging over the disorder, one ends up with the expression (29) in the limit $n \to 0$. Therefore, for the mean field spin glass problem, the Parisi approach predicts

$$\langle Y_k \rangle_{\text{spin glass}} = \lim_{n \to 0} \langle Y_k \rangle_{\text{replica}} = \frac{\Gamma(k-\mu)}{\Gamma(1-\mu)\,\Gamma(k)} \tag{31}$$

and the value of μ which enters in this expression is a complicated function of the temperature T as it is the optimal value of μ when one tries to find the extremum of the function $g(\{q_{\alpha\beta}\}, n)$.

One can repeat this calculation for all the correlation functions between the Y_k: for example, if we have n replica grouped into n/μ blocks of μ replica each, the probability that k of them are together in a block and k' of them are also together in a block is given by:

$$\frac{n(\mu-1)(\mu-2)\cdots(\mu-k-k'+1)}{n(n-1)(n-2)\cdots(n-k-k'+1)} \tag{32}$$

$$+ \frac{n(\mu-1)(\mu-2)\cdots(\mu-k+1)\ (n-\mu)(\mu-1)(\mu-2)\cdots(\mu-k'+1)}{n(n-1)(n-2)\cdots(n-k-k'+1)}.$$

The first term represents situations where the $k+k'$ replica are all in the same block whereas the second term represents the situations where the first k replica are in one block and the other k' replica are in a different block. As before, one can rewrite (32) as

$$\langle Y_k Y_{k'} \rangle_{\text{replica}} = \frac{\Gamma(k+k'-\mu)\ \Gamma(1-n)}{\Gamma(1-\mu)\ \Gamma(k+k'-n)} + (\mu-n)\frac{\Gamma(k-\mu)\ \Gamma(k'-\mu)\ \Gamma(1-n)}{[\Gamma(1-\mu)]^2\ \Gamma(k+k'-n)}. \quad (33)$$

Then the correlation function $\langle Y_k Y_{k'} \rangle$ for the spin glass problem is given by

$$\langle Y_k Y_{k'} \rangle_{\text{spin glass}} = \lim_{n \to 0} \langle Y_k Y_{k'} \rangle_{\text{replica}} = \frac{\Gamma(k+k'-\mu)}{\Gamma(1-\mu)\ \Gamma(k+k')} + \mu\frac{\Gamma(k-\mu)\ \Gamma(k'-\mu)}{[\Gamma(1-\mu)]^2\ \Gamma(k+k')}. \quad (34)$$

Remark 1: We see (31), (34) that the results are identical to the expressions (11), (12) of section 2, obtained for sums of random variables. Therefore, the statistical properties of the weights of the pure states of spin glasses as given by the replica theory are the same as what one obtains in the case of sums of random variables with long tails.

Remark 2: The simplest way of understanding the relation between spin glasses and sums of random variables is probably to consider the random energy model [7, 15, 16], for which the calculation of the moments Y_k can be done either by the replica method or by an approach very similar to the one presented in section 2. This was first pointed out to me by Jean-Philippe Bouchaud who noticed that the non-integer moments of the partition function of the random energy model [17] have identical expressions as the moments of random variables distributed according to Lévy distributions.

Remark 3: We will see in the next two sections that the expressions (29) and (33) give also the moments of the Y_k when one chooses for n values other than $n = 0$.

RANDOM MAP MODELS

We are now going to see that the moments and the correlations of the Y_k have very similar expressions in several other examples. We will discuss in this section the case of random map models, which appear as very simplified mathematical problems related to the theory of random networks of automata such as the Kauffman model [18, 19] and in the next section the problem of the returns to the origin of a random walk.

I shall consider here two cases: first, the random map model [20] without any constraint, and then the area preserving random map model.

The random map model

One considers a system which consists of M possible configurations. The dynamics is determined by a random map F in this phase space: if the system is in configuration C_t at time t, then its configuration C_{t+1} at time $t+1$ is given by

$$C_{t+1} = F(C_t). \quad (35)$$

By definition of the model the function F is totally random, i.e. for each C, $F(C)$ is chosen at random among the M configurations with equal probability. As the map is deterministic and phase space is finite, all the trajectories converge to periodic orbits (or fixed points). Thus phase space can be decomposed into the basins of attraction of the different periodic orbits.

One can define the weight W_α of a given orbit α as the fraction of phase space belonging to its basin. This means that

$$W_\alpha = \frac{\Omega_\alpha}{M} \tag{36}$$

where Ω_α is the number of points in the basin of the attractor α. Once the weights W_α are defined, one can consider their moments Y_k with the usual relation (1)

$$Y_k = \sum_\alpha (W_\alpha)^k . \tag{37}$$

From this expression, it is easy to see that Y_k is the probability that k randomly chosen configurations belong to the same basin. If we choose k random initial configurations $C^1, C^2, \cdots C^k$ converging to the same attractor and if we call m_1 the number of different configurations in the trajectory in the trajectory of C^1, m_2, the number of different configurations in the union of the trajectories of C^1 and C^2, $\cdots$, m_λ, the number of different configurations in the union of the trajectories of $C^1, \cdots C^\lambda$, one has [20]

$$\langle Y_k \rangle = \sum_{m_k=1}^{\infty} \sum_{m_{k-1}=1}^{m_k} \cdots \sum_{m_1=1}^{m_2} \left\{ \prod_{i=0}^{m_k-1} \left(1 - \frac{i}{M}\right) \right\} m_1 \frac{m_1 \, m_2 \, m_3 \cdots m_{k-1}}{M^k} \tag{38}$$

For large M, the sums become integrals and if one makes the change of variables $m_\lambda = \sqrt{M} \, x_\lambda$, one gets

$$\langle Y_k \rangle = \int_0^\infty dx_k \int_0^{x_k} dx_{k-1} \cdots \int_0^{x_2} dx_1 \, x_1^2 \, x_2 \, x_3 \cdots x_{k-1} \, e^{-x_k^2/2} \tag{39}$$

One can perform these integrals and one gets

$$\langle Y_k \rangle_{\text{random map}} = \frac{\Gamma(k) \, \Gamma\left(\frac{3}{2}\right)}{\Gamma\left(k + \frac{1}{2}\right)} . \tag{40}$$

The calculation of the correlations between the Y_k can be done in a very similar way. For example, one can use the fact that

$$\langle Y_k Y_{k'} \rangle = \sum_\alpha (W_\alpha)^{k+k'} + \sum_{\alpha \neq \beta} (W_\alpha)^k (W_\beta)^{k'} . \tag{41}$$

The first term corresponds to $k + k'$ configurations belonging to the same basin and the second term to the first k configurations belonging to one basin and the other k' configurations belonging to a different basin. The calculation follows the same steps as before and one obtains:

$$\langle Y_k Y_{k'} \rangle_{\text{random map}} = \frac{\Gamma(k+k') \, \Gamma\left(\frac{3}{2}\right)}{\Gamma\left(k+k'+\frac{1}{2}\right)} + \frac{1}{2} \frac{\Gamma(k) \, \Gamma(k') \, \Gamma\left(\frac{3}{2}\right)}{\Gamma\left(k+k'+\frac{1}{2}\right)} . \tag{42}$$

Remark 1: we see from (40) and from (42) that in general $\langle Y_k^2 \rangle \neq \langle Y_k \rangle^2$. Thus for the random map model too, the moments Y_k and the weights W_α are non-self-averaging quantities (for example $\langle Y_2 \rangle = 2/3$ whereas $\langle Y_2^2 \rangle = 52/105 \neq \langle Y_2 \rangle^2$).

Remark 2: It is interesting to notice as was first pointed out to me by Giorgio Parisi that the expressions (40) and (42) are exactly the same as those obtained in (29) and (33) in the limit $\mu = 0$ and $n = -1/2$. So the non-self-averaging effects are exactly the same for the random map model as what one gets from the replica calculation in the case of $-1/2$ replica blocked into infinitely many blocks of 0 replica each. The reason for this identity between the results obtained in the random map model and the replica approach for this special choice of n and μ is so far a coincidence. Of course, it would be interesting to try to develop a replica approach to random map models in order to see whether these values of n and μ come out in a natural way.

The area preserving random map model

A simple extension of the previous model is the area preserving random map model. In that case, the map F is random as before except that it is a one-to-one map. So F is a random permutation of the M points of phase space. In that case, obviously the basin of attraction of an attractor and the attractor itself coincide. So each point of phase space belongs to an attractor and the period of the attractor α is equal to the number of points Ω_α of this attractor. Given a point chosen at random in phase space, the probability $P(\Omega)$ that it belongs to an attractor of Ω points is

$$P(\Omega) = \frac{1}{M!} \frac{(M-1)!}{(M-\Omega)!(\Omega-1)!} (M-\Omega)! \, (\Omega-1)! \, . \tag{43}$$

Each term in this product can be understood easily: the first term corresponds to the total number of functions F; the second to choosing the $\Omega - 1$ other points of the attractor among the $M-1$ other points of phase space; the third term is the number of permutations of the remaining $M-\Omega$ points of phase space; the fourth term corresponds to the number of ways that the $\Omega - 1$ other points of the attractor are visited by the dynamics. This expression is easy to simplify to give

$$P(\Omega) = \frac{1}{M} \, . \tag{44}$$

Similarly, it is easy to show that the probability $Q(\Omega, \Omega')$ that two different points belong to two different attractors of periods Ω and Ω' is given by

$$Q(\Omega, \Omega') = \frac{1}{M(M-1)} \, . \tag{45}$$

Then one can deduce from that the expressions of $\langle Y_k \rangle$ and $\langle Y_k Y_{k'} \rangle$ by

$$\langle Y_k \rangle = \sum_{\Omega=1}^{M} P(\Omega) \, \Omega^{k-1} \tag{46}$$

and

$$\langle Y_k Y_{k'} \rangle = \sum_{\Omega=1}^{M} P(\Omega) \, \Omega^{k+k'-1} + \sum_{\Omega=1}^{M-1} \sum_{\Omega'=1}^{M-\Omega} Q(\Omega, \Omega') \, \Omega^{k-1} \Omega'^{k'-1} \, . \tag{47}$$

For large M, this leads to the following expressions

$$\langle Y_k \rangle_{\text{area preserving}} = \frac{1}{k} \tag{48}$$

$$\langle Y_k Y_{k'} \rangle_{\text{area preserving}} = \frac{1}{k+k'} + \frac{\Gamma(k) \, \Gamma(k')}{\Gamma(k+k'+1)} \, . \tag{49}$$

Remark 1: we see from (48) and from (49) that $\langle Y_k^2 \rangle \neq \langle Y_k \rangle^2$, and thus here again, the moments Y_k and the weights W_α are non-self-averaging quantities (for example $\langle Y_2 \rangle = 1/2$ whereas $\langle Y_2^2 \rangle = 7/24 \neq \langle Y_2 \rangle^2$).

Remark 2: in this case too, the expressions (48) and (49) are identical to what the replica calculation (29) and (33) would give for $n = -1$ and $\mu = 0$.

ON THE RETURNS TO THE ORIGIN OF A RANDOM WALK

Let us finally discuss a very simple problem related to the statistics of the returns of a one dimensional random walk to the origin. Consider a random walk of $2T$ steps starting and ending at the origin. Each step is either $+1$ or -1 with equal probability. There are of course

$$Z_T = \frac{(2T)!}{T!\,T!} \tag{50}$$

such walks. For each walk, one can decompose the interval $2T$ into subintervals $2t_1, 2t_2, \cdots 2t_i, \cdots$, where $2t_i$ is the time interval separating the $(i-1)$th and the ith visits of the origin by the walk. Of course, one has

$$T = t_1 + t_2 \cdots + t_i \cdots \tag{51}$$

So the interval of time $2T$ is partitionned into subintervals delimited by the successive returns to the origin. For each walk, one can define the weight W_i of the ith subinterval as

$$W_i = \frac{t_i}{T} \tag{52}$$

and the Y_k as

$$Y_k = \sum_i (W_i)^k . \tag{53}$$

Thus if one chooses k times at random between 0 and $2T$, $\langle Y_k \rangle$ is the probability that these k times fall in the same subinterval (this is because, the probability for one time to fall in the ith subinterval is W_i, and for k times to fall in the ith subintervals this probability is W_i^k). In the large T limit, one can obtain the following exact expressions:

$$\langle Y_k \rangle_{\text{random walk}} = \frac{1}{2k - 1} \tag{54}$$

$$\langle Y_k Y_{k'} \rangle_{\text{random walk}} = \frac{1}{2(k + k') - 1} + \frac{\Gamma(k - \tfrac{1}{2})\,\Gamma(k' - \tfrac{1}{2})}{2\,\Gamma(\tfrac{1}{2})\,\Gamma(k + k' + \tfrac{1}{2})} \tag{55}$$

where $\langle \rangle$ denotes the average over the Z_T walks. The derivation of (54) and (55) is not difficult. One needs the expression of the number Q_T of walks starting at the origin and returning to the origin for the first time after $2T$ steps.

$$Q_T = 2\,\frac{(2T - 2)!}{T!\,(T - 1)!} . \tag{56}$$

One can show that

$$\langle Y_k \rangle = \frac{1}{Z_T} \sum_{t=0}^{T} \sum_{\tau=0}^{T-t} Z_\tau\,Q_t\,Z_{T-t-\tau} \left(\frac{t}{T}\right)^k . \tag{57}$$

The generic term is this sum represents a subinterval of length $2t$ starting at time 2τ and it gives to the probability over all walks that this subinterval contributes to (53). Using the asymptotic forms of Z_T and Q_T for large T

$$Z_T \simeq \frac{2^{2T}}{\sqrt{\pi T}} \quad \text{and} \quad Q_T \simeq \frac{2^{2T}}{2\sqrt{\pi T^3}} \tag{58}$$

one obtains for large T the expression (54).

Similarly, $\langle Y_k Y_{k'} \rangle$ is the probability that if one chooses $k+k'$ times at random, the first k of them fall in the same subinterval and the last k' of them fall in another (possibly the same) subinterval. If one distinguishes the two cases, i.e. the case where the two subintervals are identical and the case where they are different, one finds

$$\langle Y_k Y_{k'} \rangle = \langle Y_{k+k'} \rangle + \frac{1}{Z_T} \sum_{t_1=0}^{T} \sum_{t_2=0}^{T-t_1} \sum_{\tau_1=0}^{T-t_1-t_2} \sum_{\tau_2=0}^{T-t_1-t_2-\tau_1} Z_{\tau_1} Q_{t_1} Z_{\tau_2} Q_{t_2} Z_{T-t_1-t_2-\tau_1-\tau_2} \frac{t_1^k t_2^{k'} + t_1^{k'} t_2^k}{T^{k+k'}} \tag{59}$$

which again leads to (55) in the large T limit.

Remark: Expressions (54) and (55) correspond again to a special case of (29) and (33) for $n = -1/2$ and $\mu = 1/2$. Therefore this problem corresponds to broken replica symmetry for which $-1/2$ replica are grouped into -1 block of $1/2$ replica each.

CONCLUSION

This lecture was an attempt to show that very similar non-self-averaging effects occur in a variety of systems. We saw in particular that the prediction of the Parisi replica theory for spin glasses (31), (34) gives exactly the same answer as a much simpler problem, that of sums of random variables (11), (12). Moreover, if one allows the number n of replica to be non-zero, the replica calculation (29), (33) gives the same expressions as for the random map models (40), (42) for $n = -1/2, \mu = 0$ and (48), (49) for $n = -1, \mu = 0$ or as the random walk problem (54), (55) for $n = -1, \mu = 1/2$. Of course it would be very interesting to find other examples of systems with non-self averaging effects which would correspond to the expressions obtained by the replica method (29), (33) for other values of n and μ. This would mean that the spin glass problem, at least in its mean field version, belongs to a larger class of problems [21], and it would then be very interesting to develop a more general theory than the presently existing replica approach to treat these more general problems.

The expressions obtained by the replica approach for a number of replica $n \neq 0$ has previously been considered in the literature. First, non-integer values of n were considered to try to better understand the $n \to 0$ limit in the spin -glass problem [22, 17]. More recently, non-zero values of n became important both as a mathematical tool to calculate the exact free energy of some disordered models [23], and also for physical reasons to describe situations where the quenched variables (the interactions $\{J_{ij}\}$ in the spin glass model) are no longer quenched but they are allowed to evolve according to slow dynamics [24].

As a final remark, I should say that the examples presented in this lecture could leave the impression that all problems with non-self-averaging effects are always special cases of (29) and (33) for some value of n and μ. This in fact is not true as one can easily build examples [25] for which the expressions of the moments of the Y_k do not correspond to any value of n and μ.

Acknowledgments

What is presented in this lecture originated from useful discussions with J.P. Bouchaud, M. Evans, M. Mézard, G. Parisi and N. Sourlas, as well as from very pleasant collaborations with D. Bessis, H. Flyvbjerg, E. Gardner and G. Toulouse.

While I was writing this text, the sad news of the sudden death of Claude Kipnis reached me. I have benefited a lot over the last few years from his kind advice and his encouragement. I would like to dedicate this work to him.

REFERENCES

1. M. Mézard, G. Parisi, N. Sourlas, G. Toulouse and M. Virasoro, Replica symmetry breaking and the nature of the spin glass phase, *J. Physique*, **45**, 843 (1984).
2. M. Mézard, G. Parisi and M. Virasoro, "Spin Glass Theory and Beyond", World Scientific (1987).
3. J.P. Bouchaud and A. Georges, Anomalous diffusion in disordered media: Statistical mechanisms, models and physical applications, *Phys. Rep.*, **195**, 127 (1990).
4. G. Parisi, The order parameter for spin glasses: A function on the interval 0-1, *J. Phys. A*, **48**, 1101(1980).
5. D. Sherrington and S. Kirkpatrick, Solvable model of a spin glass, *Phys. Rev. Lett.*, **35**, 1792 (1975).
6. S. Kirkpatrick and D. Sherrington, Infinite ranged models of spin glasses, *Phys. Rev.*, **B17**, 4385 (1978).
7. B. Derrida, Random energy model, an exactly solvable model of disordered systems, *Phys. Rev.*, **B24**, 2613 (1981).
8. B. Derrida and E. Gardner, Magnetic properties and the function $q(x)$ of the generalised random energy model, *J. Phys. C*, **19**, 5783 (1986).
9. B. Derrida, Directed polymers in a random medium, *Physica*, **163**, 71 (1990).
10. J.L. van Hemmen and R.G. Palmer, The replica method and a solvable spin glass model, *J. Phys. A*, **12**, 563 (1979).
11. D. Sherrington, Ising replica magnet, *J. Phys. A*, **13**, 637 (1980).
12. A. Blandin, Theories versus experiments in the spin glass systems, *J. Physique*, **C6**, 1499 (1978).
13. J.R.L. de Almeida and D.J. Thouless, Stability of the Sherrington-Kirkpatrick solution of a spin glass model, *J. Phys. A*, **11**, 983 (1978).
14. G. Parisi, Order parameter for spin glasses, *Phys. Rev. Lett.*, **50**, 1946 (1983).
15. B. Derrida and G. Toulouse, Sample to sample fluctuations in the random energy model, *J. Physique Lett.*, **46**, L223 (1985).
16. M. Mézard, G. Parisi and M.A. Virasoro, Random free energies in spin glasses, *J. Physique Lett.*, **46**, L217 (1985).
17. E. Gardner and B. Derrida, The probability distribution of the partition function of the random energy model, *J. Phys. A*, **22**, 1975 (1989).
18. S.A. Kauffman, "The Origin of Order", Oxford University Press (1993).
19. B. Derrida and H. Flyvbjerg, Multivalley structure in Kauffman's model: Analogy with spin glasses, *J. Phys. A*, **19**, L1003 (1986).
20. B. Derrida and H. Flyvbjerg, The random map model, a disordered model with deterministic dynamics, *J Physique*, **48**, 971 (1987).

21. B. Derrida and H. Flyvbjerg, Statistical properties of randomly broken objects and of multivalley structures in disordered systems, *J. Phys. A*, **20**, 5273 (1987).
22. I. Kondor, Parisi's mean field solution for spin glasses as an analytic continuation in the replica number, *J. Phys. A*, **16**, L127 (1993).
23. E. Buffet, A. Patrick and J.V. Pulé, Directed polymers on a tree: A martingale approach, *J. Phys. A*, **26**, 1823 (1983).
24. R.W. Penney, A.C. Coolen and D. Sherrington, Coupled dynamics of fast spins and slow interactions in neural networks and spin systems, *J. Phys. A*, **26**, 3681 (1993).
25. B. Derrida and D. Bessis, Statistical properties of valleys in the annealed random map model, *J. Phys. A*, **21**, L509 (1988).

QUANTUM ADIABATIC EVOLUTION

Alain Joye[1] and Charles-Edouard Pfister[2]

1. Centre de Physique Théorique
   CNRS, Luminy Case 907
   F-13288 Marseille France

2. Ecole Polytechnique Fédérale de Lausanne
   Département de Mathématiques
   CH-1015 Lausanne Switzerland

ADIABATIC THEOREM OF QUANTUM MECHANICS

The notion of adiabatic evolution or adiabatic process is an important theoretical concept, which occurs at several places in Physics. The main feature of this concept is that although the process is very slow, global changes can take place without local changes. Adiabaticity is at the border between dynamics and statics. This concept was introduced by Boltzmann in Classical Mechanics through the notion of adiabatic invariants. In Thermodynamics adiabatic processes play an important role. In Quantum Mechanics, if the state of the system is an eigenfunction $\psi(t_0)$ for the eigenvalue $e(t_0)$ at $t = t_0$, then in the adiabatic limit the state of the system at time $t = t_1$ is an eigenfunction $\psi(t_1)$ for the eigenvalue $e(t_1)$, provided the energy-level $e(t)$ remains isolated during the time-interval $[t_0, t_1]$. Even if $H(t_0) = H(t_1)$, the eigenfunction $\psi(t_1)$ is generally different from $\psi(t_0)$ by a phase which can be decomposed into a dynamical phase related to the energy-level $e(t)$ and a geometric phase related to the spectral subspaces visited during the adiabatic process. This is the fundamental observation of Berry[1] which gave rise to extensive developments during the last ten years[2,3,4].

The adiabatic theorem in Quantum Mechanics has been proven very early by Born and Fock[5]. It was refined at several occasions. We mention in particular the papers of Kato[6] and Garrido[7] which brought new ideas into the subject. In this short paper we describe recent mathematical results obtained during the last four years, mainly in the case of analytic time-dependent quantum systems, where few rigorous results were known before. Let $H(t)$ be the Hamiltonian of a time-dependent system. We suppose that $H(t)$ is a self-adjoint operator, uniformly bounded from below, defined on a dense domain D, independent of t, of a Hilbert space $\mathcal{H}$. We suppose that $H(t)$ is at least of class C^2. The time evolution of the system is governed by the evolution operator $U(s,s')$, solution of the Schrödinger equation ($\hbar = 1$)

$$i\frac{d}{ds}U(s,s') = H(\varepsilon s)U(s,s'), \quad U(s',s') = I\!I \tag{1}$$

where $1/\varepsilon$ is the typical time-scale over which the Hamiltonian changes significantly. This equation is conveniently studied using a rescaled time, $t = \varepsilon s$, which leads to

$$i\varepsilon\frac{d}{dt}U(t,t') = H(t)U(t,t'), \quad U(t',t') = I\!I \tag{2}$$

where, with an abuse of notations, we have again denoted the evolution by U. It should be stressed, although we do not write it explicitly, that $U(t,t')$ is a function of ε. The adiabatic limit corresponds to the limit ε tending to zero which is a singular limit. The adiabatic theorem states that there exists an approximate solution $V(t,t')$ of equation (2), called adiabatic evolution, such that

$$\sup_{t\in[t_0,t_1]} \|U(t,t_0) - V(t,t_0)\| = \mathcal{O}(\varepsilon) \tag{3}$$

The main assumptions needed to prove this result are the smoothness of the Hamiltonian and the

Gap Condition: *there exists $g > 0$ so that for each t the spectrum $\sigma(t)$ of $H(t)$ can be separated into two closed subsets $\sigma_1(t)$, $\sigma_2(t)$ and*

$$\text{dist}(\sigma_1(t), \sigma_2(t)) \geq g$$

By convention $\sigma_1(t)$ is always a bounded subset of $\sigma(t)$. We denote the spectral projector associated with $\sigma_1(t)$ by $Q(t)$. The main feature of $V(t,t')$ is that it is *compatible with the decomposition, or follows the decomposition,* of the Hilbert space $\mathcal{H}$ into

$$\mathcal{H} = Q(t)\mathcal{H} \oplus (I\!I - Q(t))\mathcal{H} \tag{4}$$

which is expressed by the intertwining property

$$V(t,t')Q(t') = Q(t)V(t,t') \tag{5}$$

Let us consider now the consequences of (3) and (5). Let ψ_0 be an initial wave-function at $t = t_0$, such that $Q(t_0)\psi_0 = \psi_0$. At time $t > t_0$ the wave-function is $\psi(t) = U(t,t_0)\psi_0$, and $\phi(t) = V(t,t_0)\psi_0$ is an approximate solution. Because of property (5) $Q(t)\phi(t) = \phi(t)$, i.e. $\phi(t)$ belongs to the spectral subspace $Q(t)\mathcal{H}$ of $H(t)$. This information, combined with (3), implies that the probability that the system makes a transition from the subspace $Q(t_0)\mathcal{H}$ to the subspace $(Q(t)\mathcal{H})^\perp$ is of order ε^2. In the special case $\dim Q(t) \equiv 1$, ψ_0 is an eigenfunction of $H(t_0)$ for the eigenvalue $e(t_0) = \text{tr}Q(t_0)H(t_0)$ and $\phi(t)$ an eigenfunction of $H(t)$ for the eigenvalue $e(t) = \text{tr}Q(t)H(t)$. We can give a more explicit expression for $\phi(t)$,

$$\phi(t) = V(t,t_0)\psi_0 = e^{-i/\varepsilon \int_{t_0}^t e(s)ds}\varphi(t) \tag{6}$$

where $\varphi(t)$ is the unique eigenfunction of eigenvalue $e(t)$ which is determined by the conditions $\langle \frac{d}{ds}\varphi(s)|\varphi(s)\rangle = 0$, $t_0 \leq s \leq t$, and $\varphi(t_0) = \psi_0$.

The adiabatic evolution $V(t,t')$ is solution of the equation

$$i\varepsilon\frac{d}{dt}V(t,t') = \left(H(t) + i\varepsilon\left[\frac{d}{dt}Q(t), Q(t)\right]\right)V(t,t'), \quad V(t',t') = \mathit{I\!I} \tag{7}$$

The result (3) can be improved when $H(t)$ is of class C^{q+2}, $q > 0$. For $j = 1, 2, \ldots, q$ we can construct decompositions of the Hilbert space $\mathcal{H}$ into

$$\mathcal{H} = Q_j(t)\mathcal{H} \oplus (\mathit{I\!I} - Q_j(t))\mathcal{H} \tag{8}$$

and evolutions compatible with these decompositions, which are better approximate solutions of (2). Before doing this let us recall the main result about smooth evolutions (not necessarily unitary), which are compatible with a given decomposition of the Hilbert space (4) [8,9]. There is a particular evolution $W(t,t')$ satisfying (5), which is a solution of

$$i\frac{d}{dt}W(t,t') = i\left[\frac{d}{dt}Q(t), Q(t)\right]W(t,t'), \quad W(t',t') = \mathit{I\!I} \tag{9}$$

This evolution has a geometric interpretation given below, and it depends only on $Q(t)$. All other smooth evolutions $\widehat{V}$ satisfying the intertwining property (5) are solutions of equations

$$i\frac{d}{dt}\widehat{V}(t,t') = \left(B(t) + i\left[\frac{d}{dt}Q(t), Q(t)\right]\right)\widehat{V}(t,t'), \quad \widehat{V}(t',t') = \mathit{I\!I} \tag{10}$$

where $B(t)$ commutes with $Q(t)$. The structure of equation (7) is now clear, and the best choice [10] of $B(t)$ with respect to (3) is $B(t) = \varepsilon^{-1}H(t)$. It is not difficult to show that the eigenfunction $\varphi(t)$ in (6) is simply $W(t,t')\psi_0$.

We come to the geometric interpretation of W, and give at the same time an illustration of the adiabatic theorem. Let $H(m)$ be a family of Hamiltonians $H(m)$, satisfying the gap condition for each value of the parameter m belonging to some manifold M, and let $\sigma_1(m)$ consist of a single eigenvalue $e(m)$ with $\dim Q(m) \equiv p$. We also suppose that the family of all subspaces $Q(m)\mathcal{H}$, $m \in M$, forms a vector bundle F of base M, whose points are all (m, ϕ) with $m \in M$ and $\phi \in Q(m)\mathcal{H}$, and whose fiber above m' is the set of all points (m', ϕ), $\phi \in Q(m')\mathcal{H}$. The wave-function ψ_0 is an eigenfunction of $H(m_0)$ for the eigenvalue $e(m_0)$. We decide to change the value of the parameter m_0 into m_1 in an adiabatic way, and ask what is the wave-function ψ_1 of the system after this change. To do that we choose a path γ in M, parametrized by $t \in [0, 1]$, with $\gamma(0) = m_0$, and $\gamma(1) = m_1$. The speed of the change of the parameter m is controlled by the parameter $\varepsilon > 0$. Using the *rescaled time* $\psi_1 = \psi(1)$ is the solution of

$$i\varepsilon\frac{d}{ds}\psi(s) = H(\gamma(s))\psi(s), \quad \psi(0) = \psi_0 \tag{11}$$

The adiabatic approximation of ψ_1 is $V(1,0)\psi_0$ and has an expression like (6),

$$V(1,0)\psi_0 = e^{-i/\varepsilon \int_0^1 e(\gamma(s))ds} W(1,0)\psi_0 \tag{12}$$

The phase $\varepsilon^{-1}\int_0^1 e(\gamma(s))ds$ is the *dynamical phase*. It depends on the parametrization of the path γ, and it records how long is the process for changing the value of the parameter. In this approximation $W(s,0)\psi_0$ is an eigenfunction of $H(\gamma(s))$ for all $s \in [0,1]$. The operator $W(1,0)$ depends only on the path γ, and not on its parametrization. Since the bundle F is constructed with the $Q(m)$, we automatically get a natural connection, because we have a natural decomposition of the tangent spaces into a vertical part and an horizontal part. In order to simplify the notations let us suppose that M is embedded in $\mathbb{R}^n$, so that F is embedded in $\mathbb{R}^n \times \mathcal{H}$. Let $f = (m, \phi) \in F$. Any tangent vector v_f at f can be viewed as a velocity vector of a curve $c(t) = (c_1(t), c_2(t))$ in $\mathbb{R}^n \times \mathcal{H}$ with $c(t) \in F$, and $c(0) = f$, i.e. $v_f = (\dot{c}_1(0), \dot{c}_2(0))_f$. The vertical vectors at f are velocity vectors of curves $c(t)$ with $c_1(t) \equiv m$. They are of the form $(0, \dot{c}_2(0))_f$ with $\dot{c}_2(0) = Q(m)\dot{c}_2(0)$. Using the projection $Q(m)$ we have a decomposition of v_f into a vertical vector $(0, Q(m)\dot{c}_2(0))_f$ and an *horizontal* vector $(\dot{c}_1(0), (I - Q(m))\dot{c}_2(0))_f$, and thus we have a connection. The operator $W(1,0)$ in (12) is *the holonomy operator describing the parallel transport corresponding to the above connection along the path* γ. To show this we must verify that the velocity vector to the curve $(\gamma(s), W(s,0)\psi_0)$, $s \in [0,1]$, is horizontal for any s and any $\psi_0 \in Q(m_0)$. This is immediate since

$$Q(\gamma(s))\left(\frac{d}{ds}Q(\gamma(s))\right)Q(\gamma(s)) = 0 \tag{13}$$

as a consequence of $Q(\gamma(s))^2 = Q(\gamma(s))$.

Let us now construct new decompositions (8) of $\mathcal{H}$. We define the projections Q_j as approximate solutions of the Heisenberg equation

$$i\varepsilon \frac{d}{dt}A(t) = [H(t), A(t)] \tag{14}$$

They are obtained by an iteration process [11],[12]. We set $Q_0 := Q$, $K_{-1} = 0$ and $K_j := i\left[\frac{d}{dt}Q_j(t), Q_j(t)\right]$, $j = 0, \ldots$. For each t let Γ be a simple closed path in the complex plane, counterclockwise oriented, and encircling $\sigma_1(t)$. We suppose that the distance from Γ to the spectrum $\sigma(t)$ is at least $g/2$. Clearly $Q_0(t)$ is an approximate solution of (14) up to an error term $\mathcal{O}(\varepsilon)$. Let us suppose that we have constructed $Q_k(t)$, $k = 0, \ldots, j-1$. Then $Q_j(t)$ is the spectral projection of the operator $H_j(t) := H(t) - \varepsilon K_{j-1}(t)$, defined by the Riesz formula

$$Q_j(t) = -\frac{1}{2\pi i}\oint_\Gamma (H_j(t) - \lambda)^{-1}d\lambda \tag{15}$$

Provided ε is small enough Q_j is well defined up to $j = q$ if H is C^{q+2}. Using regular perturbation theory it is easy to show that $Q_j(t)$ is an approximate solution of (14) up to an error term $\mathcal{O}(\varepsilon^{j+1})$. If it happens that all derivatives of the resolvent of $H(t_0)$ are zero up to order p, then $Q_j(t_0) = Q(t_0)$, $j \leq p$. We know how to find evolutions V_j compatible with (8) by solving equation (10). The best choice, if we want V_j to give the best approximate solution of (2), which is compatible with the decomposition (8) of $\mathcal{H}$, is

$$B(t) = \varepsilon^{-1}H_j(t) + Q_j(t)K_{j-1}(t)Q_j(t) + (I - Q_j(t))K_{j-1}(t)(I - Q_j(t)) \tag{16}$$

Notice that $V_0 = V$.

Adiabatic Theorem

A) *Let $H(t)$ be of class C^{q+2}, $q \geq 0$, and let the gap condition be satisfied for all $t \in [t_0, t_1]$. Then there exists $\varepsilon_j > 0$ so that for all $\varepsilon < \varepsilon_j$, the evolution V_j, $0 \leq j \leq q$ is defined in $[t_0, t_1]$, and*

$$\sup_{t \in [t_0,t_1]} \|U(t,t_0) - V_j(t,t_0)\| = \mathcal{O}\left(\varepsilon^{j+1}\right)$$

B) *Let $H(t)$ have an analytic continuation in the strip $S_\alpha = \{t + is \in \mathbb{C} : |s| < \alpha\}$. Let the gap condition be satisfied for all $t \in \mathbb{R}$, and let $c(t)$ be an integrable function, with $\lim_{t \to \pm\infty} c(t) = 0$, and H^+, H^- be two self-adjoint operators such that for all $\varphi \in D$*

$$\sup_{|s|<\alpha} \|(H(t+is) - H^\pm)\varphi\| \leq c(t)(\|\varphi\| + \|H^\pm \varphi\|), \quad \pm t \gg 1$$

Then there exist $\varepsilon^* > 0$ and a positive constant κ, such that for all $\varepsilon < \varepsilon^*$ and for all $t, t' \in \mathbb{R}$ we have a decomposition of $\mathcal{H}$ into $\mathcal{H} = Q_*(t)\mathcal{H} \oplus (\mathbb{I} - Q_*(t))\mathcal{H}$ and a compatible evolution $V_*(t,t')$ such that,

$$\|U(t,t') - V_*(t,t')\| = \mathcal{O}\left(e^{-\kappa/\varepsilon}\right)$$

In both cases

$$\|Q(t) - Q_j(t)\| = \mathcal{O}(\varepsilon), \quad \|Q(t) - Q_*(t)\| = \mathcal{O}(\varepsilon)$$

Moreover $\lim_{t \to \pm\infty} \|Q(t) - Q_*(t)\| = 0$.

Comments:

1) The proof [12] of the theorem is to give an explicit lower bound for ε_j or ε^* by proving the following estimate. By hypothesis we can find a, b and c such that for all $\lambda \in \Gamma$

$$\left\|\frac{d^p}{dt^p}(\lambda - H(t))^{-1}\right\| \leq ac^p \frac{p!}{(1+p^2)}, \quad \left\|\frac{d^p}{dt^p} K_0(t)\right\| \leq bc^p \frac{p!}{(1+p^2)} \tag{17}$$

with $p = 0, \ldots, q+1$ in case A), all p in case B), and in that case $b = b(t)$ is a bounded integrable function. Then there exist $\varepsilon(a, \|b\|_\infty)$, a constant $d = d(a, \|b\|_\infty)$ such that for all j, $j = 0, \ldots, q$ in case A) or $j \leq q^* = [1/ecd\varepsilon]$ in case B), all $\varepsilon \leq \varepsilon(a, \|b\|_\infty)$,

$$\|K_j(t) - K_{j-1}(t)\| \leq b\varepsilon^j d^j c^j \frac{j!}{(1+j^2)} \tag{18}$$

In case A), using this information we can repeat the proof of Theorem 3.1 in [14]. In case B) we set $Q_*(t) := Q_{q^*}(t)$ and choose as evolution V_* the solution of (10) with $B(t) = \varepsilon^{-1} H_{q^*}$. By writing the differential equation for $V_*^{-1}(t,t') U(t,t')$ the proof is immediate. Notice that we have an explicit simple upper bound for the error term in that case

$$\|U(t,t') - V_*(t,t')\| \leq \left|\int_{t'}^{t} b(s) ds\right| 2(\pi[1/ecd\varepsilon])^{1/2} \exp(-2[1/ecd\varepsilon]) \tag{19}$$

The condition on the asymptotic behaviour of $H(t)$ is not needed to establish (19), but is used in the application of remark 2) below. Of course, if it is not verified, then $b(t)$ is in general not integrable.

2) In case B) if $\psi(t)$ is a normalized solution of the Schrödinger equation such that $\lim_{t \to -\infty} \|(\mathbb{I} - Q(t))\psi(t)\| = 0$, then the transition probability to the spectral subspace $(\mathbb{I} - Q(+\infty))\mathcal{H}$ at time $t = +\infty$ is exponentially small

$$\lim_{t \to +\infty} \|(\mathbb{I} - Q(t))\psi(t)\|^2 = \mathcal{O}\left(e^{-2\kappa/\varepsilon}\right) \tag{20}$$

This result was first obtained in [13] in the case of matrices, and in the general case in [14]. The method of the proof is quite different than the above mentioned method. The evolution is studied along special complex paths in the complex time-plane. The method gives good estimations of κ in the case of matrices and even optimal values of κ in the case of 2×2 matrices. Recently Martinez [15] has rederived result (20) by a microlocal analysis. We also mention [16], however the results are weaker.

3) Nenciu [17] gave a closely related proof of B), which precedes the above sketched proof [12], and Sjöstrand [18] obtained similar results by a microlocal analysis.

4) The evolution V_* is called *superadiabatic* because of the exponential estimate.

TWO-LEVEL SYSTEMS

We consider here only the case of analytic systems with two energy levels $e_1(t)$ and $e_2(t)$, which have the same permanent degeneracy. As above $Q(t)$ is the spectral projection associated with $e_1(t)$. The paradigm model when $\dim Q(t) \equiv 1$ is a spin 1/2 in an external time-dependent magnetic field [13], and when $\dim Q(t) \equiv 2$ it is a spin 3/2 in a time-dependent quadrupole electric field [19]. We suppose that $\mathrm{tr} H(t) \equiv 0$, so that $H(t)^2 = \rho(t) \mathbb{I}$ with $\rho(t) \geq g^2/4$ in order that the gap condition be satisfied. We have

$$e_1(t) = -\sqrt{\rho(t)}, \quad \sqrt{1} = 1 \quad \text{and} \quad Q(t) = \frac{1}{2}\left(\mathbb{I} + \frac{H(t)}{e_1(t)}\right) \tag{21}$$

We suppose that the conditions of part B) of the adiabatic theorem are satisfied. For these simple systems the adiabatic evolution can be written explicitly as

$$\begin{aligned} V(t,t') &= \exp\left(-i/\varepsilon \int_{t'}^{t} e_1(s)ds\right) W(t,t')Q(t') \\ &+ \exp\left(-i/\varepsilon \int_{t'}^{t} e_2(s)ds\right) W(t,t')(\mathbb{I} - Q(t')) \end{aligned} \tag{22}$$

where $W(t,t')$ is given by (9).

Let $\chi_- \in Q(-\infty)\mathcal{H}$ and $\chi_+ \in (\mathbb{I} - Q(+\infty))\mathcal{H}$ be two normalized eigenvectors of H^- and H^+. The transition probability from χ_- to χ_+ is

$$\mathcal{P}(\varepsilon, \chi_-, \chi_+) = \lim_{\substack{t_0 \to -\infty \\ t_1 \to +\infty}} |\langle \chi_+ | U(t_1, t_0) \chi_- \rangle|^2 \tag{23}$$

In order to derive an asymptotic formula for $\mathcal{P}(\varepsilon, \chi_-, \chi_+)$, we observe that the evolution U is analytic in the strip S_α, whereas V is in general multivalued and has singularities at the zeros of ρ in S_α. Indeed, let us assume that there exists a complex eigenvalue crossing at z_j (and by symmetry at $\overline{z_j}$), i.e. $\rho(z_j) = 0$. The eigenvalues and eigenprojections have a singularity at z_j and $\overline{z_j}$ (see (21)). We further suppose that z_j is a zero of order one for $\rho(z)$ with $\mathrm{Im} z_j > 0$. We can continue analytically $e_i(t)$, $i = 1, 2$, and $Q(t)$ in the punctured strip S_α, which is defined by removing the zeros of ρ from S_α, but the analytic continuation is not unique. We fix a point t of the real axis and make the analytic continuation of e_1 and Q along a clockwise oriented simple loop γ_j based at t, encircling z_j and no other zero of ρ. After analytic continuation we obtain the values $e_1(t|\gamma_j) = e_2(t)$, and $Q(t|\gamma_j) = (\mathbb{I} - Q(t))$. A simple computation shows that

$$i\left[\frac{d}{dt}Q(t), Q(t)\right] = \frac{i}{4\rho(t)}\left[\frac{d}{dt}H(t), H(t)\right] \tag{24}$$

and therefore the generator of W is a single-valued analytic function in the punctured strip. W itself is multi-valued and $W(t|\gamma_j)$, the analytic continuation of $W(t,t)$ along γ_j, is such that

$$W(t|\gamma_j)Q(t) = (I\!\!I - Q(t))W(t|\gamma_j) \tag{25}$$

The operator $W(t|\gamma_j)$, which describes the transport of vectors along γ_j, is not unitary anymore since γ_j is a path in the complex plane and the projection $Q(z)$ is not self-adjoint when $z \notin I\!\!R$. If we start with a vector in $Q(t)$, then, at the end of the loop γ_j, we get a vector in the orthogonal subspace. When the eigenvalues are not degenerate, we can thus define a *complex* phase $\theta_j(t)$ by the relation

$$W(t|\gamma_j)\phi_1(t) = \exp(-i\theta_j(t))\phi_2(t) \tag{26}$$

where $\phi_i(t)$ is a (normalized) eigenvector of $H(t)$ for the eigenvalue $e_i(t)$, $i = 1, 2$. The important point is that $W(t|\gamma_j)$ and $\theta_j(t)$ depend only on the homotopy class of the closed path γ_j based at t in the punctured strip. Moreover, $\text{Im}\theta_j(t)$ is independent of the base point t of γ_j since t is *real*.

It is natural to write the evolution $U(t,0)$ as $U(t,0) := V(t,0)A(t)$. Using the fact that the evolution U is analytic in the strip, i.e. path independent, whereas V is path dependent, we can derive a useful identity for A by considering the Schrödinger equation along two different paths, the real axis and a path η_j in the complex plane going above z_j [19]. This identity allows us to express the transition probability (23) in terms of the analytic continuations of the eigenvalues and eigenprojections along the path γ_j and of A along the path η_j [19]. We can perform the asymptotic analysis of the transition probability (23) as $\varepsilon \to 0$ whenever we have a good control of the analytic continuation of A along η_j. At this step it is necessary to impose further conditions on the behaviour of the multi-valued function $\Phi(z) := -2\int_t^z \sqrt{\rho(z')}dz'$. These conditions specify the global behaviour in the strip of the so-called Stokes lines, which are the level-lines $\text{Im}\Phi(\cdot) = \text{Im}\Phi(z_j)$, where z_j is a zero of ρ [13,19]. The results of the analysis are given in Table 1 when the above mentioned conditions are satisfied for one (and then necessarily unique) simple zero z_j of ρ. This case is generic. In the table we have dropped the index j.

Remarks:

We can express the decay rate as

$$2\,\text{Im}\int_{\gamma_j} e_1(z)dz = 2\text{Im}\int_t^{z_j}(e_1(z) - e_2(z))dz = 2\,\text{Im}\Phi(z_j) < 0 \tag{27}$$

with the integration path from t to z inside γ_j. There is a geometric interpretation of the Stokes lines as particular geodesics in a metric naturally associated with the quadratic differential $\rho(z)d^2z$ [13]. In this metric $|\text{Im}\Phi(z_j)|$ is the distance of z_j to the real axis, and always in this metric it is the nearest complex-eigenvalue crossing in the upper half-plane to the real axis. Except for the real case and the Landau-Zener formula, we have a nontrivial prefactor of geometric nature which is

$$0 \leq |\langle \chi_+ | W(+\infty, t) W(t|\gamma) W(t, -\infty) \chi_- \rangle|^2 \leq \|W(t|\gamma)\|^2 \tag{28}$$

and which reduces to $\exp(2\text{Im}\theta)$ when $\dim Q(t)\mathcal{H} = 1$. The geometric nature of the prefactor is evident in (28): we transport the vector χ_- along the real axis from $-\infty$ to t, then along the path γ around the complex-eigenvalue crossing, and finally again along the real axis up to $+\infty$. We get a vector in $(I\!\!I - Q(+\infty))\mathcal{H}$ and we compare this transported vector with χ_+. The geometric prefactor was measured recently in a spin experiment[22]. The case of the avoided crossing leading to the Landau-Zener formula is important in Physics. The last row of the Table 1 yields an asymptotic formula accurate up to exponentially small *relative* errors. The expressions e_1^* and θ^* are defined as e_1 and θ with H_{q^*} in place of H (see comment 1 on the adiabatic theorem).

Table 1. Asymptotic transition probability in the generic case.

Characteristics of $H(t)$	Transition probability $\mathcal{P}(\varepsilon,\chi_-,\chi_+)$
General formula[19] Complex hermitian $\dim Q(t)\mathcal{H} > 1$	$\left(\lvert\langle\chi_+\rvert W(+\infty,t)W(t\lvert\gamma)W(t,-\infty)\chi_-\rangle\rvert^2 + \mathcal{O}(\varepsilon)\right) \times \exp\left(\frac{2}{\varepsilon}\operatorname{Im}\int_\gamma e_1(z)dz\right), \quad \varepsilon \to 0$
Dykhne formula[20] Real symmetric $\dim Q(t)\mathcal{H} = 1$	$(1+\mathcal{O}(\varepsilon))\exp\left(\frac{2}{\varepsilon}\operatorname{Im}\int_\gamma e_1(z)dz\right), \quad \varepsilon \to 0$
Geometrical prefactor[21,13,22] Complex hermitian $\dim Q(t)\mathcal{H} = 1$	$(1+\mathcal{O}(\varepsilon))\exp\left(\frac{2}{\varepsilon}\operatorname{Im}\int_\gamma e_1(z)dz + 2\operatorname{Im}\theta\right), \quad \varepsilon \to 0$
Landau-Zener formula[23,24] Complex hermitian $\dim Q(t)\mathcal{H} = 1$ Avoided crossing: $e_2(t) - e_1(t) \overset{\delta\to 0}{\simeq} \sqrt{a^2t^2 + 2ct\delta + b^2\delta^2}$	$(1+\mathcal{O}(\delta)+\mathcal{O}(\varepsilon))\exp\left(-\frac{\pi\delta^2}{2\varepsilon}\left(\frac{b^2}{a} - \frac{c^2}{a^3}\right)(1+\mathcal{O}(\delta))\right), \quad \varepsilon,\delta \to 0$
Accurate formula[25] Complex hermitian $\dim Q(t)\mathcal{H} = 1$	$\left(1+\mathcal{O}\left(e^{-\kappa/\varepsilon}\right)\right)\exp\left(\frac{2}{\varepsilon}\operatorname{Im}\int_\gamma e_1^*(z,\varepsilon)dz + 2\operatorname{Im}\theta^*(\varepsilon)\right), \quad \kappa > 0,\ \varepsilon \to 0$

In Table 2 we have given the results in the non generic case, when there is an eigenvalue crossing, which is a zero of $\rho(z)$ of higher order and/or when several eigenvalue crossings (say q) determine the asymptotics of the transition probability. These cases give rise to different prefactors and/or interference phenomena in the transition probability. The above outlined method must be slightly modified. The conditions on the Stokes lines, which are needed to prove the results below are given in the quoted references.

Table 2. Asymptotic transition probability in non generic cases.

Characteristics of $H(t)$	Transition probability $\mathcal{P}(\varepsilon,\chi_-,\chi_+)$
High order degeneracy[26],[27] Real symmetric $\dim Q(t)\mathcal{H}=1$ $\rho(z) \simeq (z-z_1)^n$ $H(z) \simeq (z-z_1)^m,\ 2m \leq n$	$\left(4\sin^2\left(\frac{\pi(n-2m)}{2(n+2)}\right) + \mathcal{O}(\varepsilon^p)\right) \times$ $\exp\left(\frac{2}{\varepsilon}\mathrm{Im}\int_t^{z_1}(e_1(z)-e_2(z))dz\right),\ p>0,\ \varepsilon\to 0$
Stückelberg oscillations[28],[25] Complex hermitian $\dim Q(t)\mathcal{H}=1$ q simple zeros z_j of $\rho(z)$ contribute	$\left\|\sum_{j=1}^q \exp\left(-i\theta_j - \frac{i}{\varepsilon}\int_{\gamma_j}e_1(z)dz\right)\right\|^2 +$ $\mathcal{O}(\varepsilon)\exp\left(-\frac{2}{\varepsilon}\|\mathrm{Im}\int_{\gamma_1}e_1(z)dz\|\right),\ \varepsilon\to 0$
Conjugated effects[26] Real symmetric $\dim Q(t)\mathcal{H}=1$ q zeros z_j characterized by (n_j,m_j) as above contribute	$\left\|\sum_{j=1}^q A_j(n_j,m_j)\exp\left(-\frac{i}{\varepsilon}\int_t^{z_j}(e_1(z)-e_2(z))dz\right)\right\|^2 +$ $\mathcal{O}(\varepsilon^p)\exp\left(-\frac{2}{\varepsilon}\|\mathrm{Im}\int_t^{z_1}(e_1(z)-e_2(z))dz\|\right),$ $A_j(n_j,m_j) = 2\sigma_j\sin\left(\frac{\pi(n_j-2m_j)}{2(n_j+2)}\right),\ \sigma_j=\pm 1,\ \varepsilon\to 0$
Accurate formula[25] Complex hermitian $\dim Q(t)\mathcal{H}=1$ q simple zeros z_j of $\rho(z)$ contribute	$\left\|\sum_{j=1}^q \exp\left(-i\theta_j^*(\varepsilon) - \frac{i}{\varepsilon}\int_{\gamma_j}e_1^*(z,\varepsilon)dz\right)\right\|^2 +$ $\mathcal{O}\left(e^{-\kappa/\varepsilon}\right)\exp\left(-\frac{2}{\varepsilon}\|\mathrm{Im}\int_{\gamma_1}e_1(z)dz\|\right),\ \kappa>0,\ \varepsilon\to 0$

Remarks:

For the precise definition of σ_j, as well as the paths of integration in the first and third line see [26]. Notice also that in the second and third lines

$$\mathrm{Im}\int_{\gamma_j}e_1(z)dz < 0\ ,\ \mathrm{Im}\int_t^{z_j}(e_1(z)-e_2(z))dz < 0 \qquad (29)$$

are independent of $j = 1,\ldots,q$. In the last line the energy e_1^*, or the phases θ_j^* are defined using H_{q^*}, and we have

$$\mathrm{Im}\int_{\gamma_j}e_1^*(z,\varepsilon)dz = \mathrm{Im}\int_{\gamma_j}e_1(z)dz + \mathcal{O}(\varepsilon^2) \qquad (30)$$

In the discussion the eigenvalue-crossings in the upper half-plane only play a role, because we consider the transition from the lower energy-level to the higher energy-level.

Finally we mention that these methods, the iteration scheme and the analysis in the complex plane, can be applied to study similar problems, like the semi-classical limit of the above barrier reflection coefficient for the stationary Schrödinger equation, or the adiabatic invariant of a time-dependent classical oscillator [25].

REFERENCES

1. M.V. Berry, "Quantal phase factors accompanying adiabatic changes," *Proc.Roy.Soc.Lond.A* **392**: 45 (1984).
2. A. Shapere and F. Wilczek, "Geometric Phases in Physics", World Scientific, Singapore, New Jersey, London, Hong Kong (1989).
3. J.W. Zwanziger, M. Koenig and A. Pines, Berry's phase, *Ann.Rev.Phys.Chem.*, **41**, 601 (1990).
4. C.A. Mead, The geometric phase in molecular systems, *Rev.Mod.Phys.*, **64**, 51 (1992).
5. M. Born and V. Fock, Beweis des Adiabatensatzes, *Zeit.f.Phys.*, **51**, 165 (1928).
6. T. Kato, On the adiabatic theorem of quantum mechanics, *J. Phys. Soc. Japan*,**5**, 435 (1950).
7. L.M. Garrido, Generalized adiabatic invariance, *J. Math. Phys.*, **5**, 335 (1964).
8. T. Kato, "Perturbation theory for linear operators", Springer Berlin, Heidelberg, New-York (1966).
9. S.G. Krein, "Linear differential equations in Banach spaces", Providence, R.I.: Transl. Math.Mon. 29 (1971).
10. J.E. Avron, R. Seiler and L.G. Yaffe, Adiabatic theorems and applications to the quantum Hall effect, *Comm. Math. Phys.*, **110**, 33 (1987).
11. A. Joye and C.-E. Pfister, Full asymptotic expansion of transition probabilities in the adiabatic limit, *J. Phys. A*, **24**, 753 (1991).
12. A. Joye and C.-E. Pfister, Superadiabatic evolution and adiabatic transition probability between two non-degenerate levels isolated in the spectrum, *J. Math. Phys.*, **34**, 454 (1993).
13. A. Joye, H. Kunz and C.-E. Pfister, Exponential decay and geometric aspect of transition probabilities in the adiabatic limit, *Ann. Phys.*, **208**, 299 (1991).
14. A. Joye and C.-E. Pfister, Exponentially small adiabatic invariant for the Schrödinger equation, *Comm. Math. Phys.*, **140**, 15 (1991).
15. A. Martinez, Precise exponential estimates in adiabatic theory, Preprint (1993).
16. V. Jaksic and J. Segert, Exponential approach to the adiabatic limit and the Landau-Zener formula, *Rev. Math. Phys.*, **4**, 529 (1992).
17. G. Nenciu, Linear adiabatic theory. Exponential estimates, *Comm. Math. Phys.*, **152**, 479 (1993).
18. J. Sjöstrand, Remarque sur des projecteurs adiabatiques du point de vue pseudodifférentiel Preprint (1993).
19. A. Joye and C.-E. Pfister, Non-abelian geometric effect in quantum adiabatic transitions, *Phys. Rev. A* to appear (1993).
20. J.-T. Hwang and P. Pechukas, The adiabatic theorem in the complex plane and the semi-classical calculation of non-adiabatic transition amplitudes, *J. Chem. Phys.*, **67**, 4640 (1977).
21. M.V. Berry, Geometric amplitude factors in adiabatic quantum transitions, *Proc. Roy. Soc.London A*, **430**, 405 (1990).
22. J.W. Zwanziger, S.P. Rucker and G.C. Chingas, Measuring the geometric component of the transition probability in a two-level system, *Phys. Rev. A*, **43**, 323 (1991).
23. A. Joye and C.-E. Pfister, Absence of geometrical correction to the Landau-Zener formula, *Phys. Lett. A* , **169**, 62 (1992).
24. A. Joye, Proof of the Landau-Zener formula, *Asymptotic Analysis* to appear (1993).
25. A. Joye and C.-E. Pfister, Semi-classical asymptotics beyond all orders for simple scattering systems, Preprint (1993).
26. A. Joye, Non-trivial prefactors in adiabatic transition probabilities induced by high order complex degeneracies *J. Phys. A* to appear (1993).
27. M.V. Berry and R. Lim, Universal transition prefactors derived by superadiabatic renormalization, Preprint (1993).
28. A. Joye, G. Mileti and C.-E. Pfister, Interferences in adiabatic transition probabilities mediated by Stokes lines, *Phys. Rev. A* **44**, 4280 (1991).

SEMI–CLASSICAL INELASTIC S–MATRIX FOR ONE–DIMENSIONAL N–STATES SYSTEMS

Ph. A. Martin[1] and G. Nenciu[2]

1. Institut de Physique Théorique, Ecole Polytechnique Fédérale de Lausanne PHB–Ecublens, CH–1015 Lausanne, Switzerland

2. Department of Theoretical Physics, University of Bucharest, Bucharest, Romania

INTRODUCTION

The quasi classical theory of inelastic atomic and molecular collisions has a long history going back at least to the seminal paper by Stückelberg [1], and a large body of the literature is devoted to the subject (see e.g.[2]). The calculation of inelastic scattering S–matrix elements (at least in the spherical symmetric case) proceeds usually in two steps: i) The reduction of the second order coupled channel equations to a first order "common trajectory" model by means of semi–classical treatments. ii) The approximate calculation of the inelastic probability transitions within the common trajectory model. A case of particular interest is the crossing or almost crossing of electronic Born–Oppenheimer surfaces where a Stückelberg–Landau–Zener type formula is obtained.

A discussion of the range of validity of the derivations involved in steps i) and ii) has been presented in several papers (see e.g.[2–6]). One of the main points is that, while the derivation of the common trajectory model requires a high energy assumption in order to assure that the differences between the momenta of various levels are small compared with the average momentum, the Landau–Zener formula is expected to be valid at moderate energies in order to assure that the transitions take place mainly at the avoided crossings, so that the assumptions in steps i) and ii) are somewhat conflicting. Due to the complexity of the problem (in particular the application of semiclassical methods to n–dimensional differential system), it is often difficult to turn the arguments involved in i) and ii) into real proofs. The aim of the letter is to report on a fully controlled theory in a simple setting. We treat a one–dimensional scattering system at energies such there are no classically forbidden regions. Moreover, the electronic levels have only one (almost) crossing over the whole range of the nuclear coordinate. Complete proofs will be given elsewhere.

We follow essentially the route i) and ii). Using recently developed tools in the adiabatic reduction theory [7,8] we first reduce the second order n-channel system to a first order n-level effective evolution problem up to exponentially small errors in the adiabatic parameter $\epsilon = \hbar/\sqrt{2M}$ (M is the nuclear mass). This effective evolution is governed by an effective Hamiltonian $H^{eff}(\epsilon)$, whose expansion in ϵ is explicitely provided by the adiabatic theory. At lowest order ($\epsilon = 0$) and sufficiently high energy, one recovers the standard common trajectory model. Next, for a two level system having a single avoided crossing with gap δ, we establish directly from the effective evolution a Dykne type formula valid for general (but small) ϵ and δ. Then, by perturbation theory, we calculate the inelastic probability transition to the precision requested to describe the cross–over regime between avoided crossing (Landau–Zener formula) and real crossing. This derivation does not require the high energy assumption needed for the common trajectory model.

REDUCTION TO AN EFFECTIVE EVOLUTION PROBLEM

We start with the generalized eigenfunctions equation at fixed energy E:

$$-\epsilon^2 \frac{d^2}{dx^2}\psi(x) + v(x)\psi(x) = E\psi(x), \quad \epsilon^2 = \frac{\hbar^2}{2M}, \tag{1}$$

where, for $x \in \mathbb{R}$, $\psi(x)$ is a n–component vector and $v(x)$ is a $n \times n$ hermitian matrix. We assume that $v(x)$ has an analytic continuation into a strip $\{x+iy, |y| < a\}$ and that $v(x+iy)$ tends to an hermitian limit v_0 as $x \to \pm\infty$ sufficiently fast. (For simplicity we assume that the limits at $\pm\infty$ coincide.) Moreover the energy E is fixed in such a way that $w(x) = E - v(x)$ is a strictly positive hermitian matrix for all $x \in \mathbb{R}$, i.e. if $v_j(x)$ and $|n_j(x)>$ are the eigenvalues and the eigenvectors of $v(x)$ respectively then

$$w(x) = E - v(x) = \sum_{j=1}^{n} e_j(x) |n_j(x)><n_j(x)|, \tag{2}$$

with $e_j(x) = E - v_j(x) \geq \Delta$ for some $\Delta > 0$.

The solutions of (1) have the asymptotic form

$$\psi(x) \approx \frac{1}{\sqrt{2}}(w_0)^{-1/4}\left(\exp[\frac{i}{\epsilon}(w_0)^{1/2}x]\, a^{in} + \exp[-\frac{i}{\epsilon}(w_0)^{1/2}x]\, a^{out}\right), \quad x \to -\infty$$

$$\psi(x) \approx \frac{1}{\sqrt{2}}(w_0)^{-1/4}\left(\exp[\frac{i}{\epsilon}(w_0)^{1/2}x]\, b^{out} + \exp[-\frac{i}{\epsilon}(w_0)^{1/2}x]\, b^{in}\right), \quad x \to \infty. \tag{3}$$

In (3) $w_0 = E - v_0$ and the powers $(w(x))^q$ of $w(x)$ are defined from (2) by

$$(w(x))^q = \sum_{j=1}^{n}(e_j(x))^q |n_j(x)><n_j(x)|, \quad (e_j(x))^q > 0. \tag{4}$$

The mapping of ingoing to outgoing data is given by the $2n \times 2n$ unitary scattering matrix S

$$S\begin{pmatrix} a^{in} \\ b^{in} \end{pmatrix} = \begin{pmatrix} S_{++} & S_{+-} \\ S_{-+} & S_{--} \end{pmatrix}\begin{pmatrix} a^{in} \\ b^{in} \end{pmatrix} = \begin{pmatrix} b^{out} \\ a^{out} \end{pmatrix}. \tag{5}$$

The $n \times n$ matrix S_{++} gives the transmission coefficients for an ingoing wave with positive momentum and S_{-+} gives the corresponding reflection coefficients, whereas S_{--} and S_{+-} give the similar quantities for an ingoing wave with negative momentum.

The problem is to study the behaviour of S as $\epsilon = \hbar/2M \to 0$, that is in the regime which corresponds to the semi–classical limit for the nuclei in the theory of inelastic atomic collisions. As in the one channel case one expects the reflection coefficients to be exponentially small, and indeed we prove that as $\epsilon \to 0$ (see also [9] for related results)

$$S_{+-} = O(\exp(-\frac{c_\Delta}{\epsilon})) \tag{6}$$

where, for $\Delta > 0$, c_Δ is a strictly positive constant independent of ϵ.

The main result of our note is that the transmission coefficient, i.e. S_{++}, is given up to exponentially small corrections by

$$S_{++} = \tilde{S} + O(\exp(-\frac{c_\Delta}{\epsilon})) \tag{7}$$

where $\tilde{S}$ is the $n \times n$ scattering matrix associated with the n–dimensional quantum mechanical evolution generated by an effective hermitian Hamiltonian $\widetilde{H}_\epsilon(x)$:

$$\tilde{S} = \lim_{x \to \infty} \left(\exp[-\frac{i}{\epsilon}(w_0)^{1/2}x] \right) \tilde{U}_\epsilon(x, -x) \left(\exp[-\frac{i}{\epsilon}(w_0)^{1/2}x] \right)$$

$$i\epsilon \frac{d}{dx}\tilde{U}_\epsilon(x, x_0) = \widetilde{H}_\epsilon(x) \tilde{U}_\epsilon(x, x_0), \quad \tilde{U}_\epsilon(x_0, x_0) = I. \tag{8}$$

At the lowest orders in ϵ, one has

$$\widetilde{H}_\epsilon(x) = -(w(x))^{1/2} + \frac{i\epsilon}{2}\left[\frac{d}{dx}(w(x))^{1/4}, (w(x))^{-1/4}\right] + O(\epsilon^2). \tag{9}$$

To establish (6)–(9) we introduce, as usual, the first order $2n$–dimensional first order system fully equivalent to (1):

$$i\epsilon \frac{d}{dx}\psi(x) = K(x)\psi(x) \tag{10}$$

with

$$\psi(x) = \begin{pmatrix} \psi(x) \\ \epsilon\frac{d}{dx}\psi(x) \end{pmatrix}, \quad K(x) = i\begin{pmatrix} 0 & I \\ -w(x) & 0 \end{pmatrix}.$$

The solutions of (10) define in turn a $2n \times 2n$ "propagator" $U(x, x_0)$

$$\psi(x) = U(x, x_0)\psi(x_0), \quad U(x_0, x_0) = I \tag{11}$$

and the S–matrix (5) can be expressed in terms of $U(x, x_0)$ (in the limit $x \to \infty$, $x_0 \to -\infty$).

The $2n$ eigenvalues of the (non hermitian) matrix $K(x)$ are $\pm(e_j(x))^{1/2}$, $j = 1, \ldots, n$. Due to (2), the sets of positive and negative eigenvalues of $K(x)$ are at a distance larger or equal to $2\sqrt{\Delta}$. This gap condition together with the analyticity of $v(x)$ enables to apply the adiabatic reduction theory, relatively to the n–dimensional subspace of positive eigenvalues of $K(x)$, as developed in [7]. Then the results in [7] augmented with some consequences of the symplectic structure of $K(x)$ lead to (6)–(9).

Now for (8) one can use the powerful methods of the standard adiabatic limit [10, 7, 8]. To proceed further we restrict our attention to a two level system with a simple avoided crossing. The model is defined by taking $v(x) = v_0(x) + \delta u(x)$ where the eigenvalues of $v_0(x)$ have a single linear crossing at $x = 0$ and δ is a small positive parameter, and $\lim_{x \to \pm \infty} v_0(x) = v_0$, $\lim_{x \to \pm \infty} u(x) = 0$. Consequently, for $x \neq 0$, the corresponding effective Hamiltonian $\widetilde{H_{\epsilon,\delta}}(x)$ has an expansion in ϵ and δ:

$$\widetilde{H_{\epsilon,\delta}}(x) = -(w_0(x))^{1/2} + \delta \tilde{v}^{0,1}(x) + \epsilon \tilde{v}^{1,0}(x) + \cdots \tag{12}$$

$$(w_0(x))^{1/2} = \sum_{j=1}^{2} (e_j^0(x))^{1/2} |n_j^0(x)\rangle\langle n_j^0(x)|, \quad e_j^0(x) = E - v_j^0(x) \tag{13}$$

$$\tilde{v}^{0,1}(x) = \sum_{j,k=1}^{2} \frac{\langle n_j^0(x)|u(x)|n_k^0(x)\rangle}{(e_j^0(x))^{1/2} + (e_k^0(x))^{1/2}} |n_j^0(x)\rangle\langle n_k^0(x)| \tag{14}$$

$$\tilde{v}^{1,0}(x) = \frac{i}{2} \left[\frac{d}{dx}(w_0(x))^{1/4}, (w_0(x))^{-1/4} \right] \tag{15}$$

where $v_j^0(x)$, $|n_j^0(x)\rangle$ are the eigenvalues and eigenvectors of $v_0(x)$ respectively. For further reference we note that $\tilde{v}^{0,1}(x)$ and $\tilde{v}^{1,0}(x)$ have definite limits as $x \to 0$ and moreover

$$\tilde{v}^{1,0}(0) = 0. \tag{16}$$

We conclude this section by the remark that one can easily recover the common trajectory model from (12-14). For example: restricting ourselves to the case

$$v(x) = \begin{pmatrix} v_{11}(x) & 0 \\ 0 & v_{22}(x) \end{pmatrix} + \delta \begin{pmatrix} 0 & v_{12}(x) \\ v_{21}(x) & 0 \end{pmatrix} \tag{17}$$

defining the level momenta by

$$p_j(x) = [2M(E - v_{jj}(x))]^{1/2}, \quad j = 1, 2 \tag{18}$$

and neglecting the terms of order $O(\epsilon)$, $O(\delta^2)$ and $O(p_1(x) - p_2(x)/p_1(x) + p_2(x))$ one finds that the effective evolution (8) is given (up to an overall phase factor) by

$$i\hbar \frac{d}{dt} \tilde{\psi}(t) = v(x(t))\tilde{\psi}(t) \tag{19}$$

where the trajectory $x(t)$ obeys the equation

$$M \frac{d}{dt} x(t) = \frac{1}{2}(p_1(x(t)) + p_2(x(t))). \tag{20}$$

A REFINED LANDAU–ZENER FORMULA

We consider now in more detail the two level model defined by in the previous section. The labelling of states for $-(w_0(x))^{1/2}$ is shown in Figure 1.

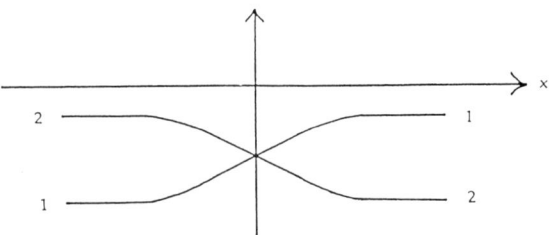

Figure 1. Eigenvalues of $-(w_0(x))^{1/2}$

Since at $x \to \pm\infty$, $\widetilde{H_{\epsilon,\delta}}(x)$ approaches $-(w_0(x))^{1/2}$, we take the same asymptotic labelling of the states of $\widetilde{H_{\epsilon,\delta}}(x)$ as for $-(w_0(x))^{1/2}$. What one has to compute are the probabilities $|\widetilde{S_{1j}}|^2$, $j=1,2$, for the system to be in the states 1 and 2 respectively at $x \to \infty$ if it was with probability one in the state 1 as $x \to -\infty$. As is well known, for small but fixed δ, the leading term as $\epsilon \to 0$, is given by the famous Landau–Zener formula

$$|\widetilde{S_{11}}|^2 \approx \exp\left[-\frac{2\pi\delta^2}{a\epsilon}|\tilde{v}_{12}|^2\right] \tag{21}$$

where

$$\tilde{v}_{12} = <n_1^0(0)|\tilde{v}^{0,1}|n_2^0(0)>, \quad a = \lim_{x\to 0}\frac{(e_2^0(x))^{1/2} - (e_1^0(x))^{1/2}}{x} > 0. \tag{22}$$

Notice that due to our labelling, $|\widetilde{S_{11}}|^2$ is what is usually called the probability of nonadiabatic transitions.

Recently, the Landau–Zener formula has been rigorously proved, and more important, bounds on the error term have been obtained (see [11,12] and references therein). Applied to our case the results in [11] give

$$|\widetilde{S_{11}}|^2 = \exp\left[-\frac{2\pi\delta^2}{a\epsilon}|\tilde{v}_{12}|^2(1+O(\delta))\right](1+O(\delta)+O(\epsilon)). \tag{23}$$

Notice, however, that the formula (23) is not sufficiently precise at very small δ; for example if $\delta \approx \epsilon$ it only gives $|\widetilde{S_{11}}|^2 = 1 + O(\epsilon)$ and then by unitarity $|\widetilde{S_{12}}|^2 = O(\epsilon)$. On the other hand for $\delta = 0$, the refinements of the adiabatic theorem for crossing eigenvalues (see [13] and references therein) imply that

$$|\widetilde{S_{11}}|^2 = 1 - \frac{2\pi}{a}\epsilon|h_{12}|^2 + o(\epsilon) \tag{24}$$

where

$$h_{12} = <n_1^0(x)|\frac{d}{dx}n_2^0(x)>\bigg|_{x=0} \tag{25}$$

(The term $\tilde{v}^{1,0}(x)$ gives no contribution to order ϵ due to (16)). Let us remark at this point that the proofs of (23) and (24) are quite different and the main difficulty is to obtain a uniform control in ϵ, δ of the error term.

It turns out that both (23) and (24) are particular cases of a refined form of the Landau–Zener formula for which we have a rigorous and fairly simple proof

$$|\widetilde{S_{11}}|^2 = \exp\left[-\frac{2\pi}{a\epsilon}|\delta\tilde{v}_{12} - i\epsilon h_{12}|^2(1+O(\epsilon+\delta))\right]. \tag{26}$$

The proof goes as follows. As usual we integrate the effective evolution given by $\widetilde{H_{\epsilon,\delta}}(x)$ along a dissipative path γ (whose existence is provided by the results in [14,8]) in the upper half plane, which amounts to consider the adiabatic expansion for nonself–adjoint Hamiltonians (see e.g. [15,16,8]). The result is that

$$|\widetilde{S_{11}}|^2 = \lim_{\substack{z_0 \to -\infty \\ z \to \infty}} \left|\exp\left\{-\frac{i}{\epsilon}\left[\int_{z_0}^{z}\sum_{j=0}^{2}\phi_j(u,\delta)\epsilon^j\,du + O(\epsilon^3)\right]\right\}\right|^2 \tag{27}$$

where the integration is taken along γ and the term $O(\epsilon^3)$ is uniform in z, z_0. In (27), we have not distinguished between the contributions of the so called geometrical

153

and dynamical phases: $\widetilde{\phi_j}(u,\delta)$ incorporates both quantities. Since for small δ the eigenvalues of $\widetilde{H_{0.\delta}}(x)$ are nondegenerate on γ one can obtain the expansion of $\phi_j(u,\delta)$ in powers of δ by using (12)–(15). As a result one obtains:

$$|\widetilde{S_{11}}|^2 = \lim_{\substack{z_0 \to -\infty \\ z \to \infty}} \left|\exp\left\{-\frac{i}{\epsilon}\left[\int_{z_0}^{z} \sum_{j+k\leq 2} \phi_{j,k}(u)\, \epsilon^j\, \delta^k\, du + O((\epsilon+\delta)^3)\right]\right\}\right|^2 \qquad (28)$$

The important point is that $\phi_{j,k}(u)$ are analytic in a strip around the real axis, some of them having a simple pole at $u = 0$. Moreover, they are real for $x \in \mathbb{R}$, $x \neq 0$. Then (26) follows from (28) by shifting the path to the real axis and using the standard formula $\frac{1}{u} = P - i\pi\delta(0)$. In other words if $\phi_{j,k}(u) = \xi_{j,k}(u)/u$ with regular $\xi_{j,k}(u)$ then $\phi_{j,k}(u)$ contributes to (26) with a factor $\exp[-2\pi\xi_{j,k}(0)\epsilon^{j-1}\delta^k]$.

We end up by writing the Landau–Zener factor (21) in terms of the data of the problem i.e. $v_j^0(x)$, $|n_j^0(x)>$ and $u(x)$. Using (14) one gets

$$|\widetilde{S_{11}}|^2 \approx \exp\left[-\frac{2\pi}{\hbar}\frac{\delta^2}{v_{cl}(0)\,b(0)}|u_{12}(0)|^2\right] \qquad (29)$$

with

$$v_{cl}(0) = \sqrt{\frac{2}{M}(E - v_1^0(0))}, \quad b(0) = \left|\frac{d}{dx}(v_1^0(x) - v_2^0(x))\right|_{x=0} \qquad (30)$$

which coincides, of course, with the standard Stückelberg–Landau–Zener formula [17].

REFERENCES

1. E.C.G. Stückelberg, *Helv. Phys. Acta*, **5**, 369 (1932).
2. E.E. Nikitin and S.Ya Umanskü, "Theory of Slow Atomic Collisions", Springer, Berlin (1984).
3. D.S.F. Crothers, *Adv. Phys.*, **20**, 405 (1971).
4. J.B. Delos, W.R. Thorson and S.K. Knudson, *Phys. Rev. A*, **6**, 709 (1972).
5. J.B. Delos and W.R. Thorson, *Phys. Rev. A*, **6**, 720 (1972).
6. W.R. Thorson, J.B. Delos and S.A. Boorstein, *Phys. Rev. A*, **4**, 1052 (1979).
7. G. Nenciu, Linear adiabatic theory. Exponential estimates. *Commun. Math. Phys.*, **153**, 479 (1993).
8. A. Joye, "Geometrical and Mathematical Aspects of the Adiabatic Theorem of Quantum Mechanics", Ph.D. Thesis, EPF-Lausanne (1992).
9. M. Fedoriouk, "Méthodes Asymptotiques pour les Equations Différentielles Ordinaires Linéaires", MIR, Moscou (1987).
10. J.-T. Hwang and P. Pechukas, *J. Chem. Phys.*, **67**, 4640 (1977).
11. A. Joye and C.-E. Pfister, *Phys. Lett. A*, **169**, 62 (1992).
12. G.A. Hagedorn, *Commun. Math. Phys.*, **136**, 433 (1991).
13. G.A. Hagedorn, *Ann. Phys.*, **196**, 278 (1989).
14. A. Joye, H. Kunz and C.-E. Pfister, *Ann. Phys.*, **208**, 299 (1991).
15. G. Nenciu and G. Rasche, *J. Phys. A*, **25**, 5749 (1992).
16. A. Joye and C.-E. Pfister, *J. Phys. A*, **24**, 753 (1991).
17. N.F. Mott and H.S.W. Massey, "The Theory of Atomic Collisions", Clarendon, Oxford (1971).

GIBBSIAN VERSUS NON-GIBBSIAN MEASURES: SOME RESULTS AND SOME QUESTIONS IN RENORMALIZATION GROUP THEORY AND STOCHASTIC DYNAMICS

A.C.D. van Enter[1*], R. Fernández[2] and A.D. Sokal[3]

1. Instituut voor Theoretische Natuurkunde, Rijksuniversiteit Groningen, Nijenborgh 4, NL 9747 AG Groningen, The Netherlands

2. Institut de Physique Théorique, EPF Lausanne PHB Ecublens, CH 1015 Lausanne, Switzerland

3. Department of Physics, New York University 4 Washington Place, New York, NY 10003 USA

We discuss some problems which arise if one tries to implement renormalization group transformations as maps from Hamiltonians to Hamiltonians. We provide various examples, involving systems not necessarily in the vicinity of a phase transition, where this can not be done, because Gibbs measures under the action of various real-space transformations become non-Gibbsian. We mention some related issues occurring in non-equilibrium problems.

Renormalization-group (RG) theory allows one to connect the different levels of description of statistical-mechanical systems. For example, one of its major accomplishments is the explanation of the phenomenon of "universality", the fact that different *microscopic* interactions give rise to the same *mesoscopic* or *macroscopic* behavior. Despite its many successful applications, the mathematical foundations of the theory are still on somewhat shaky ground. We will illustrate this by presenting some results which contradict various aspects of the folklore of RG theory. This contribution is based on [1], to which we refer for background, detailed proofs, more complete references and various related results. For a good recent treatment of RG theory on a physical level of rigor, we refer to [2].

We will discuss here the simplest case of discrete-spin classical lattice models.

Definition 1: The *configuration space* is $\Omega = \{\Omega_0\}^{\mathbb{Z}^d}$ where the single-spin space Ω_0 contains finitely many elements.

Definition 2: An *interaction* Φ is a translation invariant family of functions $\Phi_X, X \subset \mathbb{Z}^d$, where

$$\Phi_X : \{\Omega_0\}^X \to \mathbb{R}$$

---

*Speaker at the conference

The interaction is *regular* if

$$\sum_{0 \in X} ||\Phi_X||_\infty < \infty \qquad (1)$$

Definition 3: A probability measure μ on Ω is a *Gibbs measure* for a (regular) interaction Φ if its conditional probabilities satisfy

$$\frac{\mu(\sigma_1^\Lambda \mid \sigma^{\Lambda^c})}{\mu(\sigma_2^\Lambda \mid \sigma^{\Lambda^c})} = \exp - \beta \left[\sum_{X \cap \Lambda \neq \emptyset} \{\Phi_X(\sigma_1) - \Phi_X(\sigma_2)\} \right] \qquad (2)$$

for all $\Lambda \subset \mathbb{Z}^d$, σ_1^Λ and $\sigma_2^\Lambda \in \{\Omega_0\}^\Lambda$, and $\sigma^{\Lambda^c} \in \{\Omega_0\}^{\mathbb{Z}^d \setminus \Lambda}$. Here $\sigma_{1,2}$ denote the configurations that coincide with $\sigma_{1,2}^\Lambda$ in Λ and with σ_{Λ^c} in Λ^c and the functions Φ_X act on the coordinates of $\sigma_{1,2}$ in X. A standard reference on Gibbs measures is [3].

Block-spin (or real-space RG) transformations are obtained by first dividing $\mathbb{Z}^d$ into cubic, non-overlapping blocks of linear size L, denoted $B_j^L, j \in \mathbb{Z}^d$ and then prescribe either the value or (more generally) the distribution of each block-spin σ_j' in a way which only depends on the spin configuration in B_j^L.

This includes both deterministic and stochastic transformations. The block spin can take either the same or a different number of values as the original spins. Some examples of these transformations are:

a) Decimation (the σ_j' are the spins on some periodic sublattice).

b) Majority rule ($\sigma_j' = \mathrm{sgn} \sum_{i \in B_j^L} \sigma_i$; toss a coin when there is a tie).

c) Block averaging ($\sigma_j' = \sum_{i \in B_j^L} \sigma_i$).

d) Kadanoff variational transformations ($T(\sigma' \mid \sigma) = \mathrm{const} \exp(p \sigma_j' \sum_{i \in B_j^L} \sigma_i)$).

These transformations map probability measures on probability measures. If the original measure μ is a Gibbs measure for an interaction Φ, and the transformed measure μ' is a Gibbs measure for an interaction Φ', the map $R : \Phi \to \Phi'$ corresponds to what physicists usually consider to be a RG transformation. In RG theory one then studies properties of this map R (fixed points, eigenvalues of its linearization near these fixed points, etc.). The questions we addressed are the most basic ones:

I) Does R exist?

II) Is R smooth (and single-valued)?

The answers we find are that *if* R exists, it is smooth (Lipschitz continuous), but in various examples R does *not* exist.

The smoothness (and the related result that R cannot be multivalued) is proven by what a physicist would call a "droplet argument" and a mathematician a "large deviation argument" [1, section 3], combined with the strict convexity of pressure on the space of regular interactions [4].

To prove that R does not exist, i.e. that μ' is not a Gibbs measure, we demonstrate that μ' lacks the "almost Markov" [5,6] or "quasilocality" property [3], which is shared by all Gibbs measures. What happens is that each block-spin configuration acts as a constraint on the original system, and for some of these configurations σ' the constrained system has long-range order.

The unconstrained degrees of freedom thus operate as "hidden variables" which are responsible for the "non-local correlations" (the violation of quasilocality for μ'). These "hidden variables" correspond to the degrees of freedom which, according to usual physicists' terminology, are integrated out. In [1] this is shown to occur for low temperatures and small magnetic fields in the case of decimation or Kadanoff transformations (zero field for majority-rule transformations). For averaging transformations, the non-Gibbsian property is shown to hold for *all* magnetic field strengths. Our approach is based upon and extends the fundamental work of Griffiths-Pearce [7,8] and Israel [9].

On the other hand we know that R *does* exist at high temperatures in many examples [9,10,11]. At low temperatures, but in a (strong) magnetic field, the situation is more subtle. As we mentioned before, in [1] we showed that for averaging transformations, μ' is non-Gibbsian for all field strengths. On the other hand, for decimation and Kadanoff transformations, μ' *is* Gibbsian and R *does* exist for strong fields [7,8]. The recent work of Martinelli-Olivieri [12] on decimation transformations shows that in *any* non-zero field iterating the transformation sufficiently often brings the transformed measures back into the Gibbsian fold. For majority rule transformations, however, it was found that in strong fields the transformed measure is non-Gibbsian [13]. We emphasize that this pathological behavior occurs within the region of "complete analyticity"; hence the point of view of the originators of this concept [14] -eloquently expressed by the Tolstoy quote "all happy families are alike" [15]- seems less appropriate here than a more sceptical 20th century view [16]: "All animals are equal but some animals are more equal than others." Indeed, these examples show that the intuition that "good" properties which hold at high temperature should also hold at strong fields (=low density), or, even more drastically, in the whole one-phase regime, does not apply.

Let us also mention some further recent results which were obtained since we first submitted [1].

a) Israel [17] has proven that the property of being a Gibbs measure is exceptional in category sense, even within the class of ergodic, nonnull measures.

b) Not much is rigorously known in the regions near β_c, despite this being the regime of greatest physical interest. Kennedy [18] obtained some partial results for Ising models, suggesting that majority-rule transformation might be well-defined in an interval around β_c. On the other hand, for decimation applied to Potts models at high q, for which there is a first-order transition, the transformed measures around β_c are non-Gibbsian [13].

c) Schonmann's [19] result that the restriction of a 2-dimensional low temperature Ising Gibbs measure to the 1-dimensional lattice is non-Gibbsian has been extended to higher dimension [20,21], using results on the wetting transition [22,23,24]. The observation that there is a wetting phenomenon underlying Schonmann's proof was already made in [1]. In general, the operation of restricting a Gibbs measure yields more possibilities than renormalization. Let us consider for example the 2-dimensional Ising-type model with (formal) Hamiltonian

$$\sum_{i,j \in \mathbb{Z}^2} |i_1 - j_1|^{-\alpha} \delta_{i_2,j_2} \sigma_i \sigma_j \qquad , 1 < \alpha \leq 2 \qquad (3)$$

Thus one has a Dyson interaction in the horizontal direction and zero interaction in the vertical direction. The restrictions of low temperature Gibbs measures to a horizontal line are just the Gibbs measures for this Dyson interaction. The restrictions to vertical lines however are product measures and combinations thereof. Thus the restriction of μ^+ is Gibbsian for a single-site interaction (a positive magnetic field), and similarly the restriction of μ^- is a Gibbs measure for a negative magnetic field (and no further interactions). These two interactions are different, a possibility which is excluded in the RG set-up [1 section 3], as well as for stationary measures for local stochastic dynamics [20]. We note that Schonmann's restriction example (and higher dimensional generalizations) can be viewed as the invariant measure for some discrete time Markov process, due to the Global Markov Property [25]. The generator of this Markov process is the transfer matrix, but the Markov process has to be non-local in some sense [20].

Some Open Problems

1) What happens for non-equilibrium stationary states? The conjecture of Liggett [26,IV.7.5] that invariant measures of spin-flip processes with rates bounded away from zero are all Gibbsian is still unproven. We notice that Lebowitz and Schonmann [27] have the opposite intuition, that non-Gibbsian states should be prevalent in non-equilibrium statistical mechanics. Some non-Gibbsian examples are the invariant measures of the Voter model, the Martinelli-Scoppola cluster dynamics, and the totally asymmetric exclusion process [28,29,30].

2) What happens near β_c, and what happens if one iterates RG maps? As we mentioned before, in some cases there are arguments that R might exist [18], that is the transformed measure is Gibbsian. The results of [12] seem to confirm Israel's intuition [9] that if the decimation transformation is iterated, the pathological (non-Gibbsianness) region tends to the full coexistence line $\{\beta > \beta_c, h = 0\}$: that is the region shrinks in the field direction but expands in the temperature direction toward β_c. Iterating the averaging transformations is a different story; when the block-averaged spins are chosen to be zero, all field-dependence is removed from the constrained system. As this is the block-spin configuration used to prove non-quasilocality, the non-Gibbsianness region might actually increase to the infinite half plane $\{\beta \underset{(=)}{>} \beta_c, \text{field arbitrary}\}$. Because fixing the total spin to be zero in a large block will force phase separation in this block, also close to but below the critical temperature [31,32], a phase transition of the constrained system similar to the one in [1, section 4.3.5] might well persist up to the critical temperature.

3) Can one make RG well defined at the level of Hamiltonian-like objects? One might hope that this could be accomplished by a change of variables (consider contour or field variables instead of spin variables for example). This would probably be only a partial cure, as the appropriate variables would differ between different systems.

4) Can one restore (by weakening) the notion of Gibbsianness in some of these examples? For some ideas on this issue see [20,33,34].

5) Connections to statistics, image processing and "hidden Markov models". Some of these issues are discussed in [35].

Acknowledgments

A.C.D. van Enter thanks the organizers for the invitation to an inspiring conference. The scientific interaction with the Leuven group "wiskundige natuurkunde", and in particular the many useful and enjoyable discussions and exchanges of information with Christian Maes and Koen Vande Velde on non-Gibbsian and Gibbsian measures in non-equilibrium problems, has been much appreciated. Also we would like to thank Chuck Newman for his extremely valuable questions, remarks and suggestions during the many occasions he has been exposed to one of us lecturing on this topic. The research of A.C.D. van Enter was made possible by a fellowship of the KNAW (Royal Netherlands Academy of Arts and Sciences). The research of RF was supported by the Fonds National Suisse. The research of ADS was supported in part by the U.S. N.S.F. grant DMS-9200719.

REFERENCES

1. A.C.D. van Enter, R. Fernández and A.D. Sokal, Regularity properties and pathologies of position-space renormalization-group transformations: Scope and Limitations of Gibbsian theory, *J. Stat. Phys.*, **72**, 879 (1993).
2. N. Goldenfeld, "Lectures on Phase Transitions and the Renormalization Group", Addison-Wesley, Frontiers in Physics **85** (1992).
3. H.-O. Georgii, "Gibbs Measures and Phase Transitions", Walter de Gruyter, De Gruyter Studies in Mathematics **9** (1988).
4. R.B. Griffiths and D. Ruelle, Strict convexity ("continuity") of the pressure in lattice systems, *Comm. Math. Phys.*, **23**, 169 (1971).
5. W.G. Sullivan, Potentials for almost Markovian random fields, *Comm. Math. Phys.*, **33**, 61 (1973).
6. O.K. Kozlov, Gibbs description of a system of random variables, *Prob. Inform. Transmission*, **10**, 258 (1974).
7. R.B. Griffiths and P.A. Pearce, Position-space renormalization-group transformations: Some proofs and some problems, *Phys. Rev. Lett.*, **41**, 917 (1978).
8. R.B. Griffiths and P.A. Pearce, Mathematical properties of position-space renormalization-group transformations, *J. Stat. Phys.*, **20**, 499 (1979).
9. R.B. Israel, Banach algebras and Kadanoff transformations, *in:* "Random Fields (Esztergom 1979)" **II**, 593, North-Holland (1981).
10. I.A. Kashapov, Justification of the renormalization-group method, *Theor. Math. Phys.*, **42**, 184 (1980).
11. C. Cammarota, The large block spin interaction, *Nuovo Cim. B*, **96**, 1 (1986).
12. F. Martinelli and E. Olivieri, Some remarks on pathologies of renormalization group transformations for the Ising model, *J. Stat. Phys.*, **72** (1993).
13. A.C.D. van Enter, R. Fernández and R. Kotecký, in preparation.
14. R.L. Dobrushin and S.B. Shlosman, Completely analytical interactions: Constructive description, *J. Stat. Phys.*, **46**, 983 (1987).
15. L.N. Tolstoy, "Anna Karenina", translated by A. and L. Maude, Everyman's Library, Knopf (1992).
16. G. Orwell, "Animal Farm", Secker & Warburg (1945).
17. R.B. Israel, private communication.
18. T. Kennedy, Some rigorous results on majority rule renormalization group transformations near the critical point, *J. Stat. Phys.*, **72**, 15 (1993).

19. R.H. Schonmann, Projections of Gibbs measures may be non-Gibbsian, *Comm. Math. Phys.*, **124**, 1 (1989).
20. C. Maes and K. Vande Velde, The (non-)Gibbsian nature of states invariant under stochastic transformations, Leuven preprint (1993).
21. R. Fernández and C.-E. Pfister, in preparation.
22. P. Holický and M. Zahradník, On entropic repulsion in low temperature Ising models, *in:* "Proceedings of the NATO Advances Studies Institute Workshop on Cellular Automata and Cooperative Systems (Les Houches (1992))", 275 Kluwer (1993).
23. J. Fröhlich and C.-E. Pfister, Semi-infinite Ising model I, *Comm. Math. Phys.*, **109**, 493 (1987).
24. J. Fröhlich and C.-E. Pfister, Semi-infinite Ising model II, *Comm. Math. Phys.*, **112**, 51 (1987).
25. S. Albeverio and B. Zegarlinski, Global Markov property in quantum field theory and statistical mechanics: A review on results and problems *in:* "Ideas and Methods in Quantum and Statistical Physics", 331, Cambridge University Press (1992).
26. T.M. Liggett, "Interacting Particle Systems", Springer (1985).
27. J.L. Lebowitz and R.H. Schonmann, Pseudo-free energies and large deviations for non-Gibbsian FKG measures, *Prob. Th. Rel. Fields*, **77**, 49 (1988).
28. J.L. Lebowitz, C. Maes and E.R. Speer, Statistical mechanics of cellular automata, *J. Stat. Phys.*, **59**, 117 (1990).
29. F. Martinelli and E. Scoppola, A simple stochastic cluster dynamics: Rigorous results, *J. Phys. A*, **24**, 3135 (1991).
30. K. Vande Velde kindly informed us that both he and E.R. Speer have proven the non-Gibbsianness of the stationary states of the totally asymmetric exclusion process.
31. C.-E. Pfister, Large deviations and phase separation in the two-dimensional Ising model, *Helv. Phys. Act*, **64**, 953.
32. D. Ioffe, Large deviations for the 2D Ising model, Courant Institute preprint (1993). We thank Chuck Newman for informing us about this work and providing us with a preprint
33. J. Lörinczi, these proceedings.
34. J. Lörinczi and M. Winnink, Some remarks on Almost Gibbs states, *in:* "Proceedings of the NATO Advanced Studies Institute Workshop on Cellular Automata and Cooperative Systems (Les Houches (1992))", 423, Kluwer (1993).
35. A.C.D. van Enter, R. Fernández and A.D. Sokal, Renormalization transformations as a source of examples and problems in probability and statistics, Proceedings of V CLAPEM, Saõ Paulo (1993), to appear.

STABILITIES AND INSTABILITIES IN CLASSICAL LATTICE GAS MODELS WITHOUT PERIODIC GROUND STATES

Jacek Miękisz

Instituut voor Theoretische Fysica
Katholieke Universiteit Leuven
Celestijnenlaan 200-D
B-3001 Leuven, Belgium

INTRODUCTION

We present a criterion of the stability of nonperiodic ground states. It plays a role of the Peierls condition in models without periodic ground-state configurations. We discuss lattice gas models with stable and unstable nonperiodic ground states. The crystal problem is an attempt to deduce, within statistical mechanics, periodic order in systems of many interacting particles. Our model with a unique stable nonperiodic ground state constitutes a generic counterexample to that problem.

EXAMPLE 1: CLASSICAL LATTICE GAS MODEL WITHOUT PERIODIC GROUND-STATE CONFIGURATIONS

To construct our model we use the Robinson tiles [1,2]. These are unit squares with bumps and dents such that using an infinite number of their copies one can tile the plane only in a nonperiodic fashion. In any tiling, centers of the squares form the square lattice. In our lattice gas model, every site of the square lattice can be occupied by one of the 56 particles corresponding to tiles. Two nearest-neighbour particles which do not match as tiles have a positive interaction energy, say 1; otherwise, the energy is zero. We obtain a lattice gas model with nearest-neighbour translation-invariant interactions.

Any periodic configuration corresponds to a periodic configuration of tiles. Therefore it has a nonzero density of pairs of particles with the positive energy and hence a nonzero energy density. On the other hand, nonperiodic configurations corresponding to tilings have the zero energy density. It follows that the model does not have any periodic ground-state configurations.

It was proven in [3] that nonperiodic ground-state configurations of the above model are not stable against small perturbations of chemical potentials. Every ground-state configuration contains arbitrarily long sequences of particles of type A and sequences of particles of type B. Let us introduce a small negative potential h for A particles. When we change a sequence of length l of B particles into a sequence of A particles, we lower the energy by $l|h|$ along the sequence. The energy increases by 1 only at two endpoints of the sequence. Hence, no matter how small $|h|$ is, we can always lower the energy of a ground-state configurations of the unperturbed model.

CRITERION OF THE STABILITY OF NONPERIODIC GROUND STATES

Let Y and X be two lattice configurations - assignments of particles to lattice sites. Y is a local excitation of X, $Y \sim X$, if Y differs from X only on a finite region.

Particles at lattice sites $\mathbf{a}$ and $\mathbf{b}$ interact through translation-invariant potentials $\Phi(\mathbf{a} - \mathbf{b})$ and we assume that $\Phi(\mathbf{a} - \mathbf{b}) = 0$ if $dist(\mathbf{a}, \mathbf{b}) > r$.

The relative Hamiltonian is defined as follows:

$$H(Y, X) = \sum_{\{a,b\} \cap \Lambda \neq \emptyset} (\Phi(\mathbf{a} - \mathbf{b})(Y) - \Phi(\mathbf{a} - \mathbf{b})(X)) \ for \ Y \sim X.$$

$X \in \Omega$ is a **ground-state configuration** of H if

$$H(Y, X) \geq 0 \ for \ any \ Y \sim X,$$

i.e., we cannot lower the energy of a ground-state configuration by changing it locally.

We restrict ourselves to systems in which, although there are many ground-state configurations, there is a unique translation-invariant probability measure (called a **ground state**) supported by them. This is in analogy with the Ising antiferromagnet, where there are two alternating ground-state configurations but only one translation-invariant ground-state measure which assigns probability $1/2$ to both of them.

In our models in every ground-state configuration all interactions attain simultaneously their minima (we chose them to be equal to zero). Hence, if Y is not a ground-state configuration, it contains at least one pair of particles with a nonminimal energy of interaction (equal to 1 in all our examples) - the so-called **broken bond**. Denote by $B(Y)$ the number of broken bonds in Y. Let ar be a local arrangement of particles, $Y \sim X$, and $n_{ar}(Y, X)$ the difference of the number of the appearances of the arrangement ar in Y and the number of its appearances in X.

We say that a model satisfies the **strict boundary condition for local excitations** if

$$|n_{ar}(Y, X)| < C_{ar} B(Y),$$

for any ground-state configuration X and $Y \sim X$, where $C(ar)$ is a constant depending only on the arrangement ar.

Theorem If the above condition is satisfied for all ar with $diam(ar) < D$, then the unique ground state is stable against small perturbations of range smaller than D.

Proof: Assume first that the strict boundary condition does not hold for a ground-state configuration X and a local arrangement ar. Hence, for any $C > 0$ there exists $Y \sim X$ such that $n_{ar}(Y, X) > CB(Y)$ (the case $n_{ar}(Y, X) < -CB(Y)$ can be treated analogously). We introduce a small perturbation, an interaction which assigns a negative energy $E = -1/C$ to the arrangement ar. Then

$$H(Y, X) = B(Y) - n_{ar}(Y, X)/C < 0$$

so X is not a ground-state configuration for the perturbed interaction.

Assume now that $|n_{ar}(Y, X)| < C_{ar} B(Y)$ for any arrangement ar of diameter smaller than D and any local excitation $Y \sim X$ of a ground-state configuration X. Let $H' = H + \Phi$ be a perturbed Hamiltonian, where Φ is an interaction of range smaller than D. Then

$$H'(Y, X) = B(Y) + \sum_{ar} n_{ar}(Y, X) \Phi(ar)$$

$$> B(Y) - B(Y) \sum_{ar} C_{ar} |\Phi(ar)| > 0$$

if

$$\sum_{ar} C_{ar} |\Phi(ar)| < 1. \qquad (1)$$

Hence, X is a ground-state configuration for any sufficiently small perturbation. $\square$

It is easy to see that the model in Example 1 does not satisfy the strict boundary condition for local excitations.

EXAMPLE 2: STABLE NONPERIODIC GROUND STATE

We constructed a classical lattice gas model with nearest and next nearest-neighbour translation-invariant interactions. It has a unique nonperiodic ground state satisfying the strict boundary property for any two-particle arrangements. Therefore the ground state is stable against small two-body perturbations.

The main idea of the construction is to replace two translation-invariant sequences of A and B particles responsible for forcing nonperiodicity in Example 1 by translations of a periodic sequence. For details we refer the reader to [4].

REFERENCES

1. R. M. Robinson, Undecidability and nonperiodicity for tilings of the plane, *Invent. Math.*, **12**, 177 (1971).
2. B. Grünbaum and G. C. Shephard, "Tilings and Patterns", Freeman, New York (1986).
3. J. Miękisz and C. Radin, The unstable chemical structure of quasicrystalline alloys, *Phys. Letts.*, **119A**, 133 (1986).
4. J. Miękisz, Stable quasicrystalline ground states, preprint.

ONE DIMENSIONAL ANOMALY OF THE FERMI SURFACE

G. Gallavotti

Dipartimento di Fisica, Università La Sapienza
P.le Moro 2; 00185, Roma, Italia

Consider a N-fermions system with hamiltonian, see [1]:

$$H = \sum_{i=1}^{N} \frac{-\Delta_{x_i} - p_F^2}{2m} + \lambda \sum_{i<j} v(x_i - x_j) \tag{1}$$

where m, p_F are two fixed parameters, λ is the coupling constant and v is the pair potential, supposed smooth and with short range p_0^{-1}.

The fermions will be supposed also to have spin 0 and to be enclosed in a periodic box of size L. Thus in the grand canonical ensemble the pair Schwinger function at $\xi = (x,t) \in R^2$ is defined by:

$$S(\xi) = \lim_{L\to\infty} \lim_{\beta\to\infty} \frac{\mathrm{Tr}\, e^{-(\beta-t)H} \varphi_x^- e^{-tH} \varphi_0^+}{\mathrm{Tr}\, e^{-\beta H}} \tag{2}$$

where $\varphi_x^\pm$ are the creation and annihilation operators.

The above system is particularly interesting in the light of recently proposed connections with the theory of 2–dimensional Fermi gases which, in some cases, might exhibit similar properties, see [2].

The free system, $\lambda = 0$, is such that:

$$S(\xi) = S^0_{m,p_F}(\xi) = \frac{1}{(2\pi)^2} \int dk_0 dk \, \frac{e^{-i(k_0 t + kx)}}{-ik_0 + (k^2 - p_F^2)/2m} \simeq$$

$$\simeq_{\xi\to\infty} -\frac{1}{\pi} \frac{\sin(p_F x - \mathrm{arctg}\, v_F t/x)}{(x^2 + (v_F t)^2)^{1/2}}, \qquad v_F \equiv \frac{p_F}{m} \tag{3}$$

if arctg $v_F t/x$ is bounded away from $\pi/2$;

And $S(x, 0^-)$ has a Fourier transform $1 - \hat{n}(k)$ in x, with $\hat{n}(k) = \langle a_k^+ a_k^- \rangle$ being the average occupation density of the state with momentum k. It is well known that $\hat{n}(k) = 1$ if $|k| < p_F$ and $\hat{n}(k) = 0$ otherwise.

We address the question: *which is the asymptotic behaviour of $S(\xi)$ when $\lambda \neq 0$?* answering the question will give us a way to define the *Fermi momentum* and the *mass* of the particles in presence of interaction.

In fact we shall use the following definition:

Definition: *The hamiltonian (1) describes a system with particles of mass m' and Fermi momentum p'_F if there exist constants Z, η such that:*

$$S(\xi) = \frac{1}{Z} \frac{S^0_{m', p'_F}(\xi)}{|\xi|^{2\eta}} \tag{4}$$

where $|\xi|^2 = p_0^2(x^2 + v'_F{}^2 t^2)$.

One says that the system (1) is a normal Fermi liquid if $\eta = 0$: in this case Z^{-1} is the jump of the Fermi distribution $\hat{n}(k)$ at $k = p'_F$. Otherwise the system is anomalous and the momentum distribution is less singular at $k = p'_F$, behaving proportionally to $||k| - p'_F|^{2\eta}$.

Of course one expects that the quantities m', p'_F depend on λ and are different from the parameters m, p_F in (1). It is therefore convenient, if one wants to study the above problem by perturbation theory, to introduce a hamiltonian with more parameters giving us the possibility of fixing *a priori* the values m, p_F which describe the singularity of $S(\xi)$. Hence we shall consider the hamiltonian:

$$H = \sum_{i=1}^{N} \frac{-\Delta_{x_i} - p_F^2}{2m} + \lambda \sum_{i<j} v(x_i - x_j) + \sum_{i=1}^{N} \frac{-\alpha \Delta_{x_i} - \nu p_F^2}{2m} \tag{5}$$

and call $m_0 = m/(1+\alpha)$ the *bare mass* and $\mu = -(1+\nu)p_F^2/2m$ the *chemical potential*. We shall then ask the question: is it possible to fix $\alpha = \alpha(\lambda), \nu = \nu(\lambda)$ in terms of λ so that (4) holds with $(m', p'_F) = (m, p_F)$?

Or: can one fix the bare mass and bare chemical potential so that the interacting system has prescribed values of the mass and the Fermi momentum?

The above is not the only possible definition of mass and Fermi momentum: there are several others; for an example of an alternative one see [1].

That the above analysis and definitions are non empty and have a non trivial meaning is shown by the main result of[1].

Theorem: *Suppose that $a \equiv \hat{v}(0) - \hat{v}(2p_F) \neq 0$, and $|\lambda|$ small enough; then there exist functions $\alpha(\lambda), \nu(\lambda), \eta(\lambda)$ analytic in λ near $\lambda = 0$ and divisible by λ^2 such that, fixing $\alpha = \alpha(\lambda), \nu = \nu(\lambda)$ in (5) it is:*

$$S(\xi) = \frac{S^0_{m, p_F}(\xi)}{|\xi|^{2\eta}} + \frac{A_\xi(\lambda)}{|\xi|^{1+2\eta}} \tag{6}$$

with $\eta = \eta(\lambda)$ and $|A_\xi(\lambda)| \leq C|\lambda|$ for a suitable constant C and all ξ. Furthermore $\eta(\lambda) \neq 0$.

In other words it is possible to fix $\alpha = \alpha(\lambda), \nu = \nu(\lambda)$ so that the system describes fermions with Fermi momentum p_F and mass m. However it will be, generically (*i.e.* if $a \neq 0$ at least), a system with anomalous Fermi surface.

The above theorem was discussed by Tomonaga, [3], who gave very strong arguments in favour of it: the arguments led to the formulation of the Luttinger model, which was conjectured, [4], to show the same behaviour, for the Schwinger functions, of the system (1) (hence to show an anomalous Fermi surface, by [3]), and to be exactly soluble. The model was indeed solved exactly by [5] and gave the expected results (although quantitatively different from the exact conjecture in [4]).

The above theorem closes the missing gaps by proving that the "realistic" system (1) shows anomalous behaviour as predicted by [3].

Let me spend a few words about the strategy followed in our theory of (1).

Imagine to fix $L, \beta < \infty$. Then, ($d = 1$):
1) the functions $S(\xi)$ are analytic in λ, α, ν near 0, with a radius of convergence at least of order $O(L^{-1})$, unformly in β.
2) the Schwinger functions are given, for small λ, α, ν by a functional integral:

$$S(\xi_1,\ldots,\xi_n) = \frac{\int P(d\psi)\,e^{-V(\psi)}\,\psi^-_{\xi_1}\cdots\psi^+_{\xi_n}}{\int P(d\psi)e^{-V(\psi)}} \qquad (7)$$

where:

$$V(\psi) = \lambda \int v(x-y)\psi^+_{x,t}\psi^-_{x,t}\psi^+_{y,t}\psi^-_{y,t}\,dxdydt + \alpha \int \psi^+_{x,t}\frac{-\Delta}{2m}\psi^-_{x,t}\,dxdt + \nu \int \psi^+_{x,t}\psi^-_{x,t}\,dxdt$$
$$(8)$$

and $\psi^\pm_\xi$ are grassmanian variables on R^2 and $P(d\psi)$ is a grassmanian integrations with propagator:

$$g(\xi - \eta) = \langle \psi^-_\xi \psi^+_\eta \rangle \equiv \frac{1}{(2\pi)^2}\int dk_0 dk\,\frac{e^{-i\kappa(\xi-\eta)}}{-ik_0 + (k^2 - p_F^2)/2m} \qquad (9)$$

where $\kappa = (k, k_0)$, $\int dk\cdot \equiv \frac{2\pi}{L}\sum\cdot$, $\int dk_0 \cdot \equiv \frac{2\pi}{\beta}\sum\cdot$ with the sums running over the

$k = \frac{2\pi}{L}n$ and $k_0 = \frac{2\pi}{\beta}(n_0 + \frac{1}{2})$ with n_0, n integers. [1]
3) it is possible to *rearrange* the series in $\vec{r}_0 = (\lambda, \alpha, \nu)$ into power series in a new family of parameters $\vec{r}_0, \vec{r}_{-1}, \vec{r}_{-2},\ldots$, with $\vec{r}_j \in R^3$, so that the disk of convergence

---

[1] this means that the numerator and denominators of (7) are defined by developing the exponential in powers and by computing the integrals via the Wick rule. It is well known since the work [6], see also [7], [1], that the resulting series are entire functions of λ, α, ν and, for λ, α, ν small enough, of order $O(L^{-1})$ at least, the denominator does not vanish and (7) holds.

for the series expressing the Schwinger functions in powers of such parameters is at least $|\vec{r}_h| \leq \varepsilon_0$ with $\varepsilon_0 > 0$ and β, L independent. The new expansion parameters, *the running coupling constants*, are analytic functions of $\vec{r}_0$ with radius of analyticity of $O(L^{-1})$.

Comment: in other words we can "push" the problem of computing the Schwinger functions into that of studying the true dependence of a certain sequence of scalar functions of $\vec{r}_0$, the running couplings. Of course the abstract formulation looks just a rephrasing of the original problem and no gain at all. Its interest depends on the actual definition of the resummation rule and of the running couplings.

4) the $\vec{r}_h$ verify, if $|\vec{r}_h| \leq \varepsilon_0$, a recurrence relation:

$$\vec{r}_{h-1} = \Lambda \vec{r}_h + B(\vec{r}_h) + \mathcal{B}_h(\vec{r}_h, \vec{r}_h - \vec{r}_{h+1}, \vec{r}_h - \vec{r}_{h+2}, \ldots, \vec{r}_h - \vec{r}_0) \qquad (10)$$

with Λ linear and diagonal, and $B, \mathcal{B}_j$ analytic in their arguments in the polydisk of radius ε_0. Furthermore $\mathcal{B}_j$ are functions with *short memory*; this means that there exist functions $D_j^k(\vec{x}_k, \vec{x}_{k+1}, \ldots, \vec{x}_{-1}, \vec{x}_0)$ such that for some $b, d > 0$:

$$\mathcal{B}_j(\vec{x}_j, \vec{x}_{j+1}, \ldots, \vec{x}_{-1}, \vec{x}_0) = \sum_{k=j}^{-1} D_j^k(\vec{x}_j, \vec{x}_{k+1}, \ldots, \vec{x}_{-1}, \vec{x}_0) \qquad (11)$$

$$|D_j^k| \leq 2^{-(k-j)b} |\vec{x}_j| \left(\sup_{k'>k} |\vec{x}_{k'}|\right) d$$

Comment: this has the practical consequence that the dynamical system without memory defined by the map: $\vec{r}' = \Lambda \vec{r} + B(\vec{r})$ and the one in (10) behave essentially in the same way near $\vec{r}_h = \vec{0}$.

5) if $|\vec{r}_j| \leq \varepsilon_0$ and $\vec{r}_j \xrightarrow[j \to -\infty]{} \vec{r}_{-\infty}$ then there exists an analytic function $\eta(\vec{r})$ of $\vec{r}$, defined for $|\vec{r}| \leq \varepsilon_0$, and such that $\eta = \eta(\vec{r}_{-\infty})$ is the anomalous exponent for the Schwinger functions:

$$S(\xi) = \frac{S_{m,p_F}(\xi)}{|\xi|^{2\eta}} + \lambda O(|\xi|^{-1+2\eta}) \qquad (12)$$

with $|\xi|^2 = p_0^2(x^2 v_F^2 t^2)$, $v_F = p_F/m$. And finally:

6) if $\vec{r} = (\lambda, \alpha, \nu)$ then the first two components of the *beta function* $B(\vec{r})$, vanish *identically* for $\vec{r} = (\lambda, \delta, 0)$

The last property will be the vault key of the whole theory: should it have failed it would have been easy to see that *no initial* $\vec{r}_0 = (\lambda, \alpha, \nu)$ *could be found generating, via (10), a sequence of running couplings* $\vec{r}_h$ *such that* $|\vec{r}_h| \leq \varepsilon_0$ *for all* $h \leq 0$, and the whole discussion would be inconclusive.

It remains to get down to work to explain the "resummation procedure" with the above remarkable properties. This is obviously a very essential key point: the renormalization group is a well known method for resummations, but there are "many" possible resummations and it is by no means clear that there is one with the above properties. It is in fact not difficult to exhibit many resummations with all the above properties except the last. They are non trivial, nevertheless they are useless, see [1] for an example.

In order to understand the right resummation procedure or, in other words, *the right renormalization group transformation*, we must first introduce the notion of *quasi particles fields*. It is in fact by regarding the functional integrals in (7) as functional integrals over quasi particle fields that one can immediately recognize the analogy between the above Fermi surface problem and the well known theory of the anomalous dimension in $4 - \varepsilon$ dimensions of [8]. Once the analogy is recognized it is only a "technical problem" to work out the details to obtain the above properties 1–6).

We consider the grassmanian propagator and decompose it as follows; let $\varepsilon(k_0, k)^{-1} \equiv -ik_0 + (k^2 - p_F^2)/2m$ and use the identity $1 \equiv \left(1 - e^{-\varepsilon(k_0,k)^2 p_0^{-2}}\right) + e^{-\varepsilon(k_0,k)^2 p_0^{-2}}$ where p_0 is a convenient momentum scale, equal to the inverse range of the potential; then:

$$g(\xi) = g^{(>0)}(\xi) + g^{(\leq 0)}(\xi) \equiv \frac{1}{(2\pi)^2} \int \frac{dk dk_0}{\varepsilon(k_0, k)} \left(\left[1 - e^{-\varepsilon(k_0,k)^2 p_0^{-2}}\right] + \left[e^{-\varepsilon(k_0,k)^2 p_0^{-2}}\right] \right) \quad (13)$$

with the obvious identification of the two terms, that we call the *ultrviolet* and the *infrared* parts of the propagator g..

It is then possible to write the infrared part as:

$$g^{(\leq 0)}(\xi) = \sum_{\omega = \pm 1} e^{-i\omega p_F x} \bar{g}^{(\leq 0)}(\xi, \omega), \qquad \bar{g}^{(\leq n)}(\xi) = \sum_{h=-\infty}^{n} 2^h \tilde{g}^{(h)}(2^h x, \omega) \quad (14)$$

with $\tilde{g}^{(h)}(\xi, \omega)$ essentially independent of h, for $-h$ large, see [1].

This means that:

$$\psi_\xi^\pm \equiv \psi_\xi^{(>0)\pm} + \psi_\xi^{(\leq 0)\pm} \equiv \psi_\xi^{(>0)\pm} + \sum_{\omega=\pm 1} e^{\pm i\omega p_F x} \psi_{\xi,\omega}^{(\leq 0)\pm} \quad (15)$$

where $\psi_\xi^{(>0)}$ and $\psi_{\xi,\omega}^{(\leq 0)}$, $\omega = \pm$, are *independent* grassmanian variables, (*i.e.* the crossed propagators vanish).

We call the $\psi_{\xi,\omega}^{(\leq 0)}$ the quasi particle fields and the integrals can be written

$\int P(d\psi^{(>0)}) P(d\psi^{(\leq 0)}) e^{-V(\psi^{(\leq 0)})}$.

Forgetting the integral over $\psi^{(>0)}$ (see [1]), I shall explain here the definition of the running couplings (*i.e.* the definition of the resummation procedure) for the easier problem of the evaluation of the partition function: [2]

$$\Xi = \int P(d\psi^{(\leq 0)})e^{-V(\psi)} \tag{16}$$

We already mentioned that we use the renormalization group methods, and that the proper formulation is to think that the interaction V and the integral in (16) are over quasi particle fields. The reason is that the quasi particle fields propagator $\bar{g}^{(\leq 0)}(\xi,\omega)$ have *no intrinsic length scale*: they have in fact a Fourier transform with singularity at the origin. *This is not the case for the particle fields* which contain, in the propagator, a fixed length scale p_F^{-1}, see (14).

The quasi particles fields, therefore, arise naturally to recover the scale invariance that is lost by the presence of the Fermi momentum.

We can write the interaction V as an interaction between quasi particles and define the relevant part and the irrelevant parts of it *via the same prescription that would be used in the renormalization group approach to scalar fields*, (see [8],[9]). By a power counting analysis one identifies the relevant and marginal operators and then one defines the relevant part of the interaction to be $\mathcal{L}V$ where $\mathcal{L}$ is a *localization operator* defined by:

$$\mathcal{L}\psi^+_{\xi_1,\omega_1}\psi^+_{\xi_2,\omega_2}\psi^-_{\xi_3,\omega_3}\psi^-_{\xi_4,\omega_4} = \psi^+_{\xi_1,\omega_1}\psi^+_{\xi_1,\omega_2}\psi^-_{\xi_1,\omega_3}\psi^-_{\xi_1,\omega_4}$$
$$\mathcal{L}\psi^+_{\xi_1,\omega_1}\psi^-_{\xi_2,\omega_2} = \psi^+_{\xi_1,\omega_1}\left(\psi^-_{\xi_1,\omega_2} + (x_2-x_1)\partial\psi^-_{x_1,\omega_2}\right) \tag{17}$$

as it would be done in quantum field theory of the Gross–Neveu model, (see [7], and [10]).

The *effective potential* will then be defined recursively as follows. Let $Z_0 = 1$ and denote $P^0_{Z_0}$ the integration with propagator $Z_0^{-1}g^{(\leq 0)}(\xi)$. We shall write such functional integration formally as:

$$P^0_{Z_0}(d\psi) \propto e^{-Z_0(\psi^+,[(-\Delta-p_F^2)/2m]Q^0\psi^-)}\,d\psi \tag{18}$$

where $Q^0(\partial)$ has Fourier transform $Q^0(k) = e^{\varepsilon(k)^2 p_0^{-2}}$, where again $\varepsilon(k)^2 = k_0^2 + ((k^2-p_F^2)/2m)^2$. For later use we define also $Q^h(\partial)$ as $Q^h(k) = e^{2^{-2h}\varepsilon(k)^2 p_0^{-2}}$.

Then, $\Xi = \int e^{-V^{(0)}(\sqrt{Z_0}\psi^{(\leq 0)})}P^0_{Z_0}(d\psi^{(\leq 0)})$ is given by:

$$\Xi \equiv \int e^{-V^{(0)}(\sqrt{Z_0}(\psi^{(<0)}+\psi^{(0)}))}\hat{P}_{Z_0}(d\psi^{(0)})P^0_{Z_0}(d\psi^{(\leq -1)}) =$$
$$= \int P_{Z_0}(d\psi^{(\leq -1)})e^{-\tilde{V}(\sqrt{Z_0}\psi^{(\leq -1)})} \tag{19}$$

---

[2] for the treatement of the ultraviolet problem and for the (non trivial) discussion of how to connect the theory of the partition function with that of the asymptotic behaviour of the Schwinger functions see [1].

where the propagator of $\psi^{(\leq -1)}$ is $Z_0^{-1} g^{(\leq -1)}(\xi)$ and that of $\psi^{(0)}$ is $Z_0^{-1} \tilde{g}^{(0)}(\xi)$, see (14).

The integration over $\psi^{(0)}$ is performed by thinking $\psi^{(0)}$ and $\psi^{(\leq -1)}$ as a sum of quasi particle fields. The result of the integration will be expressed in terms of the function $\tilde{V}(\sqrt{Z_0}\psi^{(\leq -1)})$ which can be regarded either as a function of the particle field $\psi_\xi^{(\leq -1)}$ or of the quasi particle field $\psi_{\xi,\omega}^{(\leq -1)}$.

The $\tilde{V}$ will be written as a sum $\sum_{\omega_1,\ldots} \int w(\xi_1,\ldots,\omega_1,\ldots) \psi_{\xi_1 \omega_1}^{+(\leq -1)} \ldots d\xi_1 \ldots$ of terms and the calculation of the coefficients w of its expansion in (even) field monomials $\psi^{+(\leq -1)} \ldots$, is of course a delicate technical problem. It is obtained "trivially" by a power series expansion, which is shown to be convergent by the essential use of the Hadamard inequality on determinants following the work [7] (and the idea of [6]).

Regarding $\tilde{V}$ as a function of the quasi particle fields we can apply to it the localization operator $\mathcal{L}$ thus isolating the relevant and marginal parts from the irrelevant ones (which, of course, are the really interesting ones for the purpose of calculating the Schwinger functions and the physical properties of the system). It will result, see (17):

$$\tilde{V}(\sqrt{Z_0}\psi^{(\leq -1)}) = \overline{V}(\sqrt{Z_0}\psi^{(\leq -1)}) + Z_0^2 \lambda_0 \int \psi_{\xi,+}^{(\leq -1)+} \psi_{\xi,-}^{(\leq -1)+} \psi_{\xi,+}^{(\leq -1)-} \psi_{\xi,-}^{(\leq -1)-} d\xi +$$

$$+ Z_0 \int \left(\delta_0 \psi_{\xi,\omega}^{(\leq -1)+} \omega' \partial_x \psi_{\xi,\omega'}^{(\leq -1)-} + \zeta_0 \psi_{\xi,\omega}^{(\leq -1)+} (\partial_t + i\omega' \partial_x) \psi_{\xi,\omega'}^{(\leq -1)-} \right) e^{ip_F(\omega-\omega')x} d\xi +$$

$$+ Z_0 2^0 \nu_0 \int \psi_{\xi,\omega}^{(\leq -1)+} \psi_{\xi,\omega'}^{(\leq -1)-} e^{ip_F(\omega-\omega')x} d\xi \qquad (20)$$

where $\overline{V} \equiv (1-\mathcal{L})\tilde{V}$ and the other terms are the result of the application of $\mathcal{L}$ to $\tilde{V}$, i.e. *they are a definition of the constants* $\lambda_0, \delta_0, \zeta_0, \nu_0$. The factor $2^0 = 1$ is so written because it will be generalized, in the iterative construction of the running couplings, into a factor 2^h at the step $-h$.

Looking at the part of $\tilde{V}$ of second degree in the fields ψ (or, more generally, of any fixed degree) we realize that the dependence of the kernels in the integrals over the ξ variables and the sums over the quasi particle labels ω must be special so that it must be possible to write them as functionals of the particle fields: this explains why, in (20), the coefficients δ_0, ν_0, ζ_0 are just constants and not functions of ω, ω', as a priori, one would expect. This is a very important property called in [10] "gauge invariance".

Hence by adding and subtracting to the r.h.s. of (20) the second degree part

$W = (\zeta_0 - \delta_0) Z_0 \int \psi_{\xi,\omega}^{(\leq -1)+} \frac{\Delta_\xi}{2m} \psi_{\xi,\omega'}^{(\leq -1)-} e^{ip_F(\omega-\omega')x} d\xi$ we see, with an easy calculation,

that $\tilde{V}$ can be written as a sum of an "irrelevant part" $\overline{V} - W$ (in the sense of power counting) plus a "relevant part" given, at second degree in the fields by:

$$\int d\xi \left(Z_0 \delta_0 \, \psi_\xi^{(\leq -1)+} \frac{-\Delta_\xi - p_F^2}{2m} \psi_\xi^{(\leq -1)-} + Z_0 \zeta_0 \psi_\xi^{(\leq -1)+} (\partial_t + \frac{-\Delta_\xi - p_F^2}{2m}) \psi_\xi^{(\leq -1)-} + \right.$$
$$\left. + Z_0 2^0 \nu_0 \psi_\xi^{(\leq -1)+} \psi_\xi^{(\leq -1)-} \right)$$
(21)

The "free" integration $P_{Z_0}(d\psi^{(\leq -1)})$ will be written formally as proportional to $d\psi \exp -(\psi^{(\leq -1)+}, (\partial_t + \omega \partial_x) Q^0(\partial) \psi^{(\leq -1)-})$ so that the term with coefficient ζ_0 can be put together with the free integration and one gets, setting $Z_{-1} = Z_0(1 + \zeta_0)$:

$$\Xi = \int P_{Z_{-1}}^{-1}(d\psi^{(<-1)}) \hat{P}_{Z_{-1}}(d\psi^{(-1)}) e^{-V^{(-1)}(\sqrt{Z_{-1}}(\psi^{(\leq -2)} + \psi^{(-1)}))} \tag{22}$$

where the propagator of the integration $\hat{P}_{Z_{-1}}(d\psi^{(-1)})$ is *defined*, by difference, so that the following quadratic forms are identical:

$$Z_0(f, [\partial_t + (-\Delta - p_F^2)/2m] Q^{(-1)} f) + Z_0 \zeta_0 (f, [\partial_t + (-\Delta - p_F^2)/2m] f) =$$
$$= Z_{-1}(f, [\partial_t + (-\Delta - p_F^2)/2m] Q^{(-1)} f) + Z_{-1}(f, \tilde{Q} f) \tag{23}$$

Hence, we finally get:

$$\Xi = \int \hat{P}_{Z_0}(d\psi^{(0)}) P_{Z_0}^0(d\psi^{(<0)}) e^{-V^{(0)}(\sqrt{Z_0}(\psi^{(<0)} + \psi^{(0)}))} =$$
$$= \int \hat{P}_{Z_{-1}}(d\psi^{(-1)}) P_{Z_{-1}}^{-1}(d\psi^{(<-1)}) e^{-V^{(-1)}(\sqrt{Z_{-1}}(\psi^{(<-1)} + \psi^{(-1)}))} \tag{24}$$

One then iterates the above construction: the key remark is that at the step $-h$ the propagator of $\hat{P}_{Z_h}(d\psi^{(h)})$, regarded as a quasi particles field propagator, written as $2^h \hat{g}^{(h)}(2^h \xi, \omega)$ is essentially scale invariant (this is not completely obvious and it requires a calculation; which shows that $\hat{g}^{(h)} \neq \tilde{g}^{(h)}$ for $h < 0$, but that nevertheless $\hat{g}$ has the same approximate scale invariance as $\tilde{g}^{(h)}$). The propagator $\hat{g}^{(h)}(\xi, \omega)$ results approximately scale independent (*i.e.* h–independent) and decaying exponentially fast both in position and in momentum space.

The above analysis describes the definition of the running couplings; the approximate scale invariance of the propagators $\hat{g}^{(h)}$ is the reason for the corresponding scale invariance of the beta function, *i.e.* of the relation (10) and the h–independence of the "main term" $B(\vec{r})$. The convergence and analyticity properties of the beta function are a simple consequence of the techniques developed in [7], see [1].

The final property, item 6) above, *i.e.* the vanishing of the first two components of $B(\lambda, \delta, 0)$, is based on the comparison of the above theory with the analogous theory of the Luttinger model, [11], [12].

One shows, see [10], that the function $B(\vec{r})$ (*i.e.* the leading part of the recurrence (10)) is the same for the model (5) and for the Luttinger model. And one infers that the first two components of $B(\lambda, \delta, 0)$ must vanish by an "indirect argument", namely by showing that otherwise the exact solution of the Luttinger model would be different from what it is, see [1]. There are also arguments, see [13], suggesting that a "direct" check of the vanishing of the appropriate part of the beta function might actually be possible.

Finally the property in item 5) is based on a careful analysis of the series (now known to be convergent) for the Schwinger functions in terms of the running couplings. The analysis is somewhat delicate: see [1]: note however that heuristically the result should be expected from the closeness between the above approach and the work [8].

REFERENCES

1. G. Benfatto, G. Gallavotti, A. Procacci and B. Scoppola, Beta function and Schwinger functions for a many fermions system in one dimension. Anomaly of the Fermi surface, To appear in *Commun. Math. Phys.*
2. P. Anderson, The "infrared catastrophe": when does it trash Fermi liquid theory?, Preprint #405, february 1993, Princeton. See also: "Luttinger-liquid" behaviour of the normal metallic state of the 2D Hubbard model, *Phys. Rev. Lett.*, **64**, 1839 (1990).
3. S. Tomonaga, Remarks on Bloch's methods of sound waves applied to many fermion problems, *Prog. Theor. Phys.*, **5**, 554 (1950).
4. J. Luttinger, An exact soluble model of a many-fermions system, *J. Math. Phys.*, **4**, 1154 (1963).
5. D. Mattis and E. Lieb, Exact solution of a many fermion system and its associated boson field, *J. Math. Phys.*, **6**, 304 (1965).
6. E. Caianiello, The number of Feynman graphs and convergence, *Nuovo Cimento*, **3**, 223 (1956).
7. K. Gawedzki and A. Kupiainen, Groos-Neveu model through convergent perturbation expansion, *Commun. Math. Phys.*, **102**, 1 (1985).
8. K. Wilson and M. Fisher, Critical exponents in 3.99 dimensions, *Phys. Rev. Lett.*, **28**, 548 (1972).
9. G. Gallavotti, Renormalization theory and ultraviolet stability for scalar fields via renormalization group methods, *Rev. of Modern Phys.*, **57**, 471 (1985).
10. G. Benfatto and G. Gallavotti, Perturbation theory of the Fermi surface in a quantum liquid. A general quasiparticle formalism and one dimensional systems, *J. Stat. Phys.*, **59**, 541 (1990).
11. G. Benfatto, G. Gallavotti and V. Mastropietro, Renormalization group and the Fermi surface in the Luttinger model, *Phys. Rev. B*, **45**, 5468 (1992).
12. G. Gentile and B. Scoppola, Università di Roma, Dipartimento di Fisica, CARR preprint n. 17/92, To appear in *Commun. Math. Phys.*
13. C. Di Castro and W. Metzner, Ward Identities and the β Function in the Luttinger Liquid, *Phys. Rev. Lett.*, **67**, 3852 (1991).

QUANTUM FLUCTUATION LIMIT: EXAMPLES FROM SOLID STATE PHYSICS

A. Verbeure[1] and V.A. Zagrebnov[2]

1. K.U.Leuven, Instituut Theoretische Fysica, B-3001 Leuven

2. Ecole Nationale Supérieure des Télécommunications-IMA
   F-75634 Paris
   On leave of absence from the Bogoliubov Theoretical Laboratory
   JINR-Dubna, 141980 Dubna, CIS-Russia

QUANTUM FLUCTUATION LIMIT

The mathematical basis for the quantum fluctuation limit is given in [8]. Take any quasi-local system defined on a lattice $\mathbb{Z}^d$, let $A(x)$ be any local observable in $x \in \mathbb{Z}^d$ and ρ a state of the system. We assume that the state is space homogeneous: $\rho \cdot \tau_x = \rho$ for all $x \in \mathbb{Z}^d$, τ_x is the space translation automorphism over the distance x. We assume also that the state is time translation invariant. In the applications we take for the state ρ an equilibrium state.

We are interested in a canonical definition of the fluctuations. A local fluctuation of the observable A in the state ρ is, as usual, defined by

$$F_{0,\Lambda}^k(A) = \frac{1}{\sqrt{\Lambda}} \sum_{x \in \Lambda} (\tau_x A(o) - \rho(A)) e^{ikx}$$

for any finite volume Λ and any fixed $k \in \mathbb{R}^d$. The problem is to have a volume independent fluctuation i.e. we have to give a meaning to the limit $V \to \infty$. The basic point of [8] is to realise this limit as a central limit. The result is the intrinsic definition of an operator

$$F_0^k(A) = \lim_\Lambda F_{0,\Lambda}^k(A)$$

acting on an appropriate Hilbert space. A basic condition for the existence of this central limit is the space-clustering of the state ρ. If the state ρ is not enough clustering, then one can adapt the definition as follows

$$F_{\delta,\Lambda}^k(A) = \frac{1}{|\Lambda|^{\frac{1}{2}+\delta}} \sum_{x \in \Lambda} e^{ikx} (\tau_x A(o) - \rho(A)) \qquad (1)$$

i.e. one looks for a δ such that
$$\lim_\Lambda F^k_{\delta,\Lambda}(A)$$
does exist; we introduce δ, a measure for the degree of abnormality of the fluctuation; it turns out to be a critical exponent. The value of δ depends on the chosen observable A and the properties of the state ρ. This is easy to understand because it is well-known that e.g. the cluster properties of an equilibrium state at high temperature are quite different from its properties at low temperatures, and e.g. criticality can be detected in terms of one observable and not of an other.

An important consequence of this point of view is that we obtained the fluctuations as operators, satisfying canonical commutation relations; one can see this as follows, if A and B are local observables, then

$$[F^k_\delta(A), F^{-k}_{\delta'}(B)]$$
$$= \lim_V \frac{1}{|\Lambda|^{1+\delta+\delta'}} \sum_{x\in\Lambda} \tau_x \left(\sum_{y\in\Lambda} [A, \tau_{y-x}B] \right) e^{ik(x-y)}$$
$$= \lim_\Lambda \frac{1}{|\Lambda|^{1+\delta+\delta'}} \sum_{x\in\Lambda} \tau_x([A,B])$$
$$= \begin{cases} \rho([A,B]) & \text{if } \delta+\delta' = 0 \\ 0 & \text{if } \delta+\delta' > 0 \\ \text{undefined} & \text{if } \delta+\delta' < 0 \end{cases} \quad (2)$$

In fact the central limit theorem defines a new dynamical system in which the observables are generated by the representation of a Boson field F^k_δ of fluctuations, represented in a state $\tilde\rho$, defined by :

$$\tilde\rho((F^k_\delta(A))^m) = \lim_\Lambda \rho((F^k_{\delta,\Lambda}(A))^m)$$

and governed by a time evolution $\tilde\alpha_t$, defined by:

$$\tilde\alpha_t F^k_\delta(A) = F^k_\delta(\alpha_t A)$$

where $\alpha_t(A) = \lim_\Lambda e^{itH_\Lambda} A e^{-itH_\Lambda}$, H_Λ is the usual Hamiltonian.

We describe two applications of this mathematical theory of fluctuations in the next sections. Both applications should be situated in theoretical solid state physics. One new phenomenon has been observed, namely the phenomenon of squeezing of fluctuations at a critical point. This is somehow surprising as one generally believes that quantum effects are not visible in the presence of long range correlations. The other major contribution consists in a rigorous proof of the Kohn anomaly for electron systems on a soft lattice in interaction with the phonons.

CRITICAL QUANTUM FLUCTUATIONS AT A STRUCTURAL PHASE TRANSITION

Consider the quantum anharmonic crystal described by the Hamiltonian [1]: $H_\Lambda = T_\Lambda + V_\Lambda$, $\Lambda \in \mathbb{Z}^d$ where T_Λ describes the harmonic part

$$T_\Lambda = \sum_{l\in\Lambda} \frac{P_l^2}{2m} + \frac{a}{2} \sum_{l\in\Lambda} Q_l^2 + \frac{1}{4} \sum_{l,l'} \phi_{l,l'}(Q_l - Q_{l'})^2$$

with Q_l the displacement operator at $l \in \mathbb{Z}^d$ from the equilibrium position, and P_l, the corresponding momentum; the anharmonicity is described by

$$U_\Lambda = |\Lambda| W \left(\frac{1}{|\Lambda|} \sum_{l \in \Lambda} Q_l^2 \right).$$

W is a positive, monotonically decreasing function such that $\lim_{x \to \infty} W(x) = 0$ and $W'(0) < -a/2$ [3]. This model is of the meanfield type, with exactly soluble equilibrium states on the local observables and manifesting a displacement structural phase transition. We prove the following:

Theorem 2.1

Denote by $\lambda = \frac{\hbar}{\sqrt{m}}$ the quantum parameter. There is a critical line of temperatures $T_c(\lambda)$, depending on λ, and a critical value λ_c, such that the extremal equilibrium states ρ satisfy:

$$\rho(Q_l) = \begin{cases} 0 & \text{if } T \geq T_c(\lambda) \\ \neq 0 & \text{if } T < T_c(\lambda), \ \lambda < \lambda_c \end{cases}$$

∎

The most interesting properties of the model are found on the critical line $T_c = T_c(\lambda)$, $0 \leq \lambda \leq \lambda_c$. In particular we have the following two properties about the (quantum) fluctuations with $k = 0$.

Theorem 2.2

If $T = T_c(\lambda) > 0$, then

(a) $\lim_\Lambda F_{\delta,\Lambda}(Q) = F_\delta(Q)$ with δ a function of the dimension; the displacement fluctuation operator is Gaussian but non-normal with

$$\delta = \begin{cases} \frac{1}{3} & \text{if } d = 3 \\ \frac{1}{4} + 0 & \text{if } d = 4 \\ \frac{1}{4} & \text{if } d \geq 5 \end{cases}$$

(b) $\lim_\Lambda F_{\delta',\Lambda}(P) = F_{\delta'}(P)$ with $\delta' = 0$ i.e. the momentum fluctuation operator is normal Gaussian

∎

In view of the general commutation formula (2), as $\delta + \delta' > 0$ the algebra of fluctuation observables is abelian. This is a mathematically rigorous result in the line of the common wisdom that quantum effects are not present at criticality. However we have the following:

Theorem 2.3

If $T_c = 0$, i.e. the pure quantum limit $\lambda = \lambda_c$, then

$$\lim_\Lambda F_{\delta,\Lambda}(Q) = F_\delta(Q)$$
$$\lim_\Lambda F_{\delta',\Lambda}(P) = F_{\delta'}(P)$$

with $\delta' = -\delta$ and $\delta > 0$.

∎

177

As in this case $\delta + \delta' = 0$, the algebra of fluctuation operators is non-abelian, quantum effects are visible. One sees the phenomenon of *squeezing* of the momentum fluctuation. One observes an uncertainty relation on the level of fluctuations.

On the critical line the equilibrium state is the unique solution of the KMS or equilibrium equation. Therefore, the algebra of observables at infinity (generated by the densities) is unique. However the algebra of fluctuation observables is very sensitive to e.g. the boundary conditions. In that sense this algebra is not unique on the critical line (see [2]). For $T < T_c(\lambda)$ this fact has been discovered in [1]. For $T = T_c(\lambda)$ we communicate the following result.

Theorem 2.4

If $T = T_c(\lambda)$, consider now the following perturbed Hamiltonian:

$$H_\Lambda^\alpha(\hat{h}) = H_\Lambda - \frac{\hat{h}}{|\Lambda|^\alpha} \sum_{l \in \Lambda} Q_l \; ; \; \hat{h} \in \mathbb{R}$$

then

$$\lim_\Lambda \rho_\Lambda^\alpha \left(e^{it F_{\delta,\Lambda}(Q)} \right) = \exp -\frac{t^2}{2} s_\alpha(Q, Q)$$

and analogously for P, where ρ_Λ^α is the local equilibrium state and s_α a quadratic form. These limits exist for $\delta = -\delta'$ with for $d = 3$

$$\delta = \begin{cases} \frac{2}{5}\alpha & \text{for } 0 \leq \alpha < \frac{5}{6} \equiv \alpha_c \\ \frac{1}{3} & \text{for } \alpha \geq \frac{5}{6} \end{cases}$$

∎

Another interesting point is the long-wavelength limit and the "soft-mode" phenomenon of the dynamics at the point of the structural phase transition. Therefore one has to consider fluctuation operators of the type

$$F_{\delta,\Lambda}^k(Q) = \frac{1}{|\Lambda|^{\frac{1}{2}+\delta}} \sum_{l \in \Lambda} (Q_l - \rho(Q)) \cos q \cdot l.$$

One derives the following fluctuation dynamical equations.

Theorem 2.5

$$\tilde{\alpha}_t(F_\delta^q(Q)) = e^{it L_q} F_\delta^q(Q)$$

with $L_q F_\delta^q(Q) = \frac{1}{i} \frac{\lambda}{\sqrt{m}} F_\delta^q(P)$.

∎

The fluctuation macrodynamics of this theorem yields an explanation of the "soft-mode" behaviour on the critical line. Remark that for $T = T_c(\lambda)$ one has on the basis of Theorem 2.3: $\delta > 0$ if $q = 0$. One computes also that $\delta = 0$ if $q \neq 0$.

The dynamical equations yield

$$L_q^2 F_\delta^q(Q) = \lambda^2 \Omega_q^2 F_\delta^q(Q) \text{ for } q \neq 0$$

and

$$L_q F_\delta^0(Q) = 0 \text{ for } q = 0.$$

This equation is a manifestation of the well-known softening of the long-wavelength mode ($q \to 0$) on the critical line ($\Omega_{q=0} = 0$) [2].

THE PEIERLS-FRÖHLICH INSTABILITY AND THE KOHN ANOMALY

We consider the tight-binding model of electrons on a soft quantum lattice [4]. This model represents a linear interaction between electrons $\{a_k^\#\}_{k\in\Lambda^*}$ and phonons $\{b_k^\#\}_{k\in\Lambda^*}$. The q-mode phonon interaction is described by the model Hamiltonian:

$$H_\Lambda = \sum_{k\in\Lambda^*}(\varepsilon_k - \mu)a_k^* a_k + \sum_{k\in\Lambda^*}\Omega_k b_k^* b_k$$
$$+\lambda\left(b_q^* \frac{1}{\sqrt{N}}\sum_{k\in\Lambda^*} a_k^* a_{k+q} + h.c.\right).$$

Here $\Lambda \subset \mathbb{Z}_a$, $|\Lambda| = N$; $\varepsilon_k = -\omega\cos ka$, $k \in \Lambda^*$; a the lattice spacing, 2ω is the size of the electron band. For $\mu = 0$ we have the half-filled band, the case considered by Peierls [5] and Fröhlich [6]. In this case the Fermi surface wavevector is $k_F = \pi/2a$. We consider here only the case $\mu = 0$. The model above is also called the Mattis-Langer model [9].

If $q = \pm 2k_F$ and the lattice is soft enough or the coupling is large enough i.e. $\Omega_q/\lambda^2 \ll 1$, then the thermodynamics is described in terms of the following results:

a) there is a critical temperature $T_c > 0$.

b) the electron spectrum depends on the temperature:

$$E(k) = (\text{sign }\varepsilon_k)\sqrt{\varepsilon_k^2 + \Delta^2}$$

with Δ the gap in the spectrum

$$\Delta(T) = \begin{cases} 0 & \text{if } T \geq T_c \\ > 0 & \text{if } T < T_c \end{cases}$$

c) there is spontaneous breaking of the symmetry, crossing the critical temperature, from $\mathbb{Z}_a \to \mathbb{Z}_{2a}$, accompanied by the condensation of the phonon modes $q = \pm 2k_F$.

This condensation is connected with the "freezing" of the Ω_q-mode, a phenomenon which is called the Kohn anomaly. The physical explanation of this is based on the analysis of the dynamics of the Fourier transform of the displacement [7]:

$$Q_k = \frac{1}{\sqrt{N}}\sum_{l\in\Lambda} Q_l\, e^{ikla} \equiv \left(\frac{1}{2\Omega_k}\right)^{1/2}(b_k + b_{-k}^*).$$

This is obviously the fluctuation operator of the displacement. It is because of this point that our theory of fluctuations will turn out to be useful. But let us first sketch the heuristic physical arguments.

Consider also the Fourier transform of the electron density operator

$$\rho_k = \sum_{l\in\Lambda} a_l^* a_l e^{ikla} = \sum_{p\in\Lambda^*} a_p^* a_{p+k}$$

then one gets for the equations of motion of the lattice vibrations

$$\ddot{Q}_q = -\Omega_q^2 Q_q - \lambda(2\Omega_q)^{1/2}\rho_q \frac{1}{\sqrt{N}}. \tag{3}$$

In order to close this dynamical equation one uses [7] the following physical argumentation. The electron density ρ_q, according to the linear response theory, is proportional to the corresponding lattice deformation Q_q, i.e.

$$\frac{1}{\sqrt{N}}\rho_q \;"="\; \chi(q)\lambda(2\Omega_q)^{1/2}Q_q.$$

Here $\chi(q)$ is the linear response function, which for $q = \pm 2k_F$ at $T = T_c$ becomes negative and very large. Therefore the effective frequency

$$\tilde{\Omega}_q^2 = \Omega_q^2 + 2\lambda^2 \Omega_q \chi(q)$$

becomes non-positive or $\tilde{\Omega}_q^2 \to 0$ at $T = T_c$. This is known as the Kohn anomaly. This is the point to understand. Besides this question there is also the mathematical question to understand formula (3) in the sense that one is equating two operators ρ_q and Q_q of different nature. These operators should be embedded in the same space. That is also possible to understand on the level of the algebra of quantum fluctuations.

Let us now come to some rigorous statements.

Theorem 3.1

At $T = T_c$ the quantum fluctuation limits exist for $\delta = 1/4$ and for $q = \pm 2k_F$

$$\lim_\Lambda F^q_{1/4,\Lambda}(\rho) = F^q_{1/4}(\rho)$$
$$\lim_\Lambda F^q_{1/4,\Lambda}(b) = F^q_{1/4}(b)$$

Here $F^q_{\delta,\Lambda}(\rho) = \dfrac{1}{N^{\frac{1}{2}+\delta}} \sum_{l \in \Lambda} a_l^* a_l e^{iql}$ and $F^q_{\delta,\Lambda}(b) = \dfrac{1}{N^{\frac{1}{2}+\delta}} \sum_{l \in \Lambda} b_l e^{iql}$

∎

This means that at $T = T_c$, both these limiting fluctuation operators are abnormal with the same critical exponent $\delta = 1/4$. On the other hand, on the basis of correlation inequalities one proves the following bound:

$$\lim_\Lambda \rho_\Lambda \left(\left(b_q^* + \frac{\lambda}{\Omega_q} F^q_{0,\Lambda}(\rho) \right) \left(b_q + \frac{\lambda}{\Omega_q} F^q_{0,\Lambda}(\rho) \right) \right) \le M < \infty$$

where M is independent of the temperature. Consequently one gets the existence of the following quantum fluctuation limits:

$$\lim_\Lambda F^q_{0,\Lambda}\left(b + \frac{\lambda}{\Omega_q}\rho \right) = F^q_0\left(b + \frac{\lambda}{\Omega_q}\rho \right)$$

for all temperatures, and at $T = T_c$

$$\lim_\Lambda F^q_{\frac{1}{4},\Lambda}\left(b + \frac{\lambda}{\Omega_q}\rho \right) = 0$$

or equivalently

$$F^q_{\frac{1}{4}}(b) = -\frac{\lambda}{\Omega_q} F^q_{\frac{1}{4}}(\rho) \tag{4}$$

i.e. the boson fluctuation $F^q_{1/4}(b)$ and the Fermion density fluctuation $F^q_{1/4}(\rho)$ are linearly dependent [8] at $T = T_c$. Moreover we proved rigorously the coherence of the lattice displacement with the electron density always at $T = T_c$. There is no mathematical evidence for this if $T \neq T_c$, in contrast to what is suggested in the physical literature [7] explained above.

Let us now turn to the dynamics of the fluctuations. The generator L of the dynamics is by definition:

$$L F_\delta^q(\ ) = \lim_\Lambda F_{\delta,\Lambda}^q([H_\Lambda,\ ]).$$

It turns out that the dynamics of the fluctuation operators is given by the following infinite hierarchy of equations:

$$L F_\delta^q(b) = -\Omega_q F_\delta^q(b)^* - \lambda F_\delta^q(\rho)$$
$$L F_\delta^q(\rho) = 2\varepsilon_0 F_\delta^q(j_1)$$
$$L F_\delta^q(j_1) = \varepsilon_0^2 F_\delta^q(\rho) + \varepsilon_0^2 F_\delta^q(j_2) - 2\lambda h_0 F_\delta(\rho)$$
$$\ldots$$
$$\ldots$$
$$\ldots$$

where j_1 is the local one-step jump operator: $a_l^* a_{l+1} + h.c.$; j_2 the local two-step jump operator; etc.; h_0 is a real number.

However using (4) one gets immediately

Theorem 3.2 (Kohn anomaly)
At $T = T_c$, if $q = \pm 2k_F$,

$$L F_{1/4}^q(b) = 0\ ,\ L F_{1/4}^q(\rho) = 0$$
$$L F_{1/4}^q(j_n) = 0\ ;\ n = 1, 2, \ldots.$$

∎

At $T = T_c$ all relevant fluctuation operators are operator valued eigenvectors of the dynamics with eigenvalue equal to zero, i.e. we get a complete "freezing" of the evolution for all fluctuation operators involved in the hierarchy describing the dynamics.

Acknowledgments

The research, on which this text is based, is realised thanks to the support of the Research Council KUL, grant OT/92/2.

REFERENCES

1. A. Verbeure, V.A. Zagrebnov, *J. Stat. Phys.*, **69**, 329 (1992)
2. A. Verbeure, V.A. Zagrebnov, in preparation
3. J.L. van Hemmen, V.A. Zagrebnov, *J. Stat. Phys.*, **53**, 835 (1988)
4. J. Pulé, A. Verbeure, V.A. Zagrebnov, "Peierls-Fröhlich Instability and Kohn Anomaly", preprint-KUL-TF-93/11
5. R.E. Peierls, "Quantum Theory of Solids", London, Oxford Univ. Press (1955)
6. H. Fröhlich, *Proc. Roy. Soc.*, **A223**, 296 (1954)
7. G.A. Toombs, *Phys. Rep.*, **40**, 182 (1978)
8. D. Goderis, A. Verbeure, P. Vets, *Commun. Math. Phys.*, **128**, 533 (1990)
9. D. C. Mattis, W. D. Langer, *Phys. Rev. Lett.*, **25**, 375 (1970)

LARGE DEVIATIONS AND THE THERMODYNAMIC FORMALISM: A NEW PROOF OF THE EQUIVALENCE OF ENSEMBLES[1]

J.T. Lewis[2], C.-E. Pfister[3] and W.G. Sullivan[2,4]

2. Dublin Institute for Advanced Studies
   10 Burlington Road
   Dublin 4, Ireland

3. Ecole Polytechnique Fédérale de Lausanne
   Département de Mathématiques
   CH-1015 Lausanne Switzerland

4. University College
   Department of Mathematics
   Belfield, Dublin 4, Ireland

1. THE EQUIVALENCE OF ENSEMBLES

In statistical mechanics the problem of the equivalence of ensembles goes back to Boltzmann and Gibbs. Here it is the problem of proving that, in the thermodynamic limit, the microcanonical measures and the grand canonical measures are equivalent; making precise the meaning of "equivalent" is part of the problem. It is commonly believed that in good statistical mechanical models such an equivalence holds, even in the presence of a phase-transition. On the other hand, it is believed that equivalence of ensembles fails in mean–field models such as the Curie–Weiss model.

There is a second statement which is also known as the equivalence of ensembles: in the thermodynamic limit, the negative of the entropy and the pressure are conjugate functions in the sense of convexity theory. In statistical mechanics, the entropy function is defined directly in the microcanonical setting and the pressure in the grand canonical setting. We refer to this statement as the equivalence of ensembles at the level of thermodynamic functions. This form of the equivalence of ensembles is known to hold for good statistical models and to fail for mean–field models. One version of our main result may be stated roughly as: *equivalence of ensembles holds at the level of measures whenever it holds at the level of thermodynamic functions.*

---

[1] Lecture delivered by J.T. Lewis

The problem of the equivalence of ensembles is not confined to statistical mechanics; it can be found in other areas of applied probability theory – in information theory, for example. Here the problem is to prove that a sequence of conditioned measures is equivalent, in an appropriate sense, to a sequence of "tilted" measures. Our choice of setting is sufficiently general to cover such applications.

Probabilistic methods have been used for at least fifty years to prove results about the equivalence of ensembles: Khinchine (1943) used a local limit theorem to prove it for a classical ideal (non–interacting) gas; Dobrushin and Tirozzi (1977) proved it for lattice gas models for which they were able to establish a local central limit theorem – a restriction which ruled–out models which exhibit first–order phase transitions. Typically, local central limit theorems hold on the scale of the square–root of the volume. The right scale for the investigation of the equivalence of ensembles, however, turns out to be that of the volume itself; this is the scale on which a large deviation principles hold. Deuschel et al. (1991) and Georgii (1993) used large deviation principle for empirical measures to prove the equivalence of ensembles. One draw–back with this approach is that it is technically difficult: since it involves measures on a space of measures, there are subtle points to be settled. Another is that the connection with thermodynamic functions is obscured. Our approach is more elementary and direct: we go back to the common origin of large deviation theory and statistical mechanics, the Principle of the Largest Term, and prove a result about the specific information gain of a sequence of conditioned measures with respect to a sequence of tilted measures. This is a "soft" theorem – it uses nothing deeper than the order–completeness of the reals, but it has a wide applicability. For non–interacting systems, the equivalence of ensembles for measures then follows from an inequality relating the information gain $\mathcal{H}(\mu|\nu)$ of μ with respect to ν to the total variation norm $\|\cdot\|_{TV}$ of the difference of measures:

$$2\mathcal{H}(\mu|\nu) \geq \|\mu - \nu\|_{TV}^2. \tag{1.1}$$

For interacting systems, our "soft" theorem has to be supplemented by a "hard" theorem, proved using the combinational devices introduced in Sullivan (1973) and perfected by Preston (1976); using it, we prove the equivalence of ensembles at the level of measures for a lattice gas with translation invariant summable potentials. In order to state this result precisely, we have to describe this setting in details; this we do in § 2. In § 3 we discuss the Principle of the Largest Term and its consequences, sketching the proof of our "soft" theorem. In § 4, we give an application to the non–interacting case. In § 5, we state precisely the general result for the lattice gas. Detailed proofs will be published elsewhere.

2. CONDITIONING AND TILTING

Let $\{(\Omega_n, \mathcal{F}_n, \rho_n)\}_{n \geq 1}$ be a sequence of measure spaces; here ρ_n is a positive measure referred to as the *reference measure*, which may or may not be normalized. Let $V_\circ := \{V_n \in (0, \infty)\}_{n \geq 1}$ be a *scale*, a sequence of positive numbers diverging to $+\infty$ as $n \to \infty$. Typically, in the applications to statistical mechanics, V_n will be the volume of a region Λ_n in a Euclidean space $\mathbb{R}^d$ or the number of lattice sites in a box Λ_n in an integer lattice $\mathbb{Z}^d$, and Ω_n will be a configuration space associated with Λ_n. Let $T_\circ := \{T_n : \Omega_n \to X\}_{n \geq 1}$ be a sequence of random variables taking values in X, a closed convex subset of E, a locally convex topological vector space; we denote the Borel subsets of X by $\mathcal{B}(X)$ and the topological dual of E by E^*. In this exposition we will assume that X is compact and that $E = \mathbb{R}^k$ ($k \geq 1$). These assumptions are not necessary (for the general case, see Lewis et al. (1993)) but they simplify the proofs and yet are adequate to cover the applications we make to the lattice gas.

For $C \in \mathcal{B}(X)$ such that $0 < \rho_n[T_n^{-1}C] < \infty$ for all n sufficiently large, we define the *conditioned measures* on $\mathcal{F}_n$ by

$$\nu_n^C[d\omega] := \frac{1_{T_n^{-1}C}(\omega)\rho_n[d\omega]}{\rho_n[T_n^{-1}C]} ; \qquad (2.1)$$

for $t \in E^*$ such that $0 < \int_{\Omega_n} \exp(V_n \langle t, T_n(\omega) \rangle) \rho_n[d\omega] < \infty$ for all n sufficiently large, we define the *tilted measures* on $\mathcal{F}_n$ by

$$\gamma_n^t[d\omega] := \frac{\exp(V_n \langle t, T_n(\omega) \rangle)\rho_n[d\omega]}{\int_{\Omega_n} \exp(V_n \langle t, T_n(\omega') \rangle)\rho_n[d\omega']} . \qquad (2.2)$$

We shall compute the specific information gain $\lim_{n \to \infty} \frac{1}{V_n} \mathcal{H}(\nu_n^C | \gamma_n^t)$; recall that $\mathcal{H}(\lambda_1 | \lambda_2)$, the information gain of λ_1 with respect to λ_2, is defined by

$$\mathcal{H}(\lambda_1 | \lambda_2) := \begin{cases} \int_{\Omega_n} \ln \frac{d\lambda_1}{d\lambda_2}(\omega) \lambda_1[d\omega], & \lambda_1 \ll \lambda_2, \\ +\infty, & \text{otherwise.} \end{cases} \qquad (2.3)$$

In the statistical mechanical applications, the T_n are k-tuples of functions such as energy–per–site and magnetization–per–site; then ν_n^C is the microcanonical measure conditioned on T_n taking values in C and γ_n^t is the grand canonical measure at generalized chemical potential t. Notice that both ν_n^C and γ_n^t are absolutely continuous with respect to the reference measure ρ_n and their densities are both functions of T_n; we exploit this by using the change of variable formula in computing the specific information gain. Define the distribution $\mathbb{M}_n$ of T_n under ρ_n by $\mathbb{M}_n := \rho_n \circ T_n^{-1}$; we have

$$\nu_n^C \circ T_n^{-1} = \mathbb{M}_n[\cdot | C] := \frac{\mathbb{M}_n[\cdot \cap C]}{\mathbb{M}_n[C]} , \qquad (2.4)$$

$$\gamma_n^t \circ T_n^{-1} = \mathbb{M}_n^t[\cdot | X] := \frac{\mathbb{M}_n^t[\cdot]}{\mathbb{M}_n^t[X]} , \qquad (2.5)$$

where $\mathbb{M}_n^t[dx] := \exp(V_n \langle t, x \rangle) \mathbb{M}_n[dx]$. Thus we have

$$\mathcal{H}(\nu_n^C | \gamma_n^t) = \mathcal{H}(\mathbb{M}_n[\cdot | C] | \mathbb{M}_n^t[\cdot | X]) . \qquad (2.6)$$

We shall see that this formula is the basic manoeuvre in our treatement; it reduces an integral over Ω_n to an integral over X and relates the information gain $\mathcal{H}(\nu_n^C | \gamma_n^t)$ to the thermodynamic functions which we are about to define in this setting.

3. THE PRINCIPLE OF THE LARGEST TERM

We need to examine the behaviour as $n \to \infty$ of the measures on X defined in § 2. Since the spaces $(\Omega_n, \mathcal{F}_n, \rho_n)$ and the random variables T_n play no part in the considerations of this section it is best to start afresh. Let $\mathbb{M}_\circ := \{\mathbb{M}_n\}_{n \geq 1}$ be a sequence of locally finite positive measures on $\mathcal{B}(X)$, the Borel subsets of X, a compact convex subset of $E = \mathbb{R}^k$. Let $V_\circ$ be a scale; define set–functions $m_n, \underline{m}, \overline{m}$ on $\mathcal{B}(X)$:

$$m_n[B] := \frac{1}{V_n} \ln \mathbb{M}_n[B] , \qquad (3.1)$$

$$\underline{m} := \liminf_{n \to \infty} m_n[B] , \qquad (3.2)$$

$$\overline{m} := \limsup_{n \to \infty} m_n[B] . \qquad (3.3)$$

The following properties are straightforward consequences of the definitions

$$\underline{m}[B] \leq \overline{m}[B] \text{ for all } B \in \mathcal{B}(X) ; \quad (3.4)$$

$$\underline{m} \text{ and } \overline{m} \text{ are increasing on } \mathcal{B}(X) . \quad (3.5)$$

The next property is an abstract version of the Principle of the Largest Term, well-known in traditional accounts of statistical mechanics (see, for example, Huang (1963)). Since it is central to our development, we give a proof. (For $a, b \in \mathbb{R}$, we denote the maximum of a and b by $a \vee b$.)

Lemma 3.1 *On $\mathcal{B}(X)$, we have*

$$\overline{m}[B_1 \cup B_2] = \overline{m}[B_1] \vee \overline{m}[B_2] . \quad (3.6)$$

Proof:
For $j = 1, 2$, we have

$$\mathbb{M}_n[B_j] \leq \mathbb{M}_n[B_1 \cup B_2] \leq \mathbb{M}_n[B_1] + \mathbb{M}_n[B_2] \quad (3.7)$$

so that

$$\mathbb{M}_n[B_1] \vee \mathbb{M}_n[B_2] \leq \mathbb{M}_n[B_1 \cup B_2] \leq 2\mathbb{M}_n[B_1] \vee \mathbb{M}_n[B_2] ; \quad (3.8)$$

it follows that

$$\overline{m}[B_1 \cup B_2] = \limsup_{n \to \infty}(m_n[B_1] \vee m_n[B_2]) . \quad (3.9)$$

But for each pair $\{a_n\}_{n \geq 1}, \{b_n\}_{n \geq 1}$ of sequences of real numbers, we have

$$\limsup_{n \to \infty}(a_n \vee b_n) = (\limsup_{n \to \infty} a_n) \vee (\limsup_{n \to \infty} b_n) . \quad (3.10)$$

Thus (3.6) follows from (3.9) and (3.10). □

Define functions $\underline{\mu}, \overline{\mu}$ on X as follows:

$$\underline{\mu}(x) := \inf_{G \ni x} \underline{m}[G] , \quad G \text{ open} , \quad (3.11)$$

$$\overline{\mu}(x) := \inf_{G \ni x} \overline{m}[G] , \quad G \text{ open} . \quad (3.12)$$

The following properties are direct consequences of the definitions:

$$\underline{\mu} \text{ and } \overline{\mu} \text{ are upper semicontinuous functions} ; \quad (3.13)$$

$$\overline{m}[G] \geq \sup_{x \in G} \overline{\mu}(x) , \quad G \text{ open} , \quad (3.14)$$

$$\underline{m}[G] \geq \sup_{x \in G} \underline{\mu}(x) , \quad G \text{ open} . \quad (3.15)$$

The lower bound (3.14) for $\overline{m}$ on open sets is rarely used; of greater importance is the following upper bound for $\overline{m}$ on compact sets, a consequence of the Principle of the Largest Term (3.6)

$$\overline{m}[K] \leq \sup_{x \in K} \overline{\mu}(x) , \quad K \text{ compact} . \quad (3.16)$$

Our first application of (3.16) is to the *concentration of measures*. Let $\mathbb{M}_\circ$ be a sequence of probability measures on $\mathcal{B}(X)$; if $\mathbb{M}_\circ$ converges weakly to a Dirac measure δ_x at some point $x \in X$, we say $\mathbb{M}_\circ$ obeys a weak law of large numbers (WLLN). In the absence of a first-order phase transition, a WLLN holds in the grand canonical ensemble. We require a substitute for a WLLN which holds regardless of phase transitions.

We say that a sequence $\mathbb{M}_\circ$ of probability measures on $\mathcal{B}(X)$ is *eventually concentrated on a set* A if, for each open neighbourhood G of A, we have

$$\lim_{n \to \infty} \mathbb{M}_n[G] = 1 \ . \tag{3.17}$$

[If $A = \{x\}$ and $\mathbb{M}_\circ$ is eventually concentrated on A, then $\mathbb{M}_\circ$ converges weakly to the Dirac measure δ_x.] We shall need the following

Lemma 3.2 *Let $\mathbb{M}_\circ$ be a sequence of probability measures on $\mathcal{B}(X)$ which is eventually concentrated on a set A; if $f : X \to \mathbb{R}$ is lower semicontinuous and bounded below on X, then*

$$\inf_{x \in A} f(x) \leq \liminf_{n \to \infty} \int_X f(x) \mathbb{M}_n[dx] \ . \tag{3.18}$$

[There is an obvious complementary upper bound; together they yields the usual characterization of the WLLN in terms of bounded continuous functions when A reduces to a single point.]

The function $\overline{\mu}$, defined at (3.12) for the pair $(\mathbb{M}_\circ, V_\circ)$, enables us to determine a concentration–set for the sequence $\mathbb{M}_\circ$. (How useful it is depends on how well we have chosen the scale $V_\circ$.) Notice that, for probability measures, the function $\overline{\mu}$ is bounded above by zero; in fact, it always attains this bound and the set on which it attains it is a concentration–set for $\mathbb{M}_\circ$. Let $N_{\overline{\mu}}$ be the set defined by

$$N_{\overline{\mu}} := \{x \in X : \overline{\mu}(x) = 0\} \tag{3.19}$$

Lemma 3.3 *Let $\mathbb{M}_\circ$ be a sequence of probability measures and $V_\circ$ a scale. Then*
(a) $N_{\overline{\mu}}$ is compact and non–empty;
(b) the sequence $\mathbb{M}_\circ$ is eventually concentrated on $N_{\overline{\mu}}$.

The proofs of both (a) and (b) make use of the bound (3.16)

Let $\overline{\mu}^t$, $\underline{\mu}^t$ be the upper and lower functions determined by the pair $(\mathbb{M}_\circ^t, V_\circ)$; they are related to $\overline{\mu}$ and $\underline{\mu}$ as follows:

$$\overline{\mu}^t(x) = \overline{\mu}(x) + \langle t, x \rangle \ , \tag{3.20}$$
$$\underline{\mu}^t(x) = \underline{\mu}(x) + \langle t, x \rangle \ . \tag{3.21}$$

These relations are a consequence of the continuity of the function $x \mapsto \langle t, x \rangle$. We are now ready for our third application of the bound (3.16): we prove a special case of Varadhan's Theorem (see Varadhan (1966)). If $\overline{\mu}(x) = \underline{\mu}(x)$ for all $x \in X$, we say the *Ruelle–Lanford function (RL–function)* μ exists for the pair $(\mathbb{M}_\circ, V_\circ)$ and is given by

$$\mu(x) := \underline{\mu}(x) = \overline{\mu}(x) \ . \tag{3.22}$$

When the RL–function exists, the bounds (3.15) and (3.16) can be restated as

$$\overline{m}[K] \leq \sup_{x \in K} \mu(x) \ , \quad K \text{ compact} \ , \tag{3.23}$$
$$\underline{m}[G] \geq \sup_{x \in G} \mu(x) \ , \quad G \text{ open} \ ; \tag{3.24}$$

When (3.23) and (3.24) hold, we say (following Varadhan (1966)) that a large deviation principle (LDP) holds with rate–function $I = -\mu$ for the pair $(\mathbb{M}_\circ, V_\circ)$. This means that the sequence $m_\circ$ of set–functions m_n, defined at (3.1), converges to the set–function

$$B \mapsto \sup_{x \in B} \mu(x) \tag{3.25}$$

in *exactly the same sense* that a sequence of probability measures $\mathbb{M}_\circ$ converges to a measure δ_x in a WLLN (remember that X is assumed to be compact). [We have given μ the name "Ruelle–Lanford function" because, in the setting of a lattice gas with translation-invariant summable potentials, our definition coincides with the definition of entropy given by Ruelle (1965) and Lanford (1973). Ruelle and Lanford understood that giving precise meaning to Boltzmann's formula

$$S = k \ln W , \tag{3.26}$$

relating the entropy S of a macroscopic equilibrium state to the number W of corresponding microscopic states is the *same problem* as that of making sense of the convergence of the sequence $m_\circ$ to the set–function (3.25); by so doing, they introduced a new technique to the theory of large deviations (compare Bahadur and Zabel (1979)).]

We are now ready to begin the calculation of the specific information gain using (2.6). First we have a result which is proved using (3.23) and (3.24):

Lemma 3.4 *Suppose the RL–function μ exists for the pair $(\mathbb{M}_\circ, V_\circ)$ and the set $C \in \mathcal{B}(X)$ is such that*

$$-\infty < \sup_{x \in C} \mu(x) = \underline{m}[C] = \overline{m}[C] = \sup_{x \in \overline{C}} \mu(x) ; \tag{3.27}$$

then the sequence $\mathbb{M}_\circ[\,\cdot\,|C]$ of probability measures is eventually concentrated on the set

$$X_{\overline{C}} := \{x \in \overline{C} : \mu(x) = \sup_{y \in \overline{C}} \mu(y)\} . \tag{3.28}$$

Lemma 3.5 *Suppose that the RL–function μ exists for the pair $(\mathbb{M}_\circ, V_\circ)$; then*
(a) the RL–function μ^t exists for the pair $(\mathbb{M}_\circ^t, V_\circ)$;
(b) the pair $(\mathbb{M}_\circ^t, V_\circ)$ obeys an LDP:

$$\overline{m}^t[K] \leq \sup_{x \in K} \mu^t(x) , \quad K \text{ compact} , \tag{3.29}$$

$$\underline{m}^t[G] \geq \sup_{x \in G} \mu^t(x) , \quad G \text{ open} ; \tag{3.30}$$

(c) μ^t is given by

$$\mu^t(x) = \langle t, x \rangle + \mu(x) . \tag{3.31}$$

If $\overline{m}^t[X] = \underline{m}^t[X]$ for all $t \in E^*$, we say that the *scaled generating function* p exists for the pair $(\mathbb{M}_\circ, V_\circ)$ and is given by

$$p(t) := \overline{m}^t[X] = \underline{m}^t[X] . \tag{3.32}$$

(In the statistical mechanical setting, p is called the *grand canonical pressure*.) Recall that if $f : X \to \mathbb{R}$, then $f^* : E^* \to \overline{\mathbb{R}}$ is defined by

$$f^*(t) := \sup_{x \in X} \{\langle t, x \rangle - f(x)\} . \tag{3.33}$$

Corollary 3.1 *Suppose the RL–function μ exists for the pair $(\mathbb{M}_\circ, V_\circ)$; then the scaled generating function p exists and is given by*

$$p(t) = (-\mu)^*(t) \tag{3.34}$$

Proof:
Since X is both compact and open (as a topological space), we have

$$\sup_{x \in X} \mu^t(x) \leq \underline{m}^t[X] \leq \overline{m}^t[X] \leq \sup_{x \in X} \mu^t(x) . \tag{3.35}$$

□

We define the set X^t for $t \in E^*$ by

$$X^t := \{x \in X : p(t) = \langle t, x \rangle + \mu(x)\} . \tag{3.36}$$

Theorem 3.1 *Suppose the RL–function μ exists for the pair $(\mathbb{M}_\circ, V_\circ)$ and condition (3.27) holds; if $X_{\overline{C}} \subset X^t$, then the specific information gain is zero:*

$$\lim_{n \to \infty} \frac{1}{V_n} \mathcal{H}(\nu_n^C | \gamma_n^t) = 0 . \tag{3.37}$$

Proof:
By (2.6), we have

$$\frac{1}{V_n} \mathcal{H}(\nu_n^C | \gamma_n^t) = \frac{1}{V_n} \mathcal{H}(\mathbb{M}_n[\cdot | C] | \mathbb{M}_n^t[\cdot | X]) \tag{3.38}$$

$$= - \int \langle t, x \rangle \mathbb{M}_n[dy|C] + m_n^t[X] - m_n[C] .$$

By Lemmas 3.4, 3.2, Corollary 3.1 and condition (3.27), we have

$$0 \leq \limsup_{n \to \infty} \frac{1}{V_n} \mathcal{H}(\nu_n^C | \gamma_n^t) \leq - \inf_{y \in X_{\overline{C}}} \langle t, x \rangle + p(t) - \sup_{y \in X_{\overline{C}}} \mu(y) \tag{3.39}$$

$$= \sup_{y \in X_{\overline{C}}} \{p(t) - \langle t, y \rangle - \mu(y)\}$$

$$= 0$$

if $X_{\overline{C}} \subset X^t$. □

4. AN APPLICATION

To illustrate how Theorem 3.1 may be applied, we consider a case of sums of independent identically distributed random variables. We set $\Lambda_n := \{1, \ldots, n\}$, and in this example $V_n := |\Lambda_n| = n$, $\Omega_n := \{0,1\}^{\Lambda_n}$, $\mathcal{F}_n := \mathcal{P}(\Omega_n)$. For $\omega \in \Omega_n$, put $\xi_j(\omega) := \omega(j)$, $j \in \Lambda_n$, and set $\rho_n[\xi_j = 0] = \frac{1}{2} = \rho_n[\xi_j = 1]$. Then $T_n := V_n^{-1} \sum_{j \in \Lambda_n} \xi_j$, $X := [0,1]$, $E := \mathbb{R} = E^*$. Define $s : X \to [0,1]$ by

$$s(x) = -x \ln x - (1-x) \ln(1-x) , \quad x \in (0,1) , \quad s(0) = s(1) = 0 . \tag{4.1}$$

Choose $C = (c_1, c_2) \subset [0,1]$; the RL–function μ exists for the pair $(\mathbb{M}_\circ, V_\circ)$ and is given by

$$\mu(x) = s(x) - \ln 2 ; \tag{4.2}$$

the set $X_{\overline{C}} = \{x^*\}$ where

$$x^* = \begin{cases} c_1, & \frac{1}{2} \leq c_1, \\ \frac{1}{2}, & c_1 < \frac{1}{2} < c_2, \\ c_2, & c_2 \leq \frac{1}{2}; \end{cases} \tag{4.3}$$

p is given by
$$p(t) = \ln(1 + e^t) - \ln 2 ;\tag{4.4}$$
and the set $X^t = \{x_t\}$ where
$$x_t = p'(t) = \frac{e^t}{1+e^t} .\tag{4.5}$$
Given C, we can find t^* such that $X_{\overline{C}} = X^{t^*}$; thus we have
$$\lim_{n\to\infty} \frac{1}{V_n} \mathcal{H}(\nu_n^C | \gamma_n^{t^*}) = 0 \tag{4.6}$$

We can use (4.6) to obtain a result on the limit of the sequence $\{\nu_{n,\Delta}^C\}_{n\geq 1}$, where $\nu_{n,\Delta}^C$ is the restriction to a finite subset Δ of N. Notice that γ_n^t is a product measure; this has two important consequences:

1. the restriction of γ_n^t to $\Delta \subset \{1,\ldots,n\}$ is independent of n and we denote it by γ_Δ^t;

2. if Δ_1 and Δ_2 are disjoint copies of Δ such that $\Delta_1 \cup \Delta_2 \subset \{1,\ldots,n\}$, then
$$\mathcal{H}(\nu_{n,\Delta_1\cup\Delta_2}^C | \gamma_{\Delta_1\cup\Delta_2}^t) \geq \mathcal{H}(\nu_{n,\Delta_1}^C | \gamma_{\Delta_1}^t) + \mathcal{H}(\nu_{n,\Delta_2}^C | \gamma_{\Delta_2}^t) .\tag{4.7}$$

But
$$\mathcal{H}(\nu_{n,\Delta_1}^C | \gamma_{\Delta_1}^t) = \mathcal{H}(\nu_{n,\Delta_2}^C | \gamma_{\Delta_2}^t) ,\tag{4.8}$$
so that
$$\mathcal{H}(\nu_n^C | \gamma_n^t) \geq \left[\frac{V_n}{|\Delta|}\right] \mathcal{H}(\nu_{n,\Delta}^C | \gamma_\Delta^t) ;\tag{4.9}$$
hence (4.6) implies that
$$\lim_{n\to\infty} \mathcal{H}(\nu_{n,\Delta}^C | \gamma_\Delta^{t^*}) = 0 .\tag{4.10}$$
It now follows from (1.1) that $\{\nu_{n,\Delta}^C\}_{n\geq 1}$ converges in total variation norm to the product measure $\gamma_\Delta^{t^*}$.

5. THE LATTICE GAS

We consider the lattice gas model: let $\mathbb{Z}^d$ ($d \leq 1$) be an integer–lattice, let $\{\Lambda_n\}_{n\geq 1}$ be an increasing sequence of cubes in $\mathbb{Z}^d$ with $V_n := |\Lambda_n| \to \infty$ as $n \to \infty$; at each site $j \in \Lambda_n$ we have a configuration space S_j which is a copy of some fixed compact Hausdorff space S. For each $n \geq 1$, the configuration space Ω_n is the space $\Omega_n = \prod_{j\in\Lambda_n} S_j$ which we regard as a subspace of the product space $\Omega = \prod_{j\in\mathbb{Z}^d} S_j$ equipped with the product topology, hence which is compact; the σ–field $\mathcal{F}_n$ is the σ–field of Borel subsets of Ω generated by the coordinate projections $\Omega \to S_j$. For each $j \in \mathbb{Z}^d$ we have the action of $\mathbb{Z}^d$ on itself given by $i \mapsto i + j$, $i \in \mathbb{Z}^d$; this lifts to $\theta_j : \Omega \to \Omega$ given by $(\theta_j\omega)(i) = \omega(i - j)$ for each configuration $\omega \in \Omega$. On each S_j we define a reference measure ρ^j, a copy of a fixed positive measure on S with $\rho^j(S_j) = 1$; on Ω we define the product measure $\rho = \prod_{j\in\mathbb{Z}^d} \rho^j$ and we take ρ_n to be the restriction of ρ to $\mathcal{F}_n$. The interaction in the model is given by a k–dimensional vector of translation–invariant absolutely summable potentials with either free or fixed boundary conditions. Using these potentials, we define mappings $T_n : \Omega_n \to X$ which give the energy per site of a configuration; here X is a compact convex subset of $E = \mathbb{R}^k$. We now define the conditioned measures ν_n^C and the tilted measures γ_n^t as in § 2; in this setting, the measure ν_n^C is the microcanonical measure on the cube Λ_n condition on T_n taking values in C (if C is an open neighbourhood of a point in X, then $T_n^{-1}C$ is what is sometimes called a "thin energy-shell" in Ω_n) and γ_n^t is a Gibbs measure on Λ_n with generalized chemical potential $t \in E^* = \mathbb{R}^k$. Using standard methods, we prove that $\overline{\mu}$ and $\underline{\mu}$ are independent of boundary conditions. Let $B_\varepsilon(x)$ be an open ball of radius ε and centre x in X; we prove, in the case of free boundary conditions, the following result.

Lemma 5.1 *Let x_0, x_1, $x_2 \in X$ satisfy $x_0 + x_1 = 2x_2$ and let $0 < \varepsilon' < \varepsilon$; then*

$$2\underline{m}[B_\varepsilon(x_2)] \geq \overline{m}[B_{\varepsilon'}(x_0)] + \overline{m}[B_{\varepsilon'}(x_1)] . \tag{5.1}$$

From this and the independence of $\overline{\mu}$ and $\underline{\mu}$ on the boundary conditions, we deduce the

Corollary 5.1 *The RL-function μ exists for the pair $(\mathbb{M}_\circ, V_\circ)$ and is concave on X.*

We have reserved the name "entropy" for the RL-functions which are concave; henceforth in this section, we refer to μ as the entropy of the pair $(\mathbb{M}_\circ, V_\circ)$ and to p, given by $p(t) = (-\mu)^*(t)$, as the grand canonical pressure. We now choose C to be an open convex subset of X; using convexity theory, we prove

Lemma 5.2 *Let C be an open convex subset of X; if μ is concave, then*
(a) $\sup_{x \in C} \mu(x) = \underline{m}[C] = \overline{m}[\overline{C}] = \sup_{x \in \overline{C}} \mu(x)$;
(b) *the entropy μ_C of the pair $(\mathbb{M}_\circ[\,\cdot\,|C], V_\circ)$ is given by*

$$\mu_C(x) = \begin{cases} \mu(x) - \sup_{y \in \overline{C}} \mu(y), & y \in \overline{C}, \\ -\infty, & y \in X\backslash\overline{C}. \end{cases} \tag{5.2}$$

We see from (a) that, provided C is chosen so that it contains a point at which μ is finite, condition (3.27) is satisfied. Part (b) gives an interpretation of $X_{\overline{C}}$ in this case: $X_{\overline{C}} = N_{\mu_C}$, the set on which the entropy attains its supremum. There is also an interpretation of the set X^t which follows from the concavity of μ: using convexity theory we can show that

$$X^t = \partial p(t) , \tag{5.3}$$

(∂f denotes the subgradients to a convex function f; when $\dim X = 1$, the interval $\partial p(t)$ is "a phase-transition segment" in the grand canonical ensemble; it reduces to a point in the absence of a first order transition.) We see that Theorem 3.1 now yields

Theorem 5.1 *Let μ be the entropy of a lattice gas with translation invariant summable potential. Let C be an open convex neighbourhood of a point at which μ is finite. Then there exists t^* such that*

$$\lim_{n \to \infty} \frac{1}{V_n} \mathcal{H}(\nu_n^C | \gamma_n^{t^*}) = 0 . \tag{5.4}$$

Because, in the presence of a non-trivial interaction, the Gibbs measures γ_n^t are not product measures, the subadditivity argument used in § 4 fails. There is a second difficulty: in § 4 we exploited permutation-invariance (exchangeability) at (4.8); here we must replace it by translation-invariance, but the measures ν_n^C associated with the cubes Λ_n are not translation-invariant. The way-out is to introduce translation-averages: define

$$\overline{\nu}_n^C := \frac{1}{V_n} \sum_{j \in \Lambda_n} \nu_n^C \circ \theta_j^{-1} , \tag{5.5}$$

where ν_n^C is extended to Ω in the usual way. We are able to prove

Theorem 5.2 *Suppose that (5.4) holds; then any weak limit point of the sequence $\{\bar{\nu}_n^C\}_{n\geq 1}$ is a Gibbs state with respect to the specification associated with $\{\gamma_n^{t^*}\}_{n\geq 1}$.*

The statement of this theorem make precise the sense in which the measures $\bar{\nu}_n^C$ and $\gamma_n^{t^*}$ are "equivalent" in the thermodynamic limit – something we said in § 1 was part of the problem.

Putting Theorems 5.1 and 5.2 together, we see that the entropy μ can be used to find a value t^* of the chemical potential such that any weak limit of the sequence $\{\bar{\nu}_n^C\}_{n\geq 1}$ is a Gibbs state with respect to the specification determined by $\{\gamma_n^{t^*}\}_{n\geq 1}$. This is possible because, as a consequence of the concavity of μ, we have $\mu(x) = -p^*(x)$ as well as $p(t) = (-\mu)^*(t)$; but these statements together constitute the equivalence of ensembles at the level of thermodynamic functions. It is in this sense that *equivalence of ensemble holds at the level of measures whenever it holds at the level of thermodynamic functions.*

REFERENCES

1. R.R. Bahadur and S.L. Zabell, Large deviations of the sample mean in general vector spaces, *Ann. Prob.* 7: 587 (1979).
2. J.-D. Deuschel, D.W. Stroock and H. Zessin, Microcanonical distribution for lattice gases, *Commun. Math. Phys.* 139: 83 (1991).
3. R.L. Dobrushin and B. Tirozzi, The central limit theorem and the problem of equivalence of ensembles, *Commun. Math. Phys.* 54: 173 (1973).
4. H.-O. Georgii, Large deviations and maximum entropy principle for interacting random fields on $\mathbb{Z}^d$, *Ann. Prob.* to appear (1993).
5. K. Huang, Statistical Mechanics, Wiley, New-York (1963).
6. A.Ya. Khinchin, Matematicheskie Osnovaniya Statisticheskoi Mekhaniki, Gostekhizdat, Moscow-Leningrad (1943). Translation: Mathematical Foundations of Statistical Mechanics, Dover, New-York (1949).
7. O.E. Lanford, Entropy and equilibrium states in classical mechanics, in Statistical Mechanics and Mathematical Problems, A. Lenard, ed., Lecture Notes in Physics 20, Springer (1973).
8. J.T. Lewis, C.-E. Pfister and W.G. Sullivan, DIAS preprint (1993).
9. C.J. Preston, Random Fields. Lecture Notes in Mathematics 534, Spinger, Berlin (1976).
10. D. Ruelle, Correlation functionals, *J. Math. Physics* 6: 201 (1965).
11. W.G. Sullivan, Potentials for almost markovian random fields, *Commun. Math. Phys.* 33: 61 (1973).
12. S.R.S. Varadhan, Asymptotic probabilities and differential equations, *Comm. Pure App. Math.* 19: 261 (1966).

FINITELY CORRELATED PURE STATES

R.F. Werner

FB Physik, Universität Osnabrück
49069 Osnabrück, Germany

INTRODUCTION

This paper is a report on joint work with Mark Fannes and Bruno Nachtergaele on a class of states on one-dimensional quantum spin systems, called "finitely correlated". These states can be constructed quite explicitly, which in itself is remarkable, since apart from quasi-free states or convex combinations of product states, there seem to be no states allowing a similarly complete control. One of their characteristic properties is the absence of a certain, typically quantum mechanical, "frustration" phenomenon. In this sense they do not exhibit the full complexity of quantum mechanical correlations. On the other hand, the structure of this class is very rich, and it is large enough to approximate any translationally invariant state (in the weak*-topology). Moreover, finitely correlated states appear as the exact ground states in physical models, particularly of antiferromagnetism. Therefore their study is an ideal vantage point from which to begin an exploration of the complexities of quantum mechanical correlations. The aim of the present article is to give an overview of some of the results obtained so far, emphasizing motivations and intuitive ideas, and referring to the journal articles for the technical aspects, and further references.

Quantum Frustration

The phenomenon of frustration is very well known from classical statistical mechanics. Let us consider the simplest possible example: an antiferromagnetic Ising spin system on a triangle. Thus, at each of the vertices labelled $i = 1, 2, 3$, we have a spin variable $\sigma_i = \pm 1$, and the Hamiltonian of the system is

$$H(\sigma) = \sigma_1\sigma_2 + \sigma_2\sigma_3 + \sigma_3\sigma_1 \quad . \tag{1}$$

We are looking for the ground state of this system, i.e. a probability measure on the 2^3 configurations minimizing the expectation of H. Let us suppose, as we may, that the measure should also be invariant under the global spin flip $\sigma_i \mapsto -\sigma_i$. Then for just two points there is the obvious unique solution of assigning equal weight to the two configurations $(+,-)$ and $(-,+)$. So it is natural to start from this partial solution of the problem, and to look for a measure on three spins having the two site solution as its marginal for any pair. Phrased differently, we are looking for a common extension of the probability measures given for each of the three edges. The phenomenon of frustration is now that such a *common extension may fail to exist*.

In fact, we have a simple necessary condition for extendibility, which in this case turns out to be sufficient as well. Our setting is precisely that considered by Bell in the first version of his famous inequality[1], which just says that the above Hamiltonian is bounded below by -1. A common extension of the states for the pairs would give expectation -1 to each term, hence -3 to H, and is hence impossible.

Clearly, in a quantum system the same kind of frustration may occur, as we can see by considering the Ising spins as one spin component of a spin-1/2 particle. But worse things can happen. Consider again three points, with a quantum spin at each site, described by observable algebras $\mathcal{A}_1, \mathcal{A}_2$, and $\mathcal{A}_3$, respectively. Typically, these would be finite dimensional matrix algebras. Consider states ω_{12} and ω_{23} defined on the subalgebras $\mathcal{A}_1 \otimes \mathcal{A}_2$ and $\mathcal{A}_2 \otimes \mathcal{A}_3$, which have the same restriction to the middle algebra $\mathcal{A}_2$. We are looking for a common extension of these states to a state ω_{123} on $\mathcal{A}_1 \otimes \mathcal{A}_2 \otimes \mathcal{A}_3$. This is almost the exact quantum translation of the previous problem with the only significant difference that we make no assumptions about a state given on $\mathcal{A}_3 \otimes \mathcal{A}_1$. In the classical case this modification makes the state extension problem trivially solvable. However, in the quantum case it does not: in general the *common extension may fail to exist*[2]. Hence a quantum system may be frustrated even if its underlying graph contains no loops.

Let us again take the simplest example, the Heisenberg antiferromagnetic chain of three spins 1/2. The operator in $\mathcal{A} \otimes \mathcal{A}$ describing this nearest neighbour interaction is the flip operator taking $\phi \otimes \psi$ to $\psi \otimes \phi$, with eigenvalues ± 1. Hence for the two site problem we have a unique state minimizing the energy, given by the (up to a factor unique) antisymmetric vector $\Phi \in \mathbb{C}^2 \otimes \mathbb{C}^2$. Consider a state on the three site system with these restrictions. Then all eigenvectors of its density matrix with non-zero eigenvalues would have to be in the subspace $\Phi \otimes \mathbb{C}^2 \subset \mathbb{C}^2 \otimes \mathbb{C}^2 \otimes \mathbb{C}^2$, as well as the subspace $\mathbb{C}^2 \otimes \Phi$. Since the transpositions $(1 \leftrightarrow 2)$ and $(2 \leftrightarrow 3)$ generate the permutation group such a vector would have to be completely antisymmetric, which is impossible.

It is clear that these problems become worse in the infinite chain, and indeed the one-dimensional Heisenberg antiferromagnet still presents a great challenge to rigorous treatment. From the notion of frustration one might expect that the complexity of one-dimensional quantum systems is comparable to that of two-dimensional classical systems, and indeed, as the contribution of Bruno Nachtergaele[3] shows, there is a rigorous interpretation of this comparison.

It is suggestive, then, to look for those Hamiltonians on a quantum spin chain which are *not* frustrated. This would be a translationally invariant nearest neighbour interaction which admits a ground state with the special property that it minimizes each interaction term separately. The first detailed example of such a system appeared in a paper by Affleck, Kennedy, Lieb, and Tasaki[4]. This paper has been a paradigm for the further development, and its generalization led directly to the notion of finitely correlated states.

Locally exposed states

Let us put the notion of "frustration-free" states in more precise terms. In order to describe a quantum spin chain we first have to fix the description of the single spins, by choosing an observable algebra $\mathcal{A}$. For a spin-s system this will be the algebra $\mathcal{M}_{2s+1}$ of $(2s+1)$-dimensional square matrices. With each of the sites $i \in \mathbb{Z}$ on the chain we associate an isomorphic copy $\mathcal{A}_i \cong \mathcal{A}$. With a finite set $\Lambda \subset \mathbb{Z}$ of sites we associate the observable algebra $\mathcal{A}_\Lambda = \bigotimes_{i \in \Lambda} \mathcal{A}_i$. The translation $\tau_x, x \in \mathbb{Z}$ takes $\mathcal{A}_i$ into $\mathcal{A}_{i+x}$ by composing the isomorphisms $\mathcal{A}_i \cong \mathcal{A} \cong \mathcal{A}_{i+x}$, and consequently takes $\mathcal{A}_\Lambda$ into $\mathcal{A}_{\Lambda+x}$. If we want to let the translations act as a group, we have to go to the infinite system, whose observable algebra is the C*-inductive limit $\mathcal{A}_\mathbb{Z}$ of the algebras $\mathcal{A}_\Lambda$ for finite $\Lambda = \{-L, \ldots, L\}$.

States on the chain are by definition positive normalized linear functionals $\omega : \mathcal{A}_\mathbb{Z} \to \mathbb{C}$. Equivalently, a state is given by the family of all its restrictions to finite segments $\mathcal{A}_\Lambda$, also called the system of correlation functions associated with the state. Since each $\mathcal{A}_\Lambda$ is a finite dimensional matrix algebra, these restrictions are given by density matrices, and a state is the same thing as a family of local density matrices, consistent with respect to the operation of restriction to subchains. A state is called *translationally invariant* if $\omega(\tau_x(A)) = \omega(A)$ for all $A \in \mathcal{A}_\mathbb{Z}$, or equivalently for $A \in \mathcal{A}_\Lambda$ for arbitrary Λ. The set of translationally invariant states will be denoted by $\mathcal{T}$.

A *finite range interaction* is given by a hermitian element $h \in \mathcal{A}_{\{1,\ldots,\ell\}}$, where ℓ is the range of the interaction. The total Hamiltonian is then, formally,

$$H = \sum_{x \in \mathbb{Z}} \tau_x(h) \ . \tag{2}$$

For a finite subsystem the Hamiltonian $H_\Lambda \in \mathcal{A}_\Lambda$ is the sum of all terms in this sum such that $\{1, \ldots, \ell\} + x \subset \Lambda$. Without affecting thermodynamic or ground state properties we can also modify this by arbitrary "boundary" terms with a norm bound going to zero when divided by the number of sites. For any hermitian operator H we denote by "inf $\sigma(H)$" the infimum of the spectrum of H, i.e. the smallest eigenvalue when H is in a finite dimensional algebra. The ground state energy of the model with interaction h is now

$$E_0(h) := \liminf_{L \to \infty} \frac{1}{L} \inf \sigma(H_{\{-L,\ldots,L\}}) = \inf \left\{ \omega(h) \Big| \omega \in \mathcal{T} \right\} \ . \tag{3}$$

$\omega(h)$ depends only on the restriction of ω to $\mathcal{A}_{\{1,\ldots,\ell\}}$. What makes the second infimum non-trivial is that not every state on this subalgebra can be extended to a translation invariant state. If we drop the constraint $\omega \in \mathcal{T}$ in the infimum, we simply get $\inf \sigma(h)$. Hence $E_0(h) \geq \inf \sigma(h)$, and quantum frustration in this language is the phenomenon that this inequality is strict. In the following Definition of the frustration free states we fix $\inf \sigma(h) = 0$ by adding an irrelevant constant to h.

Definition 1 *(Ref.6) A state ω on a quantum spin chain $\mathcal{A}_\mathbb{Z}$ is called* **locally exposed**, *if there is some integer ℓ, and a hermitian $h \in \mathcal{A}_{\{1,\ldots,\ell\}}$ such that $h \geq 0$, and ω is uniquely determined by the property*

$$\omega(\tau_x(h)) = 0 \ , \quad \text{for all } x \in \mathbb{Z} \ . \tag{4}$$

The term "exposed" is used in convex geometry to describe points in a convex set which are *unique* minimizers of a linear functional. Such points are automatically extremal, because every convex component of a minimizer is necessarily a minimizer, and because of the uniqueness required by the definition of exposedness. By the same token it is immediately clear that a locally exposed state is extremal in the state space of $\mathcal{A}_{\mathbb{Z}}$. Moreover, every translate of ω will satisfy the definition, too. Hence, again by uniqueness, a locally exposed state is automatically translationally invariant. Note also that ω is uniquely defined in terms of h, but, due to frustration, there are many interactions h which do not expose any state.

We can give a more geometric picture of locally exposed states in terms of subspaces of tensor products of Hilbert spaces. Since the one-site algebra $\mathcal{A}$ is a matrix algebra, we can consider it as the algebra of linear operators on a finite dimensional Hilbert space $\mathcal{H}$. The restriction of a state ω to a piece of the chain of length L is then given by a density operator D_ω on $\mathcal{H} \otimes \mathcal{H} \otimes \cdots \otimes \mathcal{H} = \mathcal{H}^{\otimes L}$. For a positive operator H to have vanishing expectation in a state ω it is necessary and sufficient that the support of D_ω is contained in the eigenspace of H for eigenvalue zero. That is, all eigenvectors of D_ω for non-zero eigenvalues must be in that subspace. Let us denote by $\mathcal{G} \subset \mathcal{H}^{\otimes \ell}$ the eigenspace of the interaction h for eigenvalue zero. Then the condition that $\omega(\tau_x(h)) = 0$ is equivalent to saying that the local density matrix for the chain segment $\{1, \ldots, L\}$ is supported by $\mathcal{H}^{\otimes x} \otimes \mathcal{G} \otimes \mathcal{H}^{\otimes(L-\ell-x)}$. Hence for a locally exposed state the support of the length L density matrix must be in

$$\mathcal{G}_L^\cap := \bigcap_{x=0}^{L-\ell} \mathcal{H}^{\otimes x} \otimes \mathcal{G} \otimes \mathcal{H}^{\otimes(L-\ell-x)} \quad , \tag{5}$$

and, conversely, if the support of all density matrices for all finite length segments are in $\mathcal{G}_L^\cap$, the state satisfies $\omega(\tau_x(h)) \equiv 0$. Hence $\mathcal{G}$ contains all the information needed to define ω.

Let us pause here to look at frustration in terms of these subspaces: the ground state eigenspace $\mathcal{G}$ is defined for any finite range interaction $h \in \mathcal{A}_{\{1,\ldots,\ell\}}$. What happens for a frustrated interaction is that the intersection (5) becomes $\{0\}$ for large L. In our example of the Heisenberg antiferromagnet we saw this happen for $\ell = 2$, and $L = 3$. In fact, one generically expects $\mathcal{G}_{\ell+1} = \{0\}$ for $\dim \mathcal{G} < (1/2) \dim(\mathcal{H}^{\otimes \ell})$ from a simple dimension count based on the assumption that the two spaces $\mathcal{G} \otimes \mathcal{H}$ and $\mathcal{H} \otimes \mathcal{G}$ are in general position. From this point of view the notion of locally exposed states belongs to the general subject of "geometry of tensor products", and it is not surprising that it also comes up in areas quite remote from the theory of spin chains. For example, in a certain non-commutative differential calculus[5], the non-vanishing of $\mathcal{G}_L^\cap$ for some L is equivalent to the existence of non-zero differential forms of order L.

VBS AND FC STATES

We will now look at a construction that yields many examples of locally exposed states, and in fact, all the examples we know. The first examples were in papers of Affleck[7], and states obtained in this way have come to be known as "*valence bond solid*" (VBS) states. Associated with each site there will be two finite dimensional Hilbert spaces $\mathcal{K}$ and $\overline{\mathcal{K}}$ describing two kinds of "orbitals", pointing to the left and to the right, respectively. A "valence bond" connecting two sites is described by a linear combination of product orbitals, i.e. by a vector $\phi \in \overline{\mathcal{K}} \otimes \mathcal{K}$. The periodic repetition of these bonds at every site of the chain suggested the name. In order to construct a state on the given chain with one-site Hilbert space $\mathcal{H}$ we need to connect the pair of orbitals pointing away from the site with $\mathcal{H}$, which is done by an operator $S : \mathcal{K} \otimes \overline{\mathcal{K}} \to \mathcal{H}$, the second ingredient for defining our state. Then we arrive at the following diagram describing the construction of a vector in $\mathcal{H}^{\otimes L}$ for any chain length L:

$$
\begin{array}{ccccccccc}
\mathcal{H} & \otimes & \mathcal{H} & & \cdots & & \mathcal{H} & \otimes & \mathcal{H} \\
\uparrow S & \otimes & \uparrow S & & \cdots & & \uparrow S & \otimes & \uparrow S \\
\underbrace{\mathcal{K} \otimes \overline{\mathcal{K}}} & \otimes & \underbrace{\mathcal{K} \otimes \overline{\mathcal{K}}} & & \cdots & & \underbrace{\mathcal{K} \otimes \overline{\mathcal{K}}} & \otimes & \underbrace{\mathcal{K} \otimes \overline{\mathcal{K}}} \\
\chi_L & \otimes & \phi & \otimes & \phi & \cdots & \phi & \otimes & \phi \otimes \chi_R
\end{array}
\tag{6}
$$

Here $\chi_R \in \mathcal{K}$ and $\chi_L \in \overline{\mathcal{K}}$ are arbitrary vectors describing "boundary conditions". By $\mathcal{G}_L \in \mathcal{H}^{\otimes L}$ we will denote the subspace spanned by vectors of this form, and call it the space of *VBS-vectors* constructed from ϕ and S. We can paraphrase the construction by saying that VBS vectors are *product vectors, shifted by half a site*. This would suggest that VBS vectors also factorize at finite separation; but since S need not be an isometry, one can have correlations also at long distance (as we shall see).

The subspaces $\mathcal{G}_L$ have just the geometric properties required of the ground state eigenspaces of an exposing interaction. In order to see why, we expand one of the vectors ϕ in the above diagram into product vectors:

$$
\begin{aligned}
\chi_L \otimes \phi \cdots \otimes \quad & \phi \quad \otimes \phi \otimes \cdots \otimes \chi_R \\
\in \mathrm{lin}\{\chi_L \otimes \phi \cdots \otimes \ & \chi'_R \otimes \chi'_L \ \otimes \phi \otimes \cdots \otimes \chi_R\}
\end{aligned}
\tag{7}
$$

Mapping this observation into a statement about vectors in $\mathcal{H} \otimes \cdots \mathcal{H}$ by a suitable tensor power of S, we get the intersection property

$$
\mathcal{G}_{n+m} \subset \mathcal{G}_n \otimes \mathcal{G}_m \ . \tag{8}
$$

Hence if we take $\mathcal{G} = \mathcal{G}_\ell$, we find that $\mathcal{G}_L^\cap \supset \mathcal{G}_L \neq \{0\}$. We have thus met the first requirement for a locally exposed state: the spaces $\mathcal{G}_L^\cap$ do not shrink to zero. This implies the existence of states satisfying equation (4): it suffices to take a w*-cluster point of any sequence of states such that $\omega(\tau_x(h)) = 0$ for more and more $x \in \mathbb{Z}$, i.e. a w*-limit of states given by VBS vectors $\Psi_L \in \mathcal{G}_L$.

The existence of states satisfying (4) is not in itself an interesting property of $\mathcal{G}$ (think of $\mathcal{G} = \mathcal{H}^{\otimes \ell}$). We thus turn to the second requirement for locally exposed states, i.e. uniqueness. In the VBS scheme we always have the choice of the boundary vectors χ_L and χ_R, and we would at least have to verify that the limit will be independent of this choice. In order to answer such questions it is useful to reorganize the VBS construction in a *transfer matrix* like fashion. In this way, we will be able to study how the effect of a particular choice of χ_L decays with the distance from the boundary. At the same time this will permit us to obtain further interesting properties of the VBS states, such as correlation functions. We therefore introduce the operator $V: \mathcal{K} \to \mathcal{H} \otimes \mathcal{K}$ by

$$
V\chi = (S \otimes \mathrm{id}_\mathcal{K})(\chi \otimes \phi) \ . \tag{9}
$$

We can then construct iterates $V^{(n)}: \mathcal{K} \to \mathcal{H}^{\otimes n} \otimes \mathcal{K}$, by repeatedly applying V to the last factor, thereby generating VBS vectors in the first factor, as can be seen from the following diagram:

$$
\begin{array}{cccccccc}
\mathcal{H} & & \mathcal{H} & & \mathcal{H} & & \mathcal{H} & \\
\uparrow S \ \ \diagdown & & \diagdown & & \diagdown & & \diagdown & \\
\ \ \ \ \ \otimes & & \otimes & & \otimes & & \otimes & \cdots \\
\ \ \ \ \ \diagdown & \nearrow V & \diagdown & \nearrow V & \diagdown & \nearrow V & \diagdown & \\
\underbrace{\mathcal{K} \otimes \overline{\mathcal{K}} \otimes \mathcal{K}} & & \mathcal{K} & & \mathcal{K} & & \mathcal{K} & \\
\chi_L \quad \phi & & & & & & &
\end{array}
\tag{10}
$$

Here the lower left corner represents eq.(9), the definition of V, and the rest shows how to iterate V, mapping χ_L into $\mathcal{H} \otimes \mathcal{H} \otimes \cdots \mathcal{H} \otimes \mathcal{K}$.

Since V is an operator between different vector spaces, it makes no sense to speak of its spectrum. Hence V is not yet the transfer matrix we are looking for. To get the transfer matrix we have to move one level up, from the level of Hilbert spaces to the level of observable algebras. Let $\mathcal{B} = \mathcal{L}(\mathcal{K})$ denote the algebra of operators on $\mathcal{K}$, and define $\mathbb{E} : \mathcal{A} \otimes \mathcal{B} \to \mathcal{B}$ by $\mathbb{E}(X) = V^*XV$, where we consider $\mathcal{A} \otimes \mathcal{B}$ as the algebra of operators on $\mathcal{H} \otimes \mathcal{K}$. We introduce the notation $\mathbb{E}_A B = \mathbb{E}(A \otimes B)$, i.e. we consider $\mathbb{E}$ as a family of maps $\mathbb{E}_A : \mathcal{B} \to \mathcal{B}$ indexed by, and depending linearly on A. The map $\mathbb{E}$ is now the central ingredient for constructing a finitely correlated state (more precisely, after Definition 2.4 in Ref.6, a "C*-finitely correlated state"). We will allow also maps of the general form $\mathbb{E}(X) = \sum_\alpha V_\alpha^* X V_\alpha$ with $V_\alpha : \mathcal{K} \to \mathcal{H} \otimes \mathcal{K}$. This form is equivalent to $\mathbb{E}$ being "completely positive". The special form with only one term plays an important role, however. It is equivalent to $\mathbb{E}$ being "pure", that is not decomposable as a sum of other completely positive maps.

Definition 2 *Let $\mathcal{A}$ and $\mathcal{B}$ be finite dimensional C*-algebras, $\mathbb{E} : \mathcal{A} \otimes \mathcal{B} \to \mathcal{B}$ a completely positive map, and ρ a state on $\mathcal{B}$, such that $\mathbb{E}_\mathbb{1} \mathbb{1}_\mathcal{B} = \mathbb{1}_\mathcal{B}$, and $\rho \circ \mathbb{E}_\mathbb{1} = \rho$. Then the state ω on the chain $\mathcal{A}_\mathbb{Z}$, defined by*

$$\omega(A_1 \otimes \cdots A_n) = \rho\Big(\mathbb{E}_{A_1}\mathbb{E}_{A_2}\cdots \mathbb{E}_{A_n}(\mathbb{1}_\mathcal{B})\Big) \quad . \tag{11}$$

*is called the **finitely correlated state** generated by $\mathcal{B}, \mathbb{E}$, and ρ. If $\mathcal{A} = \mathcal{M}_d$ and $\mathcal{B} = \mathcal{M}_k$ are matrix algebras, and $\mathbb{E}$ is of the form $\mathbb{E}(X) = V^*XV$ for an isometry $V : \mathbb{C}^k \to \mathbb{C}^d \otimes \mathbb{C}^k$, ω will be called **purely generated**.*

Formula (11) gives an explicit expression for all finite volume restrictions of the state ω. That these finite volume states indeed satisfy the necessary consistency conditions for a state on the whole chain follows immediately from the observation that in the formula one can add a tensor factor $\mathbb{1}_\mathcal{A}$ to the right or left of the product of A_i's without changing the value of the expectation. By the same token the state thus constructed on the infinite chain is automatically *translationally invariant*. The positivity of ω is guaranteed by the positivity of ρ, and the complete positivity of $\mathbb{E}$. The reason for calling such states "finitely" correlated is *not* that correlations extend only over finite distances in such a state (this is typically false), but that the correlations between two halves of the chain can be modelled by the finite dimensional space $\mathcal{B}$. In general there may be different $(\mathcal{B}, \mathbb{E}, \rho)$ generating the same state. In order to exclude the most trivial sorts of non-uniqueness (e.g. multiplicity) we sometimes assume $\mathcal{B}$ to be *minimal* in the sense that $\mathcal{B}$ is the smallest subalgebra which contains $\mathbb{1}$, and is invariant under each operator $\mathbb{E}_A$.

An immediate consequence of the formula (11) is the following formula for *correlation functions*:

$$\omega(A \otimes \underbrace{\mathbb{1} \otimes \cdots \otimes \mathbb{1}}_{n \text{ times}} \otimes B) = \rho\Big(\mathbb{E}_A(\mathbb{E}_\mathbb{1})^n \mathbb{E}_B(\mathbb{1}_\mathcal{B})\Big) \quad . \tag{12}$$

Hence the computation of correlation functions reduces to the computation of the powers of the operator $\mathbb{E}_\mathbb{1}$, acting on the finite dimensional space $\mathcal{B}$. Hence the decay of correlation functions of finitely correlated states is always exponential.

Because of the explicit formula (11) we can use finitely correlated states as trial states for determining the ground state energy (3). One can easily see that finitely correlated states are w*-dense in $\mathcal{T}$, so we are sure that with sufficiently high dimensional $\mathcal{B}$ we can get a good approximation to $E_0(h)$. Of course, if the true ground state has power law correlations we know that we cannot reach E_0 exactly. But the finitely correlated trial states may be still better than what is usually done, namely setting correlations equal to zero after a finite length: this happens when one makes a finite chain computation, extends the state as an infinite product state, and averages over translations. Experiments with finitely correlated trial states have been encouraging.

If the eigenvalue 1 of $\mathbb{E}_\mathbb{1}$ is simple (the eigenvector is $\mathbb{1}_\mathcal{B}$), suitable means of powers of $\mathbb{E}_\mathbb{1}$ go to the projection $B \mapsto \rho(B)\mathbb{1}$, and the state is clustering in the mean, hence ergodic[8]. Similarly the peripheral spectrum, i.e. the eigenvalues of modulus 1, determine the decomposition of ω into periodic components, and hence the breaking of translational symmetry.

Theorem 3 *Let ω be the finitely correlated state generated by $\mathcal{B}, \mathbb{E}$, and ρ. Then*
(1) if the eigenvalue 1 of $\mathbb{E}_\mathbb{1}$ is simple, ω is ergodic (extremal in $\mathcal{T}$).
(2) If $\mathbb{E}$ has trivial peripheral spectrum, ω has no proper convex decompositions into periodic states.
(3) Conversely, if ω is extremal periodic, then $\mathcal{B}, \mathbb{E}, \rho$ generating ω may be chosen such that $\mathbb{E}$ has trivial peripheral spectrum.

PURE STATES

The previous Theorem gives a simple criterion for deciding that a finitely correlated state has no decomposition into states with at least a residue of translational symmetry (a finite period). Of course, most states have decompositions into non-periodic states. To cite an example from classical systems, we can condition a stationary stochastic process (=translationally invariant state) on the process at time t, and write the given state as an integral over these conditioned states. This will give us a proper decomposition, unless the value of the process at t (i.e. the spin at t, in spin chain language) is fixed with probability 1. Put differently, a pure translationally invariant state for a classical system is concentrated on a single constant trajectory.

Similar decompositions can be done for finitely correlated states, provided $\mathbb{E}$ can be decomposed as $\mathbb{E} = \mathbb{E}^{(1)} + \mathbb{E}^{(2)}$, with two completely positive maps $\mathbb{E}^{(i)} : \mathcal{A} \otimes \mathcal{B} \to \mathcal{B}$, $i = 1, 2$. We can then substitute this decomposition for any one factor in formula (11), to get a decomposition $\omega = \omega^{(1)} + \omega^{(2)}$. We can rule out this sort of decomposition by demanding that $\mathbb{E}$ has no such decompositions, i.e. $\mathbb{E}$ is "pure", and hence of the form $\mathbb{E}(X) = V^*XV$. On the other hand, we have seen via the VBS construction that this condition implies the non-vanishing of the intersections (5), and hence at least one of the requirements for local exposedness. Locally exposed states are pure, so it seems that purity of ω and purity of $\mathbb{E}$ are indeed connected. These are the intuitions behind the following Theorem[6,9]. We state it here in a simple form in which we assume that all decompositions into ergodic or periodic components have already been performed, and that the generating triple $(\mathcal{B}, \mathbb{E}, \rho)$ is chosen to reflect this fact.

Theorem 4 *Let $\mathcal{A} = \mathcal{M}_d$ a matrix algebra, and let ω be a finitely correlated state on $\mathcal{A}_\mathbb{Z}$, generated by $\mathcal{B}, \mathbb{E}, \rho$, such that $\mathcal{B}$ is minimal, and $\mathbb{E}$ has trivial peripheral spectrum. Then the following conditions are equivalent:*
(1) ω is locally exposed.

(2) ω is pure.
(3) ω has vanishing mean entropy.
(4) ω is the (necessarily unique) infinite volume limit of VBS vector states of the form eq.(6).
(5) ω is purely generated.

Moreover,

(A) an **interaction** h exposing ω is the projection onto the complement of $\mathcal{G}_\ell$ for some chain length ℓ.
(B) The isometry V implementing $\mathbb{E}$ in a minimal triple $(\mathcal{B}, \mathbb{E}, \rho)$ is **uniquely** determined by ω up to unitary equivalence.

The proof of this results is given in the cycle $(1) \Rightarrow (2) \Rightarrow (3) \Rightarrow (4) \leftrightarrow (5) \Rightarrow (1)$, of which we have already noted some trivial directions. There are two non-trivial steps, namely $(5) \Rightarrow (1)$, which was done in Ref.6, and yields (A) as a by-product, and $(3) \Rightarrow (5)$, which was proven in Ref.9, and yields (B) as a by-product.

A ground state ω for a Hamiltonian H is said to have a *spectral gap* γ, if for any local observable X,

$$\omega(X^*[H,X]) \geq \gamma\left(\omega(X^*X) - |\omega(X)|^2\right) . \tag{13}$$

Note that this is well-defined even though the Hamiltonian H from eq.(2) is only a formal infinite sum, since only a finite number of terms in this sum contribute to the commutator. It is easily verified that in a finite quantum system this inequality indeed characterizes the gap above the lowest eigenstate. For an infinite quantum system, ground states are usually defined[8] by the inequality with $\gamma = 0$, hence the $\gamma > 0$ is a strengthening of this property. It turns out[6] that all the states characterized by the above Theorem do have a non-zero spectral gap.

Theorem 4 gives a general construction procedure for locally exposed states in terms of the isometry V. One verifies that, for generic V, we get trivial peripheral spectrum of $\mathbb{E}$, so the Theorem applies. This can be used to show[10] that the states characterized by Theorem 4 are w*-dense in the space $\mathcal{T}$ of translationally invariant states, which means that any finite collection of expectation values of local quantities of any state may be approximated by a locally exposed state. This result is in strong contrast to the classical situation, where, as we have seen, pure translationally invariant states are concentrated on the constant trajectories, and are hence far from being dense. Geometrically we can view the density of locally exposed states as follows: $\mathcal{T}$ is a compact convex subset of the total state space $\mathcal{K}$ of the chain. It is a simplex, so one is tempted to view it as a tetrahedron sitting in a larger convex body. This finite dimensional analogy is somewhat misleading, though, since the extreme points of $\mathcal{T}$ are dense in $\mathcal{T}$. (In fact, $\mathcal{T}$ is characterized uniquely by this property among metrizable simplices[8]). The pure translationally invariant states are those vertices of the "tetrahedron", which happen to sit on the extreme boundary of $\mathcal{K}$. Most ergodic states are not pure, so most of the vertices of $\mathcal{T}$ do not have this property. What the result asserts, surprisingly, is that these rare points are even dense in $\mathcal{T}$.

A weakness of Theorem 4 is that it needs the hypothesis that the state is finitely correlated in the first place. The following is a direct characterization of the states, which we announce without proof.

Theorem 5 *Let $\mathcal{A} = \mathcal{M}_d$, and let ω be a translationally invariant state on $\mathcal{A}_\mathbb{Z}$. Then the following are equivalent:*

(1) ω is the finitely correlated, and purely generated.
(2) The GNS-representation of the half-chain algebra $\mathcal{A}_{\mathbb{N}}$ with respect to the restriction of ω to the half chain has finite dimensional commutant.

SYMMETRY GROUPS

In the previous sections we have begun filling in a translation table that allows us to deduce properties of a finitely correlated state ω from properties of the generating objects $(\mathcal{B}, \mathbb{E}, \rho)$: decompositions into translation invariant or periodic components are governed by the peripheral spectrum of $\mathbb{E}_{\mathbf{1}}$, and its multiplicities, and the purity of ω is connected with the purity of $\mathbb{E}$. In this section we will show how symmetries of $(\mathcal{B}, \mathbb{E}, \rho)$ are connected with symmetries of ω acting at each site.

Let us begin with a recipe for constructing states on the chain $\mathcal{A}_{\mathbb{Z}}$ with $\mathcal{A} = \mathcal{L}(\mathcal{H})$, with prescribed symmetry group G:

- Pick two finite dimensional unitary representations $U: G \to \mathcal{L}(\mathcal{H})$ and $U': G \to \mathcal{L}(\mathcal{K})$.

- Find an isometry $V: \mathcal{K} \to \mathcal{H} \otimes \mathcal{K}$ with the intertwining property $VU'_g = (U_g \otimes U'_g)V$ for $g \in G$.

- Choose an invariant state ρ for U', such that $\rho \circ \mathbb{E} = \rho$

Then it is clear from eq(11) that the finitely correlated state ω generated from $\mathcal{B}$, $\mathbb{E}(X) = V^*XV$, and ρ, is invariant under G, in the sense that with $\mathrm{ad}_U(A) = UAU^*$ we have

$$\omega\Big((\mathrm{ad}_{U_g} A_1) \otimes \cdots (\mathrm{ad}_{U_g} A_n)\Big) = \omega(A_1 \otimes \cdots A_n) \quad, \tag{14}$$

for all $g \in G$. The paradigm[4] of a finitely correlated state was in fact obtained in this way: in that case we have $G = \mathrm{SU}(2)$, $U = \mathcal{D}^1$, and $U' = \mathcal{D}^{1/2}$, where $\mathcal{D}^s$ stands for the spin s representation. Since U' is irreducible, the invariant state is uniquely fixed as the normalized trace. Since in the Clebsch-Gordan decomposition of $\mathcal{D}^1 \otimes \mathcal{D}^{1/2}$ contains $\mathcal{D}^{1/2}$ exactly once, the isometry V is uniquely determined up to a phase, with matrix elements given by the Clebsch-Gordan coefficients. Hence ω is fixed by purely group theoretical data.

This construction can be extended in two directions: firstly, we do not have to assume that ω is purely generated. When $\mathbb{E}$ is of the more general form $\mathbb{E}(X) = \sum_{\alpha=1}^{N} V_\alpha^* X V_\alpha$, we can consider the the isometry $V: \mathcal{K} \to \bigoplus_\alpha \mathcal{H} \otimes \mathcal{K} \equiv \mathbb{C}^N \otimes \mathcal{H} \otimes \mathcal{K}$, with $V\phi = \bigoplus V_\alpha \phi$, and look for the intertwining condition $VU = (U'' \otimes U' \otimes U)V$.

The second extension is to *quantum group*[11] symmetries: we can simply follow the first to steps of recipe, replacing only the notion of unitary representation by the one appropriate for quantum groups. Intertwiners V between such representations are still just operators between the underlying Hilbert spaces. The new[12] twist that comes with the "quantum" deformation is that the two conditions in the third step cannot be simultaneously satisfied. Hence we can either produce a translationally invariant state on $\mathcal{A}_{\mathbb{Z}}$, which is a ground state for a quantum group invariant interaction, or else a fully quantum group invariant state ω, which is only defined on a half chain $\mathcal{A}_{\mathbb{N}}$, but no state with both properties. In fact, it is unclear, even apart form the structure of finitely correlated states, how to define an action of a quantum group on the whole chain from a unitary representation U on the one-site Hilbert space.

If we restrict attention once again to proper group symmetry of finitely correlated pure states, we can ask whether the above recipe is the only way of getting invariant states. Indeed, the uniqueness statement (B) in Theorem 4 implies that any symmetry of ω as in eq.(14) should be connected with a unitary acting in $\mathcal{K}$. The following Theorem contains this statement, and two alternative characterizations of such symmetries. The importance of (4) is that it can be stated without reference to the generating objects. Again, there are analogues of this result also for non-purely generated finitely correlated states.

Theorem 6 *Let ω be a state on $\mathcal{A}_{\mathbb{Z}}$ with $\mathcal{A} = \mathcal{L}(\mathcal{H})$, satisfying the equivalent conditions of Theorem 4, and let $V : \mathcal{K} \to \mathcal{H} \otimes \mathcal{K}$ be the uniquely determined isometry from which ω is generated. Then, for a unitary $U \in \mathcal{A}$, the following conditions are equivalent:*

(1) ω is invariant under $\mathrm{ad}_U^{\otimes \infty}$ as in equation (14).
(2) There is a unitary U' on $\mathcal{K}$ such that $VU' = (U \otimes U')V$.
(3) $\mathbb{E}_U : \mathcal{L}(\mathcal{K}) \to \mathcal{L}(\mathcal{K})$ has an eigenvalue of modulus 1.
(4) There are no constants $\epsilon < 1$, and $C < \infty$, such that the estimate

$$\left| \omega(A_- \otimes U^{\otimes n} \otimes A_+) \right| \leq C \epsilon^n \;, \tag{15}$$

holds for all $A_\pm$ with $\|A_\pm\| \leq 1$.

Acknowledgments

The author is indebted to the DFG (Bonn) for financial support.

REFERENCES

1. J.S. Bell, On the Einstein-Podolsky-Rosen paradox, *Physics*, **1**, 195 (1964).
2. R.F. Werner, Remarks on a quantum state extension problem, *Lett.Math.Phys.*, **19**, 319 (1990).
3. B. Nachtergaele, Geometric aspects of quantum spin states, Article in this volume.
4. I. Affleck, T. Kennedy, E.H. Lieb, and H. Tasaki, Valence bond ground states in isotropic quantum antiferromagnets, *Commun.Math.Phys.*, **115**, 477 (1988).
5. P.E.T. Jørgensen, L.M. Schmitt, and R.F. Werner, Positive representations of general commutation relations allowing Wick ordering, Preprint Osnabrück Sept.1993.
6. M. Fannes, B. Nachtergaele, and R.F. Werner, Finitely correlated states of quantum spin chains, *Commun.Math.Phys.*, **144**, 443 (1992).
7. I. Affleck, Large-n limit of SU(n) quantum "spin" chains, *Phys.Rev.Lett.*, **54**, 966 (1985).
8. O. Bratteli and D.W. Robinson, "Operator Algebras and Quantum Statistical Mechanics," 2 volumes, Springer Verlag, Berlin, Heidelberg, New York 1979 and 1981.
9. M. Fannes, B. Nachtergaele, and R.F. Werner, Finitely correlated pure states, to appear in *J.Funct.Anal.*; archived at mp_arc@math.utexas.edu, #92-132.
10. M. Fannes, B. Nachtergaele, and R.F. Werner, Abundance of translation invariant pure states on quantum spin chains, *Lett.Math.Phys.*, **25**, 249 (1992).
11. S.L. Woronowicz, Compact matrix pseudogroups, *Commun.Math.Phys.*, **111**, 613 (1987).
12. M. Fannes, B. Nachtergaele, and R.F. Werner, States on quantum spin chains with quantum group symmetry, in preparation.

STATIONARY STATES OF HAMILTONIAN SYSTEMS WITH NOISE

József Fritz

Mathematical Institute
Hungarian Academy of Sciences
1364 Budapest, Pf. 127, Hungary

1. INTRODUCTION

The ergodic hypothesis is known as one of the most challenging problems of mathematical physics since the fundamental work of Boltzmann and Gibbs. There are several ways to formulate the requirements of ergodicity, roughly speaking we have a mechanical system, like anharmonic vibrations of a lattice $\mathbb{L}$ with canonical coordinates $\omega = (p_k, q_k)_{k \in \mathbb{L}}$ and energy H, thus its time evolution is given by Newton's equations of motion

$$\dot{p}_k = -\frac{\partial H}{\partial q_k}, \quad \dot{q}_k = \frac{\partial H}{\partial p_k}, \quad k \in \mathbb{L}. \tag{1}$$

Under some natural conditions the above system of differential equations defines a flow in an adequately chosen configuration space $(\mathcal{X}, \mathcal{F})$, and we have to describe the stationary measures of this Hamiltonian dynamics. A positive answer to the ergodic hypothesis means that every stationary state is a superposition of Gibbs distributions. This postulate plays a crucial role in the foundations of classical statistical mechanics, its physical relevance is convincingly motivated by numerous applications in equilibrium and non-equilibrium theories including the mathematical description of phase transitions and the microscopic derivation of the equations of irreversible thermodynamics. On the other hand, essentially the Sinai billiards are the only non-trivial examples for which the ergodic hypothesis has rigorously been proven.

An analytic formulation of the same problem can be given in terms of the Liouville operator $\mathcal{L}$ that is the generator of Hamiltonian dynamics,

$$\mathcal{L} \equiv \sum_{k \in \mathbb{L}} \left(\frac{\partial H}{\partial p_k} \frac{\partial}{\partial q_k} - \frac{\partial H}{\partial q_k} \frac{\partial}{\partial p_k} \right); \tag{2}$$

we have to find all solutions μ within a class of 'regular' probability measures on the configuration space of the system such that

$$\int \mathcal{L}\phi(\omega)\,\mu(d\omega) = 0 \tag{3}$$

for all test functions ϕ. A relatively strong condition of regularity is the absolute continuity of μ, the uniqueness of the absolutely continuous stationary measure implies the metric transitivity of the underlying stationary process. This condition is reasonable for finite systems only, but in view of KAM[1] theory we can not hope for the ergodicity of a generic Hamiltonian system with a finite number of degrees of freedom. Moreover, this conclusion is not sufficient for the derivation of macroscopic equations, we ought to control all translation invariant stationary states of the infinitely extended system. Any version of the ergodic hypothesis seems to be extremely hard to prove for true Hamiltonian dynamics, in the present discussion we need additional assumptions of symmetry.

The first observation of this kind goes back to Gallavotti-Verboven[11], they noticed that stationary states with independent Gaussian velocities are Gibbs. On the other hand, Gaussian distribution of velocities is ensured by a thermal noise, and an extension of the entropy argument of Holley[12] and Holley-Stroock[13] can be used to extract the necessary information even in the case of a degenerated noise, see Fritz[7] and Olla-Varadhan[15]. A thermal noise certainly violates the law of energy conservation, thus the macroscopic behavior of the system is radically influenced by the noise. A diffusive noise preserving both energy and momentum is used by Olla-Varadhan-Yau[16] to regularize the problem.

Our starting point here is the simple fact that (3) can really be solved in case of exchangeable velocities. Moreover, the exchangeability of velocities is implied by their reflection symmetry, see Fritz-Funaki-Lebowitz[9], where it is shown that every such symmetric solution is represented as a superposition of Gibbs states. For the proof it is quite relevant that the system is infinitely extended and the interaction between its components is a non-degenerated one. The necessary regularity of μ amounts to finiteness of its specific entropy. The required symmetry of a translation invariant stationary measure is implied by an additional random mechanism: reflection or exchange of velocities at random moments of time. It is remarkable that this randomness is so weak that it does not force finite systems to behave in a proper ergodic manner, that is infinite systems are 'more ergodic' than the finite ones.

The paper is organized as follows. First we develop a general entropy argument to extract information on the 'symmetric component' of a general Markov process in an infinite product space. This framework is materialized then for anharmonic vibrations with noise[9], in Section 4 we present the ideas of a simplified proof. The second model we are going to discuss is a caricature of gas dynamics, the so-called FHP cellular automata[5,10] or lattice gas.

2. ENTROPY AND MARKOV PROCESSES

In this section we summarize some more or less well known facts about entropy and Markov processes, this is a preparation for a study of interacting Markov processes in Section 3. Let $P = P(X, dY)$ denote the transition probability of a Markov process with discrete time in a measure space $(\mathcal{X}, \mathcal{F}, \lambda)$, where $(\mathcal{X}, \mathcal{F})$ is a complete separable metric space equipped with its Borel field and λ is a stationary Borel probability, i.e. $\lambda(dY) = \int \lambda(dX) P(X, dY)$. For bounded measurable functions ϕ the operator of conditional expectation is denoted as

$$\mathcal{P}\phi = \mathcal{P}\phi(X) \equiv \int P(X, dY)\phi(Y),$$

while the evolution of measures is defined by $\mu \to \mu \mathcal{P}$,

$$\mu\mathcal{P}(dY) \equiv \int \mu(dX)P(X,dY),$$

thus $\lambda = \lambda\mathcal{P}$. We assume that $\mathcal{P}$ maps $\mathbb{C}_b(\mathcal{X})$, the space of continuous and bounded functions into itself. The space of Borel probabilities on $(\mathcal{X}, \mathcal{F})$ will be denoted as $\mathbb{P}(\mathcal{X})$.

Entropy (I-divergence) of a Borel probability μ relative to λ is defined by

$$I[\mu|\lambda] \equiv \sup_{\phi \in \mathbb{C}_b(\mathcal{X})} \left(\mu[\phi] - F_\lambda(\phi)\right), \tag{4}$$

where $\mu[\phi] \equiv \int \phi \, d\mu$ and $F_\lambda(\phi) \equiv \log \lambda(e^\phi)$. The supremum on the right hand side can be extended to all measurable ϕ, i.e. $\mu[\phi] \leq I[\mu|\lambda] + F_\lambda(\phi)$ whenever $F_\lambda(\phi) > -\infty$. It is plain that $I[\mu|\lambda] \geq 0$ is a convex and weakly lower semi-continuous function of μ, and $I[\mu|\lambda] = 0$ if and only if $\mu = \lambda$. Moreover, $I[\mu|\lambda] < +\infty$ implies $\mu \ll \lambda$ and $I[\mu|\lambda] = \int \log \frac{d\mu}{d\lambda} d\mu$.

By convexity and $\lambda\mathcal{P} = \lambda$ we have $F_\lambda(\mathcal{P}\phi) \leq F_\lambda(\phi)$, thus from (4) $I[\mu\mathcal{P}|\lambda] \leq I[\mu|\lambda]$. The decrement of entropy is again an I-divergence, and we have a variational characterization in terms of the reversed process. Since λ is a stationary state, $\mathcal{P}$ is a contraction in $\mathbb{L}^2(d\lambda)$. Its adjoint $\mathcal{Q}$ can also be represented as the conditional expectation with respect to $Q = Q(Y,dX)$, the transition probability of the reversed process with initial distribution λ. In general $Q(Y,dX)$ is a Borel probability for λ-a.e. $Y \in \mathcal{X}$, we assume that $\mathcal{Q} : \mathbb{C}_b(\mathcal{X}) \mapsto \mathbb{C}_b(\mathcal{X})$. We say that $\mu \in \mathbb{P}(\mathcal{X})$ is reversible with respect to P if the joint distribution $\mu(dX)P(X,dY)$ is symmetric in X and Y, thus reversibility implies stationarity. In other words, the reversibility of μ with respect to P means that the operator $\mathcal{P}$ is self-adjoint in the Hilbert space $\mathbb{L}^2(d\mu)$, i.e. $\mathcal{P} = \mathcal{Q}$.

2.2 Lemma *For any Borel probability $\mu \in \mathbb{P}(\mathcal{X})$ of finite entropy we have*

$$\int\int \varphi(X,Y)\mu(dX)P(X,dY) \; - \int \mu\mathcal{P}(dY)\log \int Q(Y,dX)e^{\varphi(X,Y)}$$
$$\leq I[\mu|\lambda] - I[\mu\mathcal{P}|\lambda] \tag{5}$$

whenever $\phi \in \mathbb{C}_b(\mathcal{X} \times \mathcal{X})$. In particular, if $I[\mu|\lambda] = I[\mu\mathcal{P}|\lambda]$ then

$$\int\int \varphi(X,Y)\mu(dX)P(X,dY) = \int\int \varphi(X,Y)\mu\mathcal{P}(dY)Q(Y,dX). \tag{6}$$

Proof By convexity $\mathcal{P}e^\phi \geq e^{\mathcal{P}\phi}$, thus

$$\int\int \phi(X,Y)\mu(dX)P(X,dY) - \log \int\int e^{\phi(X,Y)}\lambda(dX)P(X,dY) \leq I[\mu|\lambda]$$

for all $\phi \in \mathbb{C}_b(\mathcal{X} \times \mathcal{X})$, thus choosing ϕ as

$$\phi(X,Y) = \varphi(X,Y) - \log \int Q(Y,dX)e^{\varphi(X,Y)} + \psi(Y),$$

where $\varphi \in \mathbb{C}_b(\mathcal{X} \times \mathcal{X})$ and $\psi \in \mathbb{C}_b(\mathcal{X})$, from $\lambda(dX)P(X,dY) = \lambda(dY)Q(Y,dX)$ we obtain by a direct calculation that

$$\int\int \phi(X,Y)\mu(dX)P(X,dY) - \int \mu\mathcal{P}(dY)\log\int Q(Y,dX)e^{\varphi(X,Y)}$$
$$\leq I[\mu|\lambda] + F_\lambda(\psi) - \mu\mathcal{P}[\psi],$$

which proves (5) by taking the infimum of the right hand side with respect to ψ.
Now if $I[\mu|\lambda] = I[\mu\mathcal{P}|\lambda]$ then (5) reduces to

$$\int\int \beta\varphi(X,Y)\mu(dX)P(X,dY) \leq \int \mu P(dY)\log\int Q(Y,dX)e^{\beta\varphi(X,Y)}$$

for all $\beta \in \mathbb{R}$, thus dividing by $\beta \neq 0$ and sending $\beta \to \pm 0$ we obtain (6). qed

It is easy to show that the supremum of the left hand side of (5) equals the right hand side. If μ is a stationary measure then (6) means that the backward transition probability of the process with initial distribution μ is the same, namely Q, as that of the process with initial distribution λ. Therefore if λ is reversible then $I[\mu|\lambda] = I[\mu\mathcal{P}|\lambda] < +\infty$ yields the reversibility of μ, too. In general $I[\mu|\lambda] = I[\mu\mathcal{P}|\lambda]$ does not even imply that $I[\mu\mathcal{P}|\lambda] = I[\mu\mathcal{P}^2|\lambda]$, but we have $\mu\mathcal{P}\mathcal{Q} = \mu$ and such a μ turns out to be a reversible measure of the transition operator $\mathcal{P}\mathcal{Q}$. This is the information on a stationary measure we can extract by means of an entropy argument. Here is now the point where we have to do something: to materialize the general symmetry property (6) and find the corresponding solutions to the stationary equation $\mu = \mu\mathcal{P}$. An abstract description of absolutely continuous stationary states of reversible models is available[6] in terms of the conservation laws of the system.

The case of Markov processes with continuous time is quite similar, but a bit more technical. We assume that the transition probabilities $P_t = P_t(X,dY)$, $t \geq 0$ have the same properties as P before, and the associated operators of conditional expectation, $\mathcal{P}^t$ form a strongly continuous contraction semigroup in $\mathbb{C}_b(\mathcal{X})$ with generator $\mathcal{G}$. The adjoint semigroup, $\mathcal{Q}^t$ is also strongly continuous in $\mathbb{L}^2(d\lambda)$ and it is generated by the adjoint $\mathcal{G}^*$ of $\mathcal{G}$, see Davies[2]; we assume that each $\mathcal{Q}^t$ maps $\mathbb{C}_b(\mathcal{X})$ into itself. Our favorite examples are of type $\mathcal{G} = \mathcal{L} + \Delta$, where $\mathcal{L}$ is the Liouville operator of a Hamiltonian system with energy H, λ is a Gibbs state for H, and Δ generates a reversible process. Since $\mathcal{L}$ is anti-symmetric in $\mathbb{L}^2(d\lambda)$, we have a formal identity $2\Delta = \mathcal{G} + \mathcal{G}^*$. In general we assume that $\text{Dom}\,\mathcal{G} \cap \text{Dom}\,\mathcal{G}^*$ is dense in $\mathbb{L}^2(d\lambda)$ and define Δ as the Friedrichs extension of $(\mathcal{G} + \mathcal{G}^*)/2$; this is the exact meaning of the relation $2\Delta = \mathcal{G} + \mathcal{G}^*$.

Continuous processes are more regular than the discrete ones, thus we get a little bit more information in this case. For any Markov generator $\mathcal{G}$ and Borel probability μ let $D[\mu|\mathcal{G}]$ denote the Donsker-Varadhan[4] rate function,

$$D[\mu|\mathcal{G}] \equiv \sup_{\psi \in \text{Dom}\,\mathcal{G}}\left\{\int \frac{-\mathcal{G}\psi}{\psi}\,d\mu : \inf\psi > 0\right\}. \tag{7}$$

It is plain that $D[\mu|\mathcal{G}] \geq 0$ is a convex and lower semi-continuous function of μ, and it is zero if μ is a stationary state. Indeed, from the maximum principle $\psi\mathcal{G}\log\psi \leq \mathcal{G}\psi$ if $\psi > 0$, which yields the last statement by a standard approximation procedure, the rest is obvious. There is an explicit formula[4] for D if $\mathcal{G}$ is self-adjoint in $\mathbb{L}^2(d\lambda)$ and $I[\mu|\lambda] < +\infty$, namely $D(\mu|\mathcal{G}) < +\infty$ implies $\sqrt{f} \in \text{Dom}\,\sqrt{-\mathcal{G}}$ and

$$D[\mu|\mathcal{G}] = -\int \sqrt{f}\mathcal{G}\sqrt{f}\,d\lambda, \tag{8}$$

where $f = d\mu/d\lambda$.

As a consequence of Lemma 2.2 we get

2.5 Lemma *If $I[\mu|\lambda] < +\infty$ and $2\Delta = \mathcal{G} + \mathcal{G}^*$ then*

$$I[\mu\mathcal{P}^t|\lambda] + 2\int_0^t D[\mu\mathcal{P}^s|\Delta]\,ds \leq I[\mu|\lambda]$$

for all $t > 0$, consequently $D[\mu|\Delta] = 0$ in a stationary regime.

Proof Substituting $\varphi(X,Y) = \log \psi(X)$ in (5), by convexity

$$\int \log \frac{\psi(X)}{\mathcal{P}^s \mathcal{Q}^s \psi(X)} \mu \mathcal{P}^t(dX) \leq I[\mu \mathcal{P}^t|\lambda] - I[\mu \mathcal{P}^{t+s}|\lambda]$$

for all $t, s > 0$. As a non-increasing function of time, $I[\mu \mathcal{P}^t|\lambda]$ is absolutely continuous, and we expect that

$$\partial_s \log \frac{\psi}{\mathcal{P}^s \mathcal{Q}^s \psi} = -\frac{\mathcal{G}\mathcal{P}^s \mathcal{Q}^s \psi + \mathcal{P}^s \mathcal{G}^* \mathcal{Q}^s \psi}{\mathcal{P}^s \mathcal{Q}^s \psi} \to -\frac{\mathcal{G}\psi + \mathcal{G}^* \psi}{\psi}$$

as $s \to 0$, which completes the formal part of the proof. Since we do not know that $\mathcal{G}\mathcal{Q}^t\psi$ converges strongly to $\mathcal{G}\psi$, we have to regularize the problem. First we verify $2D[\mu\mathcal{P}^t|\Delta] \leq \partial_t I[\mu\mathcal{P}^t|\lambda]$ a.s. for bounded $\mathcal{G}$, which is immediate. Then we introduce the Yoshida approximation $\mathcal{G}_\alpha \equiv \alpha \mathcal{G} \mathcal{R}_\alpha$ for $\alpha > 0$, where $\mathcal{R}$ denotes the resolvent (Laplace transform) of the semigroup, and letting α go to $+\infty$ we conclude the general statement by lower semi-continuity of I and D. qed

Let us remark that $2D[\mu\mathcal{P}^t|\Delta] \neq \partial_t I[\mu\mathcal{P}^t|\lambda]$ in general. We have lost a factor 2 in the case of diffusion process, and even the analytic form of $\partial_t I$ differs from that of D for jump Markov processes. Anyway, this universal entropy inequality can effectively be used to derive symmetry properties of stationary states. The 'reversible component' of the process on which we get information by means of the entropy argument is generated by $\Delta = (\mathcal{G} + \mathcal{G}^*)/2$, cf. the operators $\mathcal{P}\mathcal{Q}$ in the case of discrete time.

3. ENTROPY AND INTERACTING MARKOV PROCESSES

In this section we discuss an infinite dimensional version of the entropy argument of the previous one, see Holley[12], Fritz[7,8], Olla-Varadhan[15], Fritz-Funaki-Lebowitz[9] and Fritz[10]. Let $\mathbb{L} \subset \mathbb{R}^d$ be a d-dimensional lattice, that is a finitely generated nowhere dense subgroup of the additive group of $\mathbb{R}^d$ containing d linearly independent vectors from $\mathbb{R}^d$, and consider identical copies $(\mathcal{X}_k, \mathcal{F}_k)$, $k \in \mathbb{L}$ of a complete and separable metric space with its Borel field. The elements (configurations) of the product space

$$(\mathcal{X}, \mathcal{F}) = \prod_{k \in \mathbb{L}} (\mathcal{X}_k, \mathcal{F}_k)$$

are represented as $X = (X_k)_{k \in \mathbb{L}}$. We equip $\mathcal{X}$ with its natural product topology and Borel field $\mathcal{F}$. The space of continuous and bounded cylinder (local) functions will be denoted as $\mathbb{C}_b^0(\mathcal{X})$ and its closure in $\mathbb{C}_b(\mathcal{X})$ is the space of quasi-local functions $\bar{\mathbb{C}}_b^0(\mathcal{X})$. For $\Lambda \subset \mathbb{L}$ let $\mathcal{F}_\Lambda$ denote the σ-field generated by the variables $X_\Lambda \equiv (X_k)_{k \in \Lambda}$. An increasing sequence of boxes Λ^n, $n \in \mathbb{N}$ can be defined as follows. Let $e_1, e_2, ..., e_d$ be a distinguished set of linearly independent elements of $\mathbb{L}$, denote $\tilde{\Lambda}^n$ the convex hull of the vectors $\pm n e_1, \pm n e_2, ..., \pm n e_d$ and set $\Lambda^n \equiv \tilde{\Lambda}^n \cap \mathbb{L}$. The shift of configurations, sets, functions and measures by $k \in \mathbb{L}$ will be denoted as s^k.

As a reference measure we choose a translation invariant probability measure λ on $\mathcal{F}$ which is regular in the sense that the specific free energy $\bar{F}_\lambda(\phi)$ exists for every continuous local function ϕ, that is

$$\bar{F}_\lambda(\phi) \equiv \lim_{n \to \infty} \frac{1}{|\Lambda^n|} \log \int \exp\left(\sum_{k \in \Lambda^n} \phi(s^k X)\right) \lambda(dX) \tag{9}$$

for $\phi \in C_b^0(\mathcal{X})$, where $|\Lambda|$ denotes the cardinality of $\Lambda \subset \mathbb{L}$. The existence of the limit above can easily be verified for product measures and also for Gibbs states with superstable interactions. This property implies that the specific entropy $\bar{I}[\mu|\lambda]$ below exists for every translation invariant probability measure on $\mathcal{F}$,

$$\bar{I}[\mu|\lambda] \equiv \lim_{n\to\infty} \frac{I_n[\mu|\lambda]}{|\Lambda^n|}, \tag{10}$$

where $I_n[\mu|\lambda]$ is the relative entropy of the corresponding distributions of X_{Λ^n}. More precisely, $I_n[\mu|\lambda] \equiv I[\mu_n|\lambda]$, where μ_n is the extension of μ from $\mathcal{F}_{\Lambda^n}$ defined by

$$\int \phi(X)\mu_n(dX) = \int \lambda[\phi|\mathcal{F}_{\Lambda^n}]\,d\mu \quad \text{for } \phi \in C_b(\mathcal{X}); \tag{11}$$

$\lambda[\phi|\mathcal{F}_\Lambda]$ is the conditional expectation of ϕ given $\mathcal{F}_\Lambda$. In fact we have

$$\bar{I}[\mu|\lambda] = \sup_{\phi \in C_b^0(\mathcal{X})} \left(\mu[\phi] - \bar{F}_\lambda(\phi)\right). \tag{12}$$

Indeed, by convexity and Hölder's inequality

$$F_\lambda(|\Lambda^\ell|\phi) \geq \frac{|\Lambda^\ell|}{|\Lambda^n|} \log \int \exp\left(\sum_{k\in\Lambda^n} \phi(s^k X)\right) \lambda(dX)$$

if $0 < \ell < n$. On the other hand, for any $\varepsilon > 0$ we can find $\ell \in \mathbb{N}$ and an $\mathcal{F}_{\Lambda^\ell}$-measurable continuous ϕ such that $\bar{I}[\mu|\lambda] - \varepsilon \leq \mu[\phi] - |\Lambda^\ell|^{-1} F_\lambda(|\Lambda^\ell|\phi)$, whence $\bar{I}[\mu|\lambda] - 2\varepsilon \leq \mu[\phi] - \bar{F}_\lambda(\phi)$ follows by a direct calculation. The opposite inequality is an immediate consequence of the definition (4) of relative entropy.

Suppose now that λ is a stationary state of a local and translation invariant transition probability $P = P(X, dY)$ in $(\mathcal{X}, \mathcal{F})$; locality of P means that the operator P of conditional expectation maps $C_b^0(\mathcal{X})$ into the space of quasi-local functions $\bar{C}_b^0(\mathcal{X})$. In such situations we obviously have $\bar{I}[\mu P|\lambda] \leq \bar{I}[\mu|\lambda]$. To extend Lemma 2.2 to specific entropy we also need the locality and a factorization property of the backward transition probability Q of the stationary process with marginal λ, namely

$$\int Q(Y,dX)\phi_1(X)\phi_2(X) = \int Q(Y,dX)\phi_1(X) \int Q(Y,dX)\phi_2(X) \tag{13}$$

whenever ϕ_i is $\mathcal{F}_{\Lambda_i}$-measurable, $i = 1, 2$, and the distance of the finite sets Λ_1 and Λ_2 exceeds an arbitrary but fixed number $R > 0$. The above condition can be relaxed to an asymptotic property of factorization, but (13) can easily be verified for the examples of Section 5, where Lemma 3.6 is used.

3.6 Lemma *Suppose that $\mu \in \mathbb{P}(\mathcal{X})$ is translation invariant and $\bar{I}[\mu P|\lambda] = \bar{I}[\mu|\lambda] < +\infty$, then under the conditions listed above we have*

$$\int\int \phi(X)\psi(Y)\mu(dX)P(X,dY) = \int\int \phi(X)\psi(Y)\mu P(dY)Q(Y,dX)$$

for all $\phi, \psi \in C_b^0(\mathcal{X})$.

Proof From (5) of Lemma 2.2 we have

$$\int\int \varphi(X,Y)\mu_n(dX)P(X,dY) - \int \mu_n P(dY) \log \int Q(Y,dX)e^{\varphi(X,Y)}$$
$$\leq I[\mu_n|\lambda] - I[\mu_n P|\lambda] \tag{14}$$

for any $\varphi \in \mathbb{C}_b(\mathcal{X} \times \mathcal{X})$. First we show that

$$\liminf_{n \to \infty} \frac{I[\mu_n \mathcal{P}|\lambda]}{|\Lambda^n|} \geq \bar{I}[\mu \mathcal{P}|\lambda], \tag{15}$$

which is a direct consequence of locality. From the definition (4) of entropy

$$I[\mu \mathcal{P}|\lambda] \geq |\Lambda^{n-\ell}|\mu_n[\mathcal{P}\phi] - \log \int \exp\Big(\sum_{k \in \Lambda^{n-\ell}} \phi(s^k X)\Big) \lambda(dX)$$

for any continuous and $\mathcal{F}_{\Lambda^\ell}$-measurable ϕ. Since $\mu_n[\mathcal{P}\phi]$ tends to $\mu[\mathcal{P}\phi]$ as $n \to +\infty$, we get (3.8). Therefore dividing (14) by $|\Lambda^n|$ we see that the right hand side vanishes as $n \to +\infty$. On the other hand, choosing

$$\varphi(X, Y) = \sum_{k \in \Lambda^n \cap m\mathbb{L}} \phi(s^k X)\psi(s^k Y)$$

in (14), where $m\mathbb{L} \equiv \{mk : k \in \mathbb{L}\}$ and $m \in \mathbb{N}$ is so large that the factorization property (13) applies, from $\mu_n[\phi] \to \mu[\phi]$ as $n \to +\infty$ for $\phi \in \bar{\mathbb{C}}_b^0(\mathcal{X})$, and from the translation invariance of μ we obtain that

$$\int\!\!\int \phi(X)\psi(Y) \mu(dX) P(X, dY) \leq \int \mu \mathcal{P}(dY)\psi(Y) \log \int Q(Y, dX) e^{\phi(X)},$$

which completes the proof as we did in the case of (6) of Lemma 2.2.   **qed**

The description of all reversible measures of an interacting Markov process is not always immediate, each model might need its own specific treatment. The case of processes with continuous time is a little bit easier because the condition $D[\mu|\Delta] = 0$ is usually stronger than the reversibility of μ with respect to Δ.

Suppose now that $P_t = P_t(X, dY)$ is the transition probability of a strongly continuous Markov semigroup in $\mathbb{C}_b(\mathcal{X})$ preserving our reference measure λ. Besides the conditions listed above, translation invariance and locality of P_t, and those of the adjoint semigroup Q^t are also assumed. The strict factorization property (13) is not a reasonable condition in the case of continuous time, we suppose that the generator $\mathcal{G}$ is a sum,

$$\mathcal{G} = \sum_{B \subset \mathbb{L}} \mathcal{G}_B = \sum_{k \in \mathbb{L}} \mathcal{G}_k, \quad \mathcal{G}_k \equiv \sum_{B \ni k} \frac{\mathcal{G}_B}{|B|}, \tag{16}$$

where $\mathcal{G}_B$ is a translation invariant family of local generators such that $\mathcal{G}_B = 0$ if the diameter of the set B exceeds a given number $R > 0$. The locality of $\mathcal{G}_B$ means that $\mathcal{G}_B(\psi\phi) = \psi\mathcal{G}_B\phi$ whenever ψ does not depend on X_B and $\mathcal{G}_B$ maps cylinder (local) functions into local ones in the following, uniform sense. If $B \subset \Lambda$ and the distance between B and the complement Λ^c of Λ is larger than R, then $\mathcal{G}_B$ preserves measurability with respect to $\mathcal{F}_\Lambda$. For convenience we assume that each $\mathcal{G}_B$ generates a strongly continuous Markov semigroup in $\mathbb{C}_b(\mathcal{X})$ with stationary measure λ, and a dense subset $\mathbb{D} \subset \mathbb{C}_b(\mathcal{X})$ forms a common core of all $\mathcal{G}_B$ and $\mathcal{G}_B^*$, too.

In view of (8) and (16), at least for self-adjoint generators and translation invariant measures the specific value of the Donsker-Varadhan rate should be defined as $\bar{D}[\mu|\Delta] \equiv D[\mu|\Delta_0]$, where $2\Delta = \mathcal{G} + \mathcal{G}^*$ and

$$\Delta_k \equiv \sum_{B \ni k} \frac{\mathcal{G}_B + \mathcal{G}_B^*}{2|B|} \quad \text{for } k \in \mathbb{L}. \tag{17}$$

As a consequence of Lemma 2.5 we prove

3.11 Lemma *If $\mu \in \mathbb{P}(\mathcal{X})$ is a translation invariant stationary measure and $\bar{I}[\mu|\lambda] < +\infty$ then $\bar{D}[\mu|\Delta] = 0$.*

Proof Let μ_n be as in (11), then from Lemma 2.5

$$I[\mu_n \mathcal{P}^t|\lambda] + 2\int_0^t D[\mu_n \mathcal{P}^s|\Delta]\, ds \leq I[\mu_n|\lambda],$$

consequently by (16)

$$\lim_{n\to\infty} \frac{1}{|\Lambda^n|} \int_0^t D[\mu_n \mathcal{P}^s|\Delta]\, ds = 0$$

for all $t > 0$, whence by (8) $\lim D[\bar{\mu}_n|\Delta_0] = 0$ as $n \to +\infty$, where

$$\bar{\mu}_n \equiv \sum_{k \in \Lambda^n} \frac{1}{|\Lambda^n|} \int_0^1 s^k \mu_n \mathcal{P}^s\, ds.$$

Therefore

$$\limsup_{n\to\infty} \int \frac{-\Delta_0 \psi}{\psi}\, d\bar{\mu}_n \leq 0$$

if $\psi > 0$ is smooth enough, which completes the proof as $\bar{\mu}_n \to \mu$.   qed

In the next section we apply this result to a system of coupled anharmonic oscillators with weak noise.

4. COUPLED ANHARMONIC OSCILLATORS WITH NOISE

Let us consider a lattice system of coupled oscillators with a self-potential U and with a symmetric nearest neighbor interaction V. Let $p_k \in \mathbb{R}$ and $q_k \in \mathbb{R}$ denote the velocity and the position of the oscillator at site $k \in \mathbb{L} = \mathbb{Z}^d$. The formal Hamiltonian of the system can be written as

$$H = \sum_{k \in \mathbb{Z}^d} (\frac{1}{2} p_k^2 + U(q_k)) + \frac{1}{2} \sum_{|j-k|=1} V(q_k - q_j), \qquad (18)$$

and the underlying evolution is defined by an infinite system of differential equations (1). For convenience we assume that U and V are infinitely differentiable non-negative functions with bounded second derivatives. Let $\mathcal{X}$ denote the set of configurations $\omega = (p_k, q_k)_{k \in \mathbb{Z}^d}$ such that $\|\omega\|_\alpha < +\infty$ for each $\alpha > 0$, where

$$\|\omega\|_\alpha^2 = \sum_{k \in \mathbb{Z}^d} e^{-\alpha|k|} [p_k^2 + q_k^2].$$

We equip $\mathcal{X}$ with its natural product topology and Borel field $\mathcal{F}$. Since U' and V' are uniformly Lipschitz continuous, the most standard iteration procedure yields existence of uniquely defined global solutions in $\mathcal{X}$, and this flow is generated by the Liouville operator $\mathcal{L}$. For the existence of Gibbs states λ^β with energy H and inverse temperature $\beta > 0$ it is sufficient to assume that U'' is strictly positive for large values of its argument. The presence of a non-degenerated interaction means that $V''(x) = 0$ implies $V'''(x) \neq 0$ for all $x \in \mathbb{R}$. As a reference measure we choose a translation invariant Gibbs state $\lambda = \lambda^1$ of unit temperature.

We now add to this Hamiltonian dynamics some 'noise'. This consists of random reflections, or random exchanges of velocities between neighboring sites. These take place independently at each site or bond of the lattice with a constant rate $a > 0$. More precisely, we consider two randomized evolutions generated by $\mathcal{G}^R = \mathcal{L} + a\Delta^R$, and by $\mathcal{G}^E = \mathcal{L} + a\Delta^E$, respectively, where $\mathbb{L}^*$ is the set of bonds of $\mathbb{L}$ and

$$\Delta^R \phi \equiv \sum_{k \in \mathbb{L}} \left(\phi(R_k \omega) - \phi(\omega) \right),$$

$$\Delta^E \phi \equiv \sum_{b \in \mathbb{L}^*} \left(\phi(E_b \omega) - \phi(\omega) \right) \quad \text{for } \phi \in \mathbb{C}_b^0(\mathcal{X}). \tag{19}$$

The operator R_k changes the sign of the velocity p_k at site $k \in \mathbb{L}$, while E_b exchanges the velocities at the ends of bond $b \in \mathbb{L}^*$; the rest of the configuration is not altered. For a construction and regularity of such Markov processes see[9]. Materializing the argument of the previous section we obtain

4.3 Theorem *Suppose that $\mu \in \mathbb{P}(\mathcal{X})$ is a translation invariant stationary state of one of the randomized evolutions $\mathcal{G}^R$ or $\mathcal{G}^E$. If μ has finite specific entropy then it is a superposition of Gibbs states λ^β.*

The starting point of the proof is an entropy argument manifested by Lemma 3.11. For translation invariant stationary states of the evolution generated by $\mathcal{G}^R$ we conclude a property of reflection symmetry (20),

$$\int \phi(R_k \omega) \mu(d\omega) = \int \phi(\omega) \mu(d\omega) \quad \text{for } k \in \mathbb{Z}^d \text{ and } \phi \in \mathbb{C}_b(\mathcal{X}), \tag{20}$$

which implies (3) and an integration by parts formula

$$\int \frac{\partial \phi}{\partial p_k} \frac{\partial H}{\partial q_k} d\mu = \int \frac{\partial \phi}{\partial q_k} \frac{\partial H}{\partial p_k} d\mu \quad \text{for } k \in \mathbb{Z}^d \tag{21}$$

and continuously differentiable $\phi \in \mathbb{C}_b^0(\mathcal{X})$. The crucial step is to show that (21) and (3) imply (22),

$$\int \phi(E_b \omega) \mu(d\omega) = \int \phi(\omega) \mu(d\omega) \quad \text{for } b \in \mathbb{L}^* \text{ and } \phi \in \mathbb{C}_b(\mathcal{X}). \tag{22}$$

On the other hand, (22) follows also from Lemma 3.11 for the evolution $\mathcal{G}^E$.

Now we show that (3) and (22) result in a KMS condition

$$\int \frac{1}{\beta} \frac{\partial \phi}{\partial q_k} d\mu = \int \phi \frac{\partial H}{\partial q_k} d\mu \quad \text{for all } k \in \mathbb{Z}^d \tag{23}$$

and smooth cylinder functions ϕ, where $1/\beta$ equals the common conditional variance of p_k given the σ-field $\mathcal{F}_{inv}$ of translation invariant events. This integration by parts formula characterizes the distribution of positions as a mixture of Gibbs states with random temperature, and feeding back this information into (3) we get a second KMS condition

$$\int \frac{1}{\beta} \frac{\partial \phi}{\partial p_k} d\mu = \int \phi \frac{\partial H}{\partial p_k} d\mu \quad \text{for all } k \in \mathbb{Z}^d \tag{24}$$

and smooth ϕ, which completes the proof.

Our statement on $\mathcal{G}^R$ is certainly false in the case of finite systems because then the energy shell may be disconnected at low values of energy. Since the related energy threshold diminishes for large systems, and degenerated (ground) states are ruled out by the finiteness of specific entropy, we really have a proof for infinite systems. The condition on the presence of a generic interaction between neighboring components can certainly be weakened, but it is essential for a proper ergodic behavior.

5. FHP CELLULAR AUTOMATA

Now we consider a lattice gas[3,5,10] on the two-dimensional triangular lattice $\mathbb{L}$ with velocities $v \in \mathbb{V} \equiv \{e^\circ, e', e'', -e^\circ, -e', -e''\}$, where e°, e', e'' are distinguished unit vectors of $\mathbb{L}$ such that each angle between them is just $120°$. If $v \in \mathbb{V}$ then v' is obtained from v by rotating it with $+120°$, i.e. $v'' = (v')'$. The configuration space is restricted by an exclusion rule: particles at the same site must have different velocities; thus the configurations of our system are the subsets X of $\mathbb{X} \equiv \mathbb{L} \times \mathbb{V}$, i.e. $\mathcal{X} = \mathsf{E}^{\mathbb{X}}$. We equip $\mathcal{X}$ with its natural product topology and Borel field $\mathcal{F}$. Set $\eta^X(q, v) = 1$ if we have a particle of velocity $v \in \mathbb{V}$ at site $q \in \mathbb{L}$ and $\eta^X(q, v) = 0$ otherwise, the set of particles (velocities) at site $q \in \mathbb{L}$ will be denoted as $\nu_q^X \equiv \{v : \eta^X(q, v) = 1\}$.

Total number and momentum of the particles of a configuration $X \in \mathcal{X}$ at site $q \in \mathbb{L}$ are denoted by $N_q = N_q(X)$ and $P_q = P_q(X)$, respectively, where

$$N_q(X) = \sum_{v \in \mathbb{V}} \eta^X(q, v), \quad P_q(X) = \sum_{v \in \mathbb{V}} \eta^X(q, v) v. \qquad (25)$$

One step of the evolution is given by a random transformation $\mathcal{T} : \mathcal{X} \mapsto \mathcal{X}$ of type $\mathcal{T} = \mathcal{SBC}$, where $\mathcal{S}$ is the free streaming of particles, $\mathcal{C} = \mathcal{C}_\gamma$ specifies collisions governed by a random element γ, and $\mathcal{B} = \mathcal{B}_\beta$ describes a random walk on $\mathbb{L}$.

The free evolution $\mathcal{S} : \mathcal{X} \mapsto \mathcal{X}$ is a deterministic transformation defined by $\eta^{\mathcal{S}X}(q, v) = \eta^X(q - v, v)$. The collision mechanism $\mathcal{C}_\gamma : \mathcal{X} \mapsto \mathcal{X}$ changes only dipoles $\mathbb{D}_v \equiv \{v, -v\}$ and tripoles $\mathbb{T}_v \equiv \{v, v', v''\}$, it is specified by a binary sequence $\gamma = (\gamma_q)_{q \in \mathbb{L}}$, $\gamma_q = \pm 1$ as follows. For each $q \in \mathbb{L}$ we set

$$\nu_q^{\mathcal{C}_\gamma X} = \begin{cases} \mathbb{D}_{v'} & \text{if } \nu_q^X = \mathbb{D}_v \text{ and } \gamma_q = +1, \\ \mathbb{D}_{v''} & \text{if } \nu_q^X = \mathbb{D}_v \text{ and } \gamma_q = -1, \\ \mathbb{T}_{-v} & \text{if } \nu_q^X = \mathbb{T}_v \text{ and } \gamma_q = +1, \\ \nu_q^X & \text{in all other cases.} \end{cases}$$

Collisions at different sites are independent of each other with common distribution $P[\gamma_q = +1] = P[\gamma_q = -1] = 1/2$ for each $q \in \mathbb{L}$.

Random collisions of the FHP automata are not sufficient to ensure the ergodic behavior of the system, we need an additional stirring mechanism $\mathcal{B}$. In some situations we allow a particle to jump across a bond $b \in \mathbb{L}^*$ which is not parallel to its velocity. First we select a pairwise disjoint set of bonds $\omega \subset \mathbb{L}^*$ somehow, and if $b \in \omega$ then we let one of the particles of velocity $v \not\parallel b$ to jump across b. The exchange mechanism $\mathcal{B} \equiv \mathcal{B}_\beta : \mathcal{X} \mapsto \mathcal{X}$ is defined in terms of a set of random velocities $\beta = (\beta_q)_{q \in \mathbb{L}}, \beta_q \in \mathbb{V}$ as follows. Let $\omega = \omega(\beta)$ be the set of bonds $b \in \mathbb{L}^*$ for which $\beta_b \not\parallel b$ and $\beta_c \parallel c$ whenever $|b \cap c| = 1$ and set

$$\eta^{\mathcal{B}_\beta X}(q, v) = \begin{cases} \eta^X(p, v) & \text{if } \{q, p\} = b \in \omega(\beta) \text{ and } v = \beta_b, \\ \eta^X(q, v) & \text{in all other cases.} \end{cases}$$

For simplicity we assume that β is a system of independently selected velocities such that for all $b \in \mathbb{L}^*$ we have $P[\beta_b = v] = \kappa$ if $v \not\parallel b$, while $P[\beta_b = v] = 1/2 - 2\kappa$ if $v \parallel b$, where $0 < \kappa < 1/4$.

Given two completely independent sequences β^t and γ^t, $t \in \mathbb{N}$ of random elements as described above we define a Markov process in $\mathcal{X}$ by $\xi_0 = X \in \mathcal{X}$ and $\xi_t = \mathcal{SB}_{\beta^t}\mathcal{C}_{\gamma^t}\xi_{t-1}$ for $t \in \mathbb{N}$. Let us remark that in the original model[5] the reflection of tripoles is not randomized, and the mechanism of random stirring is not present at all.

In a stationary regime the random stirring mechanism maintains a uniform distribution of particles, all correlation functions depend only on the number of different kinds of participating particles. Since $\mathcal{T} = \mathcal{SBC}$ preserves both the number and total momentum of particles, product measures of type

$$\lambda^{z,u}(dX) \equiv \prod_{q \in \mathbb{L}} \exp\bigl(zN_q(X) + \langle u, P_q(X)\rangle - F(z,u)\bigr)\,\lambda(dX) \tag{26}$$

are stationary states of the modified FHP automata, where $z \in \mathbb{R}$, $u \in \mathbb{R}^2$ are given parameters, $\langle \cdot, \cdot \rangle$ is the inner product in $\mathbb{R}^2$, $\lambda \equiv \lambda^{0,0}$ denotes the uniform Bernoulli measure on $(\mathcal{X}, \mathcal{F})$ and F is the normalization. Following the ideas of Ref. 9, in the paper[10] we prove

5.3 Theorem *Every translation invariant stationary state μ of the modified FHP automata is a superposition of product measures. If $0 < \mu[\eta^X(0,v)|\mathcal{F}_{inv}] < 1$ μ-a.s. then $\mu[dX|\mathcal{F}_{inv}] = \lambda^{z,u}$ μ-a.s., where the random parameters are specified by $\mu[N_q|\mathcal{F}_{inv}] = F'_z(z,u)$ and $\mu[P_q|\mathcal{F}_{inv}] = F'_u(z,u)$.*

The proof is similar to the previous one. Since the randomized collisions and exchanges are reversible, from Lemma 3.6 we conclude that the measure μ is reversible with respect to collisions, while $\mu\mathcal{C}$ is reversible with respect to exchanges. In the present concrete situations reversibility implies the following symmetries. Let D'_q and D''_q denote the rotation of a dipole at site $q \in \mathbb{L}$ by $\pm 60°$, respectively, while T_q is the operation of reflecting a tripole. Then $\bar{I}[\mu\mathcal{C}|\lambda] = \bar{I}[\mu|\lambda]$ yields

$$\int \phi(D'_q X)\,\mu(dX) = \int \phi(D''_q X)\,\mu(dX) = \int \phi(T_q X)\,\mu(dX) = \int \phi(X)\,\mu(dX) \tag{27}$$

for all $q \in \mathbb{L}$ and $\phi \in \mathbb{C}_b(\mathcal{X})$, whence $\mu\mathcal{C} = \mu$. On the other hand, from $\bar{I}[\mu\mathcal{B}|\lambda] = \bar{I}[\mu|\lambda]$ we get

$$\int \phi(E^v_b X)\,\mu(dX) = \int \phi(X)\,\mu(dX) \tag{28}$$

for $b \in \mathbb{L}^*$, $b \not\Vert v \in V$ and $\phi \in \mathbb{C}_b(\mathcal{X})$, where E^v_b jumps a particle of velocity v across the bond b, whence $\mu\mathcal{B} = \mu$, too. In view of (28) μ decomposes into product measures whose explicit form is then specified by means of (27) and the stationary equation $\mu = \mu\mathcal{S}$.

By means of the entropy method[15,16] we can now derive the macroscopic equations of the lattice gas, we obtain $\partial_t \rho + \operatorname{div} \pi = 0$ and $\partial_t \pi + \operatorname{Div} Q(\rho, \pi) = 0$, where ρ and π are the density and momentum fields, Q is the pressure tensor. It is interesting that the macroscopic equations admit spatially inhomogeneous steady state solutions.

Acknowledgements. This paper was supported in part by Hungarian National Foundation for Scientific Research Grant No. 1902. I am indebted to Joel L. Lebowitz for valuable suggestions.

REFERENCES

1. V.I. Arnold, "Mathematical Methods of Classical Mechanics," Springer Verlag, Berlin-Heidelberg (1978).
2. E.B. Davies, "One-Parameter Semigroups," Academic Press, London (1980).
3. A. DeMasi, R. Esposito, J.L. Lebowitz, and E. Presutti, Hydrodynamics of stochastic cellular automata, *Comm. Math. Phys.*, **125**, 127 (1989).
4. M.D. Donsker and S.R.S. Varadhan, Asymptotic evaluation of certain Markov process expectations for large time I, *Comm. Pure. Appl. Math.*, **28**, 1 (1975).
5. U. Frish, B. Hasslacher and Y. Pomeau, Lattice gas automata for Navier–Stokes equation, *Phys. Rev. Letters*, **56**, 1505 (1986).
6. J. Fritz, An information-theoretical proof of limit theorems for reversible Markov processes, *in*: "Trans. Sixth Prague Conf. Information Theory, Statistical Decision Functions and Random Processes, Prague 1971, Academia, Prague (1973).
7. J. Fritz, On the stationary measures of anharmonic systems in the presence of a small thermal noise, *J.Statist.Phys.*, **44**, 25 (1986).
8. J. Fritz, On the diffusive nature of entropy flow in infinite systems: Remarks to a paper by Guo-Papanicolau-Varadhan, *Comm. Math. Phys.*, **133**, 331 (1990).
9. J. Fritz, T. Funaki and J.L. Lebowitz, Stationary states of random Hamiltonian systems, MKI preprint, Budapest 1992.
10. J. Fritz, Stationary states and hydrodynamics of FHP cellular automata, MKI preprint, Budapest 1993.
11. G. Gallavotti and M. Verboven, On the classical KMS condition, *Nuovo Cimento*, 274 (1976).
12. R. Holley, Free energy in a Markovian model of a lattice spin system, *Comm. Math. Phys.*, **23**, 87 (1971).
13. R. Holley, R. and D.W. Stroock, In one and two dimensions every stationary measure for a stochastic Ising model is a Gibbs state, *Comm. Math. Phys.*, **55**, 37 (1977).
14. T.M. Liggett, "Interacting Particle Systems", Springer-Verlag, New-York (1985).
15. S. Olla and S.R.S. Varadhan, Scaling limit for interacting Ornstein-Uhlenbeck processes, *Comm. Math. Phys.*, **135**, 355 (1991).
16. S. Olla, S.R.S. Varadhan and H.T. Yau, To appear in *Comm. Math. Phys.*
17. D. Ruelle, "Statistical Mechanics. Rigorous Results", Benjamin, (1968).
18. A. Rényi, On measures of entropy and information, *in*: "Proc. Fourth Berkeley Symp. on Mat. Stat. Probab. Vol. I., Uni. Cal. Press, Los Angeles (1960).
19. H.T. Yau, Relative entropy and the hydrodynamics of Ginzburg–Landau models, *Lett. Math. Phys.*, **22**, 63 (1991).

STOCHASTIC REGULARIZATION OF COHERENT-STATE PATH INTEGRALS AND QUANTUM HALL EFFECT

R. Alicki

Institute of Theoretical Physics and Astrophysics
University of Gdańsk
Wita Stwosza 57, PL–80–952 Gdańsk, Poland

INTRODUCTION AND OUTLINE

The aim of this lecture note is to present a brief overview of the recent results on stochastically regularized path integrals, their relations to the coherent states formalism and their applications to the Quantum Hall Effect.

In a series of papers Klauder and Daubechies[1] proposed a new procedure of quantization for classical systems described by the canonical phase-space variables (q,p). The formal and mathematically ill-defined phase-space integral

$$\mathcal{N}\int \exp\{i\int_0^T [p\,dq - h(p,q)\,dt]\}\mathcal{D}p\mathcal{D}q \qquad (1)$$

is replaced by the following limit of the well-defined path integrals

$$\lim_{\nu\to\infty} 2\pi e^{\nu T/2}\int \exp\left\{i\int_0^T \left[\frac{1}{2}(p\,dq - q\,dp) - h(p,q)dt\right]\right\} d\mu_W^\nu(p,q)$$

$$= <\phi[p'',q'']\,|\exp(-iTH)\phi[p',q']> \qquad (2)$$

On the LHS of the Eq. (2) μ_W^ν denotes a Wiener measure on continuous Brownian motion paths $p(t), q(t)$, $0 \leq t \leq T$, that are pinned so that $p(0), q(0) = p', q'$ and $p(T), q(T) = p'', q''$, and ν is a diffusion constant. The Hamiltonian H is given by

$$H = (2\pi)^{-1}\int h(p,q)\,P[p,q]dp\,dq, \qquad (3)$$

where $P[p,q]$ is the projection onto the unit vector $\phi[p,q]$ and in this case $\phi[p,q]$ is a canonical coherent state of an irreducible representation of the Weyl group on a standard Gaussian wave function.

Subsequently, this idea was considerably extended and generalized by Klauder and coworkers.[2-6] It has been proved that the limit of the path integral expression in (2) exists and defines a meaningful matrix element of a quantum unitary time evolution in the following cases:

A) Phase-spaces which are homogeneous spaces for certain groups (sphere for $SU(2)$, Lobachevsky plane for $SU(1,1)$), and with diffusions governed by the associated Beltrami-Laplace operators. In these cases, analogs of (2) and (3) exist with $\phi[p,q]$ being replaced by coherent states associated with relevant groups.[2,3]

B) Phase-spaces being Kähler manifolds which combine the symplectic two-form (analog of $dp \wedge dq$) and the Riemannian metric in a complex Hermitian bilinear form. In this case one uses Kähler diffusions and the coherent states are derived from Bergman reproducing kernels.[4]

The recent results which are reviewed in the present note deal with two other cases.

C) Flat phase space $\mathbf{R}^{2n}$, but the Wiener measure for the diffusion process is replaced by a probability measure for a general stochastic process with independent increments. In this case the vectors $\phi[p,q]$ are coherent states generated by the action of an irreducible representation of the Weyl group on a general faithful vector.[5]

D) Phase-space being an arbitrary two-dimensional Riemann surface with an arbitrary genus and with a general one-form $A_b d\xi^b$ being a generalization of the canonical one-form $pdq - qdp$. Diffusion is governed by the Laplace-Beltrami operator and the coherent states are related to certain reproducing kernels and holomorphic representations of the relevant Hilbert space.[6]

In all these cases the main idea is based on the observation that the path integral expression (2) (or its analog) can be treated as a limit $\nu \to \infty$ of an integral kernel for the one parameter semigroup acting on $L^2(\Gamma)$ (Γ - phase space) with a generator equal to

$$L_\nu = -\nu W - ih, \qquad (4)$$

where W is a certain self-adjoint operator and h is a multiplication operator given by $h(p,q)$. Now if $W \geq 0$ and if there exists a Hilbert subspace $\mathcal{H}$ of $L^2(\Gamma)$ such that

$$W\phi = 0 \iff \phi \in \mathcal{H} \qquad (5)$$

the LHS of the Eq. (2) can be written in an operator form, and the following equality holds in all the cases of interest

$$w - \lim_{\nu \to \infty} \exp[(-\nu W - ih)T] = \mathcal{P} \exp(-i\mathcal{P}h\mathcal{P}T)\mathcal{P}. \qquad (6)$$

Here, $\mathcal{P}$ is an orthogonal projection on $\mathcal{H}$. In this way we construct a Hilbert space $\mathcal{H}$ of the quantized system, and a quantization procedure for a classical observable, say the Hamiltonian h

$$h(p,q) \implies H = \mathcal{P}h\mathcal{P}. \qquad (7)$$

Moreover, the projector $\mathcal{P}$ may be represented as a reproducing kernel on $L^2(\Gamma)$, and as such, defines an overcomplete family of vectors which in $\mathcal{H}$ admit a resolution of unity ("general coherent states"). In this approach, the problem of quantization of a given classical system with a phase-space Γ, equipped with a Riemannian and a symplectic structure, consists of constructing a suitable operator W on $L^2(\Gamma)$, which admits path integral representation of the type (2) for the associate semigroup e^{-tW} and finding all solutions of (5). One can notice that the condition $W\phi = 0$ resembles the polarization condition – the heart of the program of geometric quantization –[7], but generally the two formalisms may lead to different results.[4] In case D) the operator W is essentially a quantum Hamiltonian for an electron moving on a surface Γ in an orthogonal but generally nonuniform magnetic field. This fact exhibits deep relations between the mathemathical formalism used in the description of the Quantum Hall Effect and in the problem of quantization. For the cases A), B) and D) with compact manifolds, a deeper mathematical insight is provided by the Riemann-Roch-Hirzebruch-Atiyah-Singer index theorems,[8] while for case C) the theory of stochastic processes is relevant. In the following we discuss in some details the cases C) and D).

GENERAL STOCHASTIC PATH INTEGRAL REGULARIZATION

The main ingredient of the presented quantization scheme is a one-parameter semigroup $e^{-tW}, t \geq 0$ acting on $L^2(\mathbf{R}^2)$, and given by the path integral expression for its integral kernel

$$(e^{-tW})(p'', q''; p', q') = e^{\kappa t} \int e^{\frac{i}{2} \int (p\,dq - q\,dp)} \, dP(p, q) . \tag{8}$$

In the above formula $\int (p\,dq - q\,dp)$ denotes the Stratonovich stochastic integral (midpoint rule) and κ is a tunable constant. The expression $dP_\nu(p, q)$ denotes a measure on trajectories for a rather general stochastic process on $\mathbf{R}^2$ which can be a superposition of a Wiener and a Poisson process and the trajectories are pinned at p', q' for the time equal to zero and p'', q'' for the time t. The stochastic equation corresponding to this process for the mean value $E[g(p(t), q(t))] \equiv F(p, q; t)$ with a sufficiently smooth function $g(\cdot)$ reads

$$\frac{\partial}{\partial t} F(p, q; t) = \left(\frac{a}{2} \frac{\partial^2}{\partial p^2} + \frac{b}{2} \frac{\partial^2}{\partial q^2} \right) F(p, q; t) + \int \{F(p+u, q+v; t) - F(p, q; t)\} \mu(du\,dv) . \tag{9}$$

Here $a, b \geq 0$ and $\mu(du\,dv)$ is a characteristic intensity measure for the Poisson process.

Using now standard methods involving Chernoff theorem[9] one may calculate the form of the operator W defined by (8)

$$W = \frac{a}{2} \left(i\frac{\partial}{\partial p} - \frac{q}{2} \right)^2 + \frac{b}{2} \left(i\frac{\partial}{\partial q} + \frac{p}{2} \right)^2$$
$$+ \int \mu(du\,dv) \left\{ 1 - \exp i \left[v(i\frac{\partial}{\partial q} + \frac{p}{2}) - u(i\frac{\partial}{\partial p} - \frac{q}{2}) \right] \right\} + \kappa . \tag{10}$$

One may notice that the Hilbert space $L^2(\mathbf{R}^2)$ can be written as a tensor product of two Hilbert spaces $L^2(\mathbf{R})$ on which acts an irreducible representation of the Weyl group. These two representations are generated by two pairs of operators

$$\bar{Q} = i\frac{\partial}{\partial p} - \frac{q}{2}, \quad \bar{P} = i\frac{\partial}{\partial q} + \frac{p}{2}, \quad [\bar{Q}, \bar{P}] = i ,$$

217

$$Q = i\frac{\partial}{\partial p} + \frac{q}{2}, \quad P = -i\frac{\partial}{\partial q} + \frac{p}{2}, \quad [Q,P] = i. \tag{11}$$

Hence, because W is a function of $\bar{Q}, \bar{P}$ only, it can be treated as a self-adjoint operator acting on the first Hilbert space $L^2(\mathbf{R})$, i.e.

$$W = W_1(\bar{P}, \bar{Q}) \otimes \mathbf{1}. \tag{12}$$

We restrict our attention to the operators W_1 satisfying the conditions: a) $W_1 \geq 0$;b) there exists a nondegenerate, normalized eigenvector η of W_1 (faithful vector) such that $W_1 \eta = 0$.

A large class of examples of operators W_1 is generated by the Markov processes defined by (9) with $b = 1, a = 0$ and $\mu(du\,dv) = m(u)\delta(v)du\,dv$. Then W_1 is the Hamiltonian of a nonrelativistic particle in a potential hole. Namely

$$W_1 = \frac{1}{2}\bar{P}^2 + U(\bar{Q}) + const., \quad U(x) = -\int m(u)e^{iux}du. \tag{13}$$

For a general $m(\cdot)$ there always exists a nondegenerate "ground state" η.

Having the faithful vector η and the associate one-dimensional projector P_η according to (12) and (6) (with $h = 0$) we have

$$\mathcal{P} = P_\eta \otimes \mathbf{1}. \tag{14}$$

Moreover the projector $\mathcal{P}$ can be represented as an integral reproducing kernel on $L^2(\mathbf{R}^2)$ given by the following expression

$$\mathcal{P}(p'',q'';p',q') = <\phi_\eta[p'',q'']|\phi_\eta[p',q']>, \tag{15}$$

where $\phi_\eta[p,q]$ is a canonical coherent state generated by the action of an irreducible representation of the Weyl group on a faithful vector η. The generalization of the above construction to the case of a phase-space $\mathbf{R}^{2n}$ with an arbitrary n is obvious.

Summarizing, even in the case of a flat phase-space, the regularized path integrals give a whole family of quantization procedure regularizing stochastic processes, or alternatively by coherent state representations with different faithful vectors. Such a flexible formalism may find different applications.

THE LOWEST LANDAU LEVEL

The motion of an electron (with a polarized spin) in a general magnetic field orthogonal to an arbitrary surface Γ, is described by the following Pauli Hamiltonian.

$$H[A,g] = -\frac{1}{2\sqrt{g}}(\partial_a + iA_a)g^{ab}\sqrt{g}(\partial_b + iA_b) - \frac{1}{2}s^{ab}B_{ab}. \tag{16}$$

Here g_{ab} is a metric tensor, A_b is a vector potential, $B_{ab} = \partial_a A_b - \partial_b A_a$ and $s^{ab} = \sqrt{g}\epsilon^{ab}$. A convenient system of units is applied here. The ground states of $H[A,g]$ can be easily found using the fact that for any two-dimensional surface one can always choose a (local) coordinate system, say u and v, $(u,v) \in \mathbf{R}^2$, such that the metric becomes conformally flat i.e.[10]

$$ds^2 = e^{2w(u,v)}(du^2 + dv^2) . \tag{17}$$

In this special coordinate system the matrix elements of the Hamiltonian $H[A,g]$ are given by the following expression

$$<\psi|H[A,g]|\phi> = -\frac{1}{2}\int \psi^* e^{-2w}\left[(\partial_u + iA_u)^2\phi + (\partial_v + iA_v)^2\phi\right]e^{2w}dudv$$

$$-\frac{1}{2}\int \psi^*(\partial_u A_v - \partial_v A_u)\phi\,dudv \equiv \frac{1}{2}\int dudv (\mathcal{D}\psi)^*\mathcal{D}\phi \tag{18}$$

where

$$\mathcal{D}\phi = \left[(\partial_u - i\partial_v) + i(A_u - iA_v)\right]\phi . \tag{19}$$

From (18) and (19) it follows that $H[A,g] \geq 0$ and that the ground states (the lowest Landau level) are all solutions of the following equation

$$\left[(\partial_u - i\partial_v) + i(A_u - iA_v)\right]\phi = 0 , \tag{20}$$

subject to square integrability condition with respect to the measure $\sqrt{g}d\xi^1 d\xi^2$ and to the topological constraints in the case of a compact manifold Γ. For the case of a surface with $\mathbf{R}^2$ topology it follows from the Eq.(20) that the subspace $\mathcal{H}$ of the ground states is spanned by the linearly independent functions

$$\phi_k(u,v) = (u - iv)^k e^{-F(u,v)} e^{iG(u,v)} \tag{21}$$

with $k = 0, 1, 2..., N(= D - 1) \leq \infty$, and real functions F, G satisfying the equations

$$(\partial_u^2 + \partial_v^2)F(u,v) = (\partial_u A_v - \partial_v A_u) ,$$

$$(\partial_u^2 + \partial_v^2)G(u,v) = (\partial_u A_u + \partial_v A_v) . \tag{22}$$

The square integrability of ϕ_k demands that the function $(u^2 + v^2)^k \exp[-2F(u,v) + 2w(u,v)]$ should decay at least as $(u^2 + v^2)^{-(1+\epsilon)}$ for $|u|, |v| \to \infty$ with $\epsilon > 0$. If now the following integrals are finite ($\Phi \geq 0$)

$$\Phi = \int (\partial_u A_v - \partial_v A_u)dudv , \tag{23}$$

$$\Psi = -\int (\partial_u^2 + \partial_v^2)w(u,v)dudv \tag{24}$$

then for large $|u|^2 + |v|^2$ one obtains the estimate

$$|\phi_k(u,v)|^2 \exp[2w(u,v)] \sim (u^2+v^2)^{k-(\Phi+\Psi)/2\pi} . \tag{25}$$

The expressions for Φ and Ψ can be easily transformed into a geometric, coordinate independent form

$$\Phi = \frac{1}{2}\int [\partial_a A_b(\xi) - \partial_b A_a(\xi)]d\xi^a \wedge d\xi^b , \tag{26}$$

$$\Psi = \frac{1}{2}\int R(\xi)\sqrt{g(\xi)}\,d\xi^1 d\xi^2 \tag{27}$$

where R is the scalar curvature given by the Riemann tensor of g

$$R = R^{\alpha\beta}{}_{\alpha\beta} = -2e^{-2w}(\partial_u^2 + \partial_v^2)w \ . \tag{28}$$

Finally, we see that under the above assumptions it follows from (25) that the dimension of the lowest Landau level is equal to

$$D = \text{largest integer less than} \left[\frac{1}{2\pi}\Phi + \frac{1}{2\pi}\Psi - 1\right]. \tag{29}$$

The above formula generalizes the Aharonov-Casher result obtained for a flat surface.[11] The case of a compact two-dimension arbitrary genus $g_e = 0, 1, 2, ...$, can be discussed using geometrical methods. The (normalized) integrals $\frac{1}{2\pi}\Phi$, the magnetic charge, and $\frac{1}{2\pi}\Psi$, the Euler characteristic, are now topological invariants and can take only integer values, namely

$$\frac{1}{2\pi}\Phi = n, \quad n = 0, 1, 2, ... \tag{30}$$

$$\frac{1}{2\pi}\Psi = 2(1 - g_e), \quad g_e = 0, 1, 2, ... \ . \tag{31}$$

The condition (30) is the famous Dirac condition on the monopole while the condition (31) is the Gauss-Bonnet Theorem.[10] The Riemann-Roch-Hirzebruch-Atiyah-Singer index theorem[8] gives the dimension D of the lowest Landau level as

$$D = n + (1 - g_e) = \left[\frac{1}{2\pi}(\Phi + \Psi) - 1\right] + g_e \tag{32}$$

if $n > 2 - 2g_e$. Note that Eq. (32) extends the formula (29) to compact manifolds. Concrete examples of compact manifolds topologically equivalent to a sphere and a torus are discussed in details in Ref.6.

QUANTIZATION OF 2-D PHASE-SPACES

The results of the previous Section have immediate application to the problem of quantization of a general system described by a phase-space being a Riemannian surface Γ equipped with a vector field A_b (A defines a "symplectic form" dA). The classical system is described by the action functional

$$\mathcal{A} = \int [A_b(\xi)\dot{\xi}^b - h(\xi)]\,dt \ , \tag{33}$$

and the corresponding Euler-Lagrange (Hamiltonian) equation reads

$$B_{ab}(\xi)\dot{\xi}^b = \partial_a h(\xi) \ . \tag{34}$$

Consider now an operator expression, which is a LHS of the Eq. (6) with W being $H[A,g]$ as given by (16)

$$w - \lim_{\nu\to\infty} e^{-\{\nu H[A,g] - i\hbar\}T} \tag{35}$$

and which possesses the following path integral representation

$$\lim_{\nu \to \infty} \mathcal{N} \int \exp\{i \int [A_b(\xi)\dot{\xi}^b - h(\xi)]dt\} \exp\{-\frac{1}{2\nu} \int g_{ab}(\xi)\dot{\xi}^a\dot{\xi}^b dt\}$$

$$\times \exp\{\frac{\nu}{2} \int s^{ab}(\xi)B_{ab}(\xi)dt\} \Pi \sqrt{g(\xi)} d\xi^1 d\xi^2 \equiv K(\xi \tag{36}$$

The second exponent gives a formal representation of a Wiener measure for a diffusion process on Γ and hence the path integral (36) is perfectly well-defined (Euclidean) path integral for any $\nu > 0$. Moreover according to the results of the previous Section the limit (35) exists (for a large class of h) and is given by

$$w - \lim_{\nu \to \infty} e^{-\{\nu H[A,g] - ih\}t} = \mathcal{P} \exp(-i\mathcal{P}h\mathcal{P}t)\mathcal{P} \tag{37}$$

where $\mathcal{P}$ is a projector on "the lowest Landau level" $\mathcal{H}$. Hence, the Hilbert subspace $\mathcal{H}$ of the Hilbert space $L^2(\Gamma)$ should be treated as a Hilbert space of the correspond quantized system and (36) is a (regularized) path integral expression for the unitary evolution operator. The quantum Hamiltonian H is related to its classical counterpart $h(\cdot)$ by

$$H = \mathcal{P}h\mathcal{P} = \int h(\xi)|\xi><\xi|\sqrt{g(\xi)}\,d\xi^1 d\xi^2, \tag{38}$$

where $\{|\xi>\}$ are "generalized coherent states" obtained from the reproducing kernel $\mathcal{P}(\xi''; \xi')$ by the identification

$$|\xi> \equiv \mathcal{P}(\cdot\,;\xi)\,. \tag{39}$$

Finally, the quantum propagator given by (36) can be written as

$$K(\xi'',t'';\xi',t') = <\xi''|e^{-iH(t''-t')}\xi'>\,. \tag{40}$$

With the above formulas we have achieved our goal of presenting a manifestly coordinate invariant quantization procedure appropriate to a general symplectic form and geometry of the underlying two manifolds. Because the form of the dynamics, i.e., the Hamiltonian h, does not play an essential role in this procedure, we have "quantized a manifold" equipped with a Riemannian and symplectic structure.

APPLICATIONS TO QUANTUM HALL EFFECT

In the previous Section we have applied the mathematical results concerning the structure of the lowest Landau level for a general surface and magnetic field to the problem of quantization of a system with two-dimensional phase-space manifold. In a complementary fashion, the so-obtained picture of general quantization helps to derive useful semi-classical formulas for the Hall current in the case of a filling factor equal to one, for a general surface and magnetic field and with an additional potential $h(\xi)$. In this case the effective Hamiltonian for an electron in the lowest Landau level is given (in the high magnetic field regime) by the expression (38). Hence, this electron may be treated as a quantized system corresponding to the classical one with a two dimensional phase-space and with the action functional given by (33). Let us treat, "semi-classically", such electrons as a fluid with a local surface density which may be derived from the formula (29) and a local velocity $\dot{\xi}^a$ which satisfies Hamiltonian equation (34). Firstly, from (29) we obtain the semi-classical phase-space density of states

$$dN(\xi) = \frac{1}{4\pi}(\partial_a A_b(\xi) - \partial_b A_a(\xi))d\xi^a \wedge d\xi^b + \frac{1}{4\pi}R(\xi)\sqrt{g(\xi)}d\xi^1 d\xi^2$$

$$= [\frac{1}{4\pi}(\partial_a A_b(\xi) - \partial_b A_a(\xi)) + \frac{1}{8\pi}R(\xi)\sqrt{g(\xi)}\epsilon_{ab}]d\xi^a \wedge d\xi^b . \quad (41)$$

Then using Eqs.(41) and (34) and the fact that $B_{ab} = (\sqrt{B_{rs}B^{rs}g/2})\epsilon_{ab}$ one obtains the expression for the Hall current flowing through the curve C with ends ξ', ξ''

$$J_C = \frac{1}{2\pi}\int_C B_{ab}\dot{x}^a dx^b + \frac{1}{4\pi}\int_C R(x)\sqrt{g(x)}\epsilon_{ab}\dot{x}^a dx^b$$

$$= \frac{1}{2\pi}[h(\xi'') - h(\xi')] + \frac{1}{4\pi}\int_C \frac{R(\xi)}{\sqrt{B_{ab}B^{ab}/2}}dh(\xi) . \quad (42)$$

The first term on the RHS of the Eq. (42) gives the standard expression for the quantum Hall current with the filling factor equal to one[12], while the second one is a geometric correction due to the curvature. The above results have been applied to the description of an anomalous Hall current in the case of flat surface but general magnetic field.[13]

Another interesting possibility emerges from the above picture. Rapid technological development makes possible the production of as the potential acting on electrons can be designed to some extent also. Therefore, effectively, highly nontrivial quantum systems with two dimensional phase-spaces and tunable "effective Planck constant" can be obtained and tested experimentally.

Acknowledgements

This lecture note summarizes a joint research work with John Klauder and Jerzy Lewandowski which has been done during the author's stay at the University of Florida in Gainesville.

The author is grateful to The Kosciuszko Foundation for a Research Scholarship, and to the Department of Physics, University of Florida, for hospitality. Partial support from the Polish Committee for Scientific Research (project PB 1436/2/91) is also acknowledged.

REFERENCES

1. J.R. Klauder and I. Daubechies, *Phys. Rev. Lett.*, **52**, 1161 (1984); I. Daubechies and J.R. Klauder, *J. Math. Phys.*, **26**, 2239 (1985).
2. I.Daubechies and J.R.Klauder, *J. Math. Phys.*, **28**, 85 (1987).
3. J.R.Klauder, *Ann. Phys.*, **188**, 120 (1988).
4. J.R.Klauder and E.Onofri, *Int. J. Mod. Phys. A*, **A**, 3930 (1989).

5. R. Alicki and J.R. Klauder, *J. Math. Phys.*, (1993), in print.
6. R. Alicki, J.R. Klauder and J. Lewandowski, Landau level ground state degeneracy, and its relevance for a general quantization procedure, preprint 1993.
7. N.P. Woodhouse, "Geometric Quantization", Oxford University Press, (1980).
8. R.C. Gunning, "Lectures on Riemann Surfaces", Princeton Univ. Press, (1966).
9. P.R. Chernoff, *J. Funct. Anal.*, **2**, 238 (1968); R. Alicki and D. Makowiec, *J. Phys. A: Math. Gen.*, **18**, 3319 (1985).
10. H.M. Farkas and I. Kra, "Riemann Surfaces", Springer-Verlag, New-York (1980).
11. Y. Aharonov and A. Casher, *Phys. Rev. A*, **19**, 2461 (1979).
12. R.E. Prange and S.M. Girvin (Eds.), "The Quantum Hall Effect", second edition, Springer-Verlag, New-York (1990).
13. R. Alicki and J.R. Klauder, Anomalous quantum Hall current driven by a non-uniform magnetic field, preprint 1993.

AN $ADE - \mathcal{O}$ CLASSIFICATION OF MINIMAL INCOMPRESSIBLE QUANTUM HALL FLUIDS

Jürg Fröhlich,[1] Urban M. Studer,[2] and Emmanuel Thiran[1]

[1]Institut für Theoretische Physik, ETH-Hönggerberg
8093 Zürich, Switzerland
[2]Instituut voor Theoretische Fysica, K.U. Leuven
3001 Leuven, Belgium

INTRODUCTION AND THE GENERAL CLASSIFICATION PROBLEM

The quantum Hall (QH) effect[1] is observed in two-dimensional electronic systems (2DES's) subjected to a strong, uniform, transverse external magnetic field. Experimentally, such systems are realized as inversion layers that form at the interfaces of heterostructures (e.g., $GaAs/Al_xGa_{1-x}As$) in the presence of an electric field (gate voltage) perpendicular to the structures. To develop an idea of the orders of magnitude involved in QH systems, we recall that sample sizes are typically of a few tenths of a mm times a few mm, whereas the charge carrier densities, $n = n_{electron} - n_{hole}$, are of the order of $10^{11}/cm^2$, and the magnetic fields, $\mathbf{B}_c$, range from about 0.1 T up to 30 T. Moreover, experiments are performed at very low temperatures, T, typically between 10 mK and 100 mK. An important quantity characterizing QH systems is the filling factor ν: denoting by $\Phi_o = h/e = 4.14 \cdot 10^{-11}\,\mathrm{Tcm^2}$ the magnetic flux quantum and by $B_{c,\perp}$ the component of the magnetic field $\mathbf{B}_c$ perpendicular to a 2DES, the filling factor is defined by $\nu = n\,\Phi_o/B_{c,\perp}$.

Two basic facts characterize the QH effect. First, the Hall (or transverse) resistance R_H, as a function of $1/\nu$, exhibits plateaux at heights which are rational multiples of h/e^2. Second, with the quantization of R_H, one observes a near vanishing of the longitudinal (or magneto-) resistance R_L. [Strictly speaking, this is true only if measurements are carried out in a *stationary* state and on *large distance scales*, that is, on scales larger than the phase-coherence length of the constituents of the systems. For QH systems, the phase-coherence length is of the order of $100\,\mu m$ (a so-called "mesoscopic" length scale). For a general account of "quantum interference fluctuations" in disordered, mesoscopic systems, see Ref. 2; and, in particular, for some results in mesoscopic QH systems, see Ref. 3.] The near vanishing of R_L indicates the *absence of dissipative processes* in the corresponding QH systems which, in turn, is interpreted as the presence of an *energy gap* above the ground-state energy in these many-body systems (so-called *incompressibility*). Incompressibility is a crucial input to our analysis and, in these notes, we use the term *quantum Hall fluid* to denote a QH system which exhibits incompressibility. Next we briefly review the main logical steps that lead to the general classification problem of QH fluids.

Gauge Symmetry and $\hat{u}(1)$-Current Algebra. For definiteness, let us consider a QH fluid which is confined to a domain Ω of the $(1,2)$-plane, and let the external magnetic field, $\mathbf{B}_c$, be along the 3-axis, i.e., $\mathbf{B}_c = (0, 0, B_c)$. Furthermore, let us assume that the spins of the constituent electrons and/or holes are aligned with the strong field $\mathbf{B}_c$. [For an analysis of QH fluids including the spin degrees of freedom, see Refs. 4-6.]

We denote by $Z_\Lambda(A)$ the partition function (at $T = 0$) of a 2DES confined to a space-time domain $\Lambda = \Omega \times \mathbb{R}$ and coupled to an external electromagnetic field with associated potential 1-form $A = \sum_{\mu=0}^{2} A_\mu(x) dx^\mu$, where $A_\mu = A_{\text{tot},\mu} - A_{c,\mu}$ with $\partial_1 A_{c,2} - \partial_2 A_{c,1} = B_c$ fixed. Exploiting the $U(1)$-gauge symmetry of non-relativistic Schrödinger quantum mechanics and the incompressibility ($R_L = 0$) of QH fluids, one can show[4] that, in the limit of large distance scales and low frequencies (*scaling limit*), the *effective action* of a QH fluid (i.e., the logarithm of $Z_\Lambda(A)$) takes the *universal* asymptotic form

$$S_\Lambda^{\text{as}}(A) = -\frac{\sigma_H}{2} \int_\Lambda A \wedge dA + B.T.(A|_{\partial \Lambda}) , \qquad (1)$$

where $\sigma_H = R_H^{-1}$ is the Hall conductivity of the fluid, $A \wedge dA = \sum_{\mu\nu\rho} \epsilon^{\mu\nu\rho} A_\mu \partial_\nu A_\rho$ is the abelian Chern-Simons form, and $B.T.(A|_{\partial\Lambda})$ denotes a "boundary term" depending only on $A|_{\partial\Lambda}$, the restriction of A to the boundary space-time $\partial\Lambda$. Since the Chern-Simons term in Eq. (1) exhibits a boundary $U(1)$-gauge anomaly, the gauge invariance of the total effective action, $S_\Lambda^{\text{as}}(A)$, constrains the form of the boundary term $B.T.(A|_{\partial\Lambda})$. The cancellation of the gauge anomaly of the Chern-Simons term can be achieved[4] by introducing a set of N_L and a set of N_R chiral $U(1)$ boundary currents whose coupling to the external electromagnetic field $A|_{\partial\Lambda}$ is given by $N_{L/R}$ electric charges which can be arranged in a "charge vector" $\vec{Q}_{L/R}$ with $N_{L/R}$ components. Physically, these left/right chiral boundary systems (CBS's) describe gapless, chiral electric currents of electrons/holes circulating at the edge of a QH sample. Denoting by $W_{L/R}(\vec{Q}_{L/R}; A|_{\partial\Lambda})$ the effective action for these CBS's, the boundary term $B.T.(A|_{\partial\Lambda})$ is given by

$$B.T.(A|_{\partial\Lambda}) = W_L(\vec{Q}_L; A|_{\partial\Lambda}) - W_R(\vec{Q}_R; A|_{\partial\Lambda}) + G.I.(A|_{\partial\Lambda}) , \qquad (2)$$

where $G.I.(A|_{\partial\Lambda})$ stands for a possible gauge-invariant boundary term. Then gauge-anomaly cancellation implies that the Hall conductivity σ_H is given by

$$\sigma_H = \left[(\vec{Q}_L, \vec{Q}_L) - (\vec{Q}_R, \vec{Q}_R) \right] \frac{e^2}{h} , \qquad (3)$$

where $(\vec{n}, \vec{m})$ denotes the Euclidean scalar product of two vectors $\vec{n}$ and $\vec{m}$. For details, see Refs. 4-6; and for related points of view on the role of edge currents in QH fluids, see the works in Ref. 7.

Quantum Hall Lattices. The vectors $\vec{Q}_L$ and $\vec{Q}_R$ are not the only data encoding the physics of an incompressible QH fluid in the scaling limit. Excitations above the ground state of the QH fluid are described by unitary representations of the $\hat{u}(1)$-current algebras. Among all possible unitary representations ($\simeq \mathbb{R}^{N_{L/R}}$) only a subset, in fact, a lattice, will be realized in a QH fluid. Combining known properties of these representations[8] with the physical requirement that the spectrum of local excitations of CBS's be generated by excitations with the quantum numbers of electrons and holes (i.e., by excitations with electric charges $\pm e$ that obey Fermi statistics and are relatively local) one infers[5,6] that the additional data needed to characterize the scaling limit of a QH fluid is a pair of *odd, positive integral lattices* Γ_L and Γ_R: $\Gamma_{L/R}$ is generated by $N_{L/R}$ vectors $\vec{q}_{(1)}, \ldots, \vec{q}_{(N_{L/R})}$ describing distinct one-electron or one-hole excitations above the ground state. A general vector $\vec{q}$ in $\Gamma_{L/R}$ represents a multi-electron/hole excitation of the QH fluid. In such a basis, $\Gamma_{L/R}$ is fully specified by its integral Gram matrix $K_{L/R}$

$$(K_{L/R})_{ij} = (\vec{q}_{(i)}, \vec{q}_{(j)}) \in \mathbb{Z}, \; i,j = 1, \ldots, N_{L/R}. \tag{4}$$

Fermi statistics of the elementary charge carriers implies the oddness of the lattice $\Gamma_{L/R}$ (i.e., in any basis of $\Gamma_{L/R}$, at least one basis vector has an odd squared length). The vectors in $\Gamma_{L/R}$ do not necessarily label all physical excitations of the QH fluid. Indeed, any vector in the dual lattice $\Gamma^*_{L/R}$ (generated by the vectors $\underline{h}^{(j)} = \sum_{i=1}^{N_{L/R}} (K_{L/R}^{-1})^{ji} \vec{q}_{(i)}$, $j = 1, \ldots, N_{L/R}$) describes a (multi-)quasiparticle excitation which is relatively local to all electrons and holes in the system. However, in general, such an excitation has fractional electric charge and anyonic statistics. It is the $N_{L/R}$-dimensional analogue of the celebrated Laughlin vortices.[9]

The vector $\underline{Q}_{L/R}$ assigns an electric charge to every boundary excitation. Hence it is a linear functional on the lattice $\Gamma_{L/R}$, i.e., an element of the dual lattice $\Gamma^*_{L/R}$, and $Q_{\text{el}}(\vec{q}) = \underline{Q}_{L/R} \cdot \vec{q}$ is the electric charge, in units of e, of the excitation labelled by $\vec{q}$. Thus $\pm 1 = Q_{\text{el}}(\vec{q}_{(i)}) = \underline{Q}_{L/R} \cdot \vec{q}_{(i)} = (Q_{L/R})_i$, for every basis vector $\vec{q}_{(i)} \in \Gamma_{L/R}$. This means that $\underline{Q}_{L/R}$ is what mathematicians call a *visible* vector in $\Gamma^*_{L/R}$. This has an all important consequence: $(\underline{Q}_{L/R}, \underline{Q}_{L/R})$ *is a rational number*. Eq. (3) then gives the rational quantization of the Hall conductivity σ_H, in units of e^2/h, as observed in QH fluids.

A pair $(\Gamma, \underline{Q})$ is called a *quantum Hall lattice* and represents the geometrical data characterizing a QH fluid. *Classifying universality classes of QH fluids thus amounts to classifying pairs of QH lattices* $(\Gamma_L, \underline{Q}_L)$ *and* $(\Gamma_R, \underline{Q}_R)$.

CLASSIFICATION OF QUANTUM HALL LATTICES

A lattice Γ which is the orthogonal direct sum of several sublattices is called *decomposable* (otherwise *indecomposable*). QH fluids corresponding to decomposable or indecomposable QH lattices are called *composite* or *elementary* fluids, respectively. For a composite QH fluid consisting of two components, we have $\sigma_H = \sigma_H^1 + \sigma_H^2$, where $\sigma_H^i = (Q^i, Q^i) e^2/h = Q^i \cdot K^{-1} Q^{iT} e^2/h$, with $\underline{0} \neq \underline{Q}^i \in \Gamma^*_i \subset \Gamma^*$, $i = 1, 2$. In the following, we focus our attention on *indecomposable* QH lattices. We note that QH fluids consisting of electrons (L) and holes (R) are composite and thus will not appear explicitly in the subsequent classification. Moreover, since the discussions of the two chirality subsectors (L and R) are analogous, we may focus on, say, the left one. We now drop the index L from our notation.

Writing $\sigma_H = (n_H/d_H)e^2/h$, with $\gcd(n_H, d_H) = 1$, Eq. (3) tells us that, for some positive integer l, called the *level* of the QH fluid,

$$ld_H = \Delta \equiv \det K, \quad \text{and} \quad ln_H = \gamma \equiv \vec{Q} \cdot \tilde{K} \vec{Q}^T, \tag{5}$$

where $\tilde{K}$ is the cofactor (or adjoint) matrix of K, i.e., $K^{-1} = \Delta^{-1}\tilde{K}$. Next we state two general classification results on quantum Hall lattices (for proofs, see Ref. 6).

First, for a given Hall fraction σ_H and two positive integers l and N_o, one can prove that there are only *finitely* many inequivalent QH lattices $(\Gamma, \vec{Q})$ with $\sigma_H = (\vec{Q},\vec{Q})e^2/h = \vec{Q} \cdot K^{-1}\vec{Q}^T e^2/h$, $\Delta = ld_H$, and $\dim \Gamma \leq N_o$.

Second, one can show that the *minimal, non-trivial fractional charge*, e^*, associated with a QH lattice $(\Gamma, \vec{Q})$ is given by

$$e^* \equiv \min_{\underline{n} \in \Gamma^*, Q_{el}(\underline{n}) \neq 0} |Q_{el}(\underline{n})e| = \frac{e}{\lambda d_H}, \tag{6}$$

where λ is some integer dividing the level l. In particular, if d_H is *even*, one can prove that the charge parameter λ has to be a multiple of 2 and the level l a multiple of 4. Thus, for the observed QH fluids with $\sigma_H h/e^2 = \frac{1}{2}, (\frac{3}{2})$, and $\frac{5}{2}$, a *model-independent* prediction is that e^* is a fraction of $e/4$: $e^* = (2/\lambda)\,e/4$!

Next we focus our attention on *minimal* QH lattices which are characterized by the property that their levels satisfy $l = 1$. There are several basic facts (for proofs, see Ref. 6) about the corresponding universality classes of *minimal elementary QH fluids* which make these classes particularly attractive from a theoretical, as well as from a phenomenological, point of view:

(M1) For the fractions $\sigma_H = 1/(2p+1)\,e^2/h$, $p = 1, 2, \ldots$, one can easily show that the famous Laughlin fluids[9] do correspond to the *unique* minimal indecomposable QH lattices $(\Gamma, \vec{Q})$ where: Γ is one-dimensional, it is generated by a vector $\vec{q}_{(1)}$ with length squared $(\vec{q}_{(1)}, \vec{q}_{(1)}) = 2p+1 = K_{11} = K$, the visible vector $\vec{Q}$ is given by $\vec{Q} = \underline{h}^{(1)}$ where $\underline{h}^{(1)} = 1/(2p+1)\,\vec{q}_{(1)}$ is generating the dual lattice Γ^*, and $\sigma_H h/e^2 = (\vec{Q}, \vec{Q}) = 1/(2p+1)$.

(M2) If $l = 1$ then $d_H = \det K$ has to be *odd*. Experimentally, there is ample evidence for the "odd-denominator rule".[10] For one-layer (or one-component) systems, the only firmly established exception to this rule is a QH fluid with $\sigma_H = \frac{5}{2} e^2/h$.[11] [For a discussion of even-denominator QH fluids, see Refs. 5 and 6.] Furthermore, the following relationship holds between the numerator n_H of the Hall conductivity σ_H and the dimension $N = \dim \Gamma$ of the indecomposable QH lattice:

$$\text{for } n_H \begin{Bmatrix} \text{even} \\ \text{odd} \end{Bmatrix}, N \text{ has to be } \begin{Bmatrix} \text{even} \\ \text{odd, and } N \equiv n_H \pmod 4 \end{Bmatrix}. \tag{7}$$

(M3) For minimal QH fluids, one can prove a *charge-statistics theorem*: for such fluids, the (anyonic) statistical phases, $\theta(\underline{n}) = (\underline{n}, \underline{n}) = \underline{n} \cdot K^{-1}\underline{n}^T$, of (quasiparticle) excitations, labelled by $\underline{n} \in \Gamma^*$, are completely determined by the single quantity of their (fractional) charges, $Q_{el}(\underline{n}) = (\vec{Q}, \underline{n}) = \vec{Q} \cdot K^{-1}\underline{n}^T$. [In non-minimal QH fluids, there are anyons which have the same fractional charges but different statistical phases!]

Experimentally, there is evidence[12] that the observed odd-denominator QH fluids exhibit "elementary" anyons with fractional charges given by $e^* = e/d_H$. Eq. (6) then suggests that $\lambda = 1$, which is a property automatically satisfied by minimal QH fluids where $\lambda = l = 1$!

Conway, Sloane, and Sloane[13] have compiled a classification of all positive integral quadratic forms K associated with an indecomposable (even or odd) lattice Γ for a range of determinants given by $1 \leq \Delta \equiv \det K \leq 25$ and for lattice dimensions up to a limit ranging from 18 (for $\Delta = 1$) to 7 (for $\Delta = 25$). This classification involves over 100 odd lattices of odd determinant. When combined with a systematic study of visible vectors $\vec{Q}$ in the associated dual lattices Γ^*, it yields, for a physically interesting range of Hall conductivities σ_H, all possible *minimal indecomposable QH lattices* $(\Gamma, \vec{Q})$ of dimension $N = \dim \Gamma \leq N_*(\sigma_H)$, where the dimensions $N_*(\sigma_H)$ depend on the limits in Ref. 13 and satisfy (7). We summarize our results in Table 1 below, where the dimensions $N_*(\sigma_H)$ are indicated in square brackets to the right of each fraction σ_H, except for fractions with $n_H = 1$ for which the Laughlin fluids are unique (see (**M1**)).

In Table 1, minimal indecomposable QH lattices (or, equivalently, universality classes of minimal elementary QH fluids) are specified in terms of explicit matrix and vector realizations of the pairs $(K, \underline{Q})$. [For a general discussion of the geometry of QH lattices making use of gluing theory, see Ref. 6.] The results are presented relative to lattice and dual lattice bases in Γ and Γ^*, respectively, which make the *symmetries* of the corresponding QH fluids most manifest.[4-6] The general results of Refs. 13 and 6 show that most QH lattices in the range of determinants and dimensions specified above have a Gram matrix of the form

$$K = \left(\begin{array}{c|c|c} \mathcal{O} & \begin{array}{cc} \underline{t}_1 & \underline{t}_2 \\ \underline{s}_1 & \underline{s}_2 \end{array} \\ \hline \underline{t}_1^T \; \underline{s}_1^T & C(X_1) & 0 \\ \hline \underline{t}_2^T \; \underline{s}_2^T & 0 & C(X_2) \end{array} \right), \tag{8}$$

which we abbreviate by writing $\{\mathcal{O}|^{t_1,s_1}X_1, {}^{t_2,s_2}X_2\}_N$, where $C(X_i)$ denotes the *Cartan matrix* of a simply-laced simple Lie algebra, X_i, with non-trivial center, $i = 1, 2$. For $i = 1, 2$, X_i is one of the following algebras: $A_{n-1} = su(n)$, $n = 2, 3, \ldots$, $D_n = so(2n)$, $n = 4, 5, \ldots$, E_6, or E_7. [E_8-sublattices do not appear in elementary QH fluids! For the Cartan matrices we use the explicit forms given, e.g., in Ref. 14.] Furthermore, in (8), $\mathcal{O}$ stands either for an odd integer, $\mathcal{O} = 2p+1$, $p = 1, 2, \ldots$, or for an odd, two-dimensional integral quadratic form with Gram matrix $\mathcal{O} = \begin{pmatrix} a & b \\ b & c \end{pmatrix}$, abbreviated by a^bc; $N = \mathrm{rank}\, K = \dim \Gamma$. The quantities $\underline{t}_i$ and $\underline{s}_i$ denote $(\mathrm{rank}\, X_i)$-dimensional integral vectors of the form $\underline{t}_i = (0, \ldots, 0, -1, 0, \ldots, 0)$, where -1 stands in the t_i^{th} place indicated by ${}^{t_i,s_i}X_i$, and similarly for $\underline{s}_i$, $i = 1, 2$. [The vectors $\underline{s}_1$ and $\underline{s}_2$ are understood to be absent if $\dim \mathcal{O} = 1$.] Finally, if t_i or $s_i = 0$ then $\underline{t}_i$ or $\underline{s}_i = \underline{0}$.

A particularly interesting subclass of minimal indecomposable QH lattices is formed by what we call *maximally symmetric* lattices. A maximally symmetric QH lattice has the property that the neutral sublattice Σ of Γ (which lies in the orthogonal complement of $\vec{Q}$, i.e., $Q_{\mathrm{el}}(\underline{n}) = \vec{Q} \cdot K^{-1} \underline{n}^T = 0$, for $\underline{n} \in \Sigma$) is a direct sum of A_n, D_n, and $E_{n \neq 8}$-lattices, and nothing else. In particular, they have $\dim \mathcal{O} = 1$ and, relative to the dual bases chosen in (8), their visible vectors are always of the form $\vec{Q} = (1,0,\ldots,0)$. One can show[4-6] that the corresponding maximally symmetric QH fluids exhibit *Kac-Moody algebras* at level 1 associated with the Lie algebras X_1 and X_2.

In Table 1, we use round brackets $(\mathcal{O}|^{t_1,s_1}X_1, {}^{t_2,s_2}X_2)_N$ to indicate maximally symmetric QH lattices, thereby distinguishing them from all other possible minimal indecomposable QH lattices (with $N \leq N_*(\sigma_H)$) for which we use square brackets $[\ldots]_N$. Moreover, for maximally symmetric QH lattices, we do not display the vectors $\vec{Q}$ explicitly.

It seems natural to call maximally symmetric QH fluids *"generalized Laughlin fluids"* because they can be constructed from Laughlin's $1/(2p + 1)$-fluids by adding symmetries and iterating. Let

$$K = \begin{pmatrix} K' & \underline{t} \\ \hline \underline{t}^T & 0 \\ & 0 & C(X) \end{pmatrix}, \quad \text{and} \quad \vec{Q} = (\overbrace{1,0,\ldots,0}^{\text{rank } K'},\overbrace{0,\ldots,0}^{\text{rank } X}), \quad (9)$$

similarly to (8). Then one easily proves the following relations for

$$
\begin{aligned}
{}^tX = {}^tA_{n-1} = {}^tsu(n) &: \gamma = n\gamma' \text{ and } \Delta = n\Delta' - t(n-t)\gamma', \text{ for } t=1,\ldots,n-1, \\
{}^tX = {}^tD_n = {}^tso(2n) &: \gamma = 4\gamma' \text{ and } \Delta = 4\Delta' - n\gamma', \text{ if } t = n-1 \text{ or } n, \\
{}^tX = {}^tE_6 &: \gamma = 3\gamma' \text{ and } \Delta = 3\Delta' - 4\gamma', \text{ if } t = 1 \text{ or } 5, \\
{}^tX = {}^tE_7 &: \gamma = 2\gamma' \text{ and } \Delta = 2\Delta' - 3\gamma', \text{ if } t = 6,
\end{aligned}
\quad (10)
$$

where $\sigma_H' = (\gamma'/\Delta')e^2/h$, with $\Delta' = \det K'$ and $\gamma' = \vec{Q}' \cdot \tilde{K}' \vec{Q}'^T$, and accordingly for the unprimed quantities; see also (5).

We note that, for a given value of σ_H, one can construct all maximally symmetric QH lattices, in arbitrary dimension, reproducing that value of σ_H. In Table 1, *all* minimal, maximally symmetric QH lattices are given for the physically relevant range of Hall conductivities $1/7 \leq \sigma_H h/e^2 = n_H/d_H \leq 3$ with $n_H \leq 12$ and $d_H \leq 17$, and all QH lattices that can be constructed on the basis of the classification in Ref. 13 are displayed. For additional results, see Refs. 5 and 6.

A step towards a comparison of our results with phenomenological data obtained in one-layer (or one-component) QH systems[10] is made, in Table 1, by indicating in bold type all experimentally observed fractions. Moreover, Hall fractions for which there is some evidence are typed in bold and enclosed in brackets. All other fractions (plain) have not been established experimentally in one-layer systems, so far. A discussion of our results can naturally be organized by collecting QH lattices of the same *"symmetry type"* into *"structural families"*; see Refs. 5 and 6, where a detailed discussion about physical implications of our results and comparisons with the standard Haldane-Halperin[15] and Jain-Goldman[16] hierarchy schemes can be found. Here we only mention that Table 1 comprises the Laughlin $1/(2p+1)$-fluids, the "basic" Jain fluids[17] (corresponding to the A-type fluids with $\sigma_H < \frac{1}{2}e^2/h$), and the two two-dimensional hierarchy fluids with $\sigma_H = \frac{4}{11}e^2/h$ and $\frac{6}{17}e^2/h$. All other (universality classes of) QH fluids given in Table 1 do not seem to have appeared previously in the literature. Evidence that some of these "new" QH fluids might actually have been observed in some experiments will appear in Refs. 5 and 6.

Table 1. Minimal indecomposable quantum Hall lattices with $1/7 \leq n_H/d_H = \sigma_H h/e^2 \leq 3$. $(\ldots)_N$ and $\underline{Q} = (1,0,\ldots,0)$: maximally symmetric lattices (complete list); $[\ldots]_N$ and $\underline{Q}$ as indicated: non-maximally symmetric lattices (all with $N \leq N_*(\sigma_H)$).

	1 [15]	**2** [15]	**3** [18]	**4** [15]	**5** [16]	**6** [13]	**7** [14]	**8** [15]	**9** [12]	**10** [13]	**11** [14]	**12** [13]
1 $(1)_1$												
$\frac{1}{3}$ $(3)_1$	$\frac{2}{3}$ [15] $(3\|^6E_7)_8$			$\frac{4}{3}$ [15] $(3\|^9D_9)_{10}$	$\frac{5}{3}$ [16]		$\frac{7}{3}$ [14]	$\frac{8}{3}$ [15]				
$\frac{1}{5}$ $(5)_1$	$\frac{2}{5}$ [13] $(3\|^1A_1)_2$	$\frac{3}{5}$ [14] $(3\|^1E_6)_7$		$\frac{4}{5}$ [13] $(3\|^7D_7)_8$	*	*	$\frac{7}{5}$ [14] $[3\|3\|^{9,0}D_9]_{11}$ $(1,1,0,\ldots,0)$	$\frac{8}{5}$ [13]	$\frac{9}{5}$ [12]	*	$\frac{11}{5}$ [14] $(3\|^4A_{10})_{11}$	$\frac{12}{5}$ [13] $(3\|^5D_5{}^1E_6)_{12}$ $[3^25\|^{10,9}D_{10}]_{12}$ $(1,1,0,\ldots,0)$
$\frac{1}{7}$ $(7)_1$	$\frac{2}{7}$ [13] $(5\|^6E_7)_8$	$\frac{3}{7}$ [14] $(3\|^1A_2)_3$		$\frac{4}{7}$ [13] $(3\|^5D_5)_6$	$\frac{5}{7}$ [12] $[3^24\|^{0,6}E_7]_9$ $(1,0,0,\ldots,0)$	$\frac{6}{7}$ [13] $(3\|^1A_1{}^1E_6)_8$	*	$\frac{8}{7}$ [13] $[3^13\|^{8,1}D_8]_{10}$ $(1,1,0,\ldots,0)$	$\left(\frac{9}{7}\right)$ [12] $(3\|^4A_8)_9$	$\frac{10}{7}$ [13] $(3\|^4A_4{}^6E_7)_{12}$ $[3^24\|^{1,1}A_1{}^{0,9}D_9]_{12}$ $(1,0,\ldots,0)$	$\frac{11}{7}$ [14]	$\frac{12}{7}$ [13] $(3\|^1A_2{}^7D_7)_{10}$
$\left(\frac{1}{9}\right)$ $(9)_1$	$\frac{2}{9}$ [9] $(5\|^1A_1)_2$	*		$\frac{4}{9}$ [9] $(3\|^1A_3)_4$ $(5\|^{11}D_{11})_{12}$	$\frac{5}{9}$ [8] $(3\|^2A_4)_5$	*	$\frac{7}{9}$ [10] $(3\|^3A_6)_7$	$\frac{8}{9}$ [9] $(3\|^3A_7)_8$	*	$\frac{10}{9}$ [9] $(3\|^3A_9)_{10}$	$\left(\frac{11}{9}\right)$ [10] $(3\|^3A_{10})_{11}$	*
	$\frac{2}{11}$ [9] $(7\|^6E_7)_8$	$\frac{3}{11}$ [10] $(5\|^1E_6)_7$		$\frac{4}{11}$ [9] $[3\|^14]_2$ $(1,0)$ $(5\|^9D_9)_{10}$	$\frac{5}{11}$ [8] $(3\|^1A_4)_5$	$\frac{6}{11}$ [9] $(3\|^1A_1{}^1A_2)_4$	$\frac{7}{11}$ [10] $(3\|^2A_6)_7$	$\frac{8}{11}$ [11]	$\frac{9}{11}$ [11]	$\frac{10}{11}$ [9] $[3\|3\|^{6,1}D_6]_8$ $(1,1,0,\ldots,0)$	$\frac{11}{11}$	$\frac{12}{11}$ [9] $[3\|^14]_2$ $(1,2)$ $(3\|^1A_3{}^1E_6)_{10}$ $(5\|^{11}D_{11}{}^1E_6)_{18}$
	$\frac{2}{13}$ [9] $(7\|^1A_1)_2$	$\frac{3}{13}$ [10] $(5\|^1A_2)_3$		$\frac{4}{13}$ [9] $(5\|^1D_7)_8$	$\frac{5}{13}$ [8]	$\frac{6}{13}$ [9] $(3\|^1A_5)_6$ $(5\|^1E_6{}^6E_7)_{14}$	$\frac{7}{13}$ [10] $[3^24\|^{0,1}A_1]_3$ $(1,0,0)$	$\frac{8}{13}$ [9] $[3^25\|^{5,0}E_6]_8$ $(1,1,0,\ldots,0)$	$\frac{9}{13}$ [8] $(3\|^2A_8)_9$	$\frac{10}{13}$ [9] $(3\|^1A_1{}^2A_4)_6$	$\frac{11}{13}$ [10] $[7^24\|^{0,5}D_5]_7$ $(1,0,\ldots,0)$	$\frac{12}{13}$ [9] $(3\|^1A_2{}^5D_5)_8$
				$\frac{4}{15}$ [9] $(5\|^5D_5)_6$		*	$\frac{7}{15}$ [10] $(3\|^1A_6)_7$	$\frac{8}{15}$ [9]	*	*	$\frac{11}{15}$ [10] $(3\|^2A_{10})_{11}$	*
	$\frac{3}{17}$ [10] $(7\|^1E_6)_7$			$\frac{4}{17}$ [7] $(5\|^1A_3)_4$ $(7\|^{11}D_{11})_{12}$	$\frac{5}{17}$ [8]	$\frac{6}{17}$ [7] $[3\|^16]_2$ $(1,0)$ $(5\|^1A_2{}^6E_7)_{10}$	$\frac{7}{17}$ [10] $[5^34\|^{0,1}A_1]_3$ $(1,0,0,0)$	$\frac{8}{17}$ [7] $(3\|^1A_7)_8$	$\frac{9}{17}$ [8]	$\left(\frac{10}{17}\right)$ [7] $(3\|^1A_1{}^1A_4)_6$	$\frac{11}{17}$ [10] $[3^33\|^{5,0}D_5]_7$ $(1,1,0,\ldots,0)$	$\frac{12}{17}$ [7] $(5\|^9D_9{}^1E_6)_{16}$

ACKNOWLEDGMENT

The research of U.M.S. is supported by Onderzoeksfonds K.U. Leuven, grant OT/92/9.

REFERENCES

1. K. von Klitzing, G. Dorda, and M. Pepper, Phys. Rev. Lett. **45**, 494 (1980);
   D.C. Tsui, H.L. Stormer, and A.C. Gossard, Phys. Rev. B **48**, 1559 (1982);
   for a review, see, e.g., R.E. Prange and S.M. Gervin, eds., *The Quantum Hall Effect*, Second Edition, Graduate Texts in Contemporary Physics (Springer, New York, 1990).
2. B.L. Al'tshuler and P.A. Lee, Physics Today **41** (12), 36 (1988);
   R.A. Webb and S. Washburn, *ibid.* **41** (12), 46 (1988).
3. R. Mottahedeh *et al.*, Solid State Commun. **72**, 1065 (1989);
   D. Yoshioka, J. Phys. Soc. Jpn. **62**, 839 (1993).
4. J. Fröhlich and U.M. Studer, Commun. Math. Phys. **148**, 553 (1992); Rev. Mod. Phys. **65**, 733 (1993).
5. J. Fröhlich, U.M. Studer, and E. Thiran, "Gauge symmetry, integral lattices, and the classification of quantum Hall fluids", preprint, KUL-TF-93/33.
6. J. Fröhlich and E. Thiran, "Integral quadratic forms, Kac-Moody algebras, and fractional quantum Hall effect: an $ADE - O$ classification", preprint, ETH-TH/93-22.
7. B.I. Halperin, Phys. Rev. B **25**, 2185 (1982);
   M. Büttiker, *ibid.* **38**, 9375 (1988);
   C.W.J. Beenakker, Phys. Rev. Lett. **64**, 216 (1990);
   A.H. MacDonald, *ibid.* **64**, 220 (1990);
   X.G. Wen, *ibid.* **64**, 2206 (1990); Phys. Rev. B **41**, 12838 (1990);
   J. Fröhlich and T. Kerler, Nucl. Phys. B **354**, 369 (1991);
   M. Stone, Ann. Phys. (N.Y.) **207**, 38 (1991);
   R.C. Ashoori *et al.*, Phys. Rev. B **45**, 3894 (1992);
   K. von Klitzing, Physica B **184**, 1 (1993).
8. P. Goddard and D. Olive, Int. J. Mod. Phys. A **1**, 303 (1986).
9. R.B. Laughlin, Phys. Rev. Lett. **50**, 1395 (1983); Phys. Rev. B **27**, 3383 (1983).
10. D.C. Tsui, Physica B **164**, 59 (1990), and references therein;
    H.W. Jiang *et al.*, Phys. Rev. B **44**, 8107 (1991);
    H.L. Stormer, Physica B **177**, 401 (1992), and references therein.
11. R.L. Willett *et al.*, Phys. Rev. Lett. **59**, 1776 (1987);
    J.P. Eisenstein *et al.*, *ibid.* **61**, 997 (1988); Surf. Sci. **229**, 31 (1990).
12. R.G. Clark *et al.*, Phys. Rev. Lett. **60**, 1747 (1988);
    S.W. Hwang *et al.*, Surf. Sci. **263**, 72 (1992).
13. J.H. Conway, F.R.S. Sloane, and N.J.A. Sloane, Proc. R. Soc. Lond. A **418**, 17 (1988), and references therein.
14. R. Slansky, Phys. Reports **79**, 1 (1981).
15. F.D.M. Haldane, Phys. Rev. Lett. **51**, 605 (1983);
    B.I. Halperin, *ibid.* **52**, 1583 (1984).
16. J.K. Jain and V.J. Goldman, Phys. Rev. B **45**, 1255 (1992).
17. J.K. Jain, Phys. Rev. Lett. **63**, 199 (1989); Phys. Rev. B **41**, 7653 (1990).

SIMPLE RANDOM WALKS: NEW DEVELOPMENTS

S.B. Shlosman

Department of Mathematics, UCI, Irvine, CA 92717, USA
Institute of the Information Transmission Problems
Russian Academy of Sciences, Moscow, Russia
Hebrew Maimonides University, Moscow, Russia

Partially supported by the NSF under grant DMS 9208029

INTRODUCTION

The present talk is based on a recent paper by K.M.Khanin, A.E.Mazel, S.B.Shlosman and Ya.G.Sinai[1], where the behavior of the random walk in random potential is studied. Here we will discuss the question about the properties of the intersection set of independent simple random walks in high dimension: how fast the probability that the intersection set contains k points, decays with k?

We consider simple random walks on the d-dimensional lattice $\mathbf{Z}^d$. A random path $\omega^{(n)}$ is the sequence $\{\omega^{(n)}(k), \ 0 \leq k \leq n\}$, where $\omega^{(n)}(k) \in \mathbf{Z}^d$ for all k and $||\omega^{(n)}(k+1) - \omega^{(n)}(k)|| = 1$. The norm of $\omega = (\omega_1, \ldots, \omega_d)$ is by definition $||\omega|| = \sum_{i=1}^{d} |\omega_i|$.

Let us take l independent infinite simple random walks starting at the origin, $l = 2, 3, \ldots$. Define the random variable η_l, which is equal to the number of points of intersections of those walks:

$$\eta_l = \#\{x \in \mathbf{Z}^d : \exists s_1 \geq 0, \ldots, s_l \geq 0 : \omega_1(s_1) = \ldots = \omega_l(s_l) = x\}.$$

In other words the random variable η_l is equal to the number of lattice points visited by all walks $\omega_1, \ldots, \omega_l$. It is well-known that for $d > 4$ the variables η_l are finite with probability one. So one can study their probability distribution:

$$\mathbf{P}_{\eta_l}(k) = Pr\{\eta_l \geq k\}.$$

The initial idea is very simple: With positive probability the trajectories $\omega_1,\ldots,\omega_l$ will meet only once — at the origin. The probability that they all would visit another point is therefore less that one. Let us denote it by p_l. After leaving that point we are in the same initial situation, so the probability to visit a second point together should be p_l^2, and so on. The argument presented suggests that the probability in question should decay exponentially in k.

To make the above speculation into a rigorous statement one has to prove the upper and lower bounds exponential in k. The lower bound turns out to be easy. The upper bound, however, turns out to be not only hard, but actually wrong! The surprising point of the problem is that the actual decay rate is slower than exponential. Namely, we will show below that the following statement holds:

Theorem 1. *For any $\delta > 0$, $l \geq 2$ and $d > 4$ there exists a number $k_0(\delta)$ such that for all $k \geq k_0(\delta)$*

$$\exp\{-k^{1-\frac{2}{d}+\delta}\} \leq \mathbf{P}_{\eta_l}(k) \leq \exp\{-k^{1-\frac{2}{d}-\delta}\}.$$

The estimate from below in Theorem 1 is based on the estimate of the probability for simple random walk to visit all the points of the cube $B(0,2k)$, with a side $2k$, centered at the origin. Namely, introduce the following probability P_k:

$$P_k = Pr\{\omega : \forall x \in B(0,2k) \, \exists i : \omega(i) = x\}.$$

Then the following result holds.

Theorem 2. *For any $\varepsilon > 0$ there exists a number $k_1(\varepsilon)$ such that for all $k > k_1(\varepsilon)$:*

$$\exp\{-(2k)^{d(1-\frac{2}{d}+\varepsilon)}\} \leq P_k \leq \exp\{-(2k)^{d(1-\frac{2}{d}-\varepsilon)}\}.$$

We will give an outline of the proof of the lower bound of the Theorem 2 in the next section. In Section 3 we prove the upper bound of the Theorem 1. We conclude the introduction by explaining how the lower bound of the Theorem 2 implies the one of the Theorem 1.

The key observation is the following: we obtain the lower bound of the Theorem 1 by taking the l-th power of the probability of the event that a single random walker visits all the sites of a given set consisting of k points. Indeed, the l-th power of the probability is the probability of the event that each of l independent walkers would visit all sites of the set, and so the set visited by all of them contains at least k points. Our choice of the set in question would be a box $B(0,n)$, where n is the smallest even integer such that $n^d \geq k$. Now the lower bound of the Theorem 1 follows from that of the Theorem 2.

THE CHEAPEST WAY TO WALK ALL OVER THE BOX: ROUNDTRIP!

In this section we will give the main idea of the proof of the lower bound of the Theorem 2.

Consider an infinite simple random walk w with $w(0) = 0$. Typically such a walk spends inside the box $B(0, 2k)$ the time of the order of k^2, and so visits about k^2 points of this box before leaving it for good. That tells us that if one would take a set of independent infinite simple random walks starting from the origin, then their union have a good chance to cover the whole box $B(0, 2k)$, provided the number of the walks is of the order of k^{d-2}. That, however, is not what we want, since we need a single random walk, and not a family of them.

The idea now is to replace the family of k^{d-2} infinite random walks by a family of $k^{d-2} - 1$ loops plus one infinite random walk, which union can be viewed as one infinite random walk. (Recall that a loop of the length n is a random walk $w^{(n)}$ for which $w^{(n)}(0) = w^{(n)}(n) = 0$ and $w^{(n)}(i) \neq 0$, $i = 1, \ldots, n-1$.) Two things should be checked now:

i) the behavior of the single loop near the origin (during the time of the order of k^2) is the same as for the infinite simple random walk;

ii) the probability of having $k^{d-2} - 1$ loops is not too small.

To ensure i) it is sufficient to consider loops which are long enough, say $2k^{10}$ (that length should be even). Then the statement i) is the usual statement about the "equivalence of the ensembles". As for ii), it is well-known that the probability to have a loop of length n decays as $const \cdot n^{-\frac{d}{2}}$ for even n and $d > 2$. So the overall probability is not smaller than

$$\exp\{-(5+\gamma)dk^{d-2}\ln k\}$$

for any $\gamma > 0$, provided k is large enough, which is essentially the bound we are looking for.

It is easy to see that the above arguments can be transformed into a rigorous proof (see [1] for details).

THE UPPER BOUND

The upper bound from Theorem 1 follows from the following lemma.

Lemma 3. *For $d > 4$ and $\gamma > 0$ there exists a constant $c_1 > 0$ such that for all m*

$$E(\eta_l^m) \leq c_1^m \cdot (m!)^q,$$

with $q = \frac{d}{d-2} + \gamma$.

Proof. Evidently, it is enough to prove the estimate for the case $l = 2$. In what follows, we denote by η the variable η_2. Consider the indicator function $I'(x)$ of the event that the trajectory w' visits the point $x \in \mathbf{Z}^d$ and the analogous indicator function $I''(x)$ for the trajectory w''. Obviously

$$\eta = \sum_{x \in \mathbf{Z}^d} I'(x)I''(x).$$

Hence

$$E(\eta^m) = E \sum_{\{x_1, \ldots, x_m\}} \prod_{i=1}^m I'(x_i)I''(x_i),$$

where the sum is taken over all ordered m-tuples $\{x_1,\ldots,x_m\}$, $x_i \in \mathbf{Z}^d$. Denote by t the number of different sites x_i among $x_1,\ldots,x_m$ and let D_t^m be the total number of ordered m-tuples which can be constructed from t different sites $x_j \in \mathbf{Z}^d$. Obviously, $D_t^m \le t^m$. We use the notation $[x_1,\ldots,x_t]$ for the non-ordered t-tuple, i.e. for the set of t different points of $\mathbf{Z}^d$. The expression for $E(\eta^m)$ can be rewritten now as

$$E(\eta^m) = \sum_{t=1}^{m} D_t^m \sum_{[x_1,\ldots,x_t]} \mathbf{P}^2([x_1,\ldots,x_t]),$$

where $\mathbf{P}([x_1,\ldots,x_t]) = E \prod_{i=1}^{t} I(x_i)$ is the probability that a random path visits all points of the set $[x_1,\ldots,x_t]$. Obviously, $0 \le \mathbf{P}([x_1,\ldots,x_t]) \le 1$ and $1 < q < 2$ for $d > 4$ and $0 < \gamma < \frac{1}{3}$. Now comes the crucial step: to replace the exponent 2 by q, $1 < q < 2$, in the last formula! We have then:

$$E(\eta^m) \le \sum_{t=1}^{m} D_t^m \sum_{[x_1,\ldots,x_t]} \mathbf{P}^q([x_1,\ldots,x_t]).$$

Let a permutation $\pi = (i_1,\ldots,i_t)$ denotes the order in which the trajectory visits points of the set $[x_1,\ldots,x_t]$ and let $\mathbf{P}([x_1,\ldots,x_t],\pi)$ be the corresponding probability. The advantage of a seemingly strange sacrifice above is that now we can use the well-known convexity inequality: for $q > 1$ and $a_j \ge 0$

$$\left(\sum_{j=1}^{n} a_j\right)^q \le n^{q-1} \sum_{j=1}^{n} a_j^q,$$

with equality in case when $a_1 = \ldots = a_n$. Then

$$E(\eta^m) \le \sum_{t=1}^{m} D_t^m \sum_{[x_1,\ldots,x_t]} \left(\sum_{\pi} \mathbf{P}([x_1,\ldots,x_t],\pi)\right)^q \le$$

$$\le \sum_{t=1}^{m} D_t^m \sum_{[x_1,\ldots,x_t]} (t!)^{q-1} \sum_{\pi} \mathbf{P}^q([x_1,\ldots,x_t],\pi) \le$$

$$\le \sum_{t=1}^{m} D_t^m (t!)^{q-1} \sum_{\{x_1,\ldots,x_t\}} \prod_{i=1}^{t} G^q(x_{i-1},x_i),$$

provided $x_0 = 0$. Here $G(x,y)$ is the Green function; clearly, $G^q(x,y)$ is of the order of $||x-y||^{-d-\gamma(d-2)}$ and the sum $\sum_{y \in \mathbf{Z}^d} G^q(x,y)$ is finite for positive γ. Hence

$$E(\eta^m) \le \sum_{t=1}^{m} D_t^m (t!)^{q-1} const^t \le \sum_{t=1}^{m} m^m (m!)^{q-1} const^m \le (m!)^q c_1^m.$$

∎

It follows from Lemma 3 and the Stirling formula that for $0 < \alpha < qc_1^{-\frac{1}{q}}$

$$E \exp\{\alpha \eta^{\frac{1}{q}}\} = \sum_{m=0}^{\infty} E \frac{(\alpha \eta^{\frac{1}{q}})^m}{m!} \le 1 + \sum_{m=1}^{\infty} E \frac{\alpha^m \eta^{[\frac{m}{q}]+1}}{m!} \le$$

$$\leq 1 + \sum_{m=1}^{\infty} \frac{a^m(([\frac{m}{q}]+1)!)^q c_1^{[\frac{m}{q}]+1}}{m!} \leq$$

$$\leq 1 + const \cdot \sum_{m=1}^{\infty} \frac{a^m e^m m^{(q-1)/2}([\frac{m}{q}]+1)^{q([\frac{m}{q}]+1)} c_1^{[\frac{m}{q}]+1}}{m^m e^{q([\frac{m}{q}]+1)}} \leq$$

$$\leq const \cdot \sum_{m=0}^{\infty} a^m m^{q+1} q^{-m} c_1^{\frac{m}{q}} < \infty,$$

where $[\cdot]$ denotes the integer part.

Now by the Chebyshev inequality we have for $\alpha = \frac{1}{2} q c_1^{-\frac{1}{q}}$ and k large enough

$$Pr\{\eta \geq k\} \leq \frac{E \exp\{\alpha \eta^{\frac{1}{q}}\}}{\exp\{\alpha k^{\frac{1}{q}}\}} \leq$$

$$\leq const \cdot \exp\{-\alpha \cdot k^{1-\frac{2}{d}-\gamma \frac{(d-2)^2}{d(d+\gamma(d-2))}}\} \leq \exp\{-k^{1-\frac{2}{d}-\delta}\},$$

for γ small enough. ∎

The upper bound in Theorem 2 follows directly from Theorem 1.

REFERENCES

1. K.M.Khanin, A.E.Mazel, S.B.Shlosman and Ya.G.Sinai. Loop condensation effects in the behavior of the random walks, to appear.

SYMMETRY BREAKING AND LONG RANGE ORDER IN HEISENBERG ANTIFERROMAGNETS

Tohru Koma

Department of Physics, Gakushuin University
Mejiro, Toshima-ku, Tokyo 171, Japan

1. INTRODUCTION

Néel order in quantum Heisenberg antiferromagnetic spin systems has been a subject of considerable interest. This is not only because the problem has close connections with experiments, but also it offers a prototype of nontrivial many body problems in quantum systems where the order operator and the Hamiltonian do not commute with each other. In such a system, one encounters strong "quantum fluctuation", which may lead to exotic behavior not observed in "classical" systems.

There have been considerable amount of theoretical and numerical works on Heisenberg antiferromagnets.[1] There are also rigorous results concerning the existence of a Néel order. By extending the Fröhlich-Simon-Spencer method based on the infrared bounds [1], Dyson, Lieb and Simon [2] proved the existence of a long range order in certain Heisenberg antiferromagnets. This result has been extended to various other situations [3].

The Dyson-Lieb-Simon method and its extensions prove that the long range order parameter σ [defined in (4) below] is nonvanishing at sufficiently low temperatures, or in the ground state. The quantity σ measures the long range correlation in the symmetric (non-pure) state under vanishing external field. The order parameter which has direct relevance to physical observations (in, say, a neutron scattering experiment), on the other hand, is believed to be the spontaneous staggered magnetization m_s [defined in (3) below], which measures the amount of symmetry breaking in the (presumably pure) state obtained by applying an infinitesimal symmetry breaking field.

The relation between the two order parameters σ and m_s is *a priori* not clear, and this has been a cause of some confusion and controversy. (See [4] and references therein.) In the Heisenberg ferromagnets, Griffiths [5] proved that $m_s \geq \sigma$, and Dyson, Lieb and Simon [2] refined this as $m_s \geq \sqrt{3}\,\sigma$,[2] but the proofs made full use of the fact that the order operators and the Hamiltonian do commute. Extensions to models with

---

[1] The literature is too large to catalogue here. See references quoted in [2, 3, 4].
[2] Unfortunately the Dyson-Lieb-Simon method of proving the existence of a long range order does not apply to ferromagnetic models.

non-commuting order operators and Hamiltonian were also investigated [5]. Recently, Kaplan, Horsch and von der Linden [4] proved the bound $m_s \geq \sigma$ in Heisenberg antiferromagnets at zero temperature.[3] But the problem at finite temperature remained unsettled.

In this talk,[4] we report an extension [6] of the above mentioned relation between m_s and σ to quantum systems with non-commuting order operators and Hamiltonian, including the Heisenberg antiferromagnets and the electron pair condensation problem in Hubbard-type models. Our result supplies the information about the nonvanishing long range order parameter from the Dyson-Lieb-Simon method, and proves that the spontaneous staggered magnetization in the Heisenberg antiferromagnet takes a nonvanishing value.

2. SPONTANEOUS MAGNETIZATION AND LONG RANGE ORDER

Although our results apply to general quantum systems on a lattice, we here describe them in the context of the simplest Heisenberg antiferromagnets. Consider a d-dimensional $L \times \cdots \times L$ hypercubic lattice $\Lambda \subset \mathbf{Z}^d$ with periodic boundary conditions, where L is an even integer. We denote by $N = L^d$ the total number of sites. With each site $x = (x_1, \cdots, x_d) \in \Lambda$, we associate a three component spin operator $\mathbf{S}_x = (S_x^{(1)}, S_x^{(2)}, S_x^{(3)})$ with $(\mathbf{S}_x)^2 = S(S+1)$, where $2S = 1, 2, 3, \cdots$. We consider the standard isotropic Heisenberg Hamiltonian

$$H_L = \sum_{x,y \in \Lambda; |x-y|=1} \mathbf{S}_x \cdot \mathbf{S}_y. \tag{1}$$

We define the antiferromagnetic order operators as

$$O_L^{(i)} = \sum_{x \in \Lambda} (-1)^{x_1 + \cdots + x_d} S_x^{(i)}, \tag{2}$$

for $i = 1, 2$ and 3.

We define the spontaneous staggered magnetization as

$$m_s = \lim_{B \downarrow 0} \lim_{L \uparrow \infty} \frac{1}{N} \frac{\mathrm{Tr}\left[O_L^{(1)} e^{-\beta(H_L - BO_L^{(1)})}\right]}{\mathrm{Tr}\left[e^{-\beta(H_L - BO_L^{(1)})}\right]}, \tag{3}$$

and the long range order parameter as

$$\sigma = \lim_{L \uparrow \infty} \frac{1}{N} \sqrt{\left\langle (O_L^{(1)})^2 \right\rangle_L}, \tag{4}$$

where β is the inverse temperature. In (4) the thermal expectation in the symmetric equilibrium state is

$$\langle \cdots \rangle_L = \frac{\mathrm{Tr}\left[(\cdots) e^{-\beta H_L}\right]}{\mathrm{Tr}\left[e^{-\beta H_L}\right]}. \tag{5}$$

It has been proved [2, 3] that, for any S, the long range order parameter satisfies $\sigma > 0$ at sufficiently low temperature if $d \geq 3$.

Our main result is the following.

---

[3]The bound was refined as $m_s \geq \sqrt{3}\sigma$ [6, 7].
[4]The talk is based on the joint work [6] with my colleague H. Tasaki.

Corollary 1 *For any $\beta \in [0, \infty)$, we have*

$$m_s \geq \sqrt{3}\, \sigma. \tag{6}$$

Since the bound (6) is valid for any temperature, it applies to the ground state obtained by letting $\beta \to \infty$ after the infinite volume limit. It is also possible to treat directly the expectation for the ground state in a finite volume. The treatment again leads to the bound $m_s \geq \sqrt{3}\,\sigma$ for the ground state [6, 7]. It is an improvement of the bound $m_s \geq \sigma$ proved for the ground state by Kaplan, Horsch and von der Linden [4].

To determine the ratio $r = m_s/\sigma$ has been a central issue of the controversy [4]. Although an elementary symmetry consideration leads to the value $r = \sqrt{3}$, it was suggested [4] that r take nonclassical value in the ground states of the Heisenberg antiferromagnets. Our bound $r \geq \sqrt{3}$ partially resolves this controversy.[5] For a system with the symmetry group $SO(n)$, we can prove the same bound with $\sqrt{3}$ replaced with $\sqrt{n}$.

The above bound follows by combining the following two theorems.

Theorem 2 *For any $\beta \in [0, \infty)$, we have*

$$m_s \geq \lim_{L\uparrow\infty} \frac{1}{N} \left\{ \left\langle (O_L^{(1)})^{2k} \right\rangle_L \right\}^{\frac{1}{2k}}, \tag{7}$$

for any positive integer k.

Theorem 3 *For any $\beta \in [0, \infty)$, we have*

$$\lim_{k\uparrow\infty} \lim_{L\uparrow\infty} \frac{1}{N} \left\{ \left\langle (O_L^{(1)})^{2k} \right\rangle_L \right\}^{\frac{1}{2k}} \geq \sqrt{3}\,\sigma. \tag{8}$$

Theorems 2 and 3 in the case of Heisenberg ferromagnets were proved by Griffiths [5] and Dyson, Lieb and Simon [2], respectively.

3. "ENERGY SPACE COARSE GRAINING"

We shall sketch the basic ideas behind the proofs of these theorems (see [6] for the details). Usually it is argued that intensive quantities in a quantum system should behave as "classical" quantities when the system size is large. Our proofs are based on rigorous versions of this folk-statement.

We recall that, when the order operator and the Hamiltonian commute with each other, exactly the same statement as Theorem 2 was proved by Griffiths [5]. Since $[(H_L/N), (O_L^{(1)}/N)] = O(N^{-1})$, one might suspect that the Griffith's idea may be applied to a large system where the order operator and the Hamiltonian "almost" commute. We shall make this intuition a rigorous estimate by using a "coarse graining" procedure in the energy space as follows. Take a connected interval $I \subset \mathbf{R}$

---

[5] If we consider two identical independent Heisenberg antiferromagnets, and define $O_L^{(1)}$ as the sum of corresponding quantities in each system. Then one easily finds that m_s and σ corresponding to $O_L^{(1)}$ satisfies $m_s \geq \sqrt{6}\,\sigma$. This trivial example demonstrates that we must take into account the detailed properties of the system in order to prove the inequality corresponding to (6). We also note that the argument behind $r = \sqrt{3}$ is not as unreliable as criticized in [4]. An assumtion about a "natural" pure state decomposition inevitaby leads to the equality $m_s = \sqrt{3}\,\sigma$. See Ref. [6].

which contains the whole spectrum of H_L, and decompose it into subintervals as $I = \cup_{m=1,\cdots,M}[\tilde{E}_m, \tilde{E}_{m+1})$ with the decomposition number $M \sim \sqrt{N}$. By using the projection P_m onto the space spanned by eigenstates of H_L with the eigenvalues contained in $[\tilde{E}_m, \tilde{E}_{m+1})$, we define the approximate order operator $\tilde{O}_L$ and the approximate Hamiltonian $\tilde{H}_L$ as

$$\tilde{O}_L = \sum_{m=1}^{M} P_m O_L^{(1)} P_m \, , \quad \tilde{H}_L = \sum_{m=1}^{M} \tilde{E}_m P_m \, . \tag{9}$$

By construction, we have $[\tilde{O}_L, \tilde{H}_L] = 0$. Moreover we can prove that, for each L and $\beta < \infty$, one can take $\{\tilde{E}_m\}$ so that $\tilde{O}_L$ and $\tilde{H}_L$ are good approximations of the original operators in the sense that

$$\frac{1}{N}\left\|H_L - \tilde{H}_L\right\| \leq O(N^{-1/2}) \tag{10}$$

and

$$\frac{1}{N^{2k}}\left|\left\langle (O_L^{(1)})^{2k}\right\rangle_L - \left\langle (\tilde{O}_L)^{2k}\right\rangle_L\right| \leq O(N^{-1/4}) \tag{11}$$

for any integer $k \geq 1$. Since $\tilde{O}_L$ and $\tilde{H}_L$ commute, we can apply the original argument of Griffiths with suitable technical modifications. If we denote by $P(\tilde{O}_L \geq Nm)$ the projection onto the space spanned by the eigenstates of $\tilde{O}_L$ with the eigenvalues $\geq Nm$, we can prove that for any $m > m_s$ there exists a $\delta > 0$ such that

$$\left\langle P(\tilde{O}_L \geq Nm)\right\rangle_L \leq 2e^{-\beta(\delta N - S^2\sqrt{N})} \tag{12}$$

holds for N sufficiently large. Theorem 2 follows from this bound.

4. CONSEQUENCE OF THE SYMMETRY

We shall prove Theorem 3. The bound (8) is a quantum analog of this relation. Now making use of the $SU(2)$ invariance, we write

$$\left\langle (O_L^{(1)})^{2k}\right\rangle_L = \frac{1}{4\pi}\int d\Omega \left\langle (xO_L^{(1)} + yO_L^{(2)} + zO_L^{(3)})^{2k}\right\rangle_L \tag{13}$$

where $\Omega = (x, y, z)$ is integrated over the unit sphere $x^2 + y^2 + z^2 = 1$ ($x, y, z \in \mathbf{R}$). By expanding the right-hand side and evaluating the integral, we get

$$\begin{aligned}\left\langle (O_L^{(1)})^{2k}\right\rangle_L &= G_k \left\langle \left\{(O_L^{(1)})^2 + (O_L^{(2)})^2 + (O_L^{(3)})^2\right\}^k\right\rangle_L + O(N^{2k-1}) \\ &\geq G_k \left\{\left\langle (O_L^{(1)})^2 + (O_L^{(2)})^2 + (O_L^{(3)})^2\right\rangle_L\right\}^k + O(N^{2k-1})\end{aligned} \tag{14}$$

with

$$G_k = \frac{1}{4\pi}\int d\Omega \, x^{2k}, \tag{15}$$

where we have used the Hölder inequality to obtain the second bound in (14). We note that $(G_k)^{1/(2k)} \to 1$ as $k \uparrow \infty$ since the maximum value of $|x|$ in the integral is 1. Therefore, the desired bound (8) follows by taking the $2k$-th root of (14), and letting $N \uparrow \infty$ and then $k \uparrow \infty$.

Acknowledgments

It is a pleasure to thank H. Tasaki for an enjoyable collaboration on the present topics. I also thank the organizers for providing me with an opportunity to give a talk.

REFERENCES

1. J. Fröhlich, B. Simon and T. Spencer, *Comm. Math. Phys.*, **50**, 79 (1976);
   J. Fröhlich, R. Israel, B. Simon and E. H. Lieb, *Comm. Math. Phys.*, **62**, 1 (1978) and *J. Stat. Phys.*, **22**, 297 (1980).
2. F. J. Dyson, E. H. Lieb and B. Simon *J. Stat. Phys.*, **18**, 335 (1978).
3. E. Jordão Neves and J. Fernando Perez, *Phys. Lett.*, **114A**, 331 (1986);
   T. Kennedy, E. H. Lieb and B. S. Shastry, *J. Stat. Phys.*, **53**, 1019 (1988);
   K. Kubo, *Phys. Rev. Lett.*, **61**, 110 (1988);
   T. Kennedy, E. H. Lieb and B. S. Shastry, *Phys. Rev. Lett.*, **61**, 2582 (1988);
   H. Nishimori, K. Kubo, Y. Ozeki, Y. Tomita and T. Kishi, *J. Stat. Phys.*, **55**, 259 (1989);
   K. Kubo and T. Kishi, *Phys. Rev. Lett.*, **61**, 2585 (1988);
   Y. Ozeki, H. Nishimori and Y. Tomita, *J. Phys. Soc. Jpn.*, **58**, 82 (1989);
   H. Nishimori and Y. Ozeki, *J. Phys. Soc. Jpn.* **58**, 1027 (1989);
   T. Kishi and K. Kubo, *J. Phys. Soc. Jpn.*, **58**, 2547 (1989);
   T. Kishi and K. Kubo, *Phys. Rev.*, **B43**, 10844 (1991);
   Y. Saika and H. Nishimori, *J. Phys. Soc. Jpn.*, **61**, 3086 (1992).
4. T. A. Kaplan, P. Horsch and W. von der Linden, *J. Phys. Soc. Jpn.*, **11**, 3894 (1989);
   P. Horsch and W. von der Linden, *Z. Phys.*, **B72**, 181 (1988).
5. R. B. Griffiths, *Phys. Rev.*, **152**, 240 (1966).
6. T. Koma and H. Tasaki, *Phys. Rev. Lett.*, **70**, 93 (1993) and *Comm. Math. Phys.*, to appear.
7. T. Koma and H. Tasaki, "Obscured Symmetry Breaking and Low Lying States in Quantum Many-Body Systems", preprint.

INTEGRABLE S=1/2 QUANTUM SPIN CHAINS WITH SHORT-RANGE EXCHANGE

V.I. Inozemtsev

Lab. of Theor. Phys.
JINR, 141980 Dubna, Russia

1 INTRODUCTION

The 1D lattice spin chains, being the simplest quantum many-body systems, are still considered as objects for intensive study due to their relations to more realistic models in quantum statistics. As a rule, even for these systems the spectral problem is overcomplicated. The rare cases of its reduction are associated with the quantum integrability. Most of them concern the famous situation of nearest- neighbor or "local" interaction considered first by Bethe [1] and have the solution in the form of the Bethe ansatz [1-4].

On the other hand, there are quantum integrable many-body problems on a continuous line being very far from local. They are described by the Hamiltonians [5,6]

$$H_c = -\frac{1}{2}\sum_{j=1}^{M}\left(\frac{\partial}{\partial x_j}\right)^2 + \sum_{j>l}^{M} V(x_j - x_l), \tag{1}$$

where the most general form of V is given by the elliptic Weierstraß $\wp$ function with two periods $\omega_{1,2}$,

$$V(x) = g^2 \wp_{\omega_1,\omega_2}(x). \tag{2}$$

In the case of the infinite real period ω_1 the potential becomes inverse square hyperbolic,

$$V_\infty(x) \to g^2 a^2 \sinh^{-2}(ax), \quad a = \frac{i\pi}{\omega_2}. \tag{3}$$

The S=1/2 spin analogs of (1) with the Hamiltonians of the form

$$H_s = -\frac{J}{2}\sum_{j\neq l} h(j-l)\frac{\vec{\sigma}_j \vec{\sigma}_l - 1}{2}, \tag{4}$$

where $\{\vec{\sigma}\}$ are Pauli matrices, have received attention only recently. The degenerate case of $(j-l)^{-2}$ exchange has been proposed and investigated in [12-14] while in [10,11]

I have shown that the Hamiltonian of isotropic chain of N spins with nearest-neighbor exchange originally solved by Bethe is simply a limit of (4) with $h(x) = \wp_{N,\omega_2}(x)$ if ω_2 tends to zero with the inverse of the coupling J.

The analogy between quantum spin and particle integrable 1D problems has been noticed already by Lieb and Liniger [7], McGuire [8] and Yang [9]. They applied the Bethe ansatz method for solving the many-body problem with local interaction $V(x) \sim \delta(x)$. In the case of complicated spin chains with the exchange (2) or (3) such an analogy is used in opposite direction. At first, one has to investigate the solutions to the many-body problem on a line and then to find a way of determining the eigenvectors of the spin Hamiltonians (4). The usual Bethe ansatz should appear as a limit of the more general construction.

This program now is completely realized for infinite spin chains with the exchange interaction (3). The eigenvectors of (4) are constructed as M-magnon excitations over ferromagnetic ground states $|0_\pm> = \prod_{j \in Z} \chi_j^\pm$,

$$|\psi_M^\pm> = \sum_{n \in Z^M} \psi^{(M)}(n_1...n_M) \left(\prod_{\gamma=1}^M a_{n_\gamma}^\pm\right) |0_\pm>, \quad (5)$$

where $\chi_j^\pm$ are two-component eigenvectors of σ_{jz}, the operator a_j^+ turns spin at jth site to opposite direction and ψ_M is M-magnon wave function. It is the object which resembles the usual wave function in the continuous case. The remaining part of the report is devoted to the explicit realization of this analogy and its applications.

2 THE MANY-PARTICLE PROBLEM ON A LINE

It was found [6] that at some values of coupling g^2 the Hamiltonians (1) with potential (3) are connected with the well-known objects in the theory of symmetric spaces, the radial parts of the second-order Laplace-Beltrami operators. Hence the corresponding wave functions can be obtained from zonal spherical functions (ZSF). The latter have been studied for a long time [15]. In the most important case of $g^2 = 2$ the symmetric space is $X_M^- = SL(M,H)/Sp(M)$. The corresponding ZSF $\Phi_k^{(M)}(x)$ are parametrized by the vector $k \in \mathbf{R}^M$. The symmetric eigenfunctions of H_c at $g^2 = 2$ can be written as

$$\sum_{P \in \pi_M} \chi_k^{(M)}(Px) = \left[\prod_{j>l} a^{-2} \sinh^2 a(x_j - x_l)\right] \Phi_k^{(M)}(x) = \sum_{P \in \pi_M} e^{i \sum_{\mu=1}^M k_\mu P x_\mu} \varphi^{(M)}(Px|k),$$

where π_M is the group of all permutations $\{P\}$ of M objects and Px is the vector with components $(x_{P1},..x_{PM})$. The eigenfunction $\chi_k^{(M)}$ has the form

$$\chi_k^{(M)}(x) = \exp\left\{\sum_{\mu=1}^M [ik_\mu - a(M-1)]x_\mu\right\} \left[\prod_{\mu > \nu}^M \sinh a(x_\mu - x_\nu)\right]^{-1} S_k^{(M)}(y), \quad (6)$$

where $S_k^{(M)}(y)$ is a polynomial in $y_\mu = \exp(2ax_\mu)$. The maximal power of each variable in S cannot exceed $M - 1$, $S_k^{(M)}(y) = \sum_{m \in D^M} d_{m_1...m_M}(k) \prod_{\lambda=1}^M y_\lambda^{m_\lambda}$, where D^M is the hypercube in $\mathbf{Z}^M$, $m \in D^M \leftrightarrow 0 \leq m_\beta \leq M - 1$, and $d_{\{m\}}(k)$ is the set of M^M coefficients.

The eigenvalue condition $\left[H_c - \frac{1}{2}\sum_{j=1}^{M} k_j^2\right]\chi_k^{(M)}(x) = 0$ gives

$$\sum_{\beta=1}^{M}\left[2R_\beta(R_\beta + ia^{-1}k_\beta - M + 1) + (M-1)(\frac{2M-1}{3} - ia^{-1}k_\beta)\right]S_k^{(M)} =$$

$$\sum_{\beta\neq\rho}^{M}\frac{y_\beta + y_\rho}{y_\beta - y_\rho}\left[R_\beta - R_\rho + \frac{i}{2a}(k_\beta - k_\rho)\right]S_k^{(M)}, \tag{7}$$

where $R_\beta = y_\beta \frac{\partial}{\partial y_\beta}$. The left-hand side of (7) is a polynomial in $\{y\}$. Hence this equality can be satisfied only if for each pair (β, ρ) the polynomial $\left[R_\beta - R_\rho + \frac{i}{2a}(k_\beta - k_\rho)\right]S_k^{(M)}$ is divisible by $y_\beta - y_\rho$. This condition is expressed in the form of $\frac{1}{2}(M-1)(2M-1)M^M$ linear equations for the coefficients $\{d\}$,

$$\sum_{l=l_-(\beta,\rho)}^{l_+(\beta,\rho)} d_{m_1...m_\beta+l...m_\rho-l...m_N}(k)\left[m_\beta - m_\rho + 2l + \frac{i}{2a}(k_\beta - k_\rho)\right] = 0, \tag{8}$$

where $l_-(\beta, \rho) = \max\{-m_\beta, m_\rho + 1 - M\}$, $l_+(\beta, \rho) = \min\{M - 1 - m_\beta, m_\rho\}$.

They determine all the structure of $\{d_{\{m\}}(k)\}$. The main properties of this set can be formulated as follows.

Proposition 1. $S_k^{(M)}(y)$ is an homogeneous polynomial of the degree $\frac{1}{2}M(M-1)$.

Proposition 2. Let $\{P\}$ be the set of non-coinciding numbers $\{m\}$ defined by the permutation $P \in \pi_M$ as $m_\mu = P\mu - 1$, $1 \leq \mu \leq M$. The nonvanishing $d_{\{m\}}(k)$ with coinciding values of $\{m_\mu\}$ are expressed with the use of (8) through the elements $d_{\{P\}}(k)$. The latter are determined by (8) up to some factor, $d_{\{P\}}(k) = d_0 \prod_{\lambda<\mu}^{M}\left[1 + \frac{i}{2a}(k_{P^{-1}\lambda} - k_{P^{-1}\mu})\right]$.

Proposition 3. If the differences between arguments of the χ function (6) tend to infinity such that $x_{P(\lambda+1)} - x_{P\lambda} \to +\infty$, $1 \leq \lambda \leq M - 1$, then $\lim \varphi^{(M)}(x|k) = (-1)^P 2^{\frac{M(M-1)}{2}} d_{\{P^{-1}\}}(k)$.

Despite the explicit calculation of all $d_{\{P\}}(k)$ is rather cumbersome, the information given by linear system (8) is quite sufficient for making definite assumptions about the structure of multi-magnon wave functions in analogous lattice problems.

3 THE EXCITATIONS OF AN INFINITE SPIN LATTICE

The eigenproblem $[H_s - \varepsilon_M]|\psi_M^\pm> = 0$ with the vectors $\psi_M^\pm >$ of the form (5) leads to the lattice Schrödinger equation

$$\sum_{\beta=1}^{M}\sum_{s\in Z_{[n]}} h(n_\beta - s)\psi(n_1..n_{\beta-1}, s, n_{\beta+1}..n_M) = -\psi(n_1..n_M)\times$$

$$\left[\sum_{\beta\neq\gamma}^{M} h(n_\beta - n_\gamma) + J^{-1}\varepsilon_M - M\varepsilon_0\right]. \tag{9}$$

Here $\mathbf{Z}_{[n]} = \mathbf{Z} - (n_1...n_M)$ and $\varepsilon_0 = \sum_{j \in \mathbf{Z}_{-0}} h(j)$. Let us search the solution to (9) in the form similar to (6),

$$\psi(\{n\}) = \prod_{\mu<\gamma}^{M} \frac{1}{\sinh a(n_\mu - n_\gamma)} \sum_{P \in \pi_M, m \in D^M} (-1)^P \tilde{d}_{\{m\}}(k) e^{\sum_{\lambda=1}^{M}(ik_{P\lambda} + a(2m_{P\lambda} - M + 1))n_\lambda}. \quad (10)$$

The set $\tilde{d}_{\{m\}}(k)$ has to be found from (9). The result can be formulated as

Proposition 4. Let $\chi_k^{(M)}(x)$ be the eigenfunction (6) of the M-particle Hamiltonian H_c with pair interaction (3) at $g^2 = 2$. Then the M-magnon wave function satisfying the lattice Schrödinger equation (9) with the exchange (3) can be written as

$$\psi(n_1...n_M) = \sum_{P \in \pi_M} \exp\left(i \sum_{\mu=1}^{M} k_\mu n_{P\mu}\right) \varphi^{(M)}(Pn|\tilde{k}), \quad (11)$$

where $\tilde{k}$ is the vector with the components $\tilde{k}_\mu = if_a(k_\mu)$, $f_a(k) = \frac{k}{\pi}\zeta_1\left(\frac{i\pi}{2a}\right) - \zeta_1\left(\frac{ik}{2a}\right)$, $\zeta_1(x)$ being the Weierstraßzeta function on the torus $\mathbf{C}/\mathbf{Z} + i\pi a^{-1}\mathbf{Z}$. The energy of corresponding M-magnon state of the infinite spin lattice with quasi-momenta $\{k\}$ is given by

$$\varepsilon_M = J \sum_{\beta=1}^{M} E(k_\beta), \quad E(k) = -\frac{1}{2}\wp_1\left(\frac{ik}{2a}\right) + \frac{1}{2}f_a^2(k) - \frac{2ia}{\pi}\zeta_1\left(\frac{i\pi}{2a}\right). \quad (12)$$

The asymptotic form of ψ at large distances between overturned spins now follows from the Propositions 2-3.

Proposition 5. Let $n_{P(\lambda+1)} - n_{P\lambda} \to +\infty$ for each $\lambda \in [1, M-1]$ and some fixed $P \in \pi_M$. Then the leading term in the asymptotic expansion of ψ at $k \in \mathbf{R}^M$ reads

$$\psi(n_1...n_M) \sim \sum_{Q \in \pi_M} (-1)^{QP} \exp\left(i \sum_{\lambda=1}^{M} k_{Q\lambda} n_\lambda\right) \prod_{\mu<\nu}^{M} \left\{1 - \frac{1}{2a}[f_a(k_{QP\mu}) - f_a(k_{QP\nu})]\right\}. \quad (13)$$

The multi-magnon scattering matrix is factorized as it would be for integrable models. The right-hand side of eq.(13) can be also written in the standard Bethe-ansatz form [3] with the phase shift $\phi(k, k')$ given by

$$\coth \frac{\phi(k, k')}{2} = (2ia)^{-1}[f_a(k) - f_a(k')] \quad (14)$$

As it should be expected, as $a \to \infty$ it coincides with the Bethe phase due to the relation $\lim_{a \to \infty} a^{-1}[f_a(k) - f_a(k')] = i(\cot \frac{k}{2} - \cot \frac{k'}{2})$. At finite a the exact wave functions (11) do not reduce to the superpositions of plane waves.

4 THE FINITE CHAINS

To obtain thermodynamic description of the infinite systems with short-range hyperbolic exchange (3), one should start from studying more complicated problem of the spin lattices with *finite* number of sites $N \in \mathbf{Z}_+$. Imposing periodic boundary conditions leads to the elliptic form (2) of the exchange interaction, the periods of the

Weierstraß $\wp_N$ function being determined by the number of spins and parameter a so as $\omega_1 = N, \omega_2 = i\pi a^{-1}$. It is natural to suppose that the solutions to the corresponding lattice Schrödinger equation resemble the eigenfunctions of the differential operator (1). In the two-particle case the problem reduces to Lamé equation. At $g^2 = 2$ its solutions were described by Hermite more than century ago. The analogous two-magnon wave functions can be written as [10]

$$\psi(n_1, n_2) = \exp[i(k_1 n_1 + k_2 n_2)] \frac{\sigma_N(n_1 - n_2 + \gamma)}{\sigma_N(n_1 - n_2)} + (n_1 \leftrightarrow n_2). \tag{15}$$

The right-hand side of (15) should be periodic in $\{n\}$. This condition can be expressed as

$$k_1 N - 2i\gamma \zeta_N \left(\frac{N}{2}\right) = 2\pi l_1, \quad k_2 N + 2i\gamma \zeta_N \left(\frac{N}{2}\right) = 2\pi l_2, \tag{16}$$

where $l_1, l_2 \in \mathbf{Z}$ mod N. The third relation between the parameters k_1, k_2 and γ follows from the lattice Schrödinger equation at $M = 2$ and reads

$$f(k_1, \gamma) - f(k_2, -\gamma) - 2\zeta_N(\gamma) = 0, \tag{17}$$

where $f(k, \gamma) = \zeta_1 \left(\frac{2i\zeta_N(\frac{\omega}{2})\gamma - k\omega}{2\pi}\right) + (i\pi)^{-1} \left[2\zeta_N \left(\frac{\omega}{2}\right) \zeta_1 \left(\frac{1}{2}\right) \gamma + ik\zeta_1 \left(\frac{\omega}{2}\right)\right]$.

The sets $\{l_1, l_2 \in \mathbf{Z} \bmod N, \gamma \in \mathbf{C}\}$ satisfying (16-17) can be classified as follows [17]. Let us call them equivalent if the corresponding wave functions (15) coincide up to normalization factor. One can show that the following sets are equivalent to each $\{l_1, l_2, \gamma\}$: $\{l_1, l_2, \gamma + jN\}$, $\{l_1 + j, l_2 - j, \gamma + j\omega\}$, $\{l_2, l_1, -\gamma\}$, where $j \in \mathbf{Z}$. So one can restrict γ to the torus $\mathbf{T}_{N,N\omega}$ and put $l_2 = 0$. For each $l_1 \in \mathbf{Z} \bmod N$ there are just N roots $\{\gamma\}$ of (17) since the left-hand side of this equation is elliptic function of γ with N simple poles within $\mathbf{T}_{N,N\omega}$. After the exclusion of equivalent configurations one gets just $\frac{N(N-3)}{2}$ nonequivalent sets $\{l_1, 0, \gamma\}$ which determine all the two-magnon eigenvectors of the model with the total spin $S = \frac{N}{2} - 2$, i.e. the description of two-magnon sector given by (15-17) is complete.

The investigation of the properties of the antiferromagnetic ground state ($J < 0$) till now can be done only with the use of asymptotical method of Sutherland [21] which has been claimed to be valid for all integrable models. It consists in approximate treatment of the finite systems at large values of N by considering the wave functions of *infinite* systems only in the asymptotic region $|n_\lambda - n_\nu| \to \infty$ instead of their exact values. Imposing periodic boundary conditions for the model under consideration leads to the asymptotic equations of the Bethe ansatz

$$\exp(iNk_j) = \prod_{l \neq j}^{M} \exp[i\phi(k_j, k_l)], \quad j = 1, ... M, \tag{18}$$

with the phase shift ϕ of the form (14). The energy of the state defined by the solutions to (18) is given by (12).

If one adopts the usual hypothesis about the structure of the antiferromagnetic vacuum for even N, i.e. that it ts formed by real roots of (20) at $M = N/2$, one arrives in the limit $N \to \infty$ to the integral equation of Hulthén type for the root density $\sigma_\infty(\lambda)$,

$$\sigma_\infty(\lambda) + \pi^{-1} \int_{-\infty}^{\infty} \sigma_\infty(\lambda')[1 + (\lambda - \lambda')^2]^{-1} d\lambda' = (2\pi)^{-1} \mu'(\lambda), \tag{19}$$

where $\lambda = (2ia)^{-1} f_a(k)$ and the function $\mu(\lambda)$ is determined by implicit relation $\lambda = (2ia)^{-1} f_a(\pi - \mu)$. The final result for the asymptotic ground-state energy per site reads

$$\epsilon_\infty = \lim_{N\to\infty} N^{-1} \varepsilon_{N/2} = J \int_{-\infty}^{\infty} E(k(\lambda)) \sigma_\infty(\lambda) d\lambda =$$

$$-(2\pi)^{-2} \frac{J}{2ia} \int_0^{2\pi} dk E(k) f_a'(k) \int_{-\infty}^{\infty} \frac{\exp[(2a)^{-1} pf(k)]}{1 + \exp(-|p|)} \int_0^{2\pi} dk' \exp[-(2a)^{-1} pf(k')]. \quad (20)$$

Denoting $J_0 = -J \left(\frac{\sinh a}{a}\right)^{-2}$ and taking the limit $a \to \infty$ one recovers the Hulthén result for the nearest-neighbor exchange, $\epsilon_\infty \sim -2J_0 \log 2$. The integrals in (20) at finite a cannot be calculated analytically. At large values of $q = \exp(2a)$ one can represent (20) as an asymptotic series in q^{-1},

$$J_0^{-1} \epsilon_\infty(q) = -2\log 2 - 2q^{-1}(4 + 2\log 2 - \frac{\pi^2}{12} - 3\zeta(3)) + O(q^{-2}),$$

where $\zeta(\nu)$ is the Riemann zeta function.

The dispersion of the lowest-lying excitations over the antiferromagnetic vacuum at even N, the s=1/2 kinks, is given by

$$v_0 = \lim_{k_{kink} \to 0} -\frac{d\varepsilon_{kink}}{dk_{kink}} = \frac{J_0}{2a} \lim_{\lambda \to +\infty} \frac{\int_{-\infty}^{\infty} p\, dp \frac{\exp(ip\lambda)}{1+\exp(-|p|)} \int_0^{2\pi} dk E(k) f_a'(k) \exp[-(2a)^{-1} pf(k)]}{\int_{-\infty}^{\infty} \frac{\exp(ip\lambda)}{1+\exp(-|p|)} \int_0^{2\pi} dk \exp([-(2a)^{-1} pf(k)]}.$$

As $a \to \infty$, $v_0(q) = \pi \left[1 + \left(4 - \frac{\pi^2}{2}\right) q^{-1} + O(q^{-2})\right]$. The evaluation of leading finite-size correction to ϵ_∞ shows that it obeys the standard relation $\Delta \epsilon_N = -\frac{\pi c v_0}{6N^2}$ with c=1. It gives an argument to the existence of the corresponding conformal field theory.

To summarize, it should be noted that the investigation of the one-parametric class of integrable spin models described above is far from being completed as on principal as on computation level. Their relation to the YB equation and the corresponding algebraic structures of the quantum inverse scattering method is not clear up to this time. The SU(L) generalizations of the nearest-neighbor model [18] can also be extended to the hyperbolic exchange (3). The validity of the triangle Yang-Baxter relation can be proven for the S matrix of the fermionic version of the SU(3) case which is equivalent to the generalization of narrow-band fermionic model with nearest-neighbor hopping [19] or its supersymmetric t-J counterpart [20].

REFERENCES

1. H. Bethe, Z.Phys. **71**, 205 (1931).
2. R.J. Baxter. "Exactly Solved Models in Statistical Mechanics," Academic Press, New York (1982).
3. M. Gaudin, "La Fonction d' Onde de Bethe," Masson, Paris (1983).
4. L.D. Faddeev, "Recent Advances in Field Theory and Statistical Mechanics," pp.561-608, North-Holland, Amsterdam (1984).
5. F. Calogero, Lett. Nuovo Cim.**13**, 411 (1975); J.Moser, Adv. Math. **16**, 1 (1975).
6. M.A. Olshanetsky and A.M. Perelomov, Phys. Rep. **94**, 313 (1983).
7. E.H. Lieb and W. Liniger, Phys. Rev. **130**, 1605 (1963).
8. J.B. McGuire, J. Math. Phys. **5**, 622 (1964).
9. C.N. Yang, Phys. Rev.**168**, 1920 (1968).
10. V.I. Inozemtsev, J. Stat. Phys. **59**, 1143 (1990).

11. V.I. Inozemtsev, *Comm. Math. Phys.* **148**, 359 (1992).
12. F.D.M. Haldane, *Phys. Rev. Lett.* **60**, 635 (1988); ibid **66**, 1529 (1991).
13. B.S. Shastry, *Phys. Rev. Lett.* **60**, 639 (1988).
14. F.D.M. Haldane et al., *Phys. Rev. Lett.* **69**, 2021 (1992).
15. S. Helgason, "Differential Geometry, Lie Groups and Symmetric Spaces," Academic Press, New York (1978).
16. O.A. Chalykh and A.P. Veselov, *Comm. Math. Phys.* **126**, 597 (1990).
17. V.I. Inozemtsev, ICTP preprint IC/92/172, to appear in *Lett. Math. Phys.*
18. B. Sutherland, *Phys. Rev.* B **12**, 3795 (1975).
19. P. Schlottmann, *Phys. Rev.* B **36**, 5177 (1987).
20. P. Bares and G. Blatter, *Phys. Rev. Lett.* **64**, 2567 (1990).
21. B. Sutherland, *Rocky Mountain J. Math.* **8**, 413 (1978).

WHO IS AFRAID OF GRIFFITHS' SINGULARITIES?

A. Klein[1]

Department of Mathematics
University of California
Irvine, CA 92717, USA

In 1969 Griffiths[1] considered the statistical mechanics of a random ferromagnetic Ising model, with Hamiltonian given by

$$H = -\sum_{\langle xy \rangle} J_{xy} \sigma_x \sigma_y + h \sum_x \sigma_x \quad (1)$$

where $x \in \mathbf{Z}^d$, $\langle xy \rangle$ denotes a pair of nearest neighbor sites in $\mathbf{Z}^d$, $\sigma_x = \pm 1$ and the couplings $\mathbf{J} = \{J_{xy} > 0\}_{\langle xy \rangle}$ are taken as identically distributed random variables. He pointed out that for the site diluted model, i.e., $J_{xy} = J\xi_x\xi_y$, where the independent random variables ξ_x are 1 or 0 with probability p and $1-p$ respectively, the quenched magnetization, considered as a function of $z = e^{\beta h}$, displayed a non-analytical behavior at $z = 1$ for values of the inverse temperature β at which the system has neither long-range order nor spontaneous magnetization. His arguments apply to a large class of ferromagnetic models; in particular, if the couplings $J_{xy} > 0$ are independent identically distributed random variables, which may assume with non zero probability arbitrarily large values, these singularities should occur for every value of the temperature. At the origin of this behavior is the fact that even if, with probability one, the infinite system is not ordered as whole, there are, also with probability one, infinitely many arbitrarily large regions inside which the system is strongly correlated.

This phenomenon is now recognized to be a regular feature in the statistical mechanics of disordered systems not just of the type discussed above. It has the unpleasant consequence that the usual high temperature or low activity expansions, the standard tools for obtaining exponential decay (and also existence and uniqueness of the thermodynamical limit) of correlation functions, fail to converge.

In this report I will describe some results of von Dreifus, Perez and myself[2] about such random systems. We consider a class of systems whose typical representative is an Ising model in $\mathbf{Z}^d$ whose Hamiltonian is given in a finite volume $\Lambda \subset \mathbf{Z}^d$ by

$$H_\Lambda = -\sum_{\langle xy \rangle \in \Lambda^*} J_{xy} \sigma_x \sigma_y + \mathcal{B} \sum_{x \in \Lambda} h_x \sigma_x + h \sum_{x \in \Lambda} \sigma_x \quad (2)$$

where the couplings $\mathbf{J} = \{J_{xy}, \langle xy \rangle \in \mathbf{Z}^{d*}\}$ and the external fields $\mathbf{h} = \{h_x, x \in \mathbf{Z}^d\}$ are independent families of independent identically distributed (within each family)

---
[1]Research partially supported by the NSF under grants DMS-9208029 and INT-9016926

On Three Levels, Edited by M. Fannes et al.,
Plenum Press, New York, 1994

random variables. We use the notation $\Lambda^* = \{\langle xy \rangle ; x, y \in \Lambda\}$. If $\mathcal{B} = 0$, the model may be used to describe a spin glass or a random ferromagnet; if the $J_{xy} \equiv J > 0$, we have the random field Ising model.

For such a model, von Dreifus, Perez and I prove that at high temperature or at strong field $\mathcal{B}$, in spite of the non-analyticity pointed out by Griffiths, the magnetization, or more generally all quenched correlation functions, are infinitely differentiable functions of the uniform external field h. Our results require no assumptions on the probability distributions of J_{xy} and h_x, except for the obvious requirement that $\mathbf{P}\{h_x = 0\}$ be small in the strong field situation.

Finite volume thermal averages (with free boundary conditions) of local observables, at fixed $\mathbf{J}$ and $\mathbf{h}$, are defined by

$$\langle A \rangle_\Lambda = \frac{\sum_\sigma A(\sigma)\, e^{-\beta H_\Lambda(\sigma)}}{Z_\Lambda} \quad \text{with } Z_\Lambda = \sum_\sigma e^{-\beta H_\Lambda(\sigma)}, \qquad (3)$$

where β is the inverse temperature. When necessary we will make explicit the dependence on the uniform external field h.

The truncated or connected correlation function of two observables $A(\sigma), B(\sigma)$ is defined by:

$$\langle A; B \rangle_\Lambda = \langle AB \rangle_\Lambda - \langle A \rangle_\Lambda \langle B \rangle_\Lambda . \qquad (4)$$

Given a local observable A we set $\|A\| = \sup_\sigma |A(\sigma)|$, and denote by $\text{supp}\, A$ the support of A, that is, the set of $x \in \mathbf{Z}^d$ such that $A(\sigma)$ depends non-trivially on σ_x.

We start with the high temperature case; we fix arbitrary $\mathbf{h} \in \mathbf{R}^{\mathbf{Z}^d}$ and $h \in \mathbf{R}$ in (2); only $\mathbf{J}$ is random. For a given $\delta > 0$ we set $p_\delta = \mathbf{P}\{|J_{xy}| > \delta\}$. We have[2]:

Theorem 1 *Let $\delta > 0$. For any inverse temperature β such that*

$$\bar{p}_{\delta,\beta} \equiv (e^{4\beta\delta} - 1)(1 - p_\delta) + p_\delta < \frac{1}{2d - 1}, \qquad (5)$$

we have that for any two local observables A and B

$$\sup_\Lambda \mathbf{E}(|\langle A; B \rangle_\Lambda|) \leq \frac{4d\,|\text{supp}\, A|\, \|A\|\, \|B\|}{(2d-1)(1 - (2d-1)\bar{p}_{\delta,\beta})}\, ((2d-1)\bar{p}_{\delta,\beta})^{d(A,B)} \qquad (6)$$

for any $\mathbf{h} \in \mathbf{R}^{\mathbf{Z}^d}$ and $h \in \mathbf{R}$, where $d(A, B)$ is the distance between the supports of A and B in the ℓ^1 norm.

In addition, there exists a set $\mathcal{J}$ of realizations of the random couplings with $\mathbf{P}\{\mathbf{J} \in \mathcal{J}\} = 1$, such that for all $0 < \beta$ satisfying (5) and all $\mathbf{J} \in \mathcal{J}$, $\mathbf{h} \in \mathbf{R}^{\mathbf{Z}^d}$, $h \in \mathbf{R}$, we have:

1. *For every local observable A, the thermodynamical limit*

$$\lim_{\Lambda \to \mathbf{Z}^d} \langle A \rangle_\Lambda \equiv \langle A \rangle$$

   exists and is independent of the boundary conditions used for the sequence of finite volume Hamiltonians. In particular, there is a unique Gibbs state.

2. *For any local observable A the quenched expectation $\mathbf{E}(\langle A \rangle (h))$ is an infinitely differentiable function of the uniform external field h. In particular, for each $n = 1, 2, \ldots$ we have*

$$\sum_{x_1, \ldots, x_n \in \mathbf{Z}^d} \mathbf{E}\,|\langle \sigma_{x_0}; \sigma_{x_1}; \ldots; \sigma_{x_n}\rangle| < \infty, \qquad (7)$$

and

$$\frac{\partial^n}{\partial h^n} \mathrm{E} \langle \sigma_0 \rangle = \sum_{x_1,\ldots,x_n \in \mathbf{Z}^d} \mathrm{E} \langle \sigma_0; \sigma_{x_1}; \ldots; \sigma_{x_n} \rangle \ . \tag{8}$$

We now turn to the strong field case, we set $q_0 = \mathbf{P}\{h_x = 0\}$. We have[2]:

Theorem 2 *If $q_0 < \frac{1}{2d^2-1}$, then for each $\beta > 0$ we can find $\mathcal{B}_1(\beta, d) < \infty$, monotonically decreasing in β, and $\epsilon(\beta, d) > 0$, such that for any $\mathcal{B} \geq \mathcal{B}_1(\beta, d)$ and $|h| \leq \epsilon(\beta, d)$, we have for any two local observables A and B that*

$$\sup_\Lambda \mathrm{E}(|\langle A; B \rangle_\Lambda|) \leq C \, |\mathrm{supp}\, A| \, \|A\| \, \|B\| \, e^{-m\, d(A,B)} \ , \tag{9}$$

for some $C < \infty$ and $m > 0$.

In addition, there exists a set Ω of realizations of the random parameters $(\mathbf{J}, \mathbf{h})$, with $\mathbf{P}\{(\mathbf{J},\mathbf{h}) \in \Omega\} = 1$, such that for any $\beta > 0$, $\mathcal{B} \geq \mathcal{B}_1(\beta, d)$, $|h| \leq \epsilon(\beta, d)$, and all $(\mathbf{J},\mathbf{h}) \in \Omega$, statements 1 and 2 in Theorem 1 are true.

Remark 3 *In Theorem 1 it suffices to pick $\delta > 0$ such that $p_\delta < p_c(d)$, where $p_c(d)$ is the critical probability for bond percolation in $\mathbf{Z}^d$. (Notice that (5) requires $p_\delta < \frac{1}{2d-1} < p_c(d)$.) In this case we show that we can find $\beta_1 = \beta_1(d, p_\delta) > 0$ such that the conclusions of the Theorem still hold for all $\beta < \beta_1$. We use a result of Kesten[3] to estimate (17) below, obtaining an estimate similar to (9) for $\beta < \beta_1$. Similarly, in Theorem 2 we only need $q_0 < \tilde{p}_c$, where $\tilde{p}_c > \frac{1}{2d^2-1}$ is the critical probability for site percolation on a certain lattice on a subset of $\mathbf{Z}^d$.*

The first rigorous results controlling the effect of Griffiths' singularities were obtained by Olivieri, Perez and Rosa Jr..[4], who showed exponential decay of correlation functions for small β in the Ising ferromagnet with random couplings ($J_{xy} \geq 0$, $h_x \equiv 0$, $h = 0$), if $\mathrm{E}(J_{xy}) < \infty$. Exponential decay of truncated correlation functions and uniqueness of the Gibbs state for the class of models described by (2), for small β or large $\mathcal{B}$, were obtained by Berretti[5] with strong restrictions on the probability distributions of the random parameters. Fröhlich and Imbrie[6], through an intricate analysis of partially resummed high temperature/low activity expansions, were able to obtain the same results under more general assumptions on the probability distribution of the relevant random parameters. Bassalygo and Dobrushin[7] proved uniqueness of the Gibbs state for small β with no assumptions on the probability distributions.

To prove these results von Dreifus, Perez and I developped a very simple high temperature/low activity expansion in a random environment, whose convergence can be displayed in a straightforward way. In this report I will describe how to prove (6).

We start by redefining the Hamiltonian (2) as

$$H_\Lambda = - \sum_{\langle xy \rangle \in \Lambda^*} (J_{xy}\, \sigma_x \sigma_y + |J_{xy}|) + \mathcal{B} \sum_{x \in \Lambda} h_x \sigma_x + h \sum_{x \in \Lambda} \sigma_x \ , \tag{10}$$

which differs from (2) by a harmless substraction of an overall constant. To deal with truncated correlation functions we use the duplication trick. We thus consider two non-interacting copies of the original system, i.e., a new spin system with configurations $\tilde{\sigma} = \{\tilde{\sigma}_x = (\sigma_x, \sigma'_x); x \in \mathbf{Z}^d\}$, $\sigma_x, \sigma'_x \in \{-1, +1\}$, and Hamiltonian:

$$\tilde{H}_\Lambda(\tilde{\sigma}) = H_\Lambda(\sigma) + H_\Lambda(\sigma') \ . \tag{11}$$

The expectation value $\langle\!\langle C \rangle\!\rangle_\Lambda$ of an observable $C(\tilde{\sigma})$ of the duplicated system is defined as in (3), the corresponding partition function being denoted by $\tilde{Z}_\Lambda$. Truncated correlation functions of the original system may be expressed as ordinary correlation functions of the duplicated system through the identity:

$$\langle A; B \rangle_\Lambda = \frac{1}{2} \langle\langle \tilde{A} \tilde{B} \rangle\rangle_\Lambda , \qquad (12)$$

where to every observable A of the original system we associate an observable $\tilde{A}$ of the duplicated system by setting $\tilde{A}(\tilde{\sigma}) = A(\sigma) - A(\sigma')$.

Let us fix $\delta > 0$. We will call a bond $\langle xy \rangle$ *good* if $|J_{xy}| \leq \delta$, otherwise we will call it *bad*. We will denote the collections of good and bad bonds in Λ^* by Λ_g^* and Λ_b^*, respectively. We denote by $\mathcal{S}$ the collection of sites in Λ that are endpoints of bad bonds. We will also denote by $\mathcal{S}$ the graph consisting of the set $\mathcal{S}$ and edges Λ_b^*, so we have $\tilde{H}_\mathcal{S}$, etc.

We now perform a high temperature expansion in the *good* bonds only. For a good bond $\langle xy \rangle$ we have

$$0 \leq E_{xy}(\tilde{\sigma}) \equiv e^{\beta J_{xy}(\sigma_x \sigma_y + \sigma'_x \sigma'_y) + 2\beta |J_{xy}|} - 1 \leq e^{4\beta\delta} - 1 \equiv \xi , \qquad (13)$$

the nonnegativity coming from the substraction in (10). We can write

$$\langle\langle \tilde{A}\tilde{B} \rangle\rangle_\Lambda = \frac{1}{\tilde{Z}_\Lambda} \sum_{\tilde{\sigma}} \tilde{A}\tilde{B} \, Q_{\Lambda \backslash \mathcal{S}}(\tilde{\sigma}) \, e^{-\beta \tilde{H}_\mathcal{S}(\tilde{\sigma})} \prod_{\langle xy \rangle \in \Lambda_g^*} (E_{xy} + 1)$$

$$= \frac{1}{\tilde{Z}_\Lambda} \sum_{\tilde{\sigma}} \tilde{A}\tilde{B} \, Q_{\Lambda \backslash \mathcal{S}}(\tilde{\sigma}) \, e^{-\beta \tilde{H}_\mathcal{S}(\tilde{\sigma})} \sum_{G \subset \Lambda_g^*} \prod_{\langle xy \rangle \in G} E_{xy} , \qquad (14)$$

where $Q_\Lambda(\tilde{\sigma}) = \exp\left\{-\beta \left(\mathcal{B} \sum_{x \in \Lambda} h_x (\sigma_x + \sigma'_x) + h \sum_{x \in \Lambda} (\sigma_x + \sigma'_x)\right)\right\}$.

Due to the invariance of the Hamiltonian of the duplicated system under the exchange $\sigma \leftrightarrow \sigma'$, any $G \subset \Lambda_g^*$ such that the graph $G \cup \Lambda_b^*$ does not connect the supports of A and B gives a zero contribution in (14). We can thus restrict the sum to those G of the form $G = \Gamma \cup Y_\Gamma$, where Γ and Y_Γ are disjoint subsets of Λ_g^* such that $\Gamma \cup W$ is a connected graph that connects the supports of A and B for some $W \subset \Lambda_b^*$. Thus we have

$$\left| \langle\langle \tilde{A}\tilde{B} \rangle\rangle_\Lambda \right| \leq 4 \|A\| \|B\| \frac{1}{\tilde{Z}_\Lambda} \sum_{\tilde{\sigma}} Q_{\Lambda \backslash \mathcal{S}}(\tilde{\sigma}) \, e^{-\beta \tilde{H}_\mathcal{S}(\tilde{\sigma})} \sum_\Gamma \prod_{\langle xy \rangle \in \Gamma} E_{xy} \sum_{Y_\Gamma} \prod_{\langle xy \rangle \in Y_\Gamma} E_{xy}$$

$$\leq 4 \|A\| \|B\| \frac{1}{\tilde{Z}_\Lambda} \sum_\Gamma \xi^{|\Gamma|} \sum_{\tilde{\sigma}} Q_{\Lambda \backslash \mathcal{S}}(\tilde{\sigma}) \, e^{-\beta \tilde{H}_\mathcal{S}(\tilde{\sigma})} \sum_{Y_\Gamma} \prod_{\langle xy \rangle \in Y_\Gamma} E_{xy}$$

$$\leq 4 \|A\| \|B\| \frac{1}{\tilde{Z}_\Lambda} \sum_\Gamma \xi^{|\Gamma|} \sum_{\tilde{\sigma}} Q_{\Lambda \backslash \mathcal{S}}(\tilde{\sigma}) \, e^{-\beta \tilde{H}_\mathcal{S}(\tilde{\sigma})} \prod_{\langle xy \rangle \in \Gamma} (E_{xy} + 1) \sum_{Y_\Gamma} \prod_{\langle xy \rangle \in Y_\Gamma} E_{xy}$$

$$= 4 \|A\| \|B\| \frac{1}{\tilde{Z}_\Lambda} \sum_\Gamma \xi^{|\Gamma|} \tilde{Z}_\Lambda$$

$$= 4 \|A\| \|B\| \sum_\Gamma \xi^{|\Gamma|} , \qquad (15)$$

where we used (13). But the sum in (15) is over all $\Gamma \subset \Lambda_g^*$ such that $\Gamma \cup W$ is a connected graph that connects the supports of A and B for some $W \subset \Lambda_b^*$. To remove this last condition, let us define

$$\rho_{xy} = \begin{cases} \xi & \text{if the bond } \langle xy \rangle \text{ is good}, \\ 1 & \text{if the bond } \langle xy \rangle \text{ is bad}. \end{cases} \qquad (16)$$

It clearly follows from (15) that

$$\left|\langle\langle \tilde{A}\tilde{B} \rangle\rangle_\Lambda\right| \leq 4 \|A\| \|B\| \sum_\Delta \prod_{\langle xy \rangle \in \Delta} \rho_{xy}, \tag{17}$$

where we now sum over all connected subgraphs Δ of Λ^* which connect the the supports of A and B. Since the ρ_{xy}, $\langle xy \rangle \in \mathbf{Z}^d$, are independent identically distributed random variables, we can take averages in (17) obtaining

$$\mathrm{E}\left(\left|\langle\langle \tilde{A}\tilde{B} \rangle\rangle_\Lambda\right|\right) \leq 4 \|A\| \|B\| \sum_\Delta \bar{\rho}_{\delta,\beta}^{|\Delta|}, \tag{18}$$

where $\bar{\rho}_{\delta,\beta} = \mathrm{E}(\rho_{xy})$ is given in (5).

If $\bar{\rho}_{\delta,\beta} < 2d$, it follows from (12) and (18) by a routine argument that

$$\sup_\Lambda \mathrm{E}(|\langle A; B \rangle_\Lambda|) \leq \frac{2\,|\mathrm{supp}\,A|\,\|A\|\,\|B\|}{(1 - 2d\,\bar{\rho}_{\delta,\beta})} (2d\,\bar{\rho}_{\delta,\beta})^{d(A,B)}, \tag{19}$$

which is just (6), except that $(2d-1)$ has been systematically replaced by $2d$.

To obtain (6) as stated, all we need to do is to go back to (15) and argue that the estimate is still valid if we only sum over graphs Γ such that $\Gamma \cup W$ is a *self-avoiding walk* going from the support of A to the support of B for some $W \subset \Lambda_b^*$. The same argument as before now gives (17) and (18) with the sums restricted to all *self-avoiding walks* Δ going from the support of A to the support of B. (6) then follows from (5).

Acknowledgments

I want to thank Harry Kesten for an observation that made a simple proof embarrassingly simple.

REFERENCES

1. R. Griffiths, Non-analytic behavior above the critical point in a random Ising ferromagnet, *Phys. Rev. Lett.*, **23**, 17 (1969).
2. H. von Dreifus, A. Klein and J. F. Perez, Taming Griffiths' singularities: infinite differentiability of correlation functions, *in preparation*.
3. H. Kesten, Aspects of first passage percolation, in: "Ecole d'Eté de Probabilités de Saint-Flour XIV", 125, Lecture Notes in Mathematics **1180**, Springer, Berlin (1986).
4. E. Olivieri, J.F. Perez and S.G. Rosa Jr., Some rigorous results on the phase diagram of dilute Ising systems, *Phys. Lett.*, **94A**, 309 (1983).
5. A. Berretti, Some properties of random Ising models, *J. Stat. Phys.*, **38**, 483 (1985).
6. J. Fröhlich and J. Imbrie, Improved perturbation expansion for disordered systems: beating Griffiths singularities, *Commun. Math. Phys.*, **96**, 145 (1984).
7. L. Bassalygo and R. Dobrushin, Uniqueness of a Gibbs field with random potential - an elementary approach, *Theory Probab. Appl*, **31**, 572 (1986).

MICRO SPECTRAL PROPERTIES OF CRYSTALS AND THEIR BAND STRUCTURE

P. Kurasov[1,2] and B. Pavlov[1]

1. Dept. of Math. and Comp. Physics
   St. Petersburg Univ., 198904, Russia
2. Stockholm Univ., Frescativagen 24
   10405 Stockholm, Sweden

INTRODUCTION

A Multichannel Schrödinger equation with a periodic matrix potential is investigated. Such an equation arises during the reduction of the multidimensional Schrödinger equation periodic in one direction to the set of connected one-dimensional equations. It is related to the calculations of the electron spectrum in solid state, superlattices and low dimensional lattices. Periodic structure of the multichannel Hamiltonian is considered as infinite number of the periodically situated short range perturbations of the translation invariant matrix operator. Relations between the spectral properties of the single scattering centers and the band structure of the periodic problem are investigated. The first section is devoted to the finite dimensional Schrödinger equation. Relations between the scattering on a finite crystal and the corresponding periodic problem are clarified. It is shown that the bound state energies of a single center define bands of the continuous spectrum, resonances produce spectral gaps in the spectrum. The second section is devoted to the scattering on a one-dimensional chain in the three-dimensional space, which corresponds to the infinite dimension of the matrix potential. An exactly solvable model is constructed with the help of zero range potentials.

ONE DIMENSIONAL CRYSTAL

Infinite and Finite Crystals: Scattering Matrix and Band Structure

Consider the matrix Schrödinger operator

$$\mathcal{H} = -\frac{\partial^2}{\partial x^2} + V(x) \tag{1}$$

acting in the Hilbert space $L_2(\mathbf{R}, H)$, where H is a finite dimensional space, $\dim H = m$, potential $V(x)$ is a self-adjoint matrix, bounded and piecewise continuous. We shall investigate relations between the spectral properties of the operators with periodic potential V, $V(x+1) = V(x)$ and cutoff periodic potential $V_N(x) = \begin{cases} V(x), & 0 \leq x \leq N \\ V_0, & x < 0, x > N \end{cases}$, where V_0 is a constant self-adjoint matrix with eigenvalues $\{\lambda_k\}$, $k = 1, 2, ..., m$ and corresponding eigenvectors $\{e_k\}$, $k = 1, 2, ..., m$.

The periodic problem has a band structure of the spectrum. The cutoff potential defines an operator with m branches of continuous spectrum $[\lambda_k, \infty)$ and eigenvalues, situated on the half-axis $(-\infty, \max\{\lambda_k\}]$. Complete information on the spectrum is accumulated by the scattering and monodromy matrices correspondingly.

The unitary scattering matrix for the second problem can be defined in the following way using the eigenfunctions associated with the energy λ

$$-\frac{\partial^2}{\partial x^2}\Psi(x) + V(x)\Psi(x) = \lambda\Psi(x). \tag{2}$$

These eigenfunctions are combinations of plane waves on the left and right half axes

$$\Psi(\lambda, x) = \sum_j A_{inc}^{-j} \frac{\exp(i\sqrt{\lambda-\lambda_j}x)e_j}{\sqrt[4]{\lambda-\lambda_j}} + \sum_j A_{out}^{-j} \frac{\exp(-i\sqrt{\lambda-\lambda_j}x)e_j}{\sqrt[4]{\lambda-\lambda_j}}, x < 0;$$

$$\Psi(\lambda, x) = \sum_j A_{inc}^{+j} \frac{\exp(-i\sqrt{\lambda-\lambda_j}(x-n))e_j}{\sqrt[4]{\lambda-\lambda_j}} + \sum_j A_{out}^{+j} \frac{\exp(i\sqrt{\lambda-\lambda_j}(x-n))e_j}{\sqrt[4]{\lambda-\lambda_j}}, x > n. \tag{3}$$

One can separate two orthogonal sets of eigenfunctions: left incoming waves $A_{inc}^+ = 0$ and right incoming waves $A_{inc}^- = 0$, parametrized by the amplitudes A_{inc}^- and A_{inc}^+ respectively. The number of the linearly independent eigenfunctions is thus determined by the number of the opened channels, which in turn depends on the energy. The dimension of the S-matrix $\begin{pmatrix} A_{out}^+ \\ A_{out}^- \end{pmatrix} = S \begin{pmatrix} A_{inc}^+ \\ A_{inc}^- \end{pmatrix}$ depends on the energy. The S-matrix disintegrates into the $m \times m$ matrices of reflection and transition coefficients in the case of a potential symmetric with respect to the center of the period: $V(1/2+x) = V(1/2-x), x < 1/2$:

$$S = \begin{pmatrix} R & T \\ T & R \end{pmatrix}. \tag{4}$$

The unitarity of the scattering matrix can be expressed by the following relations

$$RR^* + TT^* = 1, \quad RT^* + TR^* = 0. \tag{5}$$

The monodromy matrix $\mathcal{M}(\lambda)$ for the periodic problem is defined through the solutions of the equation (2) with periodic potential by the following equality:

$$\begin{pmatrix} \{u^j\}_{j=1}^N \\ \{\frac{\partial u^j}{\partial x}\}_{j=1}^N \end{pmatrix}\Big|_{x=m+1} = \mathcal{M}(\lambda) \begin{pmatrix} \{u^j\}_{j=1}^N \\ \{\frac{\partial u^j}{\partial x}\}_{j=1}^N \end{pmatrix}\Big|_{x=m} \tag{6}$$

The band spectrum of the operator is defined by the condition $|Tr\mathcal{M}(\lambda)| < 2$.

In order to calculate the scattering matrix from the monodromy matrix we need to introduce the matrix $Q(\lambda)$, which calculates Cauchy data for the set of plane waves from it's amplitudes

$$Q(\lambda) = \begin{pmatrix} \frac{1}{\sqrt[4]{\lambda-\lambda_1}}(e_1), & ..., & \frac{1}{\sqrt[4]{\lambda-\lambda_m}}(e_m), & \frac{1}{\sqrt[4]{\lambda-\lambda_1}}(e_1), & ... \\ \sqrt[4]{\lambda-\lambda_1}(e_1), & ..., & \sqrt[4]{\lambda-\lambda_m}(e_m), & -\sqrt[4]{\lambda-\lambda_1}(e_1), & ... \end{pmatrix}. \tag{7}$$

The monodromy matrix in the representation of plane waves is then given by

$$\tilde{\mathcal{M}}(\lambda) = Q^{-1}(\lambda)\mathcal{M}(\lambda)Q(\lambda). \tag{8}$$

The corresponding power of the matrix connects amplitudes of incoming and outgoing waves:

$$\tilde{\mathcal{M}}^N \begin{pmatrix} A_-^{inc} \\ A_-^{out} \end{pmatrix} = \begin{pmatrix} A_+^{out} \\ A_+^{inc} \end{pmatrix}$$

Above the threshold we thus have :

$$\tilde{\mathcal{M}}^N(\lambda) = Q^{-1}(\lambda)\mathcal{M}^N(\lambda)Q(\lambda) = \begin{pmatrix} T - RT^{-1}R & RT^{-1} \\ -T^{-1}R & T^{-1} \end{pmatrix}. \tag{9}$$

The equalities (9) give us a possibility to calculate the scattering matrix.

Let us examine the relations between the spectra of infinite and finite crystals in the limit of large number N. This question was earlier considered in the papers [6,9], where the ordinary one-dimensional Schrödinger equation was studied. Above the energy barrier $\lambda > \lambda_k$ on each zone of the crystal the special points μ_m were marked. These points divide the zone into N equal sectors in the quasi-momentum scale $t(\mu_m) = 2\pi m/N, m = 1, 2, ..., N-1$. The reflection coefficient for the ordinary one-dimensional equation is equal to zero at these points. One can thus see that if the number of periods of the cutoff potentials increases then the number of such points on each band increases also.

For the many mode crystal the corresponding degree of the monodromy matrix has the eigenvalue 1 at these points. Reflection and transition coefficients satisfy the following relations at the marked points μ_m only

$$\det(1 - T - R) = 0, \quad \det(1 - T + R) = 0. \tag{10}$$

The equalities (10) define a reflectionless scattering matrix in the case of a single mode crystal. In the multichannel case only some sets of the plane waves cross the finite crystal without reflection. Multiplicity of the zeroes of the determinants coincides with the multiplicity of the band spectrum.

Micro Spectral Properties and Band Structure

Relations between the spectral properties of the isolated centers forming the crystal and its band structure is investigated in this subsection using an exactly solvable model. Model problem for the many-mode one-dimensional crystal with the point interaction at some point x_0 is the differential equation $\left(-\frac{\partial^2}{\partial x^2} + V_0\right)\psi(\lambda, x) = \lambda\psi(\lambda, x)$ with the boundary conditions

$$u(x_0 - 0) = u(x_0 + 0) = D(\lambda)[\frac{\partial u}{\partial x}](x_0) \tag{11}$$

where $[*](x)$ denotes the jump of the function at point x, $D(\lambda)$ is $m \times m$ matrix with the positive imaginary part in $\Im\lambda > 0$, self-adjoint on the real line (so-called R-operator). These boundary conditions depend on the energy λ of the eigenfunction and do not define an operator in the space $L_2(\mathbf{R}, H)$. A rigorous operator theory formulation of the problem can be done using self-adjoint extensions of the operator with some extra channel of interaction (see [4,5,7,8]). Following ansatz can be used for the discrete spectrum eigenfunction, corresponding to the energy $\lambda = -\chi^2$

$$\psi_j(x) = c_j e^{i\sqrt{\lambda - \lambda_j}|x - x_0|}, \Im\sqrt{\lambda - \lambda_j} > 0.$$

Dispersion equation on the energy is obtained from the boundary conditions

$$\det\left(\mathrm{diag}\{2\imath\sqrt{\lambda-\lambda_j}\} - D^{-1}(\lambda)\right) = 0. \tag{12}$$

Negative singularities of the matrix $D(\lambda)$ produce eigenvalues of the problem. Some extra eigenvalues can be formed close to the zero energy.

Unitary scattering matrix can be calculated in the same way. The reflection and transition coefficients for $x_0 = 0$ are

$$R(k) = \frac{1}{2\imath\mathrm{diag}\{\sqrt[4]{\lambda-\lambda_j}\}D(\lambda)\mathrm{diag}\{\sqrt[4]{\lambda-\lambda_j}\} - 1}, \tag{13}$$

$$T(k) = \frac{2\imath\mathrm{diag}\{\sqrt[4]{\lambda-\lambda_j}\}D(\lambda)\mathrm{diag}\{\sqrt[4]{\lambda-\lambda_j}\}}{2\imath\mathrm{diag}\{\sqrt[4]{\lambda-\lambda_j}\}D(\lambda)\mathrm{diag}\{\sqrt[4]{\lambda-\lambda_j}\} - 1}. \tag{14}$$

Model of the periodic structure in one-dimensional space can be obtained considering the same differential equation with the boundary conditions (11) at the points $x_n = n - 1/2, n = 0, \pm 1, \pm 2, ...$ The monodromy matrix can be expressed in terms of the free monodromy matrix $\mathcal{M}_0$ and the jump matrix $J(\lambda)$, corresponding to the single point x_j

$$\mathcal{M}(\lambda) = \mathcal{M}_0^{1/2}(\lambda)J(\lambda)\mathcal{M}_0^{1/2}(\lambda). \tag{15}$$

The matrices $\mathcal{M}_0(\lambda)$ and $J(\lambda)$ are

$$\mathcal{M}_0(\lambda) = \begin{pmatrix} \mathrm{diag}(\cos\sqrt{\lambda-\lambda_j}) & \mathrm{diag}(\sin\sqrt{\lambda-\lambda_j}/\sqrt{\lambda-\lambda_j}) \\ \mathrm{diag}(\sqrt{\lambda-\lambda_j}\sin\sqrt{\lambda-\lambda_j}) & \mathrm{diag}(\cos\sqrt{\lambda-\lambda_j}) \end{pmatrix},$$

$$J(\lambda) = \begin{pmatrix} 1 & 0 \\ D^{-1}(\lambda) & 1 \end{pmatrix},$$

where 1 is the unit $m \times m$ matrix. As usual

$$\det\mathcal{M}(\lambda) = \det J(\lambda)\det\mathcal{M}_0(\lambda) = 1.$$

The spectrum of the problem is purely continuous, it has a band structure. The dispersion equation connects the quasi-momentum and the energy of eigenfunctions:

$$\det\left(\mathrm{diag}\left\{\frac{\sqrt{\lambda-\lambda_j}}{\sin\sqrt{\lambda-\lambda_j}}2(\cos t - \cos\sqrt{\lambda-\lambda_j})\right\} - D^{-1}(\lambda)\right) = 0. \tag{16}$$

Projecting the surfaces of solutions on the energy axis we obtain the band spectrum.

For one-mode crystal the dispersion equation reads

$$D(k^2)\frac{k}{\sin k}2(\cos t - \cos k) = 1.$$

where $\lambda_1 = 0$, $D(\lambda)$ is an R-function. Solutions of the dispersion equation in the limit of weak connection are situated near the zero-lines of the functions $D(\lambda)$ and $D^{ex}(\lambda, t) \equiv \frac{k}{\sin k}2(\cos t - \cos k), k = \sqrt{\lambda}$. The spectrum of the problem consists of infinitely many bands. Narrow gaps can be observed near the points $\lambda = n^2\pi^2, n = 1, 2, 3, ...$ and zeroes of the function $D(\lambda)$. Only narrow bands occur below the threshold. Every such band corresponds to an eigenvalue of the isolated center. Each positive eigenvalue of the internal operator produces a narrow gap in the band spectrum in the energy scale.

Similar investigation can be carried out in the case of the two mode crystal. Assume that the operator $D(\lambda)$ is of the form

$$D(\lambda) = \begin{pmatrix} D_1 + D_2 & D_1 - D_2 \\ D_1 - D_2 & D_1 + D_2 \end{pmatrix},$$

where $D_{1,2}(\lambda)$ are R-functions. The dispersion equation is:

$$1 - (D_1 + D_2)(D_1^{ex} + D_2^{ex}) + 4D_1 D_2 D_1^{ex} D_2^{ex} = 0,$$

$$D_j^{ex} = \frac{\sqrt{\lambda - \lambda_j}}{\sin\sqrt{\lambda - \lambda_j}} 2(\cos t - \cos\sqrt{\lambda - \lambda_j}).$$

Zeroes of the external dispersion functions are situated on the parabolas $t = \pm\sqrt{\lambda - a_j} + 2\pi m, m \in \mathbf{Z}$. Crossings of these zero lines of the external dispersion functions produce additional gaps in the continuous spectrum.

ONE DIMENSIONAL CHAIN IN THE THREE DIMENSIONAL SPACE

Isolated Center

The model problem will be constructed in the same way as in the previous subsection. The Laplace operator in $L_2(\mathbf{R}^3)$ can be defined on the set of smooth functions with the following singularity at one point x_0:

$$\Psi(x) \sim \frac{\Psi_-}{4\pi|x - x_0|} + \Psi_0 + o(1), x \to x_0, \Psi_-, \Psi_0 \in \mathbf{C}. \tag{17}$$

We consider the energy dependent boundary conditions

$$\Psi_-/\Psi_0 = D(\lambda), \tag{18}$$

with R-function $D(\lambda)$. The first model of this type with the real constant D was considered in [1-3]. The following Ansatz is valid for the continuous spectrum eigenfunctions

$$\Psi(\vec{k}, x) = \frac{1}{\sqrt{2\pi}} \left(e^{i<\vec{k},\vec{x}>} + \frac{D(k^2)}{1 - \frac{ik}{4\pi}D(k^2)} \frac{e^{ik|x-x_0|}}{4\pi|x - x_0|} \right), k = \sqrt{\lambda} \tag{19}$$

The scattering matrix and the eigenstate dispersion equation are

$$S(\lambda) = \frac{\frac{ik}{4\pi} + D^{-1}(k^2)}{\frac{ik}{4\pi} - D^{-1}(k^2)}, \quad \frac{ik}{4\pi} = D^{-1}(k^2) \tag{20}$$

Every negative zero of $D(\lambda)$ produces an eigenvalue of the operator, while the positive zeroes define resonances - singularities of the S-matrix on the nonphysical sheet of the energy.

One-dimensional Chain

The following model for a one-dimensional chain in three-dimensional space will be considered. We impose boundary conditions of the type (18) at the points : $x_j = j\vec{e}, j \in$

$\mathbf{Z}$, where $\vec{e}$ is vector of the period of the chain. The spectrum is purely continuous, corresponding to scattered waves and waveguide functions. Eigenfunctions of the first type describe process of the scattering on the chain and contain an incoming plane wave. Waveguide functions describe transport along the chain and define the band spectrum of the operator. These eigenfunctions can be written in the Bloch representation with the quasi-momentum t:

$$\Phi(t,x) = \sum_{j \in \mathbf{Z}} \frac{e^{ik|x-x_j|}}{4\pi|x-x_j|} e^{ijt}. \tag{21}$$

The energy and the quasi-momentum are connected by the dispersion equation

$$D^{-1}(E) = \tilde{B}(k,t) \equiv \frac{ik}{4\pi} + \sum_{j \neq 0} \frac{e^{ik|x_j|}}{4\pi|x_j|} e^{itj} \equiv \frac{1}{4\pi} \ln \frac{1}{2(\cos k - \cos t)} \tag{22}$$

Solutions of the equation (22) are situated in the region $E < 0$ or $t > |k|$. Narrow gaps in the spectrum are obtained near the zeroes of D, which corresponds to the resonances of the isolated center. Eigenstates of the isolated center produce small bands of continuous spectrum of the chain.

Acknowledgments

The first author wishes to thank Stockholm University for the financial support and hospitality and Prof. N. Elander for many fruitful discussions.

REFERENCES

1. S. Albeverio, F. Gestezy, R. Hoegh-Krohn and H. Holden, "Solvable Models in Quantum Mechanics," Springer-Verlag, Berlin-New-York (1988).
2. Yu.N. Demkov and V.N. Ostrovskii, "Zero-range Potentials and Their Applications in Atomic Physics," Plenum Press, New-York (1988).
3. Yu.E. Karpeshina "A decomposition theorem for the eigenfunctions of the problem of scattering by homogeneous periodic objects of chain type in three-dimensional space," *Selecta Math. Soviet*, **4**, 1231(1985).
4. P.B. Kurasov and B.S. Pavlov, Electron in the homogeneous crystal of point atoms with internal structure II, *Teor. Mat. Fiz.*,**74**, 82 (1988).
5. P.B. Kurasov, One-electron Model of a Long Molecule, *Leningrad Univ. Vestnik*, ser.4, **3** (nr18), 8 (1989) (in Russian).
6. B.S. Pavlov and N.V. Smirnov, Spectral properties of one-dimensional disperse crystals, *Proc. LOMI*, **133**, 197 (1984).
7. B.S. Pavlov, A model of zero-radius potential with internal structure, *Teor. Mat. Fiz.*, **59**, 345 (1984).
8. B.S. Pavlov, The theory of extensions and explicitly-solvable models, *Russian Math. Surveyes*, **42:6**, 127 (1987).
9. C. Rorres, Transmission coefficients and eigenvalues of a finite one-dimensional crystal, *SIAM J. Appl. Math.*, **27**, 303 (1974).

STOCHASTICALLY FORCED BURGERS EQUATION

L. Bertini[1,2], N. Cancrini[1] and G. Jona-Lasinio

1. Dipartimento di Fiscia, Università 'La Sapienza'
   P.le Aldo Moro 2 - 00185 Roma, Italy
2. Dipartimentodi Matematica, Università 'Tor Vergata'
   Via della Ricerca Scientifica - 00133 Roma, Italy

INTRODUCTION

One of the first attempts to arrive at the statistical theory of turbulent fluid motion was the proposal by Burgers of his celebrated equation, which in one space dimension is

$$\partial_t u_t(x) = \nu \partial_x^2 u_t(x) - u_t(x) \partial_x u_t(x) \tag{1}$$

where $u_t(x)$ is the velocity field and ν is the viscosity. As Burgers emphasized in the introduction of his book [1] this equation represents an extremely simplified model describing the interaction of dissipative and non linear inertial terms in the motion of the fluid. Equation (1) can be explicitly solved and, in the limit of vanishing viscosity, the solution develops shock waves.

Rigorous results have been recently established in the study of some statistical properties: random initial data are considered in [2,3], while in [4] a forcing term, which is white noise in time and a periodic function in space, is added.

The study of Burgers equation with a forcing term is interesting in view of the phenomenological character of (1). Since it represents an incomplete description of a system, to be physically acceptable it should exhibit stability under a sufficiently wide class of perturbations. A forcing term can provide a good model of the neglected effects; in particular a random perturbation may help to select interesting invariant measures. Translational invariance is preserved when (1) is perturbed by additive stochastic processes stationary in space and time.

In principle one can think of a wide variety of stationary random forcing terms. However there is a natural choice represented by white noise in time and space. Stability under this kind of perturbation means in particular that the solutions of the equation, which are supposed to describe macroscopic phenomena, are not too sensitive to fluctuations which take place at very small scales in place and time. This looks to us a reasonable physical requirement. We also note that with this choice, due to the absence of time correlations, the full Galilean invariance of (1) is preserved.

We thus consider the following equation

$$\partial_t u_t(x) = \nu \partial_x^2 u_t(x) - u_t(x)\partial_x u_t(x) + \varepsilon \eta_t(x) \qquad (2)$$

where $t \in \mathbf{R}^+$, $x \in \mathbf{R}$, ε is the noise intensity and $\eta_t(x)$ is white noise in space and time, i.e.

$$\mathbf{E}\left(\eta_t(x)\eta_{t'}(x')\right) = \delta(t-t')\delta(x-x') \qquad (3)$$

In this paper we outline how an existence Theorem for the Cauchy problem for equation (2) can be obtained. A complete discussion can be found in [5]. The result is not trivial due to the singular character of the forcing term. Furthermore the Theorem gives an explicit expression for the solution; it is thus a first step towards a systematic study of the structure of the solutions in various regimes of small viscosity and noise intensity.

COLE-HOPF TRANSFORMATION

We realize the white noise $\eta_t(x)$ as the generalized derivative of the brownian sheet, i.e. $\eta_t(x) = \partial_t \partial_x W_t(x)$. The gaussian process $W_t(x)$ has correlation function

$$\mathbf{E}^w\left(W_t(x)W_{t'}(x')\right) = t \wedge t'\, C(x,x'), \qquad C(x,x') = \theta(xx')\,|x| \wedge |x'| \qquad (4)$$

where $a \wedge b = \min\{a,b\}$ and θ is the indicator function of the set $[0,\infty)$. We denote by P^w the law of the brownian sheet.

We shall construct a solution of equation (2) using the Cole-Hopf transformation $u_t(x) = -2\nu \partial_x \log \psi_t(x)$, which reduces (2) to the following linear equation with multiplicative *half white* noise

$$\partial_t \psi_t(x) = \nu \partial_x^2 \psi_t(x) - \frac{\varepsilon}{2\nu}\psi_t(x)\partial_t W_t(x) \qquad (5)$$

In order to study rigorously (5), it has to be interpreted as a stochastic PDE. Since it contains a non trivial diffusion the stochastic differential presents the well known ambiguities. In order to obtain, via the Cole-Hopf transformation, the solution of the stochastically perturbed Burgers equation, the stochastic differential in (5) has to be interpreted in the Stratonovich sense. In the following we thus consider

$$d\psi_t(x) = \nu \partial_x^2 \psi_t(x) dt - \frac{\varepsilon}{2\nu}\psi_t(x) \circ dW_t(x) \qquad (6)$$

which can be written in term of Ito differential as

$$d\psi_t(x) = \left(\nu \partial_x^2 \psi_t(x) + \frac{\varepsilon}{2\nu}V(x)\psi_t(x)\right) dt - \frac{\varepsilon}{2\nu}\psi_t(x)dW_t(x) \qquad (7)$$

the extra term $V(x) = \frac{1}{2}C(x,x) = \frac{1}{2}|x|$ arises from the formal expression of the Stratonovich differential, see (4) above. In the next section we construct, via a generalized Feynman-Kac formula, a process which solves the mild form of (7)

$$\psi_t(x) = G_t \star \psi_0(x) + \frac{\varepsilon}{2\nu}\int_0^t G_{t-s} \star (V\psi_s ds - \psi_s dW_s)(x) \qquad (8)$$

where $\psi_0(x)$ is the initial condition, it is related to $u_0(x)$ by the relation $u_0(x) = -2\nu \partial_x \log \psi_0(x)$. We assume $u_0(x)$ is a continuous function we write in the form $u_0(x) = -\partial_x U_0(x)$ and there exist $\alpha, c > 0$ such that for all $x \in \mathbf{R}$ $|U_0(x)| \leq \alpha(1+|x|)$ and $|U_0'(x)| \leq c \exp(\alpha|x|)$. Furthermore G_t is the convolution operator with kernel

$$G_t(x) = (4\nu\pi t)^{-1/2} \exp\left\{-\frac{x^2}{4\nu t}\right\} \tag{9}$$

i.e.

$$G_t \star f(x) = \int dy\, G_t(x-y) f(y) \tag{10}$$

Equation (8) is meaningful for a wider class of functions than (7), i.e. for functions which do not possess two derivatives.

FEYNMAN-KAC FORMULA

We now introduce an auxiliary brownian motion to write a solution of (8) via a generalized Feynman-Kac formula. Let define $dP_{x,t}^\beta = dy\, dP_{y,0;x,t}^\beta$ where $dP_{y,0;x,t}^\beta$ is the conditional measure of a brownian motion with diffusion coefficient 2ν starting at time 0 in y and arriving at time t in x. We will denote by $\mathbf{E}_{x,t}^\beta$ the expectation with respect to $dP_{x,t}^\beta$. We stress that the brownian motion β is independent from the brownian sheet W.

The formal application of Feynman-Kac formula to equation (5), which has a time dependent potential, gives

$$\psi_t(x) = \mathbf{E}_{x,t}^\beta \left(\psi_0(\beta_0) e^{-\frac{\epsilon}{2\nu}\int_0^t ds\, \partial_s W_s(\beta_s)}\right) \tag{11}$$

however, due to the singular character of the half white noise, the integral in the exponent is meaningless. Nonetheless it is possible to prove the Feynman-Kac formula if the exponent is defined as a stochastic integral (with respect to the brownian sheet W) along the brownian curve $s \mapsto \beta_s$. Let $s_k = k 2^{-n} t$, $k = 0, \cdots, 2^n$ be a partition of $[0, t]$ and, given a realization β_s, introduce

$$M_\beta^n(t) = \sum_k \left(W_{s_{k+1}}(\beta_{s_k}) - W_{s_k}(\beta_{s_k})\right) \tag{12}$$

using (4) we have

$$\lim_{n \to \infty} \mathbf{E}^w (M_\beta^n(t))^2 = \lim_{n \to \infty} \sum_k C(\beta_{s_k}, \beta_{s_k})(s_{k+1} - s_k) = \int_0^t ds |\beta_s| \tag{13}$$

It is not difficult to verify that, for almost all β, $M_\beta^n(t)$ is a Cauchy sequence in $L^2(dP^w)$; we then define $M_\beta(t)$ as its limit. Note that $M_\beta(t)$ is a, dP^w-a.s., continuous gaussian process.

In [5] the following proposition is proved.

Proposition 1 Let $\psi_0 = \exp\{\frac{1}{2\nu}U_0(x)\}$, and
$$\psi_t(x) = \mathbf{E}^\beta_{x,t}\left(\psi_0(\beta_0)e^{-\frac{\varepsilon}{2\nu}M_\beta(t)}\right) \tag{14}$$
then $\forall (t,x) \in [0,T] \times \mathbf{R}$

(i) dP^w-a.s. $\psi_t(x)$ is a solution of (8).

(ii) $\forall p \geq 1$ $\psi_t(x) \in L^p(dP^w)$; $t \mapsto \psi_t(x)$ is dP^w-a.s locally Hölder continuous with exponent $\alpha < 1/2$

(iii) $x \mapsto \psi_t(x)$ is dP^w-a.s. differentiable. Its derivative is
$$\partial_x \psi_t(x) = G'_t \star \psi_0(x) + \frac{\varepsilon}{2\nu}\int_0^t G'_{t-s} \star (V\psi_s ds - \psi_s dW_s)(x) \tag{15}$$
where
$$G'_t \star f(x) = \int dy\, \partial_x G_t(x-y) f(y) \tag{16}$$

Furthermore the application $(t,x) \mapsto \partial_x \psi_t(x)$ is dP^w-a.s locally Hölder continuous with exponent $\alpha < 1/4$ in time and $\alpha < 1/2$ in space.

(iv) dP^w-a.s. $\psi_t(x) > 0$

The properties (i)-(ii) are statements about the existence of the solution of (8); they are proved through a careful application of stochastic calculus. Properties (iii)-(iv) are required to perform the Cole-Hopf transformation. The differentiability of $x \mapsto \psi_t(x)$ is proven in two steps. First, using (14), the Hölder continuity is established. Then (8) is shown to be differentiable obtaining (15). The latter step extends to the stochastic case the regularization property of the heat kernel. However, due to the non differentiability of the brownian sheet, it is not possible to prove higher order differentiability. Using (8) and (15) the Hölder continuity of $t \mapsto \psi_t(x)$ and $t \mapsto \partial_x \psi_t(x)$ can then be proved. Finally the positivity (iv) follows from the Feynman-Kac characterization (14).

SOLUTION OF BURGERS EQUATION

As $\psi_t(x)$ is not twice differentiable, equation (2) is not defined in the classical sense. However the process $u_t(x)$, defined by the Cole-Hopf transformation, is a solution of its mild form, i.e.
$$u_t(x) = G_t \star u_0(x) - \frac{1}{2}\int_0^t ds\, G'_{t-s} \star u_s^2(x) + \varepsilon \int_0^t G'_{t-s} \star dW_s(x) \tag{17}$$

From Proposition 1 we have the following Theorem which states the existence result for the solution of the stochastic Burgers equation (17).

Theorem 2 Set $u_t(x) = -2\nu \partial_x \log \psi_t(x)$, with $\psi_t(x)$ given by (14). Then $u_t = u_t(x)$ is a $C^0(\mathbf{R})$-valued process, adapted to the brownian sheet, such that $\forall (t,x) \in [0,T] \times \mathbf{R}$

(i) $\forall p \geq 1$ $u_t(x) \in L^p(dP^w)$; $(t,x) \mapsto u_t(x)$ is dP^w-a.s. locally Hölder continuous with exponent $\alpha < 1/4$ in time and $\alpha < 1/2$ in space.

(ii) $u_t(x)$ solves (17) dP^w-a.s.

The proof of (*i*) uses Jensen inequality and the results of Proposition 1. To prove (*ii*) we first introduce a regularization which allows the use of stochastic calculus. We then show the regularization can be removed. We stress that the equation satisfied by $u_t(x)$, defined through the Cole-Hopf transformation, would not have been (17) if we had interpreted (5) in the Ito sense.

REFERENCES

1. J.M. Burgers, "The Nonlinear Diffusion Equation," D. Reidel, Dordrecht (1974).
2. Ya.G. Sinai, Statistics of shocks in solutions of inviscid Burgers equation, *Comm. Math. Phys.* **148**, 601-621 (1992).
3. S. Albeverio, S.A. Molchanov, and D. Surgailis, Stratified structure of the universe and the Burgers' Equation - A Probabilistic Approach, Preprint (1993).
4. Ya.G. Sinai, Two results concerning asymptotic behavior of solution of the Burgers equation with force, *J. Stat. Phys.* **64**, 1-12 (1991).
5. L. Bertini, N. Cancrini, and G. Jona-Lasinio, The stochastic Burgers equation, Preprint (1993).

GLAUBER EVOLUTION FOR KAC POTENTIALS ANALYSIS OF CRITICAL FLUCTUATIONS: CONVERGENCE TO A NON LINEAR STOCHASTIC PDE

B. Rüdiger

Dipartimento di Matematica
Università di Roma Tor Vergata
Via della Ricerca Scientifica, 00133 Roma, Italy

In [1] we studied the long time behaviour of the Glauber dynamics in a system of ± 1 valued spins, on a one-dimensional discrete torus, interacting via a Kac potential at the critical temperature: we proved that by a suitable choice of scaling, which involves space, time and the range of interaction, the fluctuations of the magnetic density field, if suitable normalized, converge to the solution of the following local non linear SPDE, defined on a one-dimensional torus $\mathcal{T}$

$$dm_t(r) = \left(\frac{D}{2}\Delta m_t(r) - V'(m_t(r))\right)dt + dz_t(r) \tag{1.a}$$

$$m_0(r) = 0 \tag{1.b}$$

$$V'(m_t(r)) = m_t^3(r)/3 \tag{1.c}$$

where $D > 0$, and $dz_t(r)/dt$ is the two dimensional white noise. This result was already explained at an heuristic level in [2], where propagation of chaos was proved in the Lebowitz-Penrose limit (see [3]). We will describe in more details the problem and give an outline of the proof we did in [1].

We set γ^{-1} the range of the Kac-interaction; as we will explain below, in [1] we scaled space and time by

$$\gamma^{1+1/3}x = r, \quad \gamma^{-2/3}t = \tau \tag{2}$$

For any $\gamma > 0$ we studied the Glauber dynamics on a discret torus $\mathcal{T}_\gamma$, the finite interval in $\mathbf{Z}$ with identified endpoints $\pm L_\gamma$, where L_γ is the integer part of $\gamma^{-1-1/3}/2$.

Definition 1. *Glauber dynamics on a discrete torus $\mathcal{T}_\gamma$.*

The configuration space is $\mathcal{H}_\gamma = \{-1,1\}^{\mathcal{T}_\gamma}$, and the configurations are denoted by $\sigma = \{\sigma(x), x \in \mathcal{T}_\gamma\}$. For any $\gamma > 0$, the Glauber evolution with Kac interaction (and no external magnetic field) is a Markov process with generator $\mathcal{L}_\gamma$ acting on the cylinder functions f as

$$\mathcal{L}_\gamma f(\sigma) = \sum_{x \in \mathcal{T}_\gamma} c_\gamma^\beta(x,\sigma)[f(\sigma^{(x)}) - f(\sigma)] \tag{3}$$

where
$$c_\gamma^\beta(x,\sigma) = e^{-\beta\sigma(x)h_\gamma(x)}[e^{-\beta h_\gamma(x)} + e^{\beta h_\gamma(x)}]^{-1}, \quad \beta > 0 \tag{3.a}$$

$\sigma^{(x)}$ being the configuration obtained from σ by flipping the spin at x, and

$$h_\gamma(x) = \sum_{y \neq x} J_\gamma(x,y)\sigma(y) \tag{3.b}$$

$$J_\gamma(x,y) = n_\gamma \gamma J(\gamma|x-y|), \quad n_\gamma^{-1} = \gamma \sum_x J(\gamma|x|) \tag{3.c}$$

with $J \in C^4(\mathbf{R}_+)$, positive (ferromagnetic interaction), $J(r)$ has support in $[0,1]$, and $\int dr\, J(|r|) = 1$. The process generated by $\mathcal{L}_\gamma$ is in $D(\mathbf{R}_+, \mathcal{H}_\gamma)$ and its elements are denoted by σ_t, $t \geq 0$. $D(\mathbf{R}_+, \mathcal{H}_\gamma)$ is the Skorohood space with cadlag trajectories.

Before we describe the results in [1], let us consider, more in general, the Glauber dynamics in $\mathbf{Z}^d$, $d \geq 0$, (defined as above, by substituing $\mathbf{Z}^d$ instead of $\mathcal{T}_\gamma$). We introduce then the magnetic density field.

Definition 2. *The magnetic density field.*
For $\phi \in \mathcal{S}(\mathbf{R}^d)$, the Schwartz space of smooth functions with fast decay to infinity, and any $t \geq 0$ we set

$$Y_t^\gamma(\phi) = \gamma^d \sum_{x \in \mathbf{Z}^d} \phi(\gamma x)\sigma_t(x) \tag{4}$$

In [1] it is proved that propagation of chaos holds for the Glauber dynamics on $\mathbf{Z}^d$, in the limit $\gamma \to 0$. It follows that, if we consider the Glauber dynamics on $\mathbf{Z}^d$ with initial distribution a Bernouille measure with average $\tilde{m}_0(\gamma x)$, where $\tilde{m}_0(\xi)$ is a smooth function on $\mathbf{R}^d$, then $Y_t^\gamma(\phi)$ converges in probability to $\int d\xi \tilde{m}(\xi,t)\phi(\xi))$, where $\tilde{m}(\xi,t)$ is the solution of the Cauchy problem for the following deterministic equation

$$\frac{d\tilde{m}(\xi,t)}{dt} = tangh(\beta J \cdot \tilde{m}(\xi,t)) - \tilde{m}(\xi,t) \tag{5}$$

$$\tilde{m}(\xi,0) = \tilde{m}_0(\xi)$$

Close to the limit (γ small), (5) holds with an error due to fluctuations

$$\frac{d\tilde{m}(\xi,t)}{dt} \approx tangh(\beta J \cdot \tilde{m}(\xi,t)) - \tilde{m}(\xi,t) + \gamma^{d/2}\frac{dz_t(\xi)}{dt} \tag{6}$$

We fix now $d = 1$ (see the final Remark for $d \geq 2$), and $\beta = 1$, (critical temperature). $\tilde{m} = 0$ is then a stationary stable solution of (5). De Masi et al. in [2] made the hypotesis that, with the scaling in (2), the non linear term (1b), that we obtain by making a Taylor expansion around $\tilde{m} = 0$ in (6), would have a finite effect; they guessed that, the fluctuation fields, defined below in (7), converge to the solution of (1). This was proved in [1] for the Glauber dynamics on $\mathcal{T}_\gamma$; equation (1), obtained in the limit $\gamma \to 0$, is then defined on a torus $\mathcal{T}$.

Definition 3. *The fluctuation fields.*
For any $\phi \in \mathcal{S}(\mathcal{T})$ and any $t \geq 0$ we set

$$X_t^\gamma(\phi) = \gamma \sum_{x \in \mathcal{T}_\gamma} \phi(\gamma^{1+1/3}x)\sigma_{\gamma^{-2/3}t}(x) \tag{7}$$

We then introduce the space $D(\mathbf{R}_+, \mathcal{S}'(\mathcal{T}))$ and denote by $X_t(\phi)$ its canonical coordinates: namely if $\omega \equiv (\omega_s)_{s\geq 0} \in D(\mathbf{R}_+, \mathcal{S}'(\mathcal{T}))$, then

$$(X_t(\phi))(\omega) = (\omega_t, \phi), \quad \text{where } (\omega_t, \phi) \text{ is the value of } \omega_t \text{ at } \phi$$

We finally call $\mathcal{P}^\gamma$ the law in $D(\mathbf{R}_+, \mathcal{S}'(\mathcal{T}))$ which gives to the coordinates $X_t(\phi)$ the same distribution that the variables $X_t^\gamma(\phi)$ have in the Markov process with generator $\mathcal{L}_\gamma$ which starts from ν_0^γ, the product measure on $\mathcal{H}_\gamma$ such that the average of any spin $\sigma(x)$ is 0.

In (1) we proved that $\mathcal{P}^\gamma$ converges to the law $\mathcal{P}$ of a process $m(r,t)$, which solves (1). At an heuristic level, this result can be derived from (6) by defining

$$m(r,t) := \gamma^{-1/3} \tilde{m}(\gamma^{1/3}\xi, \gamma^{-2/3}t) \tag{8}$$

The main results in [1] are summerized in the following Theorem.

Theorem 4.
There is a unique process m_t, $t \geq 0$, in $C(\mathbf{R}_+, C(\mathcal{T}))$ adapted to z_t and which solves almost surely

$$(\phi, m_t) = \int_0^t ds \left\{ \frac{D}{2}(\phi'', m_s) - \frac{1}{3}(\phi, m_s^3) \right\} + (\phi, z_t) \tag{9}$$

for all $\phi \in \mathcal{S}(\mathcal{T})$ and all $t \geq 0$. We fix

$$D = \int dr\, J(|r|)r^2 \quad , \quad \int dr\, J(|r|) = 1$$

and denote by $\mathcal{P}$ the law of m_t. $\mathcal{P}$ is the unique solution of the "martingale problem" associated to (1). $\mathcal{P}^\gamma$ converges to $\mathcal{P}$ when $\gamma \to 0$.

Still at an heuristic level, we could prove the Theorem with the "martingale method": we would then show that 1) the family P^γ is tight, 2) for each real valued function f and each $\phi \in S(\mathcal{T})$ the martingale

$$M_t^\gamma(\phi) := f(X_t^\gamma(\phi)) - \gamma^{-2/3} \int_0^t ds \mathcal{L}_\gamma f(X_s^\gamma(\phi))$$

converges to the martingale

$$M_t(\phi) := \int_0^t (dz_s, \phi) f'((m_s, \phi))$$

This method works, if we make a Taylor expansion of the rate function $c_\gamma(x, \sigma)$ with respect to $h_\gamma(x)$ and if we assume that $h_\gamma(x)$ is a function of the fluctuation fields; similar to (8), this would correspond to make the heuristical assumption that $h_\gamma(x) \approx \gamma^{1/3} X_t^\gamma(\phi)$ for some $\phi \in S(\mathcal{T})$.

As the above proof is still not rigorous, in [1] we proposed a method where the Glauber dynamics is expressed as a perturbation of the voter model. The voter model is defined by the generator $\mathcal{L}_\gamma^0$ obtained from (3) by substituing the rate function $c_\gamma(x, \sigma)$ with $c_\gamma^0(x, \sigma) = 1/2(1 - \sigma(x)h_\gamma(x))$, i.e. the first order terms in the Taylor expansion of $c_\gamma(x, \sigma)$ with respect to $h_\gamma(x)$. We used the duality relationship between the voter model and the annihilation random walks to prove that the correlation functions of the voter model are smooth on a scale $\gamma^{-4/3}$: the correlation functions on a lenght γ^{-1} are equal to the correlation functions on a small macroscopic scale $\gamma^{-4/3}l$, in the limit $\gamma \to 0, l \to 0$. We could then prove that in this limit

$$\gamma^{-1/3}h_\gamma(x) \approx \gamma^{-1/3}\gamma \sum_y \phi_l(\gamma^{4/3}(x-y))\sigma_{\gamma^{-2/3}t}(x) \tag{10}$$

$$\phi_l(r) := l^{-1}\phi(l^{-1}r), \quad \phi \in S(T) \tag{10a}$$

Using (10) we could express h_γ as a function of the fluctuation fields. The main steps of our proof were then the following: 1) we used the "martingale method" to prove that the law $\mathcal{P}_0^\gamma$ of the fluctuations in the voter model converges to the law $\mathcal{P}_0$ of the free process, i.e the solution of (1) when the non linear term V' is missing; 2) we proved that the Radon Nikodim derivative (expressed in terms of the fluctuation fields) of the Glauber dynamics with respect to the voter model converges to $d\mathcal{P}/d\mathcal{P}_0$.

Remark 5. *It is not easy to give a meaning to the solution of (1) when $d \geq 2$, because the typical trajectories of the free process (the solution of (1) when $V' = 0$) are distributions. However, in $d = 2$, it is possible to define processes which "solve" (1) in some weak sense, see [4], and more recently [5,6]. In $d = 3$, to our knowledge, the problem is still open and in $d > 3$ there are serious doubts about the possibilty to define the theory.*

REFERENCES

1. L. Bertini, E. Presutti, B. Rüdiger, E. Saada, Dynamical fluctuations a the the critical point: convergence to a non linear stochastic PDE, CARR report n. 11/93 (1993).
2. A. De Masi, E. Orlandi, E. Presutti, L. Triolo, Glauber evolution with Kac potentials: I. Macroscopic equations and fluctuation theory, CARR report n. 9/92 (1992).
3. J. Lebowitz, O. Penrose, Rigorous treatment of the Van der Waals Maxwell theory of the liquid vapour transition, *J. Math. Phys.*, 7:98 (1966).
4. G. Jona-Lasinio, P.K. Mitter, On the Stochastic Quantization of Field Theory, *Commun. Math. Phys.*, 101:409-436 (1985).
5. S. Albeverio, M. Röckner, Stochastic differential equations in infinite dimension: solutions via Dirichlet forms, Probab. Theory Related Fields, 89:347-386 (1992).
6. M. Röckner, Z. Tu-Sheng, On Uniqueness of Generalized Schrödinger Operators and Applications, *J. Funct. Analy.*, 105:187-231 (1992).

LOCAL STRUCTURE OF INTERFACES IN A KAWASAKI+GLAUBER PARTICLE MODEL

Giambattista Giacomin

Math Dept. Hill Center-Busch Campus, Rutgers Univ.
New Brunswick NJ08903, USA

The hydrodynamics of Kawasaki+Glauber particle systems is rather well understood. With a particular choice of the Glauber rates we can can have a reactive and diffusive behavior with a symmetric double well potential as a reactive potential. In this case we can think of starting off *at the top of the hill* in infinite volume and look at the spatial patterns that emerges.

The model: It is the so called Kawasaki+Glauber process. Given $\epsilon > 0$ (small enough) we consider a Markov process $\{\sigma^\epsilon(t)\}_{t \geq 0}$ taking values on $X_\epsilon^d = \{-1, +1\}^{\mathbf{Z}_\epsilon^d}$, where $\mathbf{Z}_\epsilon^d = \mathbf{Z}^d \mathrm{mod}([\epsilon^{-1}|\log \epsilon|^{1/2} \log(|\log \epsilon|)])$. Given a realization $\sigma^\epsilon(\cdot)$ of the ϵ–process, call $\sigma(x,t)$ the value of the spin at (x,t). The generator of the dynamics is given by

$$L_\epsilon = \epsilon^{-2} L_0 + L_G \qquad (1)$$

in which L_0 is the generator of the simple symmetric exclusion process and L_G is the generator of a spin flip dynamics. Precisely $L_0 f(\sigma) = (1/2d) \sum_{x \in \mathbf{Z}_\epsilon^d} \sum_e [f(\sigma^{x,x+e}) - f(\sigma)]$ and $L_G f(\sigma) = \sum_{x \in \mathbf{Z}_\epsilon^d} c(x,\sigma)[f(\sigma^x) - f(\sigma)]$ in which $\{e\}$ are the unit vectors of the frame in $\mathbf{Z}^d$. As usual $\sigma^{x,y}$ denotes the configuration σ with the spins at x and y exchanged and σ^x denotes the configuration with the spin at x flipped. We choose

$$c(0,\sigma) = 1 - (\gamma/d)\sigma(0) \sum_e (\sigma(e) + \sigma(-e)) + (\gamma^2/d) \sum_e \sigma(e)\sigma(-e) \qquad (2)$$

with $\gamma \in (1/2, 1)$. We set $c(x,\sigma) = c(0, \tau_x \sigma)$.

The initial condition: The initial condition for the ϵ–process will be the Bernoulli measure μ^ϵ on $(X_\epsilon^d, \mathcal{B}(X_\epsilon^d))$ such that

$$\mu^\epsilon(\sigma(x)) = 0 \qquad (3)$$

for all $x \in \mathbf{Z}_\epsilon^d$. The law of the process will be denoted by $\mathbf{P}_{\mu^\epsilon}^\epsilon (\mathbf{E}_{\mu^\epsilon}^\epsilon)$.

The hydrodynamics: on the hydrodynamical scale (space $\propto \epsilon^{-1}$ and time $\propto 1$) the behavior of the system is that of the Reaction-Diffusion equation (see [1] and [7])

$$\begin{cases} \frac{\partial}{\partial t} m(\mathbf{r},t) = \frac{1}{2d} \Delta m(\mathbf{r},t) + \alpha m(\mathbf{r},t) - \beta m(\mathbf{r},t)^3 \\ m(\mathbf{r},0) = m_0(\mathbf{r}) \end{cases} \qquad (4)$$

On Three Levels, Edited by M. Fannes et al.,
Plenum Press, New York, 1994

where $\alpha = 2(2\gamma - 1)$ and $\beta = 2\gamma^2$. The reactive term in (4) is (minus) the derivative of the quartic symmetric double well potential with minima at $\pm m^*$ and $m^* = \sqrt{\alpha/\beta}$. In our case $m_0 \equiv 0$. We readily realize that this initial condition is stationary but unstable. Since the dynamics and the initial condition are stochastic, we expect an escape from this initial state, but this will not happen at any finite ($\equiv$hydrodynamical) time.

Beyond the hydrodynamical times (heuristics): Consider the quantity

$$\ell_\epsilon(\mathbf{r}, t; \delta_\sigma) = e^{\alpha t} \int \mathcal{G}_t(\mathbf{r} - \mathbf{r}')\sigma([\epsilon^{-1}\mathbf{r}'])\mathrm{d}\mathbf{r}' \quad (5)$$

that is the solution of (4) linearized around $m = 0$, with initial condition σ. Consider σ distributed according to μ^ϵ. In this formula $\mathcal{G}_t(\mathbf{r})$ is the heat kernel. By [1], formula (5) will be a good approximation of the behavior of the system on the hydrodynamical time scale, that is t finite (with finite we mean not scaling with ϵ). The function ℓ_ϵ will eventually (on times longer than the hydrodynamical ones) fail to describe the behavior of the system. This indeed happens when linearity breaks down. So let us concentrate on the breakdown of linearity. We observe that ℓ_ϵ contains a growth factor $e^{\alpha t}$ and that it is the average of a zero mean Bernoulli random field over approximately $(\epsilon^{-1}\sqrt{t})^d$ sites. Hence if we set $t = |\log \epsilon|\tau$ ($\tau > 0$), for typical trajectories σ we have that $|\ell_\epsilon(\mathbf{r}, \tau|\log \epsilon|; \sigma)| \propto e^{\alpha\tau|\log\epsilon|}(\epsilon/|\log \epsilon|)^{d/2} = \epsilon^{-\alpha\tau + (d/2)}|\log \epsilon|^{-d/2}$ and the spatial dependence of $\ell_\epsilon(\mathbf{r}, \tau|\log \epsilon|; \delta_\sigma)$ will be only on the scale $\mathbf{r} \propto \sqrt{|\log \epsilon|}$ ($\epsilon^{-1}\sqrt{|\log \epsilon|}$ lattice sites). It is clear that the breakdown of linearity will happen at

$$\tau = \tau_c = (d/2\alpha) \quad (6)$$

or (better) shortly thereafter. We want to stress that this is only a heuristic scenario, that gives the right escape time and the right scales, but the quantity in (5) will not be the right quantity to look at (see formula (12) and the observations we make after the main Theorem: the problem is also widely discussed in [5] and [4]).

The escape and the characteristic scales of the phenomenon: First of all the quantity ϵ is built in the dynamics and we set

$$\lambda = \sqrt{|\log \epsilon|} \quad (7)$$

As we understood, λ^2 becomes the relevant time scale. At $t_c = \tau_c \lambda^2$ the typical space length is λ. We can prove that the non-linear phenomenon that creates the interfaces (right after t_c) does not change this typical scale length. We will see that also the hydrodynamical space scale will still be relevant.

The non-linear phenomenon: The onset of interfaces happens on times $t_c + T \log \lambda$, with $T \in (0, \infty)$. What happens is that the fluctuation has grown to the critical point and now the magnetization profile will be literally squared out by the non-linearity. That is the non-linear mechanism will make the system reach one of the two phases ($\pm m^*$) according to the sign of the fluctuation in that point. Only points very close (at distance $\propto \epsilon^{-1}$ lattice sites) to the locus of zeros of the fluctuation will have a magnetization different from $\pm m^*$. So on the space scale $\lambda \epsilon^{-1}$ lattice sites the system will be described by regions of the different phases. The only places in which the local magnetization differs from the equilibrium ones is clearly in a neighborhood of the boundaries between clusters. Here indeed there is the interface, that is shrunk to a (hyper)surface if we look at it on the space scale of the clusters ($\epsilon^{-1}\lambda$ lattice sites), but it has a non-trivial structure on the hydrodynamical scale (ϵ^{-1} lattice sites).

The Gaussian field and its approximants: Define

$$Y^\epsilon(\mathbf{x}, \sigma) = c(\alpha)\lambda^{-d/2}\epsilon^{-d/2}\epsilon^{\alpha a}\sum_x \mathcal{G}_a(\mathbf{x} - [(\epsilon/\lambda)x])\sigma(\mathbf{x}) \quad (8)$$

in which $a > 0$ and $c(\alpha)$ is a constant that will turn out to be irrelevant. Moreover $\mathcal{G}_a(\mathbf{x})$ is the heat kernel at time a. Notice that $Y^\epsilon(\cdot, \sigma) : X_d^\epsilon \to C^\infty(\mathbf{R}^d)$. So if we set

$$Y^\epsilon(\mathbf{x}, \sigma^\epsilon(t_c - a\lambda^2)) = \rho^\epsilon(\mathbf{x}) \tag{9}$$

for any $k \in \mathbf{Z}^+$ we can consider the sequence of probability measures $\{\mathbf{P}^\epsilon\}_{\epsilon>0}$ on $(C^k(\mathbf{R}^d), \mathcal{B}(C^k(\mathbf{R}^d)))$ in which $\mathbf{P}^\epsilon$ is the law of $\rho^\epsilon(\mathbf{x})$ under $\mathbf{P}^\epsilon_{\mu^\epsilon}$. As a first result we have

$$\mathbf{P}^\epsilon \Rightarrow \mathbf{P} \tag{10}$$

as $\epsilon \to 0$ ('$\Rightarrow$' denotes weak convergence of measures). The measure $\mathbf{P}$ is the Gaussian random field specified by

$$\mathbf{E}(\rho(\mathbf{x})) = 0, \quad c(\mathbf{x} - \mathbf{y}) \equiv \mathbf{E}(\rho(\mathbf{x})\rho(\mathbf{y})) = \exp\left(-\frac{a(\mathbf{x} - \mathbf{y})^2}{2}\right) \tag{11}$$

Notice that

$$\rho^\epsilon(\mathbf{x}) \approx \text{const.} \lambda^{d/2} \ell_\epsilon(\mathbf{x}\lambda, t_a; \delta_{\sigma^\epsilon(t_c - a\lambda^2)}) \tag{12}$$

Indeed the two quantities differ only by the fact that the discrete sum has become an integral (and this difference vanishes fast as $\epsilon \to 0$, this is true also for the derivatives). In particular (10) and (12) show that the fluctuation at the critical time is of order $\lambda^{-d/2}$ and that its spatial dependence is on the scale λ hydrodynamical units.

Geometric properties of the trajectories of $\mathbf{P}$: In order to prove our main Theorem we will need some auxiliary results on the geometry of the Gaussian random field (11) and part of these results may be interesting in themselves. In particular we will consider the zero level set of the trajectories $\rho : \mathbf{R}^d \to \mathbf{R}$ of the field $\mathbf{P}$

$$Z_\rho \equiv \{\mathbf{x} \in \mathbf{R}^d : \rho(\mathbf{x}) = 0\} \tag{13}$$

We have that $\rho(\cdot)$ is $\mathbf{P}$–a.s. an analytic function and if $d \geq 2$ the set $Z_\rho \equiv \{\mathbf{x} \in \mathbf{R}^d : \rho(\mathbf{x}) = 0\}$ (imbedded in $\mathbf{R}^d$) is $\mathbf{P}$–a.s. an analytic $(d-1)$–manifold. If $d = 1$, then Z_ρ is a discrete set and Z_ρ separates the line into open intervals in which ρ is alternatively positive and negative. We do not need such a high regularity to make our arguments work, but $\mathbf{P}$ is an intrinsic characteristic of the model and so it is interesting in itself.

From the zero level set to the interfaces: Let us restrict to functions ρ with the regularity described above. Call $\text{tub}(\rho, r)$ the tubular neighborhood of radius r of Z_ρ. Set $\text{tub}(\rho) = \text{tub}(\rho, D(\lambda) \log \lambda/\lambda)$, in which $D(\lambda) = T(\lambda) \log T(\lambda)$ and $T(\lambda)$ is given after formula (18). Notice that $D(\lambda) \log \lambda/\lambda$ vanishes if λ goes to infinity. Consider the map $X_\rho : [-D \log \lambda/\lambda, +D \log \lambda/\lambda] \times Z_\rho \to \text{tub}(\rho)$ defined by

$$X_\rho(\delta, \mathbf{x}) = \mathbf{x} + \delta\nu(\mathbf{x}) \tag{14}$$

and $\nu(\mathbf{x}) = D\rho(\mathbf{x})/|D\rho(\mathbf{x})|$. Moreover if we restrict $X_\rho(\delta, \mathbf{x})$ to $[-D \log \lambda/\lambda, +D \log \lambda/\lambda] \times (Z_\rho \cap Q_{2L})$, where $Q_{2L} = [-2L, 2L]^d$ ($L > 0$), for λ (chosen dependent on ρ and L) sufficiently large X_ρ is one to one ($\text{tub}(\rho)$ is shrinking to Z_ρ as λ goes to infinity). We will refer to this property by saying that X_ρ is locally invertible for λ large enough. Hence restrict the image of X_ρ and consider the inverse map $X_\rho^{-1}(\mathbf{x}) \equiv (\mathbf{z}(\mathbf{x}), \delta(\mathbf{x}))$. Call $\Theta(r)$ the (unique up to $r \to -r$) non-trivial antisymmetric solution of the one dimensional problem $(1/2d)\Theta'' + \alpha\Theta - \beta\Theta^3 = 0$ that is

$$\Theta(r) = m^* \tanh(r\sqrt{\alpha d}) \tag{15}$$

Moreover choose $L > 0$ and consider the map $F^\epsilon : C^\infty \to C^0(Q_L)$ (once we set λ large enough to make X_ρ locally invertible)

$$F^\epsilon(\rho(\cdot))(\mathbf{x}) = \begin{cases} \Theta(\lambda\delta(\mathbf{x})) & \text{if } \mathbf{x} \in \text{tub}(\rho) \cap Q_L \text{ and } \delta(\mathbf{x}) \leq D(\lambda)(\log\lambda)/\lambda \\ \Theta(D(\lambda)\log\lambda)\text{sign}(\mathbf{x}) & \text{otherwise} \end{cases} \quad (16)$$

If ρ is not suitably regular set simply $F^\epsilon(\rho) \equiv 0$. Notice that if ρ is suitably regular and $\mathbf{x} \in B\rho^c$ then

$$\lim_{\epsilon \to 0} |F^\epsilon(\rho(\cdot))(\mathbf{x}) - m^*\text{sign}(\rho(\mathbf{x}))| = 0 \quad (17)$$

faster than any power of λ^{-1}. Notice moreover that F^ϵ depends only on $\text{sign}(\rho)$, not on the whole ρ.

The empirical averages: The new result will be stated in terms of empirical averages at a time slightly larger than t_c, precisely

$$t_e = t_c + T(\lambda)\log\lambda \quad (18)$$

and $T(\lambda)$ goes to infinity as slow as we want with λ (indicatively $T(\lambda) = \log\log\lambda$). We are doing this because the interface indeed develops on times $t = t_c + T\log\lambda$ ($T \in [0, \infty)$) and the onset will be completed asymptotically on this scale of time ($T \to \infty$). Then the interface starts moving by mean curvature on the time scale λ^2. In order to localize better the interface we have then chosen a very slow growth for $T(\lambda)$. Given $b \in (0, 1)$ we define the empirical average

$$A^\epsilon_\sigma(\mathbf{x}) = \frac{1}{|B(\mathbf{x}, \epsilon^{1-b})|} \int_{B(\mathbf{x},\epsilon^{1-b})} \sigma([\epsilon^{-1}\lambda\mathbf{x}'])d\mathbf{x}' \quad (19)$$

in which $B(\mathbf{x}, R)$ is the ball of radius R centered at $\mathbf{r}$ (and $|B|$ is its (Lebesgue) measure).

The main result: Our main result is

Theorem: *Consider a system of dimension $d = 1, 2$ or 3. For all $L > 0$ and a sufficiently small X_{ρ^ϵ} ($\rho^\epsilon = Y^\epsilon(\cdot, \sigma^\epsilon(t_c - a\lambda^2))$) is locally invertible with probability going to 1 as $\epsilon \to 0$, for any $q > 0$*

$$\lim_{\epsilon \to 0} \mathbf{P}^\epsilon_{\mu^\epsilon}\left(\left\{\sup_{\mathbf{x} \in Q_L} |A^\epsilon_{\sigma^\epsilon(t_e)}(\mathbf{x}) - F^\epsilon(\rho^\epsilon)| \leq \lambda^{-2+q}\right\}\right) = 1 \quad (20)$$

and

$$\lim_{\epsilon \to 0} \mathbf{P}^\epsilon_{\mu^\epsilon}\left(Z_{A^\epsilon_{\sigma^\epsilon(t_e)}} \subset \text{tub}\left(Z_{\rho^\epsilon}, A(\lambda)\frac{(\log\lambda)^2}{\lambda^2}\right)\right) = 1 \quad (21)$$

in which $A(\lambda)$ can be chosen equal to $T(\lambda)^3$ and we remind that $T(\lambda) \uparrow \infty$ as $\lambda \uparrow \infty$ as slow as we want.

So the result (20), combined with (10), says that the spatial configurations will really look like big clusters (typically λ hydrodynamical units) separated by interfaces. The interfaces are smooth surfaces on the cluster scale, but on the hydrodynamical scale they have a *thickness* and only one relevant direction of variation.

Result (21) instead gives us an actual estimate on the distance between the zero level set of the empirical average and of the approximants of the Gaussian field. The zero level set is the relevant object because of (20) and (17). Observe that the interface is localized also on the hydrodynamical scale, not only on the cluster scale. Indeed the radius of the tubular neighborhood on the hydrodynamical scale becomes $A(\lambda)(\log\lambda)^2/\lambda$, that vanishes as λ goes to zero.

Observe also that for our purposes all the information is embodied in $\sigma(\cdot, t_c - a\lambda^2)$. That is the spatial structure at time t_e is already fully determined (both on the hydrodynamical and the cluster scale) by the configuration at a time before t_c. In [3] the problem of phase separation in the case of Glauber dynamics with local mean field dynamics is taken into consideration: in that case they are able to prove that $\sigma(\cdot, \tau\lambda^2)$ (any $\tau \in (0, \tau_c)$) is enough to characterize the spatial structure after the escape.

The evolution of the clusters after the escape should be motion by mean curvature and the result we have just presented gives a very precise initial condition for the problem of cluster evolution.

The main Theorem holds for $d = 1, 2, 3$, because [4] and [5] cover only these cases. In [6] there is the proof of the main Theorem as well as of other facts quoted above. We want to stress all the techniques used in [5] work regardless of the dimension of the system. But [6] relies on [4] and [5] for the estimates on the correlation functions of the system. It is believed (and in [5] there are some hints in this direction) that these estimates can be extended to higher dimensional cases and in that case the main Theorem would extend.

The central analytic technique used in [6] is a modification of an argument by De Mottoni and Schatzman [2] concerning the development of interfaces for problem (4).

REFERENCES

1. A. De Masi, P. Ferrari, and J.L. Lebowitz, Reaction diffusion equations for interacting particle systems, *Journal of Statistical Physics*, **44**, 589–644 (1986).

2. P. De Mottoni, M. Schatzman, Development of interfaces in $\mathbf{R}^n$, *in* "Proc. Royal Soc. Edinburgh," **116A**, 207–220 (1990).

3. A. De Masi, E. Orlandi, E. Presutti and L. Triolo, Glauber evolution with Kac potential II: spinodal decomposition, Preprint Rome (1992).

4. A. De Masi, A. Pellegrinotti, E. Presutti, and M.E. Vares, A model for phase separation, *to appear in Annals of Probability*.

5. G. Giacomin, Phase separation and random domain patterns in a stochastic particle model, Preprint Rutgers Univ. (October 1992).

6. G. Giacomin, Interface formation and global spatial structure in a reaction-diffusion model, Preprint Rutgers Univ. (December 1992).

7. C. Kipnis, S. Olla, and S.R.S. Varadhan, Hydrodynamics and large deviations for simple exclusion processes, *Comm. Pure Appl. Math.* **42**, 115–137 (1989).

QUANTUM CHAOS, FRACTAL SPECTRA AND ATOMIC STABILIZATION

G. Casati

Dipartimento di Fisica, Università di Milano
Via Castelnuovo – 22100 Como Italy
Also at Istituto Nazionale di Fisica Nucleare, Sezione di Milano

INTRODUCTION

The modifications that quantum mechanics imposes on the general picture of classical chaos is a subject of growing interest for its theoretical relevance and for different physical applications as well. As a matter of fact the study of nonlinear dynamical systems has led to a much better understanding of the rich complexity of the classical motion. Of particular relevance was the discovery of the deterministic chaotic motion which is a type of motion indistinguishable from a purely random motion even if it is governed by strictly deterministic laws. This motion is characterized by exponential instability of orbits with respect to initial conditions which leads to exponential loss of memory, decay of correlations and approach to statistical equilibrium.

However the "true" mechanics of our world is Quantum Mechanics and therefore, as far as the general problem of understanding the qualitative behaviour of dynamical systems is concerned, we need to discover the extent at which the beautiful variety of classical chaotic motion persists at the microscopic level in the domain of quantum mechanics. This problem is interesting for the foundations of quantum statistical mechanics and it has several applications like the distribution of nuclear and atomic energy levels, the interaction of atoms with e.m. fields, the problem of localization in disordered solids the unimolecular reaction theory etc.

As a matter of fact our intuition of Quantum Mechanics has been largely based on few exactly solvable models and on some approximate solutions of Schrödinger's equation. Since quantum mechanics is strictly connected to classical mechanics, one may wonder if quantum mechanics is richer than simple perturbative approaches may lead us to suspect.

In order to shed some light on the above problem, we start with a very simple dynamical system which exhibits classical chaotic motion and compare its classical and quantum evolution. It is very instructive to consider the following simple periodically perturbed system, the so-called kicked rotator, which is described by the well known "standard" or "Chirikov map"

$$\begin{cases} \bar{I} = I + k \sin \vartheta \\ \bar{\vartheta} = \vartheta + \bar{I}T \end{cases} \quad (1)$$

This map has a very simple form but it contains most of the complexity of classical dynamical systems [1]. In particular, for $kT \geq 1$ two initially close orbits separate exponentially, the phases go rapidly random and the motion becomes chaotic and diffusive with diffusion coefficient $D = k^2/2$. These qualitative predictions have very well been confirmed by numerical experiments.

We turn now to the quantum description of model (1) which is obtained from the quantization rule $I \to i\hbar \partial/\partial \vartheta$. The quantum evolution is described by a unitary operator which maps the wave function $\psi(\vartheta)$ immediately after one kick to that immediately after the next kick.

$$\bar{\psi}(\vartheta) = U\psi(\vartheta) = e^{-i\hbar^{-1}k \cos \vartheta} e^{i\hbar \frac{T}{2} \frac{\partial^2}{\partial \vartheta^2}} \psi(\vartheta) \quad (2)$$

The interesting question is how the quantum motion compares with the classical motion. As already mentioned, one of the most important features of classical chaotic motion is the exponential instability which means that two initially close orbits separate exponentially with time. As a consequence the *Ehrenfest time scale* τ_E namely the time over which a minimum uncertainty packet follows a classical orbit must be very short. Indeed let us consider two orbits which are initially in a region of phase space of area $\hbar$. Their distance in phase which at time $\tau = 0$ is of order $\sqrt{\hbar}$ increases exponentially in time namely $\delta\vartheta(\tau) \sim \hbar^{\frac{1}{2}} exp(\lambda \tau)$ where λ is the so-called Lyapounov exponent. Then after a time $\tau_E \sim \ln(1/\hbar)/2\lambda$ the distance $\delta\vartheta$ is comparable with the whole interval 2π and therefore it is completely meaningless to compare the evolution of the quantum packet with the classical orbit. Therefore, for classical chaotic systems, the time τ_E is very short and increases only logarithmically as $\hbar \to 0$.

What happens at later times? One may expect that the evolution of the quantum packet will reproduce somehow the evolution of the classical ensemble initially started in an area $\sim \hbar$ in phase space. In such case the quantum excitation will mimic the classical average, diffusive behaviour. It was therefore a big surprise when several years ago it was found [2] that the quantum motion mimics the classical one only up to a given time τ_B while, for $\tau > \tau_B$, the quantum diffusive motion, unlike the classical, stops and enters a stationary oscillatory regime (Fig. 1).

We have now a quite satisfactory understanding of this phenomenon, even if a rigorous mathematical proof is still lacking. At the informal level one may say that in the classical case the evolution of the initial ensemble develops on separate independent orbits. In the quantum case instead the different dynamical possibilities interfer and this leads to the suppression of the diffusive process. A quantitative estimate of the *diffusive time scale* τ_B was obtained in ref.[3] on the assumption that the quasi-energy spectrum (i.e. the Floquet spectrum, see below) is discrete. The idea

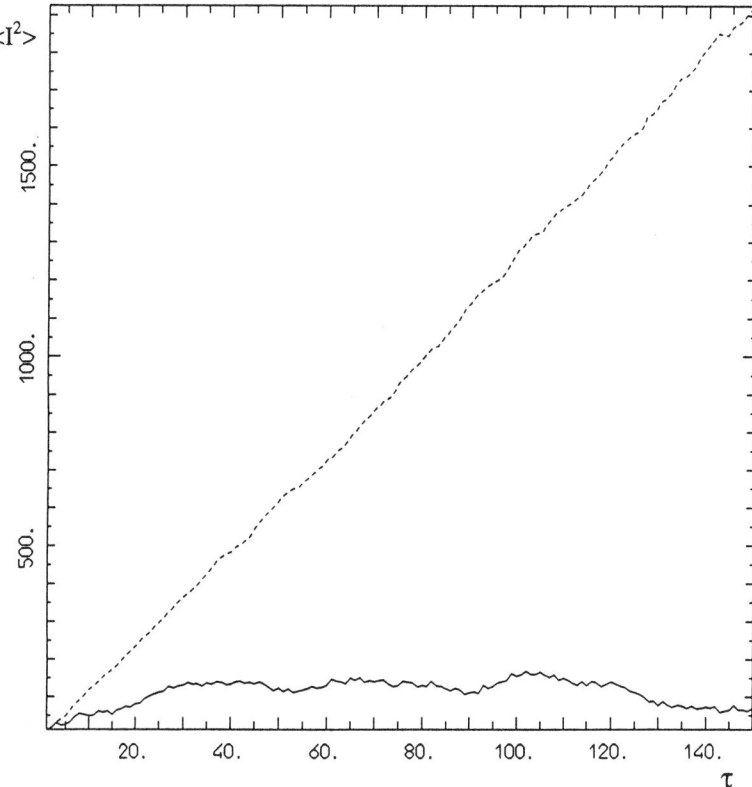

Fig. 1. Classical (dashed curve) and quantum (full curve) average value of I^2 as a function of the time τ, measured in number of map iterations. Here $k = 5$ and $T = 1 (\hbar = 1)$.

is that, according to Heisenberg's uncertainty relations, the quantum motion follows in the average the classical motion up to a time of the order of the inverse of the average q.e. levels spacing. This estimate gives $\tau \approx D/\hbar^2$.[3]

To summarize, for classical chaotic systems a coherent state follows a classical orbit only for a very short time $\tau_E \sim \ln(1/\hbar)$. For $\tau_E < \tau < \tau_D$ the quantum motion follows the classical only in the average. The diffusive time τ_D is much larger than τ_E and grows to infinite as $1/\hbar^2$.

To be more precise it is necessary to mention that the above is the "typical" behaviour. For particular (zero measure) resonant values of period T, namely when T is a rational multiple of $4\pi/\hbar$, the quasi–energy spectrum is continuous and the quantum rotator energy increases like t^2.

EXPERIMENTAL EVIDENCE

The question now is whether the interesting phenomena described above are a general occurrence in quantum mechanics or a peculiar property of the particular model chosen, namely the kicked rotator. Another interesting question is whether or not the quantum suppression of classical chaotic diffusion can be observed in real laboratory experiments.

One of the most significant cases where classical and quantum chaos confronted each other was in the explanation of an experiment on hydrogen atoms, first performed in 1974 by Bayfield and Koch [4]. Single atoms prepared in very elongated states with high principal quantum number ($n_0 \approx 63 - 69$) were injected into a microwave cavity and the ionization rate was measured. The microwave frequency was 9.9 GHz, corresponding to a photon's energy well below the ionization energy of level 66 and even lower than the transition from state 66 to 67. Much surprise therefore followed the discovery that a very efficient ionization occurred when the electric field intensity exceeded a threshold value of about 20 Volt per cm (for $n_0 = 66$), much lower than the static Stark value. A subsequent analysis [5], in classical terms, explained the threshold intensities as critical values for the onset of chaotic diffusion in action space. However, the hydrogen atom is a quantum object: The quantum mechanical evolution was investigated in the one–dimensional approximation [6] and for $\omega n_0^3 > 1$ an ionization threshold higher than the classical value was predicted in order to overcome the occurrence of quantum localization.

The dimensionless classical Hamiltonian for a hydrogen atom, interacting with a time–periodic microwave field in the dipole approximation is (in atomic units)

$$H(z, p, t,) = \frac{p^2}{2} - \frac{1}{r} + \epsilon z \cos(\omega t) \qquad (z \geq 0) \tag{3}$$

As was shown in [7], the main qualitative features of the motion are given by the one–dimensional model; moreover, since we are interested in exploring the dynamics that precedes ionization, we confine to negative energies, and introduce accordingly action–angle variables (n, θ) thus obtaining:

$$H = -\frac{1}{2n^2} + \epsilon z(n, \theta) \cos \omega t \tag{4}$$

By integrating the equations of motion over one period of the unperturbed orbit, in first order in ϵ, we obtain an approximate map for the canonical variables (ν, ϕ) where $\nu = E/\omega = -1/2n^2\omega$ is the number of photons exchanged by the atom with the field and ϕ is the field phase at perihelion [7]. A linearization of the map around the initial value $\nu_0 = -1/2n_0^2\omega$ yields, once again the Standard map [7]:

$$\bar{\nu} = \nu + k \sin \phi$$
$$\bar{\phi} = \phi + T\bar{\nu} \tag{5}$$

where $k \approx 2.6\varepsilon/\omega^{5/3}$ and $T = 6\pi\omega^2 n_0^5$.

In spite of the various simplifications so far introduced, the map (5) still gives a good description of the main behaviour of the system and leads to interesting conclusions. Indeed, the condition $kT \approx 1$ gives the threshold for transition to classical chaos leading to fast ionization:

$$\epsilon_c \approx \frac{1}{50\, n_0^5\, \omega^{1/3}} \tag{6}$$

In the corresponding quantum model we expect, similarly to the kicked rotator, exponential localization of the wave function in ν. After the relaxation time τ_D the system reaches an exponentially localized steady state with a predicted localization length in the number of photons equal to the diffusion coefficient in ν-space $l_s \approx k^2/2 \approx 3.3\epsilon^2\omega^{-10/3}$. If l_s is greater than the number of photons required to ionize the atom $l_s \geq (2\omega n_0^2)^{-1}$, then localization cannot prevent ionization. This condition leads to *quantum delocalization border*

$$\epsilon_q \approx 0.4\omega^{7/6}n_0^{-1} \tag{7}$$

Unexpected as these predictions may have been at their first appearance [6,7], they were confirmed by recent experimental results on the microwave ionization of hydrogen atoms [8,9]. It was found that experimental and numerical data fairly well agree with localization theory and at the same time appreciably deviate from classical predictions (Fig. 2).

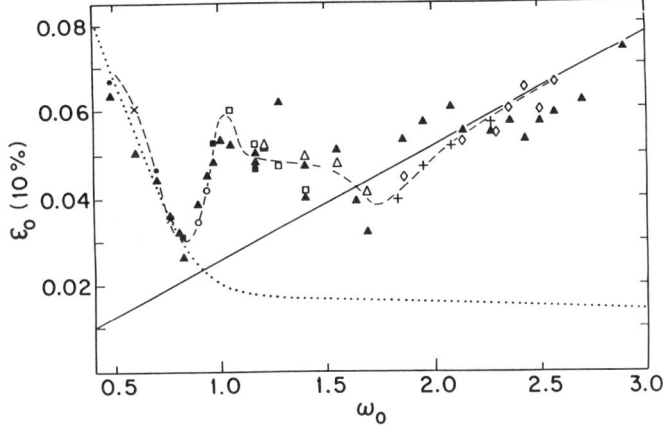

Fig. 2. A comparison at identical parameter values of experimental and quantum-mechanical values for the microwave field strength for 10% ionization probability, as a function of microwave frequency. The field and frequency are classically scaled, $\omega_0 = n_0^3\omega$ and $\varepsilon_0 = n_0^4\varepsilon$. Ionization includes excitation to states with n above $\bar{n}$. The theoretical points are shown as solid triangles. The dashed curve is one drawn through the entire experimental data set shown in Fig. 2. Values of n_0, $\bar{n}$ are •, 64, 114; x, 68, 114; o, 71, 114; ☐, 76, 114; ☐, 80, 120; △, 86, 130; +, 94, 130; ◇, 98, 130. Multiple Theoretical values at the same ω_0 are for different compensating experimental choices of n_0 and ω. The dotted curve is the classical chaos border. The solid line is the quantum 10% threshold according to localization theory for the present experimental conditions.

The experiments described in ref. [9] were precisely designed for the purpose of checking localization theory; as a matter of fact, special care was taken in order that numerical computations could simulate as closely as possible the experimental conditions. Therefore, they provide experimental evidence of the quantum suppression of the classically chaotic diffusion due to the localization phenomenon.

NEW DEVELOPMENTS

In the previous sections I have discussed a main feature of the quantum behaviour of classically chaotic systems: the quantum dynamical localization. In the following I will briefly describe two interesting developments:

a) Fractal Properties of Quantum Spectra

The quantum localization of classical chaotic diffusion is related to the discrete nature of the quasi–energy spectrum. Even though a rigorous mathematical proof is still lacking, all the empirical and analytical evidence and the general character of the argument which has initially led to the prediction of dynamical localization indicates that this is a typical feature of quantum mechanics in presence of a classical diffusion process. However, recent investigations [10] have revealed that the situation is much more complex and that the relation between classical and quantum diffusion processes is not so straightforward: actually, until now, it is unclear. A very interesting example is provided by the so–called kicked Harper model which is obtained by quantization of the following area–preserving map:

$$\bar{p} = p + K \sin x$$
$$\bar{x} = x - L \sin \bar{p} \tag{8}$$

This model describes the motion in $2\,d$ lattices in perpendicular magnetic and electric fields. The classical map is chaotic and diffusive if the parameter K is large enough. Quantization of the map (8) leads to the one–period evolution operator

$$\hat{U} = exp\left[-i\frac{L}{\hbar}\cos(\hbar\,\hat{n})\right] exp\left[-i\frac{k}{\hbar}\cos(x)\right] \tag{9}$$

where $\hat{n} = -i\partial/\partial x$. Quasi-energy eigenvalues and eigenfunctions are determined by $\hat{U}\psi_\omega = e^{-i\omega}\psi_\omega$. The quantum motion exhibits a very rich and complicated behaviour. In particular we have studied the scaling properties of the spectrum within the thermodynamic formalism for multifractals and we have found that, corresponding to classical chaotic motion, there are different quantum regimes (depending on the parameters K and L) related to different types of spectra including mixed spectra, containing both pure–point and continuous components. Moreover the investigation of the long–time quantum dynamic evolution has shown anomalous diffusion with an exponent which is related to the Hausdorff dimension of the spectrum [10].

b) The Stability of the Hydrogen Atom in a Strong External Driving Field: the Magic Mountain

From the theoretical and experimental analysis presented in Fig. 2, one naturally expects that, by increasing the intensity of the external field, a larger and larger fraction of the atoms will be ionized, until complete ionization takes place. By further increasing the field one expects this process to be even faster. However strange it may seem, this expectation turns out to be false! A careful analysis of the classical motion shows that at very high field values the atoms become stable again. For details we refer to the original papers [11,12]. It is however instructive to present an intuitive explanation of this apparently paradoxical result. If one rewrites Hamiltonian (3) in cylindrical coordinates, one obtains:

$$H = \frac{p_z^2}{2} + \frac{p_\varrho^2}{2} + \frac{m^2}{2\varrho^2} - \frac{1}{\sqrt{\varrho^2 + z^2}} + \epsilon z \cos\omega t \qquad (10)$$

For large values of ϵ and ϱ, the z–motion is dominated by the driving field $z \sim \epsilon \cos\omega t/\omega^2$. One can then perform an average over the z–motion and look at the ϱ–motion which will be described by the average Hamiltonian

$$\bar{H} = \frac{p_\varrho^2}{2} + \frac{m^2}{2\varrho^2} + \frac{2\omega^2}{\pi\epsilon} \ln\left(\frac{\varrho\omega^2}{\epsilon}\right) \qquad (11)$$

The minimum of the potential in (11) increases with the field; the physical reason is quite clear: the large amplitude of field oscillations leads to a decrease of the attractive Coulomb force, while the centrifugal potential remains the same. Obviously, the averaged Hamiltonian (11) gives a good description of the real 3D motion if the frequency $\Omega \sim \omega^2/\epsilon m$ of the oscillations in ρ is much less than the external frequency. The condition $s = \omega/\Omega \gg 1$ leads to the stabilization border

$$\epsilon > \epsilon_{stab} = \delta\frac{\omega}{m}, \qquad (12)$$

where δ is a numerical constant.

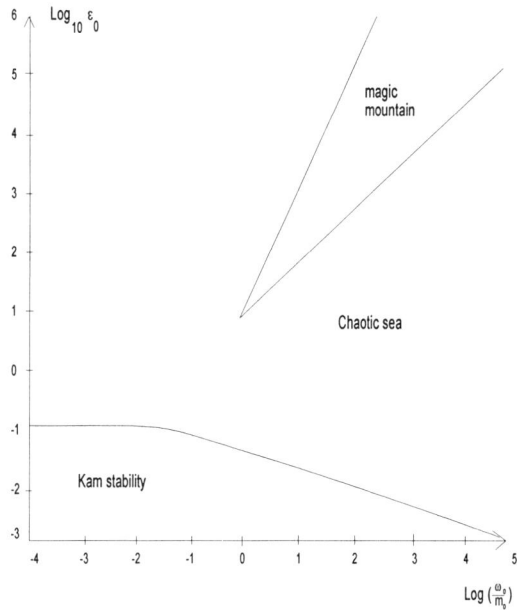

Fig. 3. Stability diagram in the ϵ_0, ω_0/m_0 plane where $\epsilon_0 = \epsilon n_0^4$, $\omega_0 = \omega n_0^3$, $m_0 = m/n_0$. The lowest curve is the usual KAM chaos border which, for small ω_0, approaches the border $\epsilon_0 \sim 0.13$ for static field ionization. Notice that the stabilization border (5) $\epsilon_0 = 12\omega_0/m_0$ is much higher than the chaos border. Notice also that the stabilization region is limited from above; this is due to the fact that in very strong fields the electron cannot be captured in the minimum of the average potential and ionization takes place.

Under condition (12), the atom is stable since the average Hamiltonian (11) is a constant of the motion with adiabatic accuracy ($\approx exp(-const \cdot s)$). For field values smaller than (12), down to the usual chaos border, ionization takes place. Numerical solutions of the exact Hamiltonian (10) confirm the estimate (12) with $\delta \approx 12$ [11,12].

The general picture of the stability diagram is shown in Fig. 3.

The most impressive result of our analysis is the "magic mountain" of stability in the upper-rightmost part of the figure which emerges from the chaotic sea and, coming from infinity, approaches, but does not touch, the familiar territory of KAM stability.

REFERENCES

1. B.V. Chirikov, *Phys. Rep.*, **52**, 623 (1979).

2. G. Casati, B.V. Chirikov, J. Ford and F.M. Izrailev *in*: Stochastic Behaviour in Classical and Quantum Hamiltonian Systems, *Lecture Notes in Physics*, **93**, 334, Springer-Verlag, Berlin (1979).

3. B.V. Chirikov. F.M. Izrailev and D.L. Shepelyansky *Sov. Sci. Rev. Sect.*, **C1**, 209 (1981).

4. J.E. Bayfield and P.M. Koch, *Phys. Rev. Lett.*, **33**, 258 (1974).

5. R. Jensen, *Phys. Rev. Lett.*, **49**, 1365 (1982); *Phys. Rev. A*, **30**, 386 (1984)
   N.B. Delone, B.P. Krainov and D.L. Shepelyansky, *Sov. Phys. Usp.*, **26**, 551 (1983).

6. G. Casati, B.V. Chirikov and D.L. Shepelyansky, *Phys. Rev. Lett.*, **53**, 2525 (1984).

7. G. Casati et al., *Phys. Rep.*, **154**, 77 (1987);
   G. Casati, I. Guarneri and D.L. Shepelyansky, *IEEE J. of Quantum Electr.*, **24**, 1420 (1988).

8. E.J. Galvez, B.E. Sauer, L. Moorman, P.M. Koch and D. Richards, *Phys. Rev. Lett.*, **61**, 2011 (1988).

9. J.E. Bayfield, G. Casati, I. Guarneri and D.W. Sokol, *Phys. Rev. Lett.*, **63**, 364 (1989).

10. R. Artuso, G. Casati and D.L. Shepelyansky, *Phys. Rev. Lett.*, **68**, 382 (1992);
    R. Artuso, F. Borgonovi, I. Guarneri, L. Rebuzzini and G. Casati, *Phys. Rev. Lett.*, **69**, 3302 (1992).

11. F. Benvenuto, G. Casati and D.L. Shepelyansky, *Phys. Rev. A*, **47**, R786 (1993).

12. F. Benvenuto, G. Casati and D.L. Shepelyansky, Classical stabilization of hydrogen atom in strong fields: The "magic mountain" in the chaotic sea, Preprint – DYSCO 025

STATISTICAL PROPERTIES OF RANDOM BANDED MATRICES: ANALYTICAL RESULTS

Yan V. Fyodorov[1,3] and Alexander D. Mirlin[2,3]

1. Weizmann Inst. of Science
   Dept. of Nuclear Physics
   Rehovot 76100, Israel

2. Institute für Theorie der Kondensierten Materie
   Universität Karlsruhe, BRD

3. Petersburg Nuclear Physics Institute
   188350, Gatchina, St. Petersburg District, Russia

Recently it was realized that there exist important classes of Random Matrices (RM)– that of *banded*[1] and *sparse*[2] type – that are appreciably different in their properties from the classical Gaussian Ensembles of RM[3] studied by Wigner, Dyson, Mehta and others. We will call the matrix *Random Banded* (RBM) if its entries H_{ij} are random variables with zero mean value and variance $\overline{|H_{ij}|^2} = A(i,j)$ decaying to zero when the distance $r = |i-j|$ from the main diagonal tends to infinity. Matrices of such a type found recently numerous applications in Solid State Physics and especially in the study of Quantum Chaos Phenomena (see papers[1,4,5] and references therein). Without much loss of generality one can put $A(i,j) = a(|i-j|)$ and in the most studied case the "shapefunction" $a(r)$ is exponentially (or faster) decaying when r exceeds some typical scale b (called the "bandwidth").

It turned out that introducing such a "structure" in RM can result in important deviations in both eigenvalue and eigenvector statistics from that typical for the Gaussian matrices. Namely, for the ensemble of RBM of infinite size N and large bandwidth $b \gg 1$ it was demonstrated numerically[4] that 1) eigenvalues are uncorrelated (Poissonic) random numbers and 2) any eigenvector has typically of the order of b^2 essentially nonzero components out of the total number of N, other components being exponentially small. Moreover, it was found that statistical properties of RBM of finite (but large) size N depend only on the *scaling parameter* $x = N/b^2$ rather then on N and b separately[4]. When x changes from zero to infinity the eigenvalue statistics gradualy deviate from being Wigner-Dyson, and the number of appreciably nonzero components of eigenvectors changes from a value of the order of N at $x \ll 1$ to a value of the order of b^2 at $x \gg 1$.

The key word allowing one to understand qualitatively the origin of such a behaviour is the Anderson localization. More precisely, one can look at RM as at the Hamiltonian of a quantum particle in a tight-binding representation. There is precisely N sites in the model with diagonal elements H_{ii} describing (random) site energies and off-diagonal elements H_{ij} describing amplitudes of random site-to-site hoppings. From this point of view a banded RM represents a generalization of a usual nearest-neighbour $1d$ tight-binding model to a situation when the typical length of hopping is of the order of bandwidth b. It has been known for a long time that the wavefunction of a quantum particle moving in a random $1d$ system of infinite size $L \to \infty$ becomes localized[6], i.e. effectively spread only over a finite space region of size ξ. The behaviour of a system of macroscopic but finite size L is assumed to be correctly described by *scaling hypothesis* stemming from the requirement that the localization length ξ is the only characteristic length of the system at given disorder and therefore all statistical properties should be dependent on the ratio ξ/L only. The eigenvalues of the random Hamiltonians are expected to be uncorrelated when the localization takes place[6]. All these facts provide a basis for understanding the observed properties of banded RM.

In the present paper we would like to give a brief summary of recent analytical results[5] concerning statistical properties of RBM for arbitrary value of the scaling parameter $x = N/b^2$.

One of the most useful and frequently used quantitative characteristics of eigenfunctions in a finite disordered sample is a so-called inverse participation ratio (IPR) defined for a given eigenstate $|\psi_\alpha\rangle$ as $P_2 = \sum_n |\langle n|\psi_\alpha\rangle|^4$. Its average value $\overline{\langle P_2\rangle}$ (the bar denotes the averaging over the distribution of random matrix elements of the Hamiltonian and brackets stand for the averaging over all eigenfunctions in a narrow window around a given point of the spectrum) is inversely proportional to a number of sites that significantly contribute to the eigenfunction normalization. In a tight-binding model the averaged value of the IPR over the whole spectrum has a simple physical meaning of the probability for a quantum particle placed initially at some site of the system to be found in the initial position after infinite time. One may wish also to introduce the whole set of generalized IPR according to the definition:

$$P_q = \sum_n |\langle n|\psi_\alpha\rangle|^{2q} \quad ; \quad q = 2, 3, \ldots \tag{1}$$

The knowledge of all these quantities can be used for the restoration of the whole probability distribution of components of eigenfunctions.

It is generally accepted nowadays that an important physical information about disordered systems of finite size can be extracted not only from average values of characteristic quantities but also from fluctuations around these average values. In particular, one can study variations of the IPR for one and the same sample by looking at different eigenfunctions corresponding to energy levels E_α in a narrow energy window around a given point E in the spectrum (level-to-level fluctuations). For this purpose it is necessary to calculate all higher moments of the IPR $m_k = \overline{\langle P_2^k\rangle}$ as well. Performing the spectral average we get the following expressions for P_q and m_k:

$$P_q = \frac{1}{\rho(E)N}\sum_1^N P_q(E,n) \quad ; \quad P_q(E,n) = \overline{\sum_\alpha |\langle n|\psi_\alpha\rangle|^{2q}\delta(E-E_\alpha)}$$
$$m_k = \frac{1}{\rho N}k! \sum_{n_1 \leq n_2 \leq \ldots \leq n_k} \Gamma(n_1,\ldots,n_k) \tag{2}$$

where

$$\Gamma(n_1,\ldots,n_k) = \overline{\sum_\alpha |\langle n_1|\psi_\alpha\rangle|^4 \ldots |\langle n_k|\psi_\alpha\rangle|^4 \delta(E-E_\alpha)} \tag{3}$$

and ρ stands for the density of states at the given point E of the spectrum.

It was realized recently that it is possible to develop a very effective analytical approach[5] to calculate averages as in eq.(2) based on a method originally devised by Efetov[7] The method operates with both commuting and anticommuting (Grassmann) auxiliary variables. Using this approach it becomes possible to map the original *stochastic* random matrix problem onto a kind of *regular* field theoretical model of interacting graded matrices (supermatrices). This model belongs to a class of $1d$ nonlinear graded σ−models. Thus, in this approach the calculation of the average of products of Green functions over the disorder amounts to performing the following supermatrix integration:

$$\overline{\langle \ ... \ \rangle} = \int \prod_{n=1}^{N} d\mu(Q_n) \mathcal{F}\{Q\} \exp\{\sum_n Str(\gamma Q_n Q_{n+1} - i\epsilon Q_n L)\} \quad (4)$$

Here Str stands for a graded trace[7], the small imaginary "frequency" $\epsilon \to 0$ is essential for the regularization and a specific form of the pre-exponent $\mathcal{F}\{Q\}$ depends on the particular object in the original stochastic model that we want to calculate. The parameters characterizing this nonlinear graded σ−model is the coupling constant γ that plays a role of an effective localization length. It is expressed in terms of the RBM parameters as follows:

$$\gamma = (\pi\rho)^2 B_2/4 \ ; \ \pi\rho = (2B_0 - E^2)^{1/2}/B_0 \ : \ B_k = \sum_{r=-\infty}^{\infty} |r|^k a(|r|)$$

The explicit parametrization of graded matrices Q_i depends on the symmetry of RBM under consideration and for the case of Hermitian matrices $H_{ij} = H^*_{ji}$ which we would like to discuss in more details these matrices belong to the graded coset space $U(1,1/2)/U(1/1) \times U(1/1)$. In order to find an explicit expression for the IPR moments m_k one should exploit a one-dimensional nearest-neighbour structure of the corresponding σ−model and to employ a transfer matrix method taking into account that the constant $\gamma \propto b^2 >> 1$ and $\epsilon \to 0$. Finally one finds the following expressions:

$$\begin{array}{l} \beta_k = \frac{k!}{(2k-2)!} x^{k-1} \frac{1}{2} \int_0^\infty dt t \int_0^x d\tau_1 \int_0^{u_1} d\tau_2 ... \int_0^{u_{k-1}} d\tau_k g^{(1)}(u_k\ ;\ t) g^{(k)}(\{\tau\}_k,\ ;\ t) \\ u_k = x - \sum_{i=1}^k \tau_i \ ; \quad g^{(p)}(\{\tau\}_p; t) \equiv g^{(p)}(\tau_p, \tau_{p-1}, ..., \tau_1; t) \\ \overline{\langle P_q \rangle} = \frac{q(q-1)}{(16\gamma)^{(q-1)}} \frac{1}{x} \int_0^\infty dt t^{2q-1} \int_0^x d\tau g^{(1)}(x-\tau;t) g^{(1)}(\tau,t) \end{array} \quad (5)$$

where the functions $g^{(p)}(\{\tau\}_p\ ;\ t)$ satisfy the reccurence equation:

$$\begin{array}{l} g^{(p)}(\{\tau\}_p;\ t) = \frac{t}{8\pi^2} \int_0^\infty d\nu_p \nu_p \sinh \pi\nu_p K_{i\nu_p}(t) e^{-\frac{1+\nu_p^2}{4}\tau_p} \int_0^\infty du u^2 K_{i\nu_p}(u) g^{(p-1)}(\{\tau\}_{p-1};\ u) \\ g^{(1)}(\tau_1\ ;\ t) = t\{K_1(t) + \frac{2}{\pi}\int_0^\infty \frac{d\nu\nu}{1+\nu^2} \sinh\{\pi\nu/2\} K_{i\nu}(t) e^{-\frac{1+\nu^2}{4}\tau_1}\} \end{array} \quad (6)$$

Here $K_{i\nu}(t)$ stands for the Macdonald functions, $x = N/4\gamma \propto N/b^2$ is the only scaling parameter characterizing the RBM ensemble and we introduced normalised moments $\beta_k \equiv m_k/P^k_{GUE}$ where $P_{GUE} = 2/N$ is a value of IPR for completely extended eigenstates typical for the Gaussian Unitary Ensemble.

Eqs.(4-5) constitute the main result of the present paper and describe in a closed form the scaling dependence of averaged moments of the eigenvector components $\overline{\langle P_q \rangle}$ and their fluctuations described by m_k on the ratio x between the matrix size N and the localization length γ. At arbitrary x, k, q these expressions should be studied numerically. However two limiting cases $x << 1$ ($x >> 1$) corresponding to delocalized (completely localized) eigenstates render themselves to detailed analytical investigation and the case $q = 2$ can be studied at any x. The main results can be summarized as follows:

1) For arbitrary value of x the IPR P_2 satisfies the following *exact* scaling relation:

$$\overline{\langle P_2 \rangle}/P_{GUE} = 1 + x/3$$

2) For small $x \ll 1$ (delocalized states):
i) the probability distribution function $\mathcal{P}_x(y)$ of the *eigenvector components* $y = N|\langle n|\psi_\alpha\rangle|^2$ is given by

$$\mathcal{P}_x(y) = e^{-y}\{1 + x(\frac{1}{3} - \frac{2}{3}y + \frac{1}{6}y^2) + o(x^2)\} \qquad (7)$$

ii) The distribution function of *inverse participation ratio* P_2 has a form of a narrow peak with the mean value $2/N$ and the relative width $\delta = \sqrt{m_2 - m_1^2}/m_1 = \frac{2}{3\sqrt{5}}x \ll 1$.

3) In the opposite case of completely localized eigenstates $x \gg 1$ the probability distribution of eigenvector components at leading order in x^{-1} is found to be equal to:

$$\mathcal{P}_x(y > xe^{-x}) = \frac{8}{x^2}\{K_1^2(2\sqrt{y/x}) + K_0^2(2\sqrt{y/x})\} \qquad (8)$$

As to the probability distribution of the inverse participation ratio P_2 it is given by the following expression:

$$\mathcal{P}^{IPR}(z = \gamma P_2) = 4\pi^2 \sum_{k=1}^{\infty}(4\pi^2 z k^4 - 3k^2)e^{-2\pi^2 k^2 z} \qquad (9)$$

It is possible to extract from this expression the asymptotic behaviour in the domain of small and large arguments:

$$\mathcal{P}^{IPR}(z \ll 1) \approx \frac{1}{(2\pi)^{1/2}} z^{-7/2} e^{-1/2z} \quad ; \quad \mathcal{P}^{IPR}(z \gg 1) \approx 16\pi^4 z e^{-2\pi^2 z} \qquad (10)$$

The most important conclusion that can be drawn from comparison of two limiting cases $x \gg 1$ and $x \ll 1$ is that at fixed disorder (i.e. at fixed localization length $\xi \propto \gamma$) a relative magnitude of fluctuations of the IPR P_2 measured by the relative width δ considerably increases from small value $\delta \propto x \ll 1$ for short samples $N \ll \gamma$ to the value of the order of one: $\delta = 1/\sqrt{5}$ for the infinitely long samples $N \gg \gamma$. The whole function $\delta(x)$ at intermediate values of $x = N/4\gamma$ can be calculated numerically by using the exact expressions in eqs.(4-5).

It is instructive to compare the strong fluctuations of the IPR in infinitely long samples with the relatively weak fluctuations of the envelope $r_n = \ln|\langle n|\psi_\alpha\rangle|$ in exponential "tails" (i.e. far from the central "bump"): the latter is shown[8] to have asymptotically the normal distribution with the relative width $\overline{(r_n - \overline{r_n})^2}^{1/2}/\overline{r_n} = (2/n)^{1/2}, n \to \infty$. This means that localized eigenfunctions are quite rigid at their "tails" but the forms of their "bumps" strongly fluctuate from one eigenfunction to another.

All results presented above were obtained for the Hermitian RBM. It is easy, however, to generalize them to the case of symmetric RBM (see 5).

Acknowledgments

The authors are grateful to A. Aronov, G. Casati, B. Chirikov, M. Feingold, S. Fishman, I. Guarnery, L. Molinari, D. Shepelyansky, U. Smilansky, H.-J. Sommers and especially to F. Izrailev for interest in their work and stimulating discussions.

The financial support from Feinberg Postdoctoral Fellowship and the program PHIT-NEDO (Y.V.F) and Alexander von Humboldt Foundation (A.D.M) is acknowledged with thanks.

REFERENCES

1. F. Izrailev, Scaling properties of localized quantum chaos, *in*: "Quantum Chaos–Quantum Measurement", Kluwer (1992);   M. Feingold, Banded random matrix ensembles, *ibid*.
2. A.D. Mirlin and Y. V. Fyodorov, Universality of level correlations in sparse random matrices, *J. Phys. A* , **24**, 2273 (1991); *Phys. Rev. Lett.*, **67**, 2049 (1991);
   S. Evangelou, A numerical study of sparse random matrices, *J. Stat. Phys*, **69**, 361 (1992).
3. O. Bohigas, Random matrices and chaotic dynamics, *in*: "Chaos and Quantum Physics", North-Holland, Amsterdam (1991).
4. G. Casati, L. Molinari and F. Izrailev, Scaling properties of band random matrices, *Phys. Rev. Lett.*, **64**, 16 (1990).
5. Y.V. Fyodorov and A. D. Mirlin, Scaling properties of localization in random banded matrices: A nonlinear σ–model approach, *Phys. Rev. Lett.*, **67**, 2405 (1991); *ibid*, **69**, 1093 (1992); *ibid*, **71**, 412 (1993); *J. Phys. A*, **26**, L551 (1993).
6. I. Lifshitz, S. Gredeskul and A. Pastur, "Introduction to the Theory of Disordered Systems", Wiley , New-York (1988).
7. K.B. Efetov, Supersymmetry and the theory of disordered metals, *Adv. Phys.*, **32**, 53 (1983).
8. Y.V. Fyodorov and A.D.Mirlin, Distribution of exponential decay rates of eigenfunctions in finite quasi 1d disordered systems, *JETP Letters*, **58**, (1993), in press.

ON THE WULFF CONSTRUCTION AS A PROBLEM OF EQUIVALENCE OF STATISTICAL ENSEMBLES

Salvador Miracle-Sole and Jean Ruiz

Centre de Physique Théorique, CNRS Luminy Case 907
F-13288 Marseille, Cedex 9, France

In this note, the statistical mechanics of SOS (solid-on-solid) 1-dimensional models under the global constraint of having a specified area between the interface and the horizontal axis, is studied. We prove the existence of the thermodynamic limits and the equivalence of the corresponding statistical ensembles. This gives a simple alternative microscopic proof of the validity of the Wulff construction for such models, first established in Ref. 1.

We consider the SOS model defined as follows: to each site i of the lattice $\mathbf{Z}$ an integer variable h_i is assigned which represents the height of the interface at this site. The energy $H_N(\mathbf{h})$ of a configuration $\mathbf{h} = \{h_0, h_1, ..., h_N\}$, in the box $0 \leq i \leq N$, of length N, is equal to the length of the corresponding interface

$$H_N(\mathbf{h}) = \sum_{i=1}^{N} (1 + |h_i - h_{i-1}|) \tag{1}$$

Its weight, at the inverse temperature β, is proportional to the Boltzmann factor $\exp[-\beta H(\mathbf{h})]$. More general Hamiltonians of the form $H = \sum P(|h_i - h_{i-1}|)$, where P is a strictly increasing function such that $P(x) \geq |x|$, when $x \to \infty$, can be treated in the same way. The proofs extend also to the case of continuous height variables.

We introduce the Gibbs ensemble which consists of all configurations, in the box of length N, with specified boundary conditions $h_0 = 0$ and $h_N = Y$. The associated partition function is given by

$$Z_1(N, Y) = \sum_{\mathbf{h}} e^{-\beta H(\mathbf{h})} \, \delta(h_0) \delta(h_N - Y) \tag{2}$$

where the sum runs over all configurations in the box and $\delta(t)$ is the dicrete Dirac delta ($\delta(t) = 1$ if $t = 0$ and $\delta(t) = 0$ otherwise). We define the corresponding free energy per site as the limit

$$\tau_p(y) = \lim_{N \to \infty} -\frac{1}{\beta N} \ln Z_1(N, yN) \tag{3}$$

where $y = -\tan\theta$, the slope of the interface, is a rational number. This free energy is called the projected surface tension. The surface tension, which represents the interfacial free energy per unit length of the mean interface, is

$$\tau(\theta) = \cos\theta\, \tau_p(-\tan\theta) \qquad (4)$$

We introduce also a second Gibbs ensemble, which is conjugate to the previous ensemble, and whose partition function, in the box of length N, is given by

$$Z_2(N, x) = \sum_{\mathbf{h}} e^{-\beta H(\mathbf{h})} e^{\beta x h_N} \delta(h_0) \qquad (5)$$

where $x \in \mathbf{N}$ replaces as a thermodynamic parameter the slope y and $h_0 = 0$. We define the associated free energy as

$$\varphi(x) = \lim_{N\to\infty} -\frac{1}{\beta N} \ln Z_2(N, x) \qquad (6)$$

Theorem 1. *Limits (3) and (6), which define the above free energies, exist. The first, τ_p, is a convex even function of y. The second, φ, is a concave even function of x. Moreover, τ_p and $-\varphi$ are conjugate convex functions, i.e., they are related by the Legendre transformations*

$$\begin{aligned} -\varphi(x) &= \sup_y [xy - \tau_p(y)] \\ \tau_p(y) &= \sup_x [xy + \varphi(x)] \end{aligned} \qquad (7)$$

Proof. The validity of the above statements is well known. See for instance Refs. 2, 3 for a proof of these results in a more general setting. □

The convexity of τ_p is equivalent to the fact that the surface tension τ satisfies a stability condition called the triangular inequality[2,3]. Relations (7) between the free energies express the thermodynamic equivalence of the two ensembles (2) and (5).

These relations imply that the curve $z = \varphi(x)$ gives, according to the Wulff construction or the equivalent the Andreev construction, the equilibrium shape of the crystal associated to our system[3].

The function $\varphi(x)$ defined by (6) is easily computed by summing a geometrical series. One introduces the difference variables

$$n_i = h_{i-1} - h_i \qquad (8)$$

for $i = 1, ..., N$, so that the partition function factorizes and one obtains

$$\varphi(x) = 1 - \beta^{-1} \ln \sum_{n\in\mathbf{Z}} e^{-\beta|n|+\beta xn} \qquad (9)$$

The explicit form of this function is

$$\varphi(x) = 1 - \beta^{-1} \ln \frac{\sinh\beta}{\cosh\beta - \cosh\beta x} \qquad (10)$$

if $-1 < x < 1$ and $\varphi(x) = -\infty$ otherwise.

We next define two new Gibbs ensembles for the system under consideration. In the first of these ensembles we consider the configurations such that $h_N = 0$, which have a specified height at the origin $h_0 = M$ and which have a specified volume V between the interface and the horizontal axis, this volume being counted negatively for negative heights:

$$V = V(\mathbf{h}) \equiv \sum_{i=0}^{N} h_i. \tag{11}$$

The corresponding partition function is

$$Z_3(N, V, M) = \sum_{\mathbf{h}} e^{-\beta H(\mathbf{h})} \, \delta(h_N) \delta(V(\mathbf{h}) - V) \delta(h_0 - M). \tag{12}$$

The second ensemble is the conjugate ensemble of the first one. Its partition function is given by

$$Z_4(N, u, \mu) = \sum_{\mathbf{h}} e^{-\beta H(\mathbf{h})} \, e^{\beta u(V(\mathbf{h})/N) + \beta \mu h_0} \, \delta(h_N), \tag{13}$$

where $u \in \mathbf{N}$ and $\mu \in \mathbf{N}$ are the conjugate variables. Our next step will be to prove the existence of the thermodynamic limit for these ensembles and their equivalence in this limit. We shall take $N = 2^n$, $n \in \mathbf{N}$. In (12), V and M must be understood as their integer part when they do not belong to $\mathbf{Z}$.

Theorem 2. *The following limits exist*

$$\psi_3(v, m) = \lim_{N \to \infty} -\frac{1}{\beta N} \ln Z_3(N, vN^2, mN) \tag{14}$$

$$\psi_4(u, \mu) = \lim_{N \to \infty} -\frac{1}{\beta N} \ln Z_4(N, u, \mu) \tag{15}$$

They define the free energies per site associated to the considered ensembles as, respectively, convex and concave functions of their variables. Moreover, ψ_3 and $-\psi_4$ are conjugate convex functions:

$$\begin{aligned} -\psi_4(u, \mu) &= \sup_{v,m} \, [uv + \mu m - \psi_3(v, m)], \\ \psi_3(v, m) &= \sup_{u,\mu} \, [uv + \mu m + \psi_4(u, \mu)]. \end{aligned} \tag{16}$$

Proof. The crucial observation is the subadditivity property given in Lemma 1 below. Then we addapt known arguments[4,5] in the theory of the thermodynamic limit. A more detailed discussion is given in the Appendix. □

Lemma 1. *The partition function Z_3 satisfies the subadditivity property*

$$Z_3(2N, 2(V' + V''), M' + M'') \geq Z_3(N, V', M') \, Z_3(N, V'', M'') \, e^{-2\beta |M''|/(2N-1)} \tag{17}$$

Proof. In order to prove this property, we associate a configuration $\mathbf{h}$ of the first system in the box of length $2N$ to a pair of configurations $\mathbf{h}'$ and $\mathbf{h}''$ of the system in a box of length N:

$$\begin{aligned} h_{2i} &= h'_i + h''_i, & i = 0, \ldots, N, \\ h_{2i-1} &= h'_{i-1} + h''_i, & i = 1, \ldots, N. \end{aligned} \tag{18}$$

Then $h_{2N} = h'_N + h''_N = 0$, $h_0 = h'_0 + h''_0 = M' + M''$ and

$$\begin{aligned} V(\mathbf{h}) &= 2 \sum_{i=1}^{N} h'_i + \sum_{i=0}^{N} h''_i + \sum_{i=1}^{N} h''_i \\ &= 2 \, [V(\mathbf{h}') + V(\mathbf{h}'')] - M''. \end{aligned}$$

This shows that the configuration **h** belongs to $Z_3(2N, 2(V' + V'') - M'', M' + M'')$. Since $H_{2N}(\mathbf{h}) = H_N(\mathbf{h}') + H_N(\mathbf{h}'')$, because $n_{2i} = n'_i$ and $n_{2i-1} = n''_i$ for $i = 1, ..., N-1$, as follows from (18), we get

$$Z_3(N, V', M') \, Z_3(N, V'', M'') \le Z_3(2N, 2(V' + V'') - M'', M' + M'')$$

Then we use the change of variables $\tilde{h}_i = h_i + [M''/(2N-1)]$ for $i = 1, ..., 2N-1$, $\tilde{h}_0 = h_0$, $\tilde{h}_{2N} = h_{2N} = 0$ which gives

$$Z_3(2N, V - M'', M) \le e^{2\beta |M''|/(2N-1)} \, Z_3(2N, V, M)$$

to conclude the proof. □

The subadditivity property and thus also Theorem 2 are satisfied when the height variables h_i are restricted to take only nonnegative values.

Theorem 3. *The functions ψ_3 and ψ_4 can be expressed in terms of the functions φ and τ_p as follows:*

$$\psi_3(v, m) = \frac{1}{u_0} \int_{\mu_0}^{\mu_0 + u_0} \tau_p(\varphi'(x)) dx, \tag{19}$$

$$\psi_4(u, \mu) = \frac{1}{u} \int_0^u \varphi(x + \mu) dx \tag{20}$$

if u_0 and μ_0 satisfy

$$\frac{1}{u_0^2} \int_0^{u_0} \varphi(x + \mu_0) dx - \frac{1}{u_0} \varphi(\mu_0 + u_0) = v, \tag{21}$$

$$\frac{1}{u_0} [\varphi(\mu_0) - \varphi(\mu_0 + u_0)] = m. \tag{22}$$

Proof. We consider again the difference variables (8) and observe that

$$V(\mathbf{h}) = \sum_{i=0}^{N} h_i = \sum_{i=1}^{N} i n_i,$$

therefore

$$Z_4(N, u, \mu) = \prod_{i=1}^{N} \left(\sum_{n_i \in \mathbb{Z}} e^{-\beta |n_i| + \beta(u/N) i n_i + \beta \mu n_i} \right).$$

Taking expression (9) into account we get that

$$Z_4(N, u, \mu) = \exp\left(-\beta \sum_{i=1}^{N} \varphi\left(\frac{u}{N} i + \mu\right)\right)$$

and

$$\psi_4(u, \mu) = \lim_{N \to \infty} \frac{1}{N} \sum_{i=1}^{N} \varphi\left(\frac{u}{N} i + \mu\right) = \lim_{N \to \infty} \frac{1}{u} \sum_{i=1}^{N} \frac{u}{N} \varphi\left(\frac{u}{N} i + \mu\right),$$

which implies expression (20).

The function ψ_3 is determined by the Legendre transform (16). The supremum over u and μ is obtained for the value u_0 and μ_0 for which the partial derivatives of the right hand side are zero: $v + (\partial \psi_4/\partial u)(u_0, \mu_0) = 0$, $m + (\partial \psi_4/\partial \mu)(u_0, \mu_0) = 0$. That is, for u_0 and μ_0 which satisfy (21) and (22).

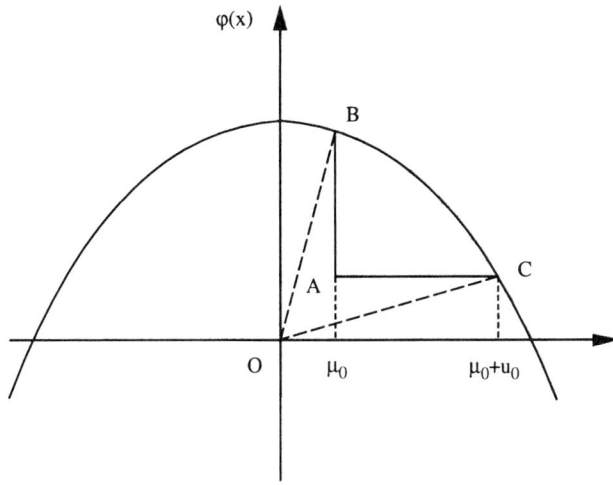

Figure 1. Graphical interpretation of Theorem 3.

Then, from (16), (20), (21) and (22), we get that

$$\psi_3(v,m) = 2\psi_4(u_0,\mu_0) - \frac{1}{u_0}[(\mu_0 + u_0)\varphi(\mu_0 + u_0) - \mu_0\varphi(\mu_0)]. \tag{23}$$

The right hand side of (23) represents twice the area of the sector OBC in Fig. 1 divided by u_0.

But, it is a known property in the Wulff construction, that twice this area is equal to the integral in (19). Indeed, by using the relation (7) in the form

$$\varphi(x) = x\varphi'(x) + \tau_p(\varphi'(x))$$

in (20) and integrating by parts $x\varphi'(x)$, we get that

$$2\psi_4(u_0,\mu_0) = \frac{1}{u_0}\int_{\mu_0}^{\mu_0+u_0} \tau_p(\varphi'(x))dx + \frac{1}{u_0}[(\mu_0 + u_0)\varphi(\mu_0 + u_0) - \mu_0\varphi(\mu_0)],$$

which together with (23) implies the expression (19). □

To interpret these relations, let us observe that the right hand side of (21) represents the area ABC, in Fig. 1, divided by $\overline{AC}^2$. Therefore, the values u_0 and μ_0, which solve (21) and (22), are obtained when this area is equal to v, with the condition coming from (22), that the slope $\overline{AB}/\overline{AC}$ is equal to m. Then, according to (19), the free energy $\psi_3(v,m)$ is equal to the integral of the surface tension along the arc BC, of the curve $z = \varphi(x)$, divided by the same scaling factor $\overline{BC} = u_0$.

We conclude that, for large N, the configurations of the SOS model, with a prescribed area vN^2, follow a well defined mean profile, the macroscopic profile given by the Wulff construction, with very small fluctuations. This follows from the fact that the probability of the configurations which deviate macroscopically from the mean profile is zero in the thermodynamic limit. The free energy associated with configurations which satisfy the above conditions, and moreover, are constrained to pass through a given point not belonging to the mean profile, can be computed with the help of Theorem 3. The corresponding probabilities decay exponentially as $N \to \infty$, as a consequence of the usual large deviations theory in statistical mechanics[6].

Acknowledgments

The authors thank Mons University, where part of this work was done, for warm hospitality and acknowledge the NATO and the ERASMUS project for financial support.

REFERENCES

1. J. De Coninck, F. Dunlop and V. Rivasseau, On the microscopic validity of the Wulff construction and of the generalized Young equation, *Commun. Math. Phys.* **121**, 401–419 (1989).
2. R. L. Dobrushin, R. Kotecký and S. B. Shlosman, "Wulff Construction: A Global Shape from Local Interactions," Am. Math. Soc., Providence, RI (1992).
3. A. Messager, S. Miracle-Solé and J. Ruiz, Convexity properties of the surface tension and equilibrium crystals, *J. Stat. Phys.* **67**, 449–470 (1992).
4. D. Ruelle, " Statistical Mechanics: Rigorous Results," Benjamin, New York Amsterdam (1969).
5. L. Galgani, L. Manzoni, and A. Scotti, Asymptotic equivalence of equilibrium ensembles of classical statistical mechanics, *J. Math. Phys.* **12**, 933–935 (1971).
6. D. Ruelle, "Hasard et Chaos" (chap. 19), Odile Jacob, Paris (1991). O. Landford, Entropy and equilibrium states in classical statistical mechanics, *in*: "Statistical Mechanics and Mathematical Problems," A. Lenard, ed., Springer, Berlin (1973).

APPENDIX

We give here a more detailed discussion of the proof of Theorem 2. We define:

$$f_n(v,m) = -\frac{1}{\beta 2^n} \ln Z_3(2^n, 2^{2n}v, 2^n m).$$

For v and m of the form $2^{-q}p$, the subadditivity property with $N = 2^n$, $V' = V'' = vN^2$ and $M' = M'' = mN$ implies that f_n is a decreasing sequence: $f_{n+1}(v,m) \leq f_n(v,m)$. Since this sequence is bounded from below, its limit exists when n tends to infinity and coincides with

$$\psi_3(v,m) = \inf_N \left[-\frac{1}{\beta N} \ln Z_3(N, vN^2, mN) \right]. \tag{24}$$

Indeed,

$$Z_3(N, V, M) \leq \sum_{\mathbf{h}} e^{-\beta H(\mathbf{h})} \delta(h_N).$$

The right hand side of the above expression is easily computed by inroducing the difference variables (8) and we get that f_n is bounded from below by $1 + (1/\beta)\ln \tanh(\beta/2)$. Let us notice that one can obtain a lower bound for $Z_3(N, vN^2, mN)$ by restricting the summation to configurations such that $(N-1)h_i = vN^2 - mN$ for $i = 1, \ldots, N-1$. This shows that f_n is bounded from above by $1 + |v - m| + |v|$.

To prove that ψ_3 is convex, we notice that the subadditivity inequality (17) with $N = 2^n$, $V' = v_1 N^2$, $V'' = v_2 N^2$ and $M' = M'' = mN^2$ gives that

$$\psi_3\left(\frac{1}{2}(v_1 + v_2), m\right) \leq \frac{1}{2}\psi_3(v_1, m) + \frac{1}{2}\psi_3(v_2, m),$$

which applied iteratively implies that

$$\psi_3\Big(\alpha v_1 + (1-\alpha)v_2, m\Big) \leq \alpha \psi_3(v_1, m) + (1-\alpha)\psi_3(v_2, m)$$

for α of the form $2^{-q}p$ and $0 \leq \alpha \leq 1$. For such α we obtain analogously that

$$\psi_3\Big(v, \alpha m_1 + (1-\alpha)m_2\Big) \leq \alpha \psi_3(v, m_1) + (1-\alpha)\psi_3(v, m_2)$$

by applying the subadditivity inequality (17) with $N = 2^n$, $M' = m_1 N^2$, $M'' = m_2 N^2$ and $V' = V'' = vN^2$. Since ψ_3 is bounded, it follows, cf. Ref. 4, that ψ_3 can be extended to a convex Lipshitz continuous function of the real variables v and m.

To prove the the existence of the limit (15) and relations (16), we introduce

$$Z_4^+(N, u, \mu) = \sup_{V, M \in \mathbf{Z}} \left[e^{\beta u(V/N) + \beta \mu M} Z_3(N, V, M) \right]$$

and proceed, as in the Appendix of Ref. 5, to study the thermodynamic limit for this quantity. We introduce the convex function

$$\psi_4^*(u, \mu) = \sup_{v, m} [uv + \mu m - \psi_3(v, m)].$$

According to (24), we have

$$e^{\beta u(V/N) + \beta \mu M} Z_3(N, V, M) \leq e^{\beta N[uv + \mu m - \psi_3(v, m)]}$$

for all $V, M \in \mathbf{Z}$, so that

$$Z_4^+(N, u, \mu) \leq e^{\beta N \psi_4^*(u, \mu)}. \tag{25}$$

On the other hand, for any $\delta > 0$ and sufficiently large N, one can find $V = vN^2$ and $M = mN$ such that

$$e^{\beta u(V/N) + \beta \mu M} Z_3(N, V, M) = e^{\beta N[uv + \mu m - \psi_3(v, m)]} e^{\beta N \psi_3(v, m)} Z_3(N, V, M)$$
$$\geq e^{\beta N[\psi_4^*(u, \mu) - \delta]}$$

and therefore

$$Z_4^+(N, u, \mu) \geq e^{\beta N[\psi_4^*(u, \mu) - \delta]}. \tag{26}$$

Inequalities (25) and (26) imply that

$$\lim_{N \to \infty} -\frac{1}{\beta N} \ln Z_4^+(N, u, \mu) = -\psi_4^*(u, \mu). \tag{27}$$

We shall now prove that the thermodynamic limit (15) exists. We follow the argument of Theorem 2 in Ref. 5. First, we notice that

$$Z_4(N, u, \mu) = \sum_{V, M \in \mathbf{Z}} e^{\beta u(V/N) + \beta \mu M} Z_3(N, V, M)$$

which implies that

$$Z_4^+(N, u, \mu) \leq Z_4(N, u, \mu). \tag{28}$$

Moreover, the inequality
$$Z_3(N,V,M) \le e^{-\beta \bar{u}(V/N) - \beta \bar{\mu} M} Z_4^+(N,\bar{u},\bar{\mu})$$
used with $\bar{u} = u'$ and $\bar{u} = u''$, $\bar{\mu} = \mu'$ and $\bar{\mu} = \mu''$ gives for any $u'' < u < u'$ and any $\mu'' < \mu < \mu'$:

$$\begin{aligned} Z_4(N,u,\mu) &\le Z^* \sum_{V \ge 0} e^{\beta(u-u')(V/N)} \sum_{M \ge 0} e^{\beta(\mu-\mu')M} \\ &+ Z^* \sum_{V \le 0} e^{\beta(u''-u)(V/N)} \sum_{M \ge 0} e^{\beta(\mu-\mu')M} \\ &+ Z^* \sum_{V \ge 0} e^{\beta(u-u')(V/N)} \sum_{M \le 0} e^{\beta(\mu''-\mu)M} \qquad (29) \\ &+ Z^* \sum_{V \le 0} e^{\beta(u''-u)(V/N)} \sum_{M \le 0} e^{\beta(\mu''-\mu)M} \\ &\le \frac{4Z^*}{(1 - e^{-\beta(\Delta u/N)})(1 - e^{-\beta(\Delta \mu)})} \end{aligned}$$

where

$$\begin{aligned} Z^* &= \sup \{Z_4^+(N,u',\mu'), Z_4^+(N,u',\mu''), Z_4^+(N,u'',\mu'), Z_4^+(N,u'',\mu'')\} \\ \Delta u &= \min\{u' - u, u - u''\} > 0, \\ \Delta u &= \min\{\mu' - \mu, \mu - \mu''\} > 0. \end{aligned}$$

Eq. (27) and the continuity of ψ_4^* together with inequalities (28) and (29) imply (15) and (16).

SCALING PROFILES OF A SPREADING DROP FROM LANGEVIN OR MONTE-CARLO DYNAMICS

F. Dunlop and M. Plapp

Centre de Physique Théorique *(CNRS – UPR14)*
Ecole Polytechnique
91128 Palaiseau, France

INTRODUCTION

When a drop spreads on a plane solid surface under the influence of surface tensions, one has a reasonable macroscopic understanding of the shape of the bulk of the drop, and a good understanding of the contact angle, which quickly approaches the equilibrium angle θ satisfying Young's equation

$$\cos\theta\ \sigma_{AB}(\theta) - \sin\theta\ \sigma'_{AB}(\theta) = \sigma_{AW} - \sigma_{BW}, \tag{1}$$

where σ_{AB}, σ_{AW} and σ_{BW} are the appropriate surface tensions, and the prime denotes a derivative with respect to the angle. What has been less studied, and is more related to non-equilibrium, is the shape of the foot of the drop, where the interface bends to match the contact angle. There the scale is intermediate between molecular and macroscopic, which justifies a statistical mechanical study.

Such a study was initiated in ref. [1] for a two-dimensional model, where the one-dimensional moving interface was represented by a continuous Solid-On-Solid model with Langevin dynamics. An exact solution was given for a Gaussian Hamiltonian, and a solution derived from a local equilibrium assumption for a generic Hamiltonian. Both solutions were in terms of a deterministic equation for scaled variables, which is remininscent of a hydrodynamic limit. Hydrodynamic limits for interfaces have been the object of much interest recently [2, 3, 4], with the common feature that such limits are in terms of motion by curvature, whenever the phases are at coexistence and the respective volumes of the phases are not conserved.

In the next section, we review those results of ref. [1], which deal with the scaling limit as $1 \ll t \ll L^2$. We then use Spohn's study of the mobility to show, by taking anisotropy into account that these results are indeed motion by curvature. The following section is devoted to Monte-Carlo dynamics for the Restricted-Solid-On-Solid model, as applied to the study of spreading in ref. [5], which we also relate here to motion by curvature.

LANGEVIN DYNAMICS OF CONTINUOUS SOLID-ON-SOLID MODELS

Variables $h_0, h_1, ... h_L$ represent the spreading lengths of layers $0, 1, ... L$ on top of the substrate. The Hamiltonian is taken as

$$H(h_0, h_1, ... h_L) = \sum_{i=1}^{L} U(h_{i-1} - h_i) - (\sigma_{AW} - \sigma_{BW}) h_0 , \qquad (2)$$

where $U(x)$ is an even function increasing at least linearly when $x \to +\infty$. The corresponding Langevin dynamics is given by

$$dh_i = -\frac{\partial H}{\partial h_i} dt + (2\beta^{-1})^{1/2} d\omega_i \qquad i = 0, 1, ..., L-1 , \qquad (3)$$

where $\omega_i(t)$ are independent Brownian motions, with a boundary condition $h_L(t) = 0\ \forall t$, and initial conditions $h_i(0) = 0\ \forall i$. As $t \to \infty$, expectation values over the Brownians converge to expectation values computed with the Gibbs measure

$$\exp\left(-\beta H(h_0, h_1, ... h_L)\right) \prod_0^{L-1} dh_i . \qquad (4)$$

The corresponding mean profile is a straight line making an angle θ_e satisfying Young's equation (1). In the Gaussian case, $U(x) = Jx^2$, the Langevin equation (3) is linear and can be solved explicitly at all times, showing relaxation to equilibrium in time $t \sim L^2$. Studying the scaling of the foot of the drop means studying the profile at times $1 \ll t \ll L^2$. There one finds, denoting by E the expectation over the Brownians,

$$\lim_{t \to \infty} \lim_{L \to \infty} t^{-1/2} E h_{[yt^{1/2}]}(t) = (\sigma_{AW} - \sigma_{BW}) \int_0^\infty \frac{\cos yx}{x^2} \left(1 - \exp(-x^2)\right) dx \equiv \psi(y) , \qquad (5)$$

so that $\psi(y)$ obeys

$$\psi(y) - y\psi'(y) = 2J\psi''(y) . \qquad (6)$$

For a general $U(x)$, the Langevin equation (3) averaged over the Brownians reads

$$\frac{d}{dt} E h_i(t) = E\beta U'(h_{i-1} - h_i) - E\beta U'(h_i - h_{i+1}) . \qquad (7)$$

At this point, the following ansatz is introduced: $EU'(h_{i-1} - h_i)$ is taken as a function of $E(h_{i-1} - h_i)$ up to an error $O(t^{-1})$, and this function is the same that gives the Gibbs average $\langle U'(h_{i-1} - h_i) \rangle$ as a function of $\langle h_{i-1} - h_i \rangle$ in the measure (4) with boundary conditions such that $\langle h_{i-1} - h_i \rangle$ is prescribed.

This ansatz is a weak but precise form of the following conjecture: as $t \to \infty$, a local equilibrium is approached; this local equilibrium can be thought of as a measure for local height differences; the conjecture is that this measure is nothing but the Gibbs measure corresponding to the average local slope.

The ansatz is convenient for computing. Indeed let $c(\tau)$ be the function defined by

$$\int_{-\infty}^{+\infty} dx (\tau - x) \exp\left(-\beta U(x) + \beta c(\tau) x\right) = 0 . \qquad (8)$$

Then applying the ansatz gives

$$\frac{d}{dt} E h_i(t) = c(E(h_{i-1} - h_i)) - c(E(h_i - h_{i+1})) + O(t^{-1}) . \qquad (9)$$

The ansatz and eq.(9) do not depend in an essential way upon boundary conditions. A scaling assumption can also be made, depending upon boundary conditions, like the existence of

$$\psi(y) \equiv \lim_{t \to \infty} \lim_{L \to \infty} t^{-1/2} E h_{[yt^{1/2}]}(t) . \tag{10}$$

Then $\psi(y)$ is found to obey

$$\psi(y) - y\psi'(y) = c'(\psi'(y))\psi''(y) . \tag{11}$$

In the next section, we show that eq.(9) corresponds to motion by curvature discretised in space, and eq.(11) to a scaling solution of motion by curvature, in both cases taking anisotropy into account. This gives a strong support, of course not a proof, to the conjecture.

MOTION BY CURVATURE

Starting point for this study is the Lifshitz equation which describes the time evolution of an interface in a system with a second-order phase transition. The normal component of velocity is given in terms of the mean curvature $K = R^{-1}$ by $v_n = \lambda \cdot K$ where λ is a constant proportional to the surface tension. This equation, stated for the isotropic case, can be generalized to take care of the anisotropy of σ: Let the (one-dimensional) interface be represented by a curve $C(l)$ parametrized by arc length. Then the curvature $K(l)$ and the angle $\theta(l)$ between the tangent at C and the x-axis can be expressed as functions of l, and we have $K(l) = d\theta(l)/dl$. We start from the formula for the surface free energy

$$F = \int_C \sigma(\theta(l)) dl$$

and compute the variation of F under a small deformation $\delta \mathbf{r}(l)$:

$$\delta F = \int_C (\sigma + \sigma'') \cdot K \cdot dl \cdot \delta \mathbf{r} \cdot \hat{\mathbf{n}} , \tag{12}$$

where $\hat{\mathbf{n}}(l)$ is the unitary vector normal to the interface. The linear response argument which is the basis of the Lifshitz equation then gives

$$v_n = -\mu \cdot (\sigma + \sigma'') \cdot K , \tag{13}$$

where the coefficient $\mu(\theta)$ called the mobility, gives a time scale. As boundary conditions, we can have fixed points or fixed contact angles with walls.

Returning to the problem of the preceding section, and looking for solutions of eq. (13) scaling as $h(x,t) = t^{1/2} \cdot \psi(x \cdot t^{-1/2})$ we get

$$\psi(y) - y\psi'(y) = \mu \cdot (\sigma + \sigma'') \cdot (\cos \theta)^2 \cdot \psi'' . \tag{14}$$

Next, we remark that the stiffness $(\sigma + \sigma'')$ gives the fluctuations at equilibrium, which can be computed using eq. (8):

$$(\sigma + \sigma'') \cdot (\cos \theta)^3 = \frac{\int dx \exp\left(-\beta U(x) + \beta c(\tan \theta)x\right)}{\beta \int dx (\tan \theta - x)^2 \exp\left(-\beta U(x) + \beta c(\tan \theta)x\right)} = c'(\tan \theta) . \tag{15}$$

Finally, using[3] $\mu = \cos\theta$ in Langevin dynamics, eqs. (11) and (14) are found to be exactly the same. Similarly, we can rewrite eq. (13) using the identities $v_n = \cos\theta \cdot \dot{h}$, $\mu = \cos\theta$, $K = h'' \cdot (\cos\theta)^3$ and (15) as

$$\dot{h} = c'(\tan\theta) \cdot h'' . \qquad (16)$$

But this, in the discrete case, is just the limit of eq. (9) for smooth interfaces. Hence, we have shown that generalized motion by curvature and continuous Langevin dynamics with a local equilibrium assumption give the same results for smooth solutions, and have the same scaling limit when boundary conditions permit.

We can furthermore use motion by curvature to achieve a better understanding of interface motion at a macroscopic level. Suppose that we stick to a point of an interface, following motion by curvature, our point always moving in the direction of the local normal. We can develop an evolution equation for the curvature, similar to advective derivation in hydrodynamics. We find for the isotropic case, using the Lifshitz equation,

$$\frac{d}{dt}K = \lambda\left(K^3 + \frac{\partial^2}{\partial l^2}K\right) , \qquad (17)$$

a nonlinear diffusion equation. This accounts very well for the different features of motion by curvature. The first term causes interface regions of constant curvature to shrink, whereas the second term describes a diffusion of curvature towards flat interface regions, as observed for interfaces near a wall or in a corner. For the anisotropic case the equation is slightly more complicated, but the structure is the same.

MONTE-CARLO DYNAMICS OF RESTRICTED-SOLID-ON-SOLID MODELS

We now return to the usual Solid-On-Solid model, where the variables $h_0...h_i...h_l$ represent the height of the spreading phase on top of the plane solid surface, here a straight solid line in two-dimensional space. The allowed configurations are $\mathbf{h} = \{h_i\}_{i=0}^{\infty} \in Z^Z$ such that $\forall i \geq 0$, $h_i \geq 0$ and $h_{i+1} - h_i \in \{-1, 0, 1\}$, and

$$\forall i \geq 0, \quad h_i = 0 \Longrightarrow h_{i+1} = 0 \quad \text{and} \quad \max\{i | h_i > 0\} = l < \infty$$

The last condition means taking the foot of a single drop. The Hamiltonian is taken as

$$H(\mathbf{h}) = a_0 h_0 + \sum_{i=0}^{\infty}\Big(J|h_{i+1} - h_i| + (J - \sigma_{AW} + \sigma_{BW})(1 - \delta_{h_i,0})\Big) . \qquad (18)$$

Letting $\mathbf{h} = (\mathbf{h}^o, \mathbf{h}^e)$ with $\mathbf{h}^o = \{h_{2j+1}\}_{j=0}^{\infty}$ and $\mathbf{h}^e = \{h_{2j}\}_{j=0}^{\infty}$, the discrete time dynamics is defined by the transition probabilities

$$Pr((\mathbf{h}^o, \mathbf{h}^e) \to (\mathbf{h}'^o, \mathbf{h}^e)) = \frac{1}{2}\frac{e^{-\beta H(\mathbf{h}'^o, \mathbf{h}^e)}}{\sum_{\mathbf{h}''^o} e^{-\beta H(\mathbf{h}''^o, \mathbf{h}^e)}},$$

$$Pr((\mathbf{h}^o, \mathbf{h}^e) \to (\mathbf{h}^o, \mathbf{h}'^e)) = \frac{1}{2}\frac{e^{-\beta H(\mathbf{h}^o, \mathbf{h}'^e)}}{\sum_{\mathbf{h}''^e} e^{-\beta H(\mathbf{h}^o, \mathbf{h}''^e)}}, \qquad (19)$$

where $\mathbf{h}'^o \neq \mathbf{h}^o$ and $\mathbf{h}'^e \neq \mathbf{h}^e$, and where the factors $1/2$ correspond to the probability of either the odd sites, or the even sites being updated. Scaling profiles of the foot of a spreading drop have been obtained numerically within this model[5]: the initial condition was a straight line $h_i \simeq h_0 - i \tan \theta_0$ making an angle θ_0 larger than the equilibrium angle θ_e obeying eq.(1). At times $1 \ll t \ll h_0^2$, a scaling limit was obtained as

$$t^{-1/2}\left(h_{l_0+xt^{1/2}}(t) - h_{l_0+xt^{1/2}}(0)\right) \longrightarrow \Phi(x) \qquad \text{as} \qquad t \to \infty. \tag{20}$$

This scaling limit looks like motion by curvature, with boundary conditions such that curvature diffuses away from the original angle at l_0. In order to make a quantitative comparison, we estimate the mobility from the linear response: consider

$$H_\epsilon(\mathbf{h}) = \sum_{i=0}^{L-1} \left(J|h_{i+1} - h_i| + c(\tan\theta)(h_{i+1} - h_i)\right) + \epsilon\chi(t \geq 0)\sum_{i=0}^{L} h_i, \tag{21}$$

and run the associated dynamics from $t = -\infty$ up to $t = 0$ with the boundary condition $h_0 = 0$, and then with free boundary condition for $t > 0$. Then

$$\mu(\theta) = \cos\theta \lim_{L \to \infty} \frac{1}{L} \sum_{i=0}^{L-1} \lim_{\epsilon \to 0} \epsilon^{-1} \lim_{t \to +\infty} \left(Eh_i(t+1) - Eh_i(t)\right). \tag{22}$$

An approximation to the mobility is obtained by taking (22) at $t = 0$, where the formula can be evaluated exactly from the Gibbs measure, giving

$$\mu(\theta) \simeq \cos\theta \; \frac{2e^{-2\beta J} + \frac{1}{2}e^{-\beta J}\left(e^{\beta c(\tan\theta)} + e^{-\beta c(\tan\theta)}\right)}{\left(1 + e^{-\beta J}\left(e^{\beta c(\tan\theta)} + e^{-\beta c(\tan\theta)}\right)\right)^2}, \tag{23}$$

with $c(.)$ defined by

$$\tau = \frac{e^{-\beta J}\left(e^{\beta c(\tau)} - e^{-\beta c(\tau)}\right)}{1 + e^{-\beta J}\left(e^{\beta c(\tau)} + e^{-\beta c(\tau)}\right)}. \tag{24}$$

This value of the mobility gives at least the right leading order at low temperatures. We have integrated numerically the corresponding equation for motion by curvature,

$$\dot{h} = \frac{\mu(\theta)}{\cos\theta} c'(\tan\theta) h'', \tag{25}$$

and also the Monte-Carlo dynamics associated to (18), with various boundary and initial conditions, and found a good agreement even for long times. As a notable feature, the dynamics of wetting in a right-angle corner, with contact angle θ_1 on one boundary and θ_2 on the other, shows spreading if $\theta_1 + \theta_2 < \pi/2$ and shrinking if $\theta_1 + \theta_2 > \pi/2$.

REFERENCES

1. D.B. Abraham, P. Collet, J. De Coninck and F. Dunlop, Langevin dynamics of spreading and wetting, *Phys. Rev. Lett.* **65**, 195 (1990); Langevin dynamics of an interface near a wall, *J. Stat. Phys.* **61**, 509 (1990).
2. L.C. Evans, H.M. Soner and P.E. Souganidis, Phase transitions and generalized motion by mean curvature, *Comm. Pure and Appl. Math.* **XLV**, 1097 (1992).

3. H. Spohn, Interface motion in models with stochastic dynamics, *J. Stat. Phys.* **71**, 1081 (1993).
4. A. De Masi, E. Orlandi, E. Presutti and L. Triolo, Motion by curvature by scaling non local evolution equations, *J. Stat. Phys.* (1993).
5. J. De Coninck, F. Dunlop and F. Menu, Spreading of a Solid-On-Solid drop, *Phys. Rev. E* **47**, 1820 (1993).

RIGOROUS CALCULATION OF COLLECTIVE EXCITATIONS IN A MEAN FIELD MODEL

B. Momont

Instituut voor Theoretische Fysica, K.U. Leuven
Celestijnenlaan 200D, B-3001 Leuven

INTRODUCTION

In this contribution a new way to study collective excitations is presented. In order for the calculations to be tractable, the approach is applied to a solvable model: the mean field Overhauser model. The dynamics of this model will be studied on the level of macroscopic fluctuations of densities. In the third section it will be made clear what is meant by: dynamics on the level of macroscopic fluctuations.

The approach is based on the theory of macroscopic fluctuations and their dynamics as it was introduced by Goderis, Verbeure and Vets[1,2]. The scheme that is put forward here can be used to tackle other problems involving collective phenomena. In the literature collective phenomena are usually studied in the random phase approximation[3,4,5] and even then one does not find a closed expression for the excitation spectrum. Here a rigorous calculation of the dispersion relation (i.e. the excitation spectrum as a function of momentum) for spin density waves in the Overhauser model is presented. A more complete exposition, including proofs and technical details is published elsewhere[6].

THE MICROSCOPIC LEVEL

In this section some earlier results[7] are briefly summarized. The Overhauser model describes a system of interacting electrons. The algebra of observables $\mathcal{A}$ is generated by the creation and annihilation operators $c_i^+(f)$, $c_i^-(f)$; $f \in L^2(\mathbb{R})$, where $i\,(=1,2)$ stands for the two possible spin states of the spin 1/2 Fermions (electrons). These operators satisfy the canonical anti-commutation relations (CAR)

$$\{c_i^-(f), c_j^+(g)\} = \langle f, g \rangle \delta_{ij}, \quad \{c_i^-(f), c_j^-(g)\} = 0, \quad (c_i^-(f))^* = c_i^+(f),$$

with $\langle f, g \rangle$ the $L^2(\mathbb{R})$ scalar product $\int_\mathbb{R} dx \,\overline{f(x)} g(x)$. The formal $c_i^\#(x)$ $(\# = \pm)$ are determined by $c_i^+(f) = \int_\mathbb{R} f(x)\, c_i^+(x)\, dx$. The Overhauser Hamiltonian for a finite volume Λ is taken to be

$$H_\Lambda = \sum_{i=1}^{2} \frac{1}{2} \int_\Lambda dx \, \nabla c_i^+(x) \nabla c_i^-(x) - \frac{c}{\Lambda} \int_\Lambda dx \, e^{iqx} c_1^+(x) c_2^-(x) \int_\Lambda dy \, e^{-iqy} c_2^+(y) c_1^-(y).$$

The group $G_0 = \{\sigma_a | a \in \mathbb{R}\}$ of automorphisms of $\mathcal{A}$ is defined by $\sigma_a(c_1^-(f)) = c_1^-(f_a)e^{-iqa/2}$ and $\sigma_a(c_2^-(f)) = c_2^-(f_a)e^{+iqa/2}$, where $f_a(x) = f(x-a)$.

Theorem 1 *For the infinite system, all β-equilibrium states ($0 \le \beta \le \infty$) with bounded densities of particle number, kinetic energy and momentum are G_0-invariant, i.e. $\omega \circ \sigma_a = \omega$; $\forall a \in \mathbb{R}$.*

From now on take a fixed, extremal G_0-invariant equilibrium state ω with order parameter $b = \omega(c_1^+(0)c_2^-(0)) \ne 0$.

The two-component version of the CAR-algebra $\mathcal{A}$ is introduced as follows; for $f = (f_1, f_2) \in L \equiv L^2(\mathbb{R}) \oplus L^2(\mathbb{R})$ define the creation operator $c^+(f)$ by

$$c^+(f) = (c_1^+(f_1), c_2^+(f_2)).$$

The time derivative is determined by $\omega - \text{weak} - \lim_{\Lambda \to \mathbb{R}}[H_\Lambda, c^+(f)] \equiv \delta_\omega(c^+(f)) = c^+(h_\omega f)$ where h_ω is given by

$$(h_\omega f)(x) = \begin{pmatrix} -\Delta/2 & -c\bar{b}e^{iqx} \\ -cbe^{-iqx} & -\Delta/2 \end{pmatrix} \begin{pmatrix} f_1(x) \\ f_2(x) \end{pmatrix}$$

with Δ the Laplacian.

Define for all $k \in \mathbb{R}$ the functions $e_\pm^k$ on $\mathbb{R}$ by

$$e_\pm^k(x) = \frac{1}{\sqrt{2\pi}} \frac{1}{\sqrt{1 + |\mu_\pm(k)|^2}} \begin{pmatrix} e^{i(k+q/2)x} \\ \mu_\pm(k) e^{i(k-q/2)x} \end{pmatrix}$$

where $\mu_\pm(k) = \frac{1}{2c\bar{b}}(kq \mp \sqrt{k^2 q^2 + 4c^2|b|^2})$. Note that $e_\pm^k \notin L$. One can check that $h_\omega e_\pm^k = \epsilon_\pm(k) e_\pm^k$ with the quasi particle spectrum given by

$$\epsilon_\pm(k) = \frac{1}{2}(k^2 + q^2/4 \pm \sqrt{k^2 q^2 + 4c^2|b|^2}).$$

Theorem 2 *The extremal G_0-invariant β-equilibrium states of bounded densities of particle number, kinetic energy and momentum are the quasi-free states ω on $\mathcal{A}$, determined by the two-point function $\omega(c^+(f)c^-(g)) = \langle g, \frac{1}{e^{\beta h_\omega}+1} f \rangle$ where the value of b in h_ω is determined as a solution of the gap equation*

$$0 = b\left(1 - \frac{c}{2\pi} \int dk \frac{1}{2(\epsilon_+(k) - \epsilon_-(k))} \frac{\sinh[\beta(\epsilon_+(k) - \epsilon_-(k))/2]}{\cosh[\beta\epsilon_+(k)/2]\cosh[\beta\epsilon_-(k)/2]}\right).$$

THE MACROSCOPIC LEVEL (FLUCTUATIONS)

Let $\mathcal{V}^2$ be the smallest complex vectorspace generated by the densities $c^+(f)c^-(g)$. $\mathcal{V}^2$ thus contains all finite sums $\sum_{i \in I} c^+(f^i)c^-(g^i)$. Define the local fluctuations of elements of $\mathcal{V}^2$ as[2]

$$\widetilde{\left(\sum_{i \in I} c^+(f^i) c^-(g^i)\right)}_\Lambda = \frac{1}{\sqrt{\Lambda}} \int_\Lambda dx \left[\sigma_x\left(\sum_{i \in I} c^+(f^i) c^-(g^i)\right) - \omega\left(\sum_{i \in I} c^+(f^i) c^-(g^i)\right)\right].$$

What we are ultimately interested in, is the dynamics of the density fluctuations in the state ω, as $\Lambda \to \infty$. In the following lemma $\mathcal{V}^2$ is equipped with a non negative sesquilinear form $\langle \cdot, \cdot \rangle_\omega$.

Lemma 1 *The definition*

$$\langle \sum_{i \in I} c^+(f^i) c^-(g^i), \sum_{j \in J} c^+(f^j) c^-(g^j) \rangle_\omega \equiv$$

$$\lim_{\Lambda \to \mathbb{R}} \omega \left[\left(\sum_{i \in I} \widetilde{c^+(f^i) c^-(g^i)} \right)_\Lambda^* \left(\sum_{j \in J} \widetilde{c^+(f^j) c^-(g^j)} \right)_\Lambda \right]$$

determines a non negative sesquilinear form on $\mathcal{V}^2$.

The kernel $\mathcal{V}_0^2 = \{ A \in \mathcal{V}^2 | \langle A, A \rangle_\omega = 0 \}$ of $\langle \cdot, \cdot \rangle_\omega$ is non-trivial, therefore consider the quotient space $\mathcal{V}^2/\mathcal{V}_0^2$. The induced scalar product on this quotient space will also be denoted by $\langle \cdot, \cdot \rangle_\omega$. Let $A, B \in \mathcal{V}^2$ and let $[A], [B]$ be the corresponding elements in the quotient space $\mathcal{V}^2/\mathcal{V}_0^2$, then

$$[A] = [B] \iff \langle A - B, A - B \rangle_\omega = \lim_{\Lambda \to \mathbb{R}} \omega[\widetilde{(A-B)}_\Lambda^* \widetilde{(A-B)}_\Lambda] = 0.$$

Therefore one can say that the observables $A, B \in \mathcal{V}^2$ that correspond to the same element $[A] = [B] \in \mathcal{V}^2/\mathcal{V}_0^2$ have "the same fluctuations" in the thermodynamic limit.

The closure of the quotient space $\mathcal{V}^2/\mathcal{V}_0^2$ will be called the Hilbert space of macroscopic density fluctuations. No new notation is introduced for the extension of the scalar product to $\left(\overline{\mathcal{V}^2/\mathcal{V}_0^2}, \langle \cdot, \cdot \rangle_\omega \right)$.

When $\mathcal{V}_0^2$ is explicitly calculated the following lemma results.

Lemma 2 *The elements of* $\mathcal{V}_0^2$ *are determined by*

$$\sum_{i \in I} c^+(f^i) c^-(g^i) \in \mathcal{V}_0^2$$
$$\iff$$
$$\sum_{i \in I} \hat{f}^i_{1,-q/2} \overline{\hat{g}^i_{1,-q/2}} = 0$$
$$\sum_{i \in I} \hat{f}^i_{1,-q/2} \overline{\hat{g}^i_{2,+q/2}} = 0$$
$$\sum_{i \in I} \hat{f}^i_{2,+q/2} \overline{\hat{g}^i_{1,-q/2}} = 0$$
$$\sum_{i \in I} \hat{f}^i_{2,+q/2} \overline{\hat{g}^i_{2,+q/2}} = 0$$

with $\hat{f}^i_{1,-q/2}(k) = \hat{f}^i_1(k + q/2), \ldots$

Let $(\mathcal{L}, \langle \cdot, \cdot \rangle_\omega)$ be the vectorspace $\oplus_{j=1}^4 \mathcal{U}$, with $\mathcal{U} = \{ P | P = \sum_{i \in I} \hat{f}^i \hat{g}^i \; ; \; f^i, g^i \in L^2(\mathbb{R}) \}$. With the scalar product defined appropiately[6] lemma 2 proves that each element $[\sum_{i \in I} c^+(f^i) c^-(g^i)] \in \mathcal{V}^2/\mathcal{V}_0^2$ can be represented by

$$\sum_{i \in I} \begin{pmatrix} \hat{f}^i_{1,-q/2} \overline{\hat{g}^i_{1,-q/2}} \\ \hat{f}^i_{1,-q/2} \overline{\hat{g}^i_{2,+q/2}} \\ \hat{f}^i_{2,+q/2} \overline{\hat{g}^i_{1,-q/2}} \\ \hat{f}^i_{2,+q/2} \overline{\hat{g}^i_{2,+q/2}} \end{pmatrix} \in \mathcal{L}$$

This mapping determines an isometric isomorphism between the vectorspaces $(\mathcal{V}^2/\mathcal{V}_0^2, \langle \cdot, \cdot \rangle_\omega)$ and $(\mathcal{L}, \langle \cdot, \cdot \rangle_\omega)$, and by extension also between the Hilbertspaces $\left(\overline{\mathcal{V}^2/\mathcal{V}_0^2}, \langle \cdot, \cdot \rangle_\omega \right)$ and $(\overline{\mathcal{L}}, \langle \cdot, \cdot \rangle_\omega)$.

The Overhauser dynamics induces a macro-dynamics on the space $\left(\overline{\mathcal{V}^2}/\mathcal{V}_0^2, \langle \cdot, \cdot \rangle_\omega\right)$. This macro-dynamics will be studied on the equivalent space $(\overline{\mathcal{L}}, \langle \cdot, \cdot \rangle_\omega)$. The spectral decomposition of the time derivative will be determined.

The Overhauser dynamics is determined by the time derivative δ_ω. The operator δ_ω on $\overline{\mathcal{V}^2}$ is given by $\delta_\omega \left(\sum_{i \in I} c^+(f^i)c^-(g^i) \right) = \sum_{i \in I} \left(c^+(h_\omega f^i)c^-(g^i) - c^+(f^i)c^-(h_\omega g^i) \right)$. The induced macro-dynamics $\tilde{\delta}_\omega$ on $\overline{\mathcal{V}^2}/\mathcal{V}_0^2$ is defined by² $\tilde{\delta}_\omega \left(\left[\sum_{i \in I} c^+(f^i)c^-(g^i) \right] \right) = \left[\delta_\omega \left(\sum_{i \in I} c^+(f^i)c^-(g^i) \right) \right]$. For $X \in \mathcal{L}$ this becomes

$$\left(\tilde{\delta}_\omega(X) \right)(k) = \begin{pmatrix} 0 & c\bar{b} & -c\bar{b} & 0 \\ \bar{c}b & kq & 0 & -c\bar{b} \\ -\bar{c}b & 0 & -kq & c\bar{b} \\ 0 & -\bar{c}b & \bar{c}b & 0 \end{pmatrix} X(k) \equiv D(k)X(k)$$

One sees that the macrodynamics $\tilde{\delta}_\omega$ is a " matrix multiplication operator " on the space $\mathcal{L}$. For a fixed $k \in \mathbb{R}$, the matrix $D(k)$ has the eigenvalues $\pm \lambda(k) = \pm \sqrt{k^2 q^2 + 4c^2 |b|^2}$ and the degenerate eigenvalue 0. Decomposing $\mathcal{L}$ according to the eigenvalues and with the weights $\rho_+, \rho_-, \rho_0^1, \rho_0^2$ appropriately defined⁶, $\overline{\mathcal{L}}$ can be identified with the following direct sum space

$$\overline{\mathcal{L}} = L^2(\mathbb{R}, \rho_+) \oplus L^2(\mathbb{R}, \rho_-) \oplus L^2(\mathbb{R}, \rho_0^1) \oplus L^2(\mathbb{R}, \rho_0^2)$$

Choosing orthonormal bases of the spaces $L^2(\mathbb{R}, \rho_\#)$, it becomes easy to prove the following

Theorem 3 *If $q \neq 0$ the spectrum of $\tilde{\delta}_\omega$ is given by*

$$\sigma(\tilde{\delta}_\omega) = \sigma_{disc}(\tilde{\delta}_\omega) \cup \sigma_{cont}(\tilde{\delta}_\omega) = \{0\} \cup (Ran(+\lambda) \cup Ran(-\lambda)).$$

The spectral decomposition of $\tilde{\delta}_\omega$ is

$$\tilde{\delta}_\omega = \int_{\mu \in Ran(+\lambda)} \mu \, dE_\mu^+ + \int_{\mu \in Ran(-\lambda)} \mu \, dE_\mu^-.$$

Where the projections $E_\mu^\pm$ are defined by

$$(E_\mu^\pm X)(k) = (P_\pm X)(k) \quad \text{if } k \in [-\sqrt{\mu^2 - 4c^2 |b|^2}/|q|, \sqrt{\mu^2 - 4c^2 |b|^2}/|q|]$$
$$= 0 \qquad \text{otherwise.}$$

The spectrum of $\tilde{\delta}_\omega$ is thus given by $]-\infty, -2c|b|] \cup \{0\} \cup [2c|b|, +\infty[$.

DISCUSSION

For β large enough, i.e. low enough temperatures there exist β-KMS-states of the Overhauser model that break the translation symmetry. These states are still G_0-invariant though (Theorem 2), i.e. they exhibit a $2\pi/q$-periodicity. The order parameter associated with the symmetry breaking is $\omega \left(c_1^+(0) c_2^-(0) \right) \equiv b \neq 0$. This can be interpreted as a pairing of electrons with opposite spin or a non trivial transition from one spin state to the other.

The quasi particles display a continuous spectrum that is given by

$$\epsilon_\pm(k) = \frac{1}{2}(k^2 + q^2/4 \pm \sqrt{k^2 q^2 + 4c^2 b^2})$$

and they obey Fermi-Dirac statistics. Fixing the particle density and the temperature T determines the chemical potential $\mu(T)$, which at zero temperature is called the Fermi energy $\mu(T=0) = \epsilon_F \equiv \epsilon_+(k_F^+) \equiv \epsilon_-(k_F^-)$. Working in the grand canonical ensemble, one could equivalently say that the particle density is determined by fixing the chemical potential μ and the temperature T.

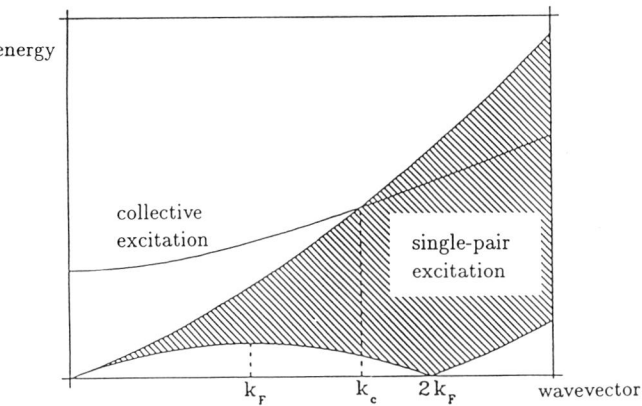

Figure 1. Excitation spectrum of the Overhauser model.

The single-pair excitations[5] can be of two types

$$\epsilon_+(p) \to \epsilon_+(p+k) \quad \text{with } |p| \leq k_F^+ \text{ and } |p+k| \geq k_F^+$$
$$\epsilon_-(p) \to \epsilon_-(p+k) \quad \text{with } |p| \leq k_F^- \text{ and } |p+k| \geq k_F^-$$

Where the restrictions are because of the Pauli principle. The single-pair excitation energies are given by

$$\epsilon_\pm(p+k) - \epsilon_\pm(p) = \frac{1}{2}\left(k^2 + 2kp \pm (\sqrt{(p+k)^2 q^2 + 4c^2|b|^2} - \sqrt{p^2 q^2 + 4c^2|b|^2})\right)$$

The single-pair excitations $\epsilon_+(p) \to \epsilon_+(p+k)$ are indicated by the shaded region in the figure.

The collective modes associated with the symmetry breaking[8] are called spin density waves. The main result of the article is the exact calculation of the spectrum of the collective modes

$$\lambda(k) = \sqrt{k^2 q^2 + 4c^2|b|^2}.$$

These collective modes are also indicated in the figure. This spectrum will always intersect the shaded region at a critical wave vector k_c, since for large k one has that $\lambda(k) \sim k$, while the single-pair excitations are of the order k^2. This can be interpreted[3,5] as the possibility for the collective excitations with wave vector $k > k_c$ to decay into a single-pair excitation.

Acknowledgments

The author would like to acknowledge A. Verbeure and V.A. Zagrebnov for pointing out the subject that was studied in this contribution.

REFERENCES

1. D. Goderis, P. Vets, Central limit theorem for mixing quantum systems and the CCR-algebra of fluctuations, *Comm. Math. Phys.* **122**, 249 (1989).
2. D. Goderis, A. Verbeure, P. Vets, Dynamics of fluctuations for quantum lattice systems, *Comm. Math. Phys.* **128**, 533 (1990).
3. P.A. Martin, F. Rothen, Problèmes a N-corps et Champs Quantiques, Presses Polytechniques et Universitaires Romandes, 1990, Chap. 4.
4. N.H. March, M. Parrinello, Collective Effects in Solids and Liquids, Adam Hilger Ltd, 1982, Chap.2.
5. D. Pines, P. Noziéres, The Theory of Quantum Liquids, W.A. Benjamin, inc., 1966.
6. B. Momont, Rigorous calculation of the collective excitation spectrum in a mean field model, to appear in *J. Math. Phys.*
7. M. Broidioi, B. Nachtergaele, A. Verbeure, The Overhauser model: equilibrium fluctuation dynamics, *J. Math. Phys.* **32** (10), 2929 (1991).
8. S. Nakajima, Y. Toyozawa, R. Abe, The Physics of Elementary Excitations, Springer Verlag, 1980, Chap.4.

THE FLUID-DYNAMICAL LIMIT FOR THE BBGKY HIERARCHY OF A DISCRETE VELOCITY MODEL

V. Gorunovich

Institute of Mathematics
Tereshchenkovskaya St 3, Kiev - 4, Ukraine

1. INTRODUCTION

We derive rigorously the fluid-dynamical equations for the one-dimensional discrete velocity model, considered in [1], directly from the BBGKY hierarchy of this model. The model we consider is not a mathematical one like the well-known Broadwell models of the Boltzmann equation [2]. It is also different from the Broadwell model of the BBGKY hierarchy considered in [3], the stochastic model dealt with in [4] and lattice gas cellular automata as well (see e.g.[5] and references therein). It is a deterministic model which can be regarded as physically realistic. The scheme that we present is apparently the first where the fluid limit for such a system is obtained in a rigorous way from the BBGKY hierarchy.

The problem we deal with is very similar to that which was solved in [6] by using a probabilistic approach. The restricted values of the velocities lead however to a fundamental difference. The limit obtained in [6] is a kinetic limit and our result agrees with that. As was shown in [7], a hydrodynamical type behavior occurs in the Tonk's gas only when its velocity distribution contains some delta-functions. From this point of view the scheme of the fluid-dynamical limit presented by us seems to be more general and could be used elsewhere.

The material of these notes is organized as follows. In Sec.2 we define our model and give the explicit form of the BBGKY hierarchy. Since our system is symmetric, the BBGKY hierarchy does not coincide with the hierarchy derived in [1]. In Sec.3 we give a non-rigorous derivation of the Euler equations. In Sec.4 the problem of determining the local equilibrium functions for our system is solved. In Sec.5 we state our main result.

2. THE MODEL

Let us consider a system of N identical hard rods with length d and unit mass, which can move along a line in both directions like beads on a string. By $x = (x_1, \ldots, x_N)$, where $x_i = (q_i, p_i)$, $i \in (1, \ldots, N)$ and $q_i \in R^1$, $p_i \in \{1, -1\}$ are the position of the center of i-th particle and its momentum, respectively, we denote a point in the N-particle phase space. We assume that the momentum of any particle (rod) takes values in a discrete set, and, for simplicity, we consider only the case where there are two possible values of velocity. By $W_N = \{(q_1, \ldots, q_N) \in R^N \mid |q_i - q_j| \leq d$ at least for one pair $(i,j); i,j \in (1, \ldots, N)\}$ we denote the set of all "forbidden" configurations of N particles. From the set of all "allowed" values of positions and momenta for N particles $(R^N \setminus W_N) \times \{1, -1\}^N$ we exclude a set M_N which consists of:

(i) the phase points associated with multiple collisions (when three or more particles collide);

(ii) the phase points associated with pasting together two or more particles (when the distance between particles having the same momentum equals to d).

(The Lebesgue measure of the set M_N is equal to zero.)

We regard the set $((R^N \setminus W_N) \times \{1, -1\}^N) \setminus M_N$ as the phase space of a statistical system which we call the one-dimensional discrete velocity model. Treating this system as a symmetrical system and following in general the scheme presented in [1], we get the BBGKY hierarchy for this system, which is of the form

$$\frac{\partial F_s(t, (x)_s)}{\partial t} = -\sum_{i=1}^{s} \{p_i \frac{\partial}{\partial q_i} F_s(t, (x)_s) + 2(\delta_{p_i,1} F_{s+1}(t, (x)_s; q_i - d, 1)$$

$$+ \quad \delta_{p_i,-1} F_{s+1}(t, (x)_s; q_i + d, -1) - \delta_{p_i,-1} F_{s+1}(t, x_1, \ldots; q_i, -1; \ldots, x_s; q_i - d, 1)$$

$$- \quad \delta_{p_i,1} F_{s+1}(t, x_1, \ldots; q_i, 1; \ldots, x_s; q_i + d, -1))\}, \quad s \geq 0 \tag{1}$$

where $\delta_{n,m}$ is the Kronecker symbol, $F(t) = \{F_s(t, (x)_s)\}_{s \geq 0}$ denotes s-particle distribution functions, $(x)_s = (x_1, \ldots, x_s)$ (see e.g. [1] and references therein).

The main result of [1] can be reformulated with the emphasis on the symmetry of the system under consideration. It says that the hierarchy (1) has a unique global solution global, in the space E_ς with the norm

$$\| f \| = \sup_{s \geq 0} \varsigma^{-s} \max_{p_1, \ldots, p_s} \operatorname{ess\ sup}_{q_1, \ldots, q_s} |f(x_1, \ldots, x_s)|, \quad \varsigma > 0 \tag{2}$$

3. THE EULER EQUATIONS

Let us consider the microscopic values $\rho(t,q) = F_1(t,q,1) + F_1(t,q,-1)$ and $m(t,q) = F_1(t,q,1) - F_1(t,q,-1)$ which we call the mean (number) density and the mean momentum, respectively, in analogy with three-dimensional systems. To obtain a closed system of evolution equations for these values we will generally follow the scheme developed in [8] for the case of a three-dimensional continuous system with a regular pair potential.

Consider a class of distribution functions

$$f(t, \varsigma) = (f_1(t, \varsigma, p_1), f_2(t, \varsigma, (x)_2), \ldots, f_s(t, \varsigma, (x)_s)),$$

where $\zeta = \varepsilon q_1$, which depend on a small parameter ε, which we regard as the ratio of typical microscopic and macroscopic lengths, so that these functions are symmetric with respect to the variables $x_1, \ldots, x_s$; the variable ζ plays a special role. Furthermore, depending on the case, we will regard this variable as εq_1 or as a smooth (macroscopic) variable ζ. Taking into account that

$$\frac{\partial F_s(t, \zeta, (x)_s)}{\partial q_1} = \frac{\partial f_s(t, \zeta, (x)_s)}{\partial q_1} + \varepsilon \frac{\partial f_s(t, \zeta, (x)_s)}{\partial \zeta} \tag{3}$$

and using the short hand notation $A[F_{s+1}(t)]$ for the part of the hierarchy (1) which is responsible for particle interaction, the discrete velocity hierarchy for functions which belong to this class get the form

$$\frac{\partial f_s(t, \zeta, (x)_s)}{\partial t} = -\sum_{i=1}^{s} p_i \frac{\partial f_s(t, \zeta, (x)_s)}{\partial q_i}$$

$$- \varepsilon p_1 \frac{\partial f_s(t, \zeta, (x)_s)}{\partial \zeta} + A[f_{s+1}(t, \zeta)]. \tag{4}$$

Below, by using this hierarchy, we will obtain a closed system of evolution equations for ρ and m.

If we differentiate ρ, expressed by the functions $f_1(t, \zeta, p_1)$, with respect to t, and then use the first equation of the hierarchy (2.2), we get

$$\frac{\partial \rho(t, \zeta)}{\partial t} = -\varepsilon \frac{\partial m(t, \zeta)}{\partial \zeta} \tag{5}$$

which is the continuity equation. Note that we regard here εq_1 as the smooth (hydrodynamical) variable ζ, and consequently, ρ and m as macroscopic values. Then, treating m similarly, we have

$$\frac{\partial m(t, \zeta)}{\partial t} = -\varepsilon \frac{\partial \rho(t, \zeta)}{\partial \zeta} + A[f_2(t, \zeta)] \tag{6}$$

Thus, the equations (5) and (6) do not form a closed system of equations for ρ and m. To transform these equations into a closed system we should somehow express the value $A[f_2(t, \zeta)]$ by ρ and m. As it is done in [8], we will realize that by using a local equilibrium distribution. But we should first carry out an expansion of $A[f_2(t, \zeta)]$. Expanding it into the Taylor series up to the first order in ε, we have

$$A[f_2(t, \zeta)] = 4\varepsilon d \frac{\partial f_2(t, \zeta; q, 1; q - d, -1)}{\partial \zeta} + O(\varepsilon^2) \tag{7}$$

by virtue of the symmetry of the functions $f(t, \zeta)$. Then, since the Euler equations take only the first order in ε into account, we should replace $f_2(t, \zeta; q, 1; q - d, -1)$ by the two-particle function which is the solution of the hierarchy (4), obtained by considering only the terms of first order in ε. At the same time such procedure is not the usual expansion in ε. For this reason we demand that such functions are solutions of the equation

$$-\mathcal{H}_\varepsilon f + Af = 0 \tag{8}$$

where $\mathcal{H}_\varepsilon f = \mathcal{H} + \varepsilon p_1 \frac{\partial}{\partial \zeta}$, and, in addition, the conditions $f_1^1(t, \zeta, 1) = 0$, $f_1^1(t, \zeta, 1) = 0$ are satisfied. Furthermore, in order to obtain functions which depend ρ and m we should solve (8) using the physical argument that these solutions are given by a local equilibrium distribution which is a Gibbs distribution. We will show in the next section that the two-particle distribution function, which should be substituted into (6) is of the form

$$f_2(\zeta; q, 1; q - d, -1) = \frac{(\rho(\zeta))^2}{1 - d\rho(\zeta)}. \tag{9}$$

Then, by substituting (9) into the right hand side of (6), we finally get the Euler equations for our model:

$$\frac{\partial \rho}{\partial t} + \varepsilon \frac{\partial m}{\partial \zeta} = 0, \tag{10}$$

$$\frac{\partial m}{\partial t} + \varepsilon \frac{\partial \rho}{\partial \zeta} + \varepsilon d \frac{\partial}{\partial \zeta}\left(\frac{\rho^2 - m^2}{1 - d\rho}\right) = 0, \tag{11}$$

which are to be solved subject to the initial conditions

$$\rho(t=0,\zeta) = \rho(0,\zeta); \; m(t=0,\zeta) = m(0,\zeta) \tag{12}$$

(since boundary conditions are of no importance for our consideration, we omit them.)

Summing up this section, we note that the initial value problem for the system (10)–(11) has an unique solution. This is a direct consequence of the statement on the existence of the solution for the equation considered in [6].

4. LOCAL EQUILIBRIUM FUNCTIONS

Using the idea given in [10], we treat the expression

$$D \sim \exp\{-\sum_{i=1}^{N}(\gamma(\varepsilon q_i)p_i + \mu(\varepsilon q_i)\}\chi_{R^N \setminus W_N}((q)_N) \tag{13}$$

as the Gibbs local equilibrium distribution of the model. Here γ, μ are thermodynamic fields; $\mu(\zeta)$ is the local chemical potential, $\gamma(\zeta)$ is the local field conjugate to the momentum density, $\chi_{R^N \setminus W_N}$ is the characteristic function of $R^N \setminus W_N$. In the local distribution we drop the local temperature, since the energy of the model is trivially conserved. By $f_l(t, \zeta)$ we denote the distribution functions constructed by this distribution. Similary as it is done for the equilibrium case in [9] one can easily obtain the integral equations for these functions, which are the local equilibrium analogues of the integral equations derived in [9]. Following a treatment similar to that in [9], we obtain the exact solutions of these equations. The two-particle function is of the form

$$f_2(\zeta, r) = \frac{\rho(\zeta)}{d} \sum_{k=1}^{\infty} T(r-k)\left(\frac{d\rho(\zeta)}{1-d\rho(\zeta)}\right)^k \frac{(r-k)^{k-1}}{(k-1)!}$$

$$\times \exp\{-\frac{d\rho(\zeta)(r-k)}{1-d\rho(\zeta)}\}; \tag{14}$$

$$r = \frac{|q_1 - q_2|}{d}; \; T(y) = \begin{cases} 0 & \text{if } y < 0 \\ 1 & \text{if } y \geq 0 \end{cases}$$

So, the value of this function at $|q_1 - q_2| = d$, which was used in previous section, is the solution of (8).

The integral equations as well as their solutions are given for the infinite volume system. However, it should be noted that carefully performing the thermodynamic limit for our model is a routine task.

5. THE MAIN CONVERGENCE THEOREM

Now we present our main theorem and give the sketch of its proof.

Let us consider the following initial value problem for the BBGKY hierarchy (4)

$$\frac{df(t,\zeta)}{dt} = \mathcal{H}_\varepsilon f(t,\zeta) + Af(t,\zeta), \tag{15}$$

$$f(t=0,\zeta) = f_l(\rho(0), m(0)), \tag{16}$$

where the functions $\{f_{l;s}(\rho(0), m(0), (x)_s)\}_{s\geq 0}$ for $s = 2$ are given by (14).

Theorem. *Assume that the initial value problem for the fluid equations (10)–(11) has a smooth solution ρ for m in some fixed time interval $[0,T]$, $T < 0$. Then there is a constant C which depends only on the initial data $\rho(0)$, $m(0)$, the solutions of (10)–(12) ρ, m and their first derivatives such that*

$$\sup_{t\in[0,T]} \sup_{\rho,m} |f_s(t,\zeta,(x)_s) - f_{l;s}(\rho(t,\zeta), m(t,\zeta), (x)_s)| \leq \varepsilon\varsigma^s C \tag{17}$$

uniformly with respect to $(x)_s$ on any compact, where $f(t,\zeta)$ is the solution of the initial value problem (15)–(16) and $\varsigma > 0$.

Remark 1. The theorem shows in which way the solutions of the BBGKY hierarchy (15)–(16) are aproximated by the solutions of fluid equations (10)–(11). The estimate (17) also justifies the validity of replacing $A[f_2(t,\zeta)]$ in the right hand side of (6) by $f_2(t,\zeta,d,1,-1)$ as ε tends to zero. Therefore the theorem completes the rigorous derivation of the fluid-dynamical equations (10)–(11).

Remark 2. The generality of this result is restricted to the special form of the initial data for the BBGKY hierarchy, that is when the evolution of the system starts with a local equilibrium state. So we do not consider here an initial layered behavior.

Sketch of the proof. Let us represent the solutions of the initial value problem (15)–(16) in the form

$$f(t,\zeta) = f_l(\rho(t,\zeta)) + \varepsilon R(t,\zeta) \tag{18}$$

Substituting (18) into (15)–(16), we obtain for the functions $R(t,\zeta)$ the following initial value problem

$$\frac{\partial R(t)}{\partial t} = -\mathcal{H}R - \varepsilon\mathcal{H}_\zeta R + \mathcal{K} \tag{19}$$

$$R(t=0,\zeta) = 0 \tag{20}$$

where $\mathcal{K}$ is some polynomial of the first order in the derivatives of $f_l(\rho, m,)$ with respect to ρ and m, and $\mathcal{H}_\zeta = \sum_{i=1}^s p_1 \frac{\partial}{\partial \zeta}$. The solutions of (19)–(20) can be represented as

$$R(t,\zeta) = \int_0^t d\tau \, T(t-\tau) U_\zeta(t-\tau) \mathcal{K}(\rho(t,\zeta), m(t,\zeta)) \tag{21}$$

where $U_\zeta(t)$ is the evolution operator constructed by $\mathcal{H}_\zeta$. $T(t)$ is a symbolic form of the iteration series for this hierarchy (see [1]). The inequality (17) can be obtained by estimating each term of the iteration series, using the boundedness of the norms of $U_\zeta(t)$, $f_l(\rho, m)$ in the space E_ς and the conditions of the Theorem on the initial data and the solutions of the fluid equations (10)–(12).

REFERENCES

1. V.I. Gerasimenko, V.V. Gorunovich, The Bogoliubov's equations for a one-dimensional particle system, *Dokl. Ak. Nauk. Ukr.*, **8**, 34 (1991).
2. K. Uchiyama, On the Boltzmann-Grad limit for the Broadwell model of Boltzmann equation, *Jour. Stat. Phys.*, **52**, 331 (1988).
3. R. Caflisch, G. Papanicolaou, The fluid-dynamical limit of of non-linear model Boltzmann equation, *Com. Pur. Appl. Math.*, **32**, 589 (1979).
4. S. Caprino, A. DeMasi, E. Presutti, M. Pulvirenti, A derivation of the Broadwell equation, *Comm. Math. Phys.*, **135**, 433 (1991).
5. M.H. Ernst, "Statistical mechanics of cellular automata fluids," Course 2, J.P. Hansen, D. Levesque and J. Zinn-Justin, eds., Elsevier Science Publishers B.V. (1991).
6. J. Boldrighini, R. Dobrushin, Yu. Suhov, One-dimensional hard rod caricature of hydrodynamics, *Journ. Stat. Phys.* **31**, 577 (1983).
7. J. Lebowitz, J. Percus, J. Sykes, Time evolution of the total distribution function of a 1-dimensional system of hard rods, *Phys. Rev.*, **721**, 224 (1968).
8. N.N. Bogoliubov, The hydrodynamics equations in statistical mechanics, *in*: "Sbo-rnik trudov", vol. 2, Nauk. dumka, Kiev (1970).
9. Z. Salsburg, R. Zwanzig, and J. Kirkwood, Molecular distribution functions in a one-dimensional fluid, *Journ. Chem. Phys.*, **21**, 1098 (1953).
10. C. Morrey, On the derivation of the equations of hydrodynamics from statistical mechanics, *Comm. Pure. Appl. Math.*, **8**, 279 (1955).

ISOSPECTRAL DEFORMATION OF DISCRETE RANDOM LAPLACIANS

Oliver Knill

Mathematikdepartement
ETH Zentrum
CH-8092 Zürich

DISCRETE RANDOM LAPLACIANS

Let $T_1, T_2, \ldots, T_d$ be commuting automorphisms of the probability space (X, μ). We call the $\mathbb{Z}^d$ action (X, T, μ) a *dynamical system* and write $T^n x = T_1^{n_1} T_2^{n_2} \ldots T_d^{n_d}(x)$ for $n \in \mathbb{Z}^d$. Denote by $\mathcal{X}$ the *crossed product* of the von Neumann algebra $\mathcal{A} = L^\infty(X, M(N, \mathbb{C}))$ with the dynamical system (X, T, μ). The group $\mathbb{Z}^d$ acts on $\mathcal{A}$ by automorphisms $f \mapsto f(T^n)$ where $f(T^n)(x) = f(T^n x)$ and the algebra $\mathcal{X}$ is obtained by completing the algebra of all polynomials in the variables $\tau_1, \ldots, \tau_d$ with coefficients in $\mathcal{A}$

$$K = \sum_{n \in F \subset \mathbb{Z}^d} K_n \tau^n, \quad (KL)_n = \sum_{l+m=n} K_l L_m(T^l) \tau^n$$

with respect to the norm $|||K||| = |\;||K(x)||\;|_\infty$, where $K(x)$ is the bounded linear operator on $l^2(\mathbb{Z}^d)$ defined by $(K(x)u)_n = \sum_m K_m(x) u_{n+m}$ and where $||\cdot||$ is the operator norm on $\mathcal{B}(l^2(\mathbb{Z}^d))$ and $|\cdot|_\infty$ the essential supremum norm. With the involution on $\mathcal{X}$ defined by

$$\left(\sum_n K_n \tau^n\right)^* = \sum_n K_n^*(T^{-n}) \tau^{-n}$$

$\mathcal{X}$ becomes a von Neumann algebra. It has the *trace*

$$\operatorname{tr}(K) = \int_X \operatorname{Tr}(L_0(x))\, d\mu(x),$$

where Tr denotes the usual trace on the matrix algebra $M(N, \mathbb{C})$. We call the elements in $\mathcal{X}$ *discrete random operators*.

In $\mathcal{X}$ lies the set of *discrete random Laplacians*

$$\mathcal{L} = \{L = a\tau + (a\tau)^* + b = \sum_{i=1}^d a_i \tau_i + (a_i \tau_i)^* + b\},$$

where we assume $b = b^*$ and a_i to be invertible. We call the vector $(a_1, \ldots a_d)$ the *gauge potential part* and b the *scalar potential part* of the Laplacian. Discrete random Laplacians are by definition self-adjoint.

Given a normal element $K \in \mathcal{X}$. The functional calculus defines $f(K)$ for every function $f \in C(\sigma(K))$, where $\sigma(K)$ denotes the spectrum of K. According to Riesz's representation theorem, the bounded linear functional $f \mapsto \text{tr}(f(K)) \in \mathbb{C}$ on $C(\sigma(K))$ defines a measure dk on $\sigma(K)$ which is called the *density of states* of K. It satisfies $\text{tr}(f(K)) = \int f(E) \, dk(E)$ for all $f \in C(\sigma(K))$.

Given $L \in \mathcal{L}$ and $d \geq 2$. The *(multiplicative) curvature* of the Laplacian $L = a\tau + (a\tau)^* + b \in \mathcal{L}$ is defined as

$$F = \sum_{i<j} F_{ij}\, \tau_i \tau_j ,$$

where the value $F_{ij} = a_i a_j(T_i) a_i(T_j)^{-1} a_j^{-1}$ can be considered as the result of the parallel transport with the connection $a\tau$ around the *plaquette* $P_{ij}x = \{x, T_i x, T_i T_j x, T_j x\}$. The *additive curvature* given by

$$\sum_{ij} F^+_{ij} \tau_i \tau_j = \sum_{ij} [a_i \tau_i, a_j \tau_j] \tau_i \tau_j = \sum_{ij} (a_i a_j(T_i) - a_j a_i(T_j)) \tau_i \tau_j .$$

is vanishing if and only if $F_{ij} = 1$. We say in this case, a Laplacian has *zero curvature*. This means that parallel transport around each closed curve in the lattice $\mathbb{Z}^d$ gives the identity.

Examples of discrete random Laplacians are:

- *Random Jacobi operators*

If $d = N = 1$, we get random Jacobi matrices $L = a\tau + (a\tau)^* + b$. Special cases are discrete random Schrödinger operators $L = \tau + \tau^* + V$. Such operators have been studied much in the last years. (See [1], [2].)

- *Harper Laplacians*

Let $d = 2$ and $a_1, a_2 \in L^\infty(X, SU(N))$ satisfy $a_1(x) a_2(T_1 x) = e^{2\pi i \alpha} a_2(x) a_1(T_2 x) \cdot 1$, where α is an irrational number. In this case

$$L = a_1 \tau_1 + a_2 \tau_2 + (a_1 \tau_1)^* + (a_2 \tau_2)^*$$

is called a *Harper Laplacian*. It has by definition a constant curvature $e^{2\pi i \alpha}$ and $2\pi\alpha$ has a physical interpretation as a normalized magnetic flux. A special case of a Harper Hamiltonian is given by $(X, T_1, T_2, \mu) = (\mathbb{T}^1, x \mapsto x + \alpha, x \mapsto x, dx)$ with $a_1 = 1$, $a_2(x) = e^{2\pi i x}$, which leads to the Hofstadter case of the *discrete Mathieu operator* $L = \tau_1 + \tau_1^* + V$, where $V(x) = 2 \cos(2\pi x)$. See [3] for recent results on the spectrum.

- *D'Alembert operator with Dirac operator*

Let $d = 4$. The operator $L = \sum_{i=1}^4 g_i(\tau_i + 2 + \tau_i^*)$ with $g_1 = -1, g_2 = g_3, g_4 = 1$ is a discrete version of the flat d'Alembert operator $\square = \sum_{i=1}^4 g_i \frac{\partial^2}{(\partial x_i)^2}$. Define the Dirac operator $D = \sum_{i=1}^4 \gamma_i(\tau_i + \tau_i^*)$, where γ_i be the Dirac matrices, satisfying $\{\gamma_i, \gamma_j\} = 2g_i \delta_{ij}$. The square

$$D^2 = \sum_{i=1}^4 g_i(\tau_i^2 + 2 + (\tau_i^*)^2)$$

is a Laplacian over the $\mathbb{Z}^d$-dynamical system $(X, T^2 = (T_1^2, T_2^2, \ldots, T_d^2), \mu)$.

• *Periodic Laplacians*
In the case $|X| = M < \infty$, the automorphisms T_i are just finite permutations of X. Ergodic T_i lead to periodic Laplacians which are mostly studied in the case $N = 1, a_i = 1, b = V$. If $d = N = 1$, we get periodic Jacobi matrices. The case $d = 2, N = 1, a_i = 1$ is the subject of the book [4].

We think about elements in $\mathcal{X}$ as discrete versions of *differential operators*. They can also be considered as the *Hamiltonian* of a quantum mechanical particle or describing the *geometry* of a discrete manifold. We think of the a_i as components of a *connection* or a *gauge potential* or a *one-form* and of the F_{ij} as the components of the *curvature* or a *gauge field* or a *two form*, always depending on geometrical, physical or algebraic preferences.

DISCRETE RANDOM DIRAC OPERATORS

The classical d'Alembert operator $\Box = \sum_{i=1}^{4} g_i \frac{\partial^2}{(\partial x_i)^2}$ can be factorized as $L = D^2$ with the Dirac operator $D = \sum_i \gamma_i \delta_{x_i}$, where the Dirac matrices γ_i satisfy the anti-commutation relation $\{\gamma_i, \gamma_j\} = \pm 2g_i \delta_{ij}$. Our aim is to construct a discrete random Dirac operator, which is by definition the square root of a discrete random Laplace operator L. For doing so, it is necessary to construct a new dynamical system from the given dynamical system (X, T, μ). We get a new probability space (Y, ν) by defining $Y = \bigcup_{I \subset \{1,\ldots,d\}} X_I$ to be the union of 2^d copies X_I of (X, μ) and letting ν be the normalized measure on Y, such that for $Z \subset X = X_I$, $\mu(Z) = 2^d \nu(Z)$. Define

$$S_i : X_I \to X_{I \Delta \{i\}}, \quad S_i(x) = \begin{cases} x, & i \notin I, \\ T_i(x), & i \in I. \end{cases}$$

$(Y, S_1, S_2, \ldots, S_d, \nu)$ is again a $\mathbb{Z}^d$ dynamical system and because $S_i^2(x) = T_i(x)$ for $x \in X_I$, we get the old system back by restricting S_i^2 on X_I. We call (Y, S, ν) a $2:1$ *integral extension* of (X, T, μ).

Define $\mathcal{Y}$ to be the crossed product of $\mathcal{B} = \mathcal{L}^\infty(\mathcal{Y}, \mathcal{M}(\mathcal{N}, \mathbb{C}))$ with the dynamical system (Y, S, ν). We will write elements in $\mathcal{Y}$ as $C = \sum_n C_n \sigma_n$, where $\sigma_i f = f(S_i) \sigma_i$. Let

$$\mathcal{D} = \{D = \sum_i d_i \sigma_i + (d_i \sigma_i)^* \mid d_i \in \mathcal{B}\} \subset \mathcal{Y}$$

be the set of discrete random Laplacians in $\mathcal{Y}$ which have zero scalar part. Define $\psi : \mathcal{Y} \to \mathcal{X}$ by $\psi(K)_n(x) = K_{2n}(x)$ for $x \in X = X_\emptyset$, where we use the notation $2n = (2n_1, \ldots, 2n_d)$. We say, an operator $D \in \mathcal{D}$ is a *discrete random Dirac operator*, if there exists $E \in \mathbb{C}$ such that $\psi(D^2 + E) \in \mathcal{L}$. In [5], we have constructed Dirac operators to every one-dimensional random Jacobi operator. In [6] we showed how the iteration of the factorization $L \mapsto D$ satisfying $L = D^2 + E$ leads to operators with spectra on Julia sets.

If $D = d\sigma + (d\sigma)^*$ is a Dirac operator to L, we can rewrite $\psi(D^2 + E) = L = a\tau + (a\tau^*) + b$ as

$$\{d_i \sigma_i, d_j \sigma_j\} = \delta_{ij} 2a_i,$$
$$\{d_i \sigma_i, (d_j \sigma_j)^*\} = \delta_{ij} b_i,$$

with $\sum_{i=1}^{d} b_i = b - E$. This implies

$$[a_i T_i, a_j T_j] = 0, \tag{1}$$
$$[a_i T_i, (a_j T_j)^*] = \delta_{ij} c_i. \tag{2}$$

and the Laplacian L must have zero curvature.

ISOSPECTRAL TODA DEFORMATIONS OF DISCRETE LAPLACIANS

Define the cones $(\mathbb{Z}^d)^+ := \{n \in \mathbb{Z}^d \mid n_i > 0\}$ and $(\mathbb{Z}^d)^- = -(\mathbb{Z}^d)^+$ in the lattice $\mathbb{Z}^d$. We define in $\mathcal{X}$ the projections

$$K = \sum_n K_n T^n \mapsto K^\pm = \sum_{n \in (\mathbb{Z}^d)^\pm} K_n T^n$$

and denote the images of these projections in $\mathcal{X}$ by $\mathcal{X}^\pm$. We remark that if $K \in \mathcal{X}^+$ then $K^* \in \mathcal{X}^-$. Denote by $C^\omega(\mathbb{R})$ the set of all entire functions $\mathbb{C} \to \mathbb{C}$ which map $\mathbb{R}$ into itself and define the space of Hamiltonians

$$C^\omega(\mathcal{X}) = \{H : \mathcal{X} \to \mathbb{C} \mid K \mapsto H(K) = \mathrm{tr}(h(K)), \ h \in C^\omega(\mathbb{R})\}.$$

To such a Hamiltonian $H \in C^\omega(\mathcal{X})$ belongs the *Toda* differential equation

$$\dot{K} = [h'(K)^+ - h'(K)^-, K]$$

which gives a globally defined isospectral flow in $\mathcal{X}$. It is isospectral because $B = h'(K)^+ - h'(K)^-$ is a skew-symmetric operator. The *first Toda flow* is obtained with the Hamiltonian $H(L) = \mathrm{tr}(L^2/2)$. The random Toda flows do not leave $\mathcal{L}$ invariant unless $d = 1$. We have shown in [7] how one can linearize in the case $d = 1$ these infinite dimensional integrable dynamical systems. For finite $|X|$, the flows are then the classical periodic Toda lattices which can be linearized explicitly using algebraic geometry [8]. In more dimension, an isospectral Toda deformation in $\mathcal{L}$ is in general not possible. Already for finite $|X|$, there is a result of Mumford [9] which says that, generically, there exist no isospectral deformations of higher dimensional real Laplacians (N=1). In the complex case, it is however trivial to deform a self-adjoint Laplacian $L = \sum_i a_i T_i + (a_i T_i)^* + b \in \mathcal{L}$. Take any curve $g_t \in L^\infty(X, SU(N))$ satisfying $g(0) = 1$. Then $L(t) = g(t) L g(t)^{-1}$ consists of self-adjoint isospectral Laplacians.

A SUFFICIENT CONDITION FOR ISOSPECTRAL DEFORMATIONS

Theorem 1 *If there exists $D \in \mathcal{D}$ and $E \in \mathbb{R}$ such that $L = \psi(D^2 + E)$, the orbit of the isospectral deformation*

$$\frac{d}{dt} L = [L^+ - L^-, L] =: [B, L]$$

stays in $\mathcal{L}$. If L is not a stationary point of this Toda flow, there exists a curve of isospectral Laplacians through L. The deformation can be written as

$$\dot{a}_i = a_i b(T_i) - b a_i,$$
$$\dot{b} = \sum_{i=1}^{d} 2(a_i a_i^* - (a_i^* a_i)(T_i^{-1})),$$

The differential equation for $L = \psi(D^2 + E)$ is equivalent to
$$\dot{D} = [(D^2)^+ - (D^2)^-, D]$$
in $\mathcal{Y}$ and this is a decoupled system of one-dimensional random Volterra systems
$$\dot{d}_i = [d_i^2, d_i^*] = d_i \cdot (d_i d_i^*)(T_i) - (d_i^* d_i)(T_i^{-1}) \cdot d_i .$$

Proof. The differential equation $\frac{d}{dt} L = [B, L]$ has a solution in $\mathcal{X}$. We want to show that for an initial condition L having a factorization $L = \psi(D^2 + E)$, the solution stays in $\mathcal{L}$. It follows from Equation (1) and (2) that

$$\begin{aligned}[L^+ - L^-, L] &= [a\tau - (a\tau)^*, a\tau + (a\tau)^* + b] \\ &= [a\tau, b] - [(a\tau)^*, b] + [a\tau, (a\tau)^*] - [(a\tau)^*, a\tau]\end{aligned}$$

is in $\mathcal{L}$. This holds true as long as $L = \psi(D^2) + E$. To see that the factorization $L = \psi(D^2) + E$ is preserved by the flow we have to show that the Clifford structure

$$\begin{aligned}\{d_i \sigma_i, d_j \sigma_j\} &= \delta_{ij} 2 a_i, \\ \{d_i \sigma_i, (d_j \sigma_j)^*\} &= \delta_{ij} b_i .\end{aligned}$$

stays invariant under the flow. The differential equation
$$\dot{D} = [(D^2)^+ - (D^2)^-, D]$$
in $\mathcal{Y}$ can be written as a decoupled set of Volterra systems
$$\dot{d}_i = [d_i^2, d_i^*] = d_i \cdot (d_i d_i^*)(T_i) - (d_i^* d_i)(T_i^{-1}) \cdot d_i .$$

We write from now on simply d instead of $d\sigma$ and d^* instead of $(d\sigma)^*$ in order to simplify the writings. With the differential equations
$$\dot{d}_i = [d_i^2, d_i^*] , \quad \dot{d}_i^* = -[(d_i^*)^2, d_i] ,$$
we get for $i \neq j$ using $\{d_i, d_j^*\} = 0$

$$\begin{aligned}\frac{d}{dt}\{d_i, d_j\} &= \dot{d}_i d_j + d_i \dot{d}_j + \dot{d}_j d_i + d_j \dot{d}_i \\ &= [d_i^2, d_i^*] d_j + d_i [d_j^2, d_j^*] + [d_j^2, d_j^*] d_i + d_j [d_i^2, d_i^*] \\ &= d_i^2 d_i^* d_j - d_i^* d_i^2 d_j + d_i d_j^2 d_j^* - d_i d_j^* d_j^2 + d_j^2 d_j^* d_i - d_j^* d_j^2 d_i + d_j d_i^2 d_i^* - d_j d_i^* d_i^2 \\ &= d_i^2 \{d_i^*, d_j\} + d_j^2 \{d_i, d_j^*\} - \{d_i^*, d_j\} d_i^2 - \{d_j^*, d_i\} d_j^2 = 0 .\end{aligned}$$

Similarly, using $\{d_i, d_j\} = \{d_i^*, d_j^*\} = 0$ for $i \neq j$, we get for $i \neq j$

$$\begin{aligned}\frac{d}{dt}\{d_i, d_j^*\} &= \dot{d}_i d_j^* + d_i \dot{d}_j^* + \dot{d}_j^* d_i + d_j^* \dot{d}_i \\ &= [d_i^2, d_i^*] d_j^* - d_i[(d_j^*)^2, d_j] - [(d_j^*)^2, d_j] d_i + d_j^* [d_i^2, d_i^*] \\ &= d_i^2 (d_i^* d_j^* + d_j^* d_i^*) - (d_j^*)^2 (d_i d_j + d_j d_i) - (d_i^* d_j^* + d_j^* d_i^*) d_i^2 + \\ &\quad (d_i d_j + d_j d_i)(d_j^*)^2 = 0 .\end{aligned}$$

$\square$

Remarks.
• The condition in the proposition is always satisfied in one dimension, where every Laplacian can be deformed.

- The zero curvature conditions $[a_i \tau_i, a_j \tau_j] = [a_i \tau_i, (a_j \tau_j)^*] = 0$, $i \neq j$ for L do not imply in general that we can factor $L = D^2 + E$. We need that the zero curvature holds true also after deformation. This would need for example that the equation

$$\frac{d}{dt}[a_i \tau_i, (a_j \tau_j)^*] = 2[[a_i \tau_i, b], (a_j \tau_j)^*] = 0$$

is valid which implies that also a condition for b must be satisfied in order to have an isospectral deformation.

- Operators with unitary a_i and constant b are stationary points of the first Toda flow. Especially random Harper operators can not be deformed in this way.

- The necessary conditions

$$\begin{aligned} a_i a_j(T_i) &= a_j a_i(T_j), \\ a_j a_i &= a_i(T_j) a_j(T_i), \quad i \neq j \end{aligned}$$

for isospectral deformations have only constant solutions if we require the a_i to be complex valued cocycles and if the maps T_i are ergodic. (To see this, we divide the first equation by the second one giving $a_j(T_i)^2 = a_j^2$ which implies that a_j is constant if T_i is ergodic.) We conclude that we need the entries of the Laplacian to be *matrix-valued* in order to obtain interesting isospectral deformations.

DISCRETE RANDOM PARTIAL DIFFERENCE EQUATIONS

Given a $\mathbb{Z}^d$ dynamical system (X, T, μ) and a *Lagrangian*

$$l(q_0, q_1, \ldots, q_d) = -\sum_{i=1}^{d} g_i \frac{(q_i - q_0)^2}{2} + \gamma \cdot V(q_0),$$

where $g_i \in \mathbb{R} \setminus \{0\}$ and $V \in C^2(\mathbb{T}^1, \mathbb{R})$ is a real-valued potential and γ is a real coupling constant. We define on the space $L^\infty(X, \mathbb{T}^1)$ the *functional*

$$S(q) = \int_X l(q(x), q(T_1 x), q(T_2 x), \ldots, q(T_d x)) \, d\mu(x)$$

on $L^\infty(X, \mathbb{T}^1)$. A critical point q of this functional satisfies the *discrete random partial difference equation*

$$-\sum_{i=1}^{d} g_i(q(T_i) - 2q + q(T_i^{-1})) - \gamma \cdot V'(q) = \delta S(q) = 0.$$

The second variation at a critical point (the Hessian) is the random Laplacian

$$L = \sum_{i=1}^{d} a_i \tau_i + (a_i \tau_i)^* + b$$

with $a_i = -g_i$ and $b = \gamma \cdot V'' - \sum_{i=1}^{d} 2g_i$.

It is a nontrivial problem to find nontrivial critical points of the above functional. For large coupling constants γ, it is however quite easy using the anti-integrable limit idea of Aubry [10]. Denote by

$$\Sigma = \{\sigma \in \mathbb{T}^1 \mid V'(\sigma) = 0, \; V''(\sigma) \text{ invertible}\}$$

the set of non-degenerate critical points of V.

Theorem 2 *Assume that Σ has at least 2 elements. Then there exists $\gamma_0 \in \mathbb{R}$ such that for coupling constants $|\gamma| > \gamma_0$, there exist nontrivial critical points of the functional S.*

Proof. Write $\gamma = 1/\epsilon$ and take the equivalent functional

$$S(q) = \int_X -\sum_{i=1}^d \epsilon g_i \frac{(q(T_i x) - q(x))^2}{2} + \gamma \cdot V(q(x)) \, d\mu(x) \,.$$

In the limit $\epsilon = 0$ every function $q \in L^\infty(X, \Sigma)$ is a critical point. The Hessian at such a point is $L = V''(q)$ which is by definition of Σ invertible. Applying the implicit function theorem, there exists also a critical point q_ϵ for small $|\epsilon|$. □

We introduced the above functional S in [7] for the one-dimensional case, where the problem of finding critical points is equivalent to embed a factor of the dynamical system in a *monotone twist map* defined by the generating function l. In the case, when the dynamical system is the irrational rotation on $\mathbb{T}^1$, the functional S is called the Percival functional. In [11], we used the functional together with a theorem of Krieger and the anti-integrable limit of Aubry to show that every ergodic dynamical system with finite metric entropy can be embedded into a monotone twist map. For two-dimensional systems $d = 2$ with $a_1 = -a_2 = 1$ and $V(q) = \cos(q)$ a critical point satisfies a discrete version of the Sine-Gordon equation.

One can not only establish the existence of critical points but also prove that there exist uncountably many different critical points. This can be done with the help of a *generalized Morse index* at a critical point q which is defined to be the value of the integrated density of states $k(0) = \int_{-\infty}^0 dk(E')$ of the Hessian $L = \delta^2(S)$ at the energy $E = 0$. This number is lying in the interval $[0, 1]$ and measures, how much of the spectrum of L is below 0. One could say that it measures the "dimension" of the infinite dimensional unstable manifold, which is passing through the critical point. If two critical points have a different index, then they must be essentially different in the sense that it is then excluded that one critical point is a translation of the other one.

Proposition 3 *If there exist two critical points σ_1, σ_2 of V with $V''(\sigma_1) > 0$ and $V''(\sigma_2) < 0$, then there exist uncountably many different critical points of S for $|\gamma|$ large enough. The Morse index at a critical point near $q \in L^\infty(X, \Sigma))$ is*

$$m(\{x \in X \mid V''(q(x)) < 0\}) \,.$$

Proof. In the anti-integrable limit, the Hessian is an invertible diagonal operator and the density of states is a Dirac measure located on the set

$$\{V''(\sigma) \mid \exists Y(\sigma) \subset X, \; m(Y(\sigma)) > 0, \; \forall x \in Y(\sigma), \; \sigma = q(x) \in \Sigma\} \,.$$

Each point σ in this set has the mass $m(Y(\sigma))$. The integrated density of states at 0 is given by $m(\{x \in X \mid V''(q(x)) < 0 \})$. Because the Hessian in the anti-integrable limit is invertible, the mass of the spectrum below 0 does not change under perturbations of the operator or (by the implicit function theorem) under perturbations of the critical point. This means that the Morse index is constant for critical points near the anti-integrable limit and it is given by the same formula. □

ZERO CURVATURE AND THE DENSITY OF STATES

We consider in this section Laplacians in $\mathcal{L}_{SU} = \{L \in \mathcal{L} \mid a_i \in L^\infty(X, SU(N)), b = 0\}$. Call
$$\mathcal{G} = \{G \in \mathcal{L} \mid G_0 \in L^\infty(X, SU(N)), G_n = 0, n \neq 0\}$$
the *group of $SU(N)$ gauge fields*. For $G \in \mathcal{G}$, the map
$$L \mapsto GLG^{-1}$$
on $\mathcal{L}_{SU}$ is called a *gauge transformation*. The $a_i \tau_i$ are transformed as
$$a_i \tau_i \mapsto G a_i G(T_i)^{-1} \tau_i \ .$$
Gauge transformations leave the set of zero-curvature Laplacians invariant. Also the density of states is invariant under such transformations so that gauge transformations are isospectral deformations. We will just see that the density of states decides also, if the operator has zero curvature or not. In the special case of a Harper Hamiltonian $d = 2, N = 1$ with constant curvature, the density of states determines the curvature and so the normalized magnetic flux:

Proposition 4 *a) The operator $L \in \mathcal{L}_{SU}$ has zero curvature if and only if*
$$\text{tr}(L^4) = \int_\mathbb{R} E^4 \ dk(E) = NP_4(d) \ ,$$
where $P_4(d)$ is the number of closed paths of length 4 starting at $0 \in \mathbb{Z}^d$.
b) For a Harper operator $L = a_1\tau_1 + a_2\tau_2 + (a_1\tau_1)^* + (a_2\tau_2)^*$ with $N = 1$ satisfying
$$a_1 a_2 (T_1) a_1(T_2)^{-1} a_2^{-1} = e^{2\pi i \alpha} \ ,$$
the density of states determines the normalized magnetic flux $2\pi i \alpha$:
$$\text{tr}(L^4) = \int_\mathbb{R} E^4 \ dk(E) = 16 + 8\cos(2\pi i \alpha) \ .$$
c) Harper Hamiltonians with different $\cos(2\pi \alpha)$ are not isospectral.

Proof.
a) If we know the density of states, we can calculate
$$\text{tr}(L^n) = \int E^n \ dk(E) \ .$$
The operator has zero curvature if and only if $a_i a_j(T_i)(a_i(T_j))^{-1}(a_j)^{-1} = 1$ and this is the case if and only if
$$\text{Tr}(a_i a_j(T_i)(a_i(T_j))^{-1}(a_j)^{-1}) = N$$

for all i,j. Because the trace of an element a in $SU(N)$ is always $\leq N$ with equality in the case of zero curvature, this is equivalent to

$$\mathrm{tr}(L^4) = NP_4(d) \,,$$

where $P_4(d)$ is the number of closed paths of length 4 in $\mathbb{Z}^d$ starting at one point.
b) L^4 contains 4 curvature terms like $a_1 a_2(T_1)(a_1(T_2)^{-1})^*(a_2^{-1})^*$ for each positively oriented plaquette at $x \in X$ and 4 curvature terms like $a_1 a_2(T_1 T_2^{-1})^* a_1(T_2^{-1})^* a_2(T_2^{-1})$ for each negatively oriented plaquette. There are additionally 16 constant summands 1 belonging to closed paths of length 4 which are not passing around a plaquette.
c) As a corollary of b), we obtain that Harper Laplacians with different $\cos(2\pi\alpha)$ are not isospectral because isospectral Laplacians have the same $\mathrm{tr}(L^4)$. □

The functional
$$S_{gauge}(L) = g \cdot (\mathrm{tr}(L^4) - NP_4(d))$$

on $\mathcal{L}_{SU}$ is the lattice action of pure lattice $SU(N)$ gauge field on a d dimensional infinite lattice. For finite X, the lattice is periodic and the functional is a finite sum. In general it is an averaged sum.

If we don't assume α to be constant for the Harper Hamiltonian, we can make the following remark about the relation between the curvature and the density of states.

Proposition 5 *Given $L \in \mathcal{L}_{SU}$. The curvature $F = a_1 a_2(T_1) a_1(T_2)^* a_2^*$ determines the density of states dk. On the other hand, the density of states dk does not determine the curvature F.*

Proof. If F is known, we can calculate $\mathrm{tr}(L^n)$ for each $n \in \mathbb{N}$ because $\mathrm{tr}(L^n)$ contains summands labeled by paths of length n and each path summand is the product of all the curvatures belonging to the plaquettes which are surrounded by this path. We can determine therefore also $\mathrm{tr}(\log(L-E))$ for $\mathrm{Im}(E) > 0$ and so the integrated density of states

$$k(E) = \mathrm{Im}(\mathrm{tr}(\log(L-E))) \,, \ E \in \mathbb{R}$$

which determines the density of states $dk = \frac{d}{dE}k(E)dE$. The curvature function $F \in L^\infty(X)$ and a translated function $F(T^n)$ belong both to the same density of states. It is thus in general not possible to determine dk from F. □

It can be seen in the same way that also for higher dimensional Laplacians $L = \sum_{i=1}^{d} a_i \tau_i + (a_i \tau_i)^*$ with $a_i \in L^\infty(X, SU(N))$, the curvature $F = \sum_{ij} F_{ij} \tau_i \tau_j$ determines the density of states.

QUESTIONS

We formulate some open points. We don't know if the existence of a Dirac operator also implies that higher Toda flows exist.

Is the sufficient condition in the Theorem 1 also necessary? In other words, can one factorize an operator if an isospectral deformation is possible?

We don't know if the property of factorization is a spectral property in the sense that the density of states decides whether there exists D with $\psi(D^2) + E = L$.

REFERENCES

1. H. L. Cycon, R. G. Froese, W. Kirsch, and B. Simon, "Schroedinger Operators," Texts and Monographs in Physics, Springer (1987).
2. R. Carmona, and J. Lacroix, "Spectral Theory of Random Schrödinger Operators," Birkhäuser, Boston (1990).
3. Y.Last, Zero measure spectrum for the almost Mathieu operator, Technion preprint March (1993).
4. Gieseker, Knörrer, and Trubowitz, "The Geometry of algebraic Fermi curves," Perspectives in Mathematics, Academic Press, San Diego (1993).
5. O. Knill, Factorization of Random Jacobi Operators and Bäcklund transformations, Comm. Math. Phys., **151**, 589-605 (1993).
6. O. Knill, Renormalization of random Jacobi operators, to be submitted.
7. O. Knill, Isospectral Deformations of Random Jacobi Operators, Comm. Math. Phys., **151**, 403-426 (1993).
8. P. v. Moerbeke, The spectrum of Jacobi Matrices, Invent. Math., **37**, 45-81 (1976).
9. P. v. Moerbeke, "About isospectral deformations of discrete Laplacians," Lecture Notes in Mathematics, No. 755, Springer (1978).
10. S. Aubry, The concept of anti-integrability: definition, theorems and applications to the standard map, *in :* "Twist mappings and their Applications", IMA Volumes in Mathematics, Vol. 44., Eds. R.Mc Gehee, K.Meyer, (1992).
11. O. Knill, Embedding abstract dynamical systems in monotone twist maps, to be submitted.

SCATTERING AND THE ROLE OF OPERATORS IN BOHMIAN MECHANICS

M. Daumer[1], D. Dürr[1], S. Goldstein[2], N. Zanghí[3]

1. Fakultät für Mathematik, Universität München, Theresienstr. 39, 80333 Munich, Germany

2. Department of Mathematics, Rutgers University, New Brunswick, New Jersey 08903, USA

3. Istituto di Fisica, Università di Genova, INFN, Via Dodecaneso 33, 16146 Genova, Italy

ABSTRACT

Using Bohmian mechanics, we analyze the problem of describing escape time, escape position and sojourn time—quantities for which the quantum formalism assigns no self-adjoint operator—for quantum systems. The large-scale behavior relevant to scattering theory is also discussed.

HOW DOES ONE HANDLE THIS?

Consider an electron with a localized initial wave function, with support inside a certain region G. Surrounding the electron are detectors, placed along the boundary of G, which measure the position and the time of "escape" of the electron from the region. What are the quantum mechanical predictions for the statistics of these quantities?

Predictions in quantum mechanics are based on a correspondence between operators and observable quantities, with the operators that correspond to classical observables arising as follows: The operators $\hat{\mathbf{q}}$, $\hat{\mathbf{p}}$ for position and momentum may be found by replacing the classical Poisson bracket by the commutator: $\{\ ,\ \} \to \frac{1}{i\hbar}[\ ,\ ]$. For a general classical observable, given by a function $f(\mathbf{q},\mathbf{p})$ on phase space, the rule is to replace $\mathbf{q},\mathbf{p}$ in f by the operators $\hat{\mathbf{q}}, \hat{\mathbf{p}}$. (This procedure is however ambiguous since it does not specify the order in which noncommmuting operators should appear in a product.) The spectral measure of the operator is then supposed to describe the statistics for the outcome of the measurement of the corresponding observable. This rule is

On Three Levels, Edited by M. Fannes *et al.*,
Plenum Press, New York, 1994

trusted to yield the correct operators in the usual "measurement situations" where the observable is measured at a specific time chosen by the experimenter.

But what are the operators for the escape time and escape position? For a classical particle with trajectory $\mathbf{Q}(t)$, the escape time (the first exit time) and the corresponding escape position from the region G are given by

$$T_e := \inf\{t | \mathbf{Q}(t) \in G^c\} \tag{1}$$

and

$$\mathbf{Q}_e = \mathbf{Q}(T_e). \tag{2}$$

One should notice at once that these expressions are not "simple" functions $f(\mathbf{q}, \mathbf{p})$ on phase space, since they depend explicitly on the trajectory, i.e., on the classical dynamics. The simple rule of replacement mentioned above would presumably lead to grotesque ambiguities if applied to these quantities. Notice also that the moment of time at which the "counter clicks" is indeed random and is not a parameter chosen by the experimenter performing the measurement. Because of this "problem of continuous observation" (see, e.g., [2, 3]), some physicists have felt the need to generalize the quantum formalism so that it may be applied in such situations [1, 2, 3].

It is not at all clear what rule should in fact be used to find the "correct" operators. Moreover, the mere existence of an operator for "time" seems to conflict with general principles: From the sizable literature on the subject of "time operators" we may cite the argument of Pauli [4] that there can be no self-adjoint operator $\hat{t}$ which is canonically conjugate to the Hamiltonian H.[1] Furthermore it is shown in [1] that no orthogonal basis of arrival or escape time eigenstates exists,[2] and hence no corresponding self-adjoint time operator.

On the other hand, by being more flexible with the quantum rules various researchers have proposed solutions to some of these problems: Since $\|P_G \psi_t\|^2$ ($P_G \psi_t$ denotes the projection of ψ_t onto G) is the probability of finding the particle at time t in G, a natural guess for the probability density of the escape time is to define $\rho_e(t) = -\frac{d}{dt}\|P_G \psi_t\|^2$. Why is it natural? Imagine a classical picture in which the particle never returns to G once it leaves. In this case $\|P_G \psi_t\|^2$ would indeed be the probability that the particle is still in G at time t, which is $(1-\ )$ the distribution function of the escape time. However, in general $\rho_e(t)$ may well be negative (the particle may return to G) and thus $\rho_e(t)$ cannot in general be interpreted as a probability density [2].

Using similar ideas, one may also arrive at a "time-operator" for the total time spent in G, called the sojourn-time or dwell-time operator [2],[5]. For this particular "continuous observation," the generalization of the rule above is straightforward. The classical expression for the total time T_s spent by the particle in the region G is such that the position $\mathbf{Q}(t)$ can be unambiguously replaced by the Heisenberg position operator

---

[1] $[\hat{t}, H] = i\hbar$ implies that the spectrum of both $\hat{t}$ and H is the whole real line, which conflicts with the semiboundedness of H.

[2] If, for example, "escape time eigenstates" $|t\rangle$ existed, states for which the electron would leave a certain region at an exactly specified time t, then any time-evolved eigenstate would have to be an eigenstate itself. That is, for times $t' > 0$, $e^{-iHt'}|t\rangle = |t - t'\rangle$, and this state must be orthogonal to $|t\rangle$. But the scalar product

$$\langle t | t - t' \rangle = \langle t | e^{-iHt'} | t \rangle$$

may be seen as a (distributional) boundary value from which it may be analytically continued (in t') into the lower half plane, Im $t' < 0$, (assuming $H \geq 0$). But by the unicity of the analytic continuation, the boundary value cannot be zero on a set of positive measure; otherwise it would have to be zero at $t' = 0$, which is surely not the case.

$\hat{q}(t)$: With χ_G denoting the indicator function of the set G

$$T_s = \int_0^\infty \chi_G(\mathbf{Q}(t))dt. \qquad (3)$$

This becomes

$$\hat{t}_s = \int_0^\infty P_G(t)dt \qquad (4)$$

where $P_G(t)$ is the Heisenberg operator for the projection onto G. Note that formally $[\hat{t}_s, H] = i\hbar$ on the subspace of wave functions which are localized in G [2].

This operator gives a mean sojourn time which is in agreement with the classical result when $\|P_G\psi_t\|^2$ is the probability for the particle to be in G at time t (see the next section). However, in view of the negative result mentioned above,[3] there are doubts as to whether this operator yields more than the "correct" mean.

While little discussion seems to have been devoted to the problem of finding an operator for the escape position, the statistics for the escape position have received some attention, having been addressed in the "scattering-into-cones" theorem [6] and the "flux-across-surfaces" theorem [7]. We shall discuss these theorems later when we consider the large-scale behaviour typical of scattering theory.

We wish to focus on the question of how one handles situations where no self-adjoint operators exist for escape and sojourn time and escape position, i.e., where the usual quantum rules for predictions do not apply. What can be done?

We shall give a systematic discussion of this issue within the context of Bohmian mechanics, which is a theory of point particles in motion—in which particles have *trajectories*—and which is known to yield the same predictions as quantum mechanics whenever the latter is unambiguous [8, 9].

THE BOHMIAN WAY

In Bohmian mechanics a particle having wave function ψ moves along a trajectory $\mathbf{Q}(t)$ determined by

$$\frac{d}{dt}\mathbf{Q}(t) = \mathbf{v}_t(\mathbf{Q}(t)) = \frac{\hbar}{m}\mathrm{Im}\frac{\nabla\psi_t}{\psi_t}(\mathbf{Q}(t)) \qquad (5)$$

where ψ_t is a solution of Schrödinger's equation

$$i\hbar\frac{\partial}{\partial t}\psi_t = (-\frac{\hbar^2}{2m}\nabla^2 + V)\psi_t. \qquad (6)$$

The wave function $\psi \in L_2(\mathbb{R}^3)$ is supposed to be sufficiently smooth, so that the dynamics exists.[4] We do not wish to limit our discussion to free particle motion. Indeed we wish to include in the description the escape of the particle for scattering states. We may therefore think of the potential V as a typical scattering potential with scattering center lying in G.

The initial position $\mathbf{Q}$ of the particle with initial wave function ψ is distributed according to the probability density $\rho = |\psi|^2$ (ψ is assumed to be normalized). The continuity equation (quantum flux equation) shows that the flux

$$(|\psi_t|^2, \mathbf{j}_t) = (|\psi_t|^2, |\psi_t|^2 \mathbf{v}_t,)$$

---

[3] The argument above does not directly apply in general to the sojourn time since the particle may leave the region G and return to it, but in cases of no return the escape time and sojourn time are the same.

[4] For the "existence of dynamics" see [11] and the contribution of K. Berndl in this volume.

is conserved, implying *equivariance*, i.e., that under the motion (5) the time-evolved probability density $\rho_t = |\psi_t|^2$ at all times.

We may now discuss our problem—in accordance with the title of this conference—on three levels.

Level 1: microscopic

The escape time T_e, the escape position $X(T_e)$ and the sojourn time T_s ((1)–(3)) are in Bohmian mechanics random variables on the probability space $\Omega = G$, the set of initial positions, equipped with the probability measure $\mathbb{P}$ given by the density $|\psi|^2$. Thus, in principle, all the statistics can be computed.

As we remarked above, the mean of T_s is readily computed from (3) applying Fubini's theorem and using equivariance (see also [10]):

$$\mathbb{E}(T_s) = \int_0^\infty \|P_G(t)\psi\|^2 dt. \qquad (7)$$

As we have said, this connects with (4) since

$$\mathbb{E}(T_s) = (\psi, \hat{t}_s \psi). \qquad (8)$$

The *distribution* of T_s, however, will depend in a rather complicated manner on ψ; in particular, there is no reason to expect it to be given by the the spectral measure of $\hat{t}_s$.

Level 2: mesoscopic

In general, the particle may well return to G after the first-exit time T_e. By the "mesoscopic level" we have in mind the situation where the particle never returns to G once it leaves. This is guaranteed by the following current positivity condition (CPC):

$$\text{CPC}: \quad \forall t \geq 0 \quad \text{and} \quad \forall \mathbf{q} \in \partial G, \quad \mathbf{j}(\mathbf{q}, t) \cdot \mathbf{n}(\mathbf{q}) \geq 0, \qquad (9)$$

where $\mathbf{n}(\mathbf{q})$ denotes the outward normal. This ensures that a trajectory may cross the surface ∂G of G at most once. For a given surface ∂G, the CPC is of course a condition on the wave function. Under the CPC, the product of the current and the surface-time element $d\sigma dt$ is the joint distribution for escape time and position:

$$\mathbb{P}((\mathbf{Q}_e, T_e) \in (d\sigma, dt)) = \mathbf{j}(\mathbf{q}, t) \cdot \mathbf{n}(\mathbf{q}) d\sigma dt \qquad (10)$$

For the sake of proper normalization, we assume that the particle leaves the region G with certainty, i.e., that

$$\lim_{t \to \infty} \|P_G(t)\psi\|^2 = 0. \qquad (11)$$

(This holds, for example, for a wave function whose spectral decomposition has only an absolutely continuous component [12].)

Integrating (10) over the surface and applying Gauss' theorem yields, by virtue of the continuity equation (the quantum flux equation), the escape-time density

$$\rho_e(t) = -\frac{d}{dt}\|P_G(t)\psi\|^2. \qquad (12)$$

This is now completely clear since the CPC implies that if the particle is in G at time t, it did not leave G before time t, i.e., that $\mathbb{P}(t_e > t) = \|P_G(t)\psi\|^2$.

We may introduce in (12) the self-adjoint operator

$$Z(t) = -\frac{d}{dt}P_G(t) = -\frac{i}{\hbar}[H, P_G(t)] \qquad (13)$$

so that

$$\rho_e(t) = (\psi, Z(t)\psi). \qquad (14)$$

$Z(t)$ is clearly a positive operator on a linear subspace of wave functions satisfying the CPC.

The expression for the probability that the escape time is within an arbitrary set Δ

$$\mathbb{P}(T_e \in \Delta) = (\psi, Z(\Delta)\psi) := (\psi, \int_\Delta Z(t)dt\psi) \qquad (15)$$

leads us now to the map $\Delta \mapsto Z(\Delta)$, which is a positive-operator-valued measure (POV) [13]. A POV is a generalization of a projection-valued measure (PV) in the sense that the operators $(Z(\Delta))$ need not be projections.[5]

As in the spectral theorem, we may associate with a POV a self-adjoint operator, namely its first moment. This association is, however, many-to-one, and the operator itself is of rather limited value: We may define $\hat{t}_e$ by

$$\hat{t}_e = \int_0^\infty tZ(t)dt = \int_0^\infty P_G(t)dt, \qquad (16)$$

where the second equality follows from (13). Thus $\hat{t}_e = \hat{t}_s$ (cf.(4)) as, of course, it must. But in general the $Z(\Delta)$ are not projections ($Z(\Delta)^2 \neq Z(\Delta)$ in general) and hence they do not define the spectral resolution of $\hat{t}_e$. Therefore the spectral measure for $\hat{t}_e$ does not in general describe the escape time distribution. (Of course this operator does yield the correct mean.)

We obtain the distribution of the escape position in the mesoscopic regime by integrating (10) over t:

$$\rho_e(\mathbf{q}) = \int_0^\infty \mathbf{j}(\mathbf{q},t) \cdot \mathbf{n}(\mathbf{q})dt, \quad \mathbf{q} \in \partial G. \qquad (17)$$

As with the escape time, one may extract from (17) a POV,[6] which will in general not be a PV.

Level 3: macroscopic

By the macroscopic level we have in mind the "scattering regime," in which the surface ∂G is very far from the scattering center so that the large-time asymptotic behavior becomes relevant.

While on the mesoscopic level we have assumed by definition that the CPC holds, we may expect that on the macroscopic level the CPC must hold for scattering states (i.e. states for which (11) holds). However, to be sure, we still assume the CPC.

For simplicity, we assume that G is the ball K_r of radius r centered at the origin, which coincides with the scattering center. Let C be a concentric cone and let $\Delta_r^C := \partial K_r \cap C$ be the surface defined by the intersection of the surface of the ball and the cone. What is the probability that the particle escapes through Δ_r^C in the limit as $r \to \infty$? For this we need to integrate (17) over Δ_r^C and take the limit. One easily finds heuristically that

---

[5] What we call a POV is in [3] called a "generalized resolution of identity" (GRI). It is introduced there in connection with the problem of the escape time operator.

[6] (17) as well as (10) indeed define a POV [15].

$$\lim_{r \to \infty} \mathbb{P}(\mathbf{Q}_e \in \Delta_r^C) = \lim_{r \to \infty} \int_{\Delta_r^C} \int_{t=0}^{\infty} \mathbf{j}_t \cdot d\sigma dt = \lim_{t \to \infty} \|P_C(t)\psi\|^2. \tag{18}$$

Together with Dollard's "scattering-into-cones theorem"[6] (which assumes the existence of the wave operators $\Omega_\pm := s - \lim_{t \to \mp\infty} e^{iHt} e^{-iH_0 t}$)

$$\lim_{t \to \infty} \|P_C(t)\psi\|^2 = \|P_C \mathcal{F} \Omega_-^\dagger \psi\|^2, \tag{19}$$

this yields the flux-across-surfaces theorem:

$$\lim_{r \to \infty} \int_{\Delta_r^C} \int_{t=0}^{\infty} \mathbf{j}_t \cdot d\sigma dt = \|P_C \mathcal{F} \Omega_-^\dagger \psi\|^2. \tag{20}$$

Here $\mathcal{F}$ denotes the Fourier transform.

We would like to emphasize that while (20) may conceivably be satisfied even without the CPC's holding (in some appropriate asymptotic sense), the CPC is crucial if we are to regard (20) as the probability that the particle first exits through Δ_r^C. Thus the CPC, together with the flux-across-surfaces theorem, appears to be the basis of scattering theory. The importance of (20) for scattering was recognized in [7], but the proof was only given for the free evolution.[7]

We wish next to point out that the POV structure of (17) characteristic of the mesoscopic regime has become, in the macroscopic regime, a PV structure: We may readily recognize on the r.h.s. of (20) the PV defined by the (direction of the) "asymptotic momentum" $\hat{\mathbf{p}}^+ = \Omega_- \hat{\mathbf{p}} \Omega_-^\dagger$, i.e., by $C \mapsto \chi_C(\hat{\mathbf{p}}^+)$.

We can also of course write down the differential cross section $\frac{d\sigma}{d\omega}$ by introducing angular variables θ_e, ϕ_e and the solid angle $d\omega$ and by writing the PV on the r.h.s. of (20) as

$$Z(d\omega) := \Omega_- \mathcal{F}^{-1} P_{C_{d\omega}} \mathcal{F} \Omega_-^\dagger = \chi_{C_{d\omega}}(\hat{\mathbf{p}}^+) \tag{21}$$

where $C_{d\omega}$ is the cone spanned by $d\omega$. Then

$$d\sigma := \lim_{r \to \infty} \mathbb{P}(\theta_e, \phi_e \in d\omega) = (\psi, Z(d\omega)\psi) \tag{22}$$

is the probability that the particle escapes asymptotically within the solid angle $d\omega$.

WHAT CAN BE OBSERVED?

The reader may now well ask the "Gretchenfrage"[8]: "Can one measure all the random variables of levels 1, 2, and 3? Can one do realistic experiments, involving apparatuses as measuring devices, which actually record the escape time and position as computed from the theory; in other words, can one *observe* these random variables?"

The answer is provided by the following straightforward and completely general result of Bohmian mechanics [15]: The statistics of quantities which are measured in an experiment are *always* given by a POV. Thus we may say the following concerning the measurability of the random variables $T_e, \mathbf{Q}_e$ and T_s:

---

[7] For the general proof see [14].
[8] "Nun sag, wie hast du's mit der Religion?
Du bist ein herzlich guter Mann
Allein ich glaub du hälst nicht viel davon." (Faust I, Goethe)

- Level 1, microscopic, the CPC is not assumed to hold.

  In general, the random variables are not distributed according to a POV and there can be no experiment to measure them.

  As a side remark, we mention that measurements of the mean sojourn time have been discussed in connection with the so-called Larmor clock: In the limit of an infinitely small magnetic field localized in G, it can be shown that the precession angle of the spin of the particle is proportional to the mean of the sojourn time [5].

- Level 2, mesoscopic, the CPC holds.

  As we have seen, in this case the statistics are always given by POV's. Moreover, the detectors around the boundary should accurately measure the random variables under discussion. (An analysis of the experiment and a discussion of the accuracy of the measurement is contained in [1].)

- Level 3, macroscopic.

  The statistics of the escape position are given by (the PV associated with) the asymptotic momentum, which of course can be measured.

WHAT ROLE DO OPERATORS PLAY?

We have remarked that Bohmian mechanics yields in very natural situations a description of what goes on which cannot be directly verified through measurements. But that turns out to be a strength rather than a weakness of the (indeed of any *complete*) theory: In Bohmian mechanics the act of measurement can be analyzed in as much detail as one wishes; in particular, Bohmian mechanics tells us precisely which quantities can be measured and which physical processes qualify as measurements of whatever it is that can be measured. This analysis reveals the status of operators in the description of nature, and allows a clear view of the range of applicability of the usual quantum mechanical formulas. The particular example of escape statistics exemplifies the general situation [15], namely, that operators as "observables" appear merely as computational tools in the phenomenology of certain types of experiments—those for which the statistics of the result are governed by a PV. However, they have no fundamental significance and do not at all reflect what, on the microscopic level, is really going on.

Acknowledgments

This work has been partially supported by NSF Grants no. DMS-9105661 and no. DMS-9305930, by DFG, and by INFN.

REFERENCES

1. G. R. Allcock, The time of arrival in quantum mechanics I-III, *Ann. of Phys.* **53**, 253 (1969).
2. H. Ekstein and A. J. F. Siegert, On a reinterpretation of decay experiments, *Ann. of Phys.* **68**, 509 (1971).

3. B. Misra and E C G. Sudarshan, The Zeno's paradox in quantum theory, *Journal of Math. Physics* **18**, 756 (1977).
4. W. Pauli, Encyclopedia of physics, S. Flügge ed., Springer, Berlin, Heidelberg, New York, 60 (1958).
5. Ph. Martin, Time delay of quantum scattering processes, *Acta Physica Austriaca, Suppl.* XXIII, 157 (1981).
6. J.D. Dollard: Scattering into cones, *Comm. math. Phys.* **12**, 193 (1969).
7. J.-M. Combes, R. G. Newton, and R. Shtokhamer, Scattering into cones and flux across surfaces, *Phys. Rev. D* **11**, 366 (1975).
8. D. Bohm, A suggested interpretation of quantum theory in terms of hidden variables I+II, *Phys. Rev.* **85**, 166 (1952).
9. D. Dürr, S. Goldstein, and N. Zanghí, Quantum equilibrium and the origin of absolute uncertainty, *J. Stat. Phys.*, **67**, 843 (1992).
10. C. R. Leavens, Transmission, reflection and dwell times within Bohm's causal interpretation of quantum mechanics, *Solid State Comm.* **74**, 923 (1990).
11. K. Berndl, D. Dürr, G. Peruzzi, S. Goldstein, and N. Zanghí, On the global existence of Bohmian mechanics, (in preparation).
12. W. Amrein, J. Jauch, and K. Sinha, Scattering Theory in Quantum Mechanics, Lecture Notes and Suppl. in Physics, W.A. Benjamin Inc., London-Amsterdam-Don Mills-Sydney-Tokyo (1977).
13. D. Davies, Quantum Theory of Open Systems, Academic Press, London-New York-San Franzisco (1976).
14. M. Daumer, PhD thesis at the LMU, Munich, (in preparation).
15. M. Daumer, D. Dürr, S. Goldstein, and N. Zanghí: On the role of operators in quantum theory, (in preparation).

TOWARDS THE EUCLIDEAN FORMULATION OF QUANTUM STATISTICAL MECHANICS

Roman Gielerak[1], Lech Jakóbczyk,[1] Robert Olkiewicz

Institute of Theoretical Physics
University of Wrocław
pl.M.Borna 9, PL-50-205 Wrocław, Poland

INTRODUCTION

Several aspects of the Quantum Statistical Mechanics in the Euclidean Region are discussed. The axiomatic approach to the purely Euclidean formulation of QSM is proposed. Some reconstruction procedures of real time structures are presented. A new functional integral representation of multi-time Green functions by a bounded complex reflection positive measures is obtained.

1. EUCLIDEAN MULTI-TIME GREEN FUNCTIONS

Let $(\mathcal{A}, \alpha_t, \omega, \beta)$ be given, where $\mathcal{A}$ is a C^* - algebra with unit, α_t is a norm - continuous one - parameter group of $*$ - automorphisms of $\mathcal{A}$ and ω is a β - KMS state on $(\mathcal{A}, \alpha_t)$ [1]. The corresponding GNS quadruplet is denoted by $(\mathcal{H}_\omega, \Omega_\omega, \pi_\omega, e^{itH_\omega})$ and the multi-time Green functions

$$G^\omega : \mathcal{A} \times \cdots \times \mathcal{A} \times \mathbf{R} \times \cdots \times \mathbf{R} \mapsto \mathbf{C}$$

are defined as

$$G^\omega_{A_1,\ldots,A_n}(t_1,\ldots,t_n) = \omega(\alpha_{t_1}(A_1)\cdots\alpha_{t_n}(A_n)) = G^\omega_{A_1,\ldots,A_n}(t_2 - t_1,\ldots,t_n - t_{n-1}) =$$

$$= <\Omega_\omega, \pi_\omega(A_1)e^{i(t_2-t_1)H_\omega}\pi_\omega(A_2)\cdots e^{i(t_n-t_{n-1})H_\omega}\pi_\omega(A_n)\Omega_\omega > . \qquad (1.1)$$

By Araki theorem [2], (1.1) can be analytically continued to the tubular domain $T^n_\beta \subset \mathbf{C}^n$ given by

$$T^n_\beta = \{(z_1,\ldots,z_n) \in \mathbf{C}^n : Im\, z_j \leq Im\, z_{j+1}; \sum_{j=1}^{n-1}(Im\, z_{j+1} - Im\, z_j) < \beta\}.$$

---

[1]Supported by KBN grant No 200609101

The Euclidean Region $T_\beta^{E,n}$ is defined as

$$T_\beta^{E,n} = \{(is_1, \ldots, is_n) : -\beta/2 < s_1 \leq \cdots \leq s_n < \beta/2\}$$

and the Euclidean Green functions $G^{E,\omega}$ are restrictions of analytically continued G^ω to $\overline{T_\beta^{E,n}}$ i.e.

$$G^{E,\omega}_{A_1,\ldots,A_n}(s_1,\ldots,s_n) = G^\omega_{A_1,\ldots,A_n}(is_1,\ldots,is_n). \tag{1.2}$$

From the Araki theorem we know that

$$|G^{E,\omega}_{A_1,\ldots,A_n}(s_1,\ldots,s_n)| \leq \|A_1\| \cdots \|A_n\|$$

pointwise on $\overline{T_\beta^{E,n}}$ and $G^{E,\omega}_{A_1,\ldots,A_n}$ are continuous in $(s_1,\ldots,s_n)$ on $\overline{T_\beta^{E,n}}$. The KMS condition can be rewritten in terms of $G^{E,\omega}$ as:

$$G^{E,\omega}_{A^*,B}(-\beta/2, \beta/2) = G^{E,\omega}_{B,A^*}(s,s) \tag{1.3}$$

for every $A, B \in \mathcal{A}$.

The real time algebra of observables $R(\mathcal{A})$ is defined as in [3] where fundamental properties of $R(\mathcal{A})$ are discussed. The Euclidean algebra of observables $E_\beta(\mathcal{A})$ is defined by the following construction. Let $\mathcal{A}_\beta$ be a free algebra generated by symbols $P = (A_1, s_1) \cdots (A_n, s_n)$, where $A_j \in \mathcal{A}$ and $-\beta/2 \leq s_1 \leq s_2 \leq \cdots \leq s_n \leq \beta/2$. The ideal generated by the relations

$$(A,s)(B,s) - (AB,s),$$

$$\lambda(A,s) + (B,s) - (\lambda A + B, s),$$

$$(A_1, s_1)(A_2, s_2) - (A_i, s_i)(A_{i^C}, s_{i^C}),$$

where $i = 1, 2$ is such that $s_i \leq s_{i^C}$, $i^C = \{1,2\} \setminus \{i\}$, and $(1,0) - (1,s)$, is denoted by $\mathcal{I}$. Then the algebra $E_\beta(\mathcal{A})$ is defined as the quotient $\mathcal{A}_\beta/\mathcal{I}$. Note that the product in $E_\beta(\mathcal{A})$ is a chronological multiplication. The positive (respectively negative) time subalgebra $E_\beta^+(\mathcal{A})$ (respectively $E_\beta^-(\mathcal{A})$) is defined as

$$E_\beta^\pm(\mathcal{A}) = \{P \in E_\beta(\mathcal{A}) : P \text{ is supported on positive (negative) euclidean times}\}$$

The reflection R is defined by:

$$R : E_\beta^\pm(\mathcal{A}) \mapsto E_\beta^\mp(\mathcal{A}),$$

$$R(A_1, s_1) \cdots (A_n, s_n) = (A_n^*, -s_n) \cdots (A_1^*, -s_1). \tag{1.4}$$

The subalgebras $E_\beta^{\leq s}(\mathcal{A})$ are defined as those generated by the symbols $P = (A_1, s_1) \cdots (A_n, s_n)$ for which $\max\{s_i\} \leq s$. Then the local shift transformation τ_α can be defined on $E_\beta^{\leq s}(\mathcal{A})$ by

$$\tau_\alpha((A_1 s_1) \cdots (A_n, s_n)) = (A_1, s_1 + \alpha) \cdots (A_n, s_n + \alpha) \tag{1.5}$$

provided that $\alpha \leq \beta/2 - s$. Given multi-time Green functions $G^{E,\omega}_{A_1,\ldots,A_n}(s_1,\ldots,s_n)$ determine a linear functional on the algebra $E_\beta(\mathcal{A})$ by the formula:

$$G^{E,\omega}((A_1, s_1) \cdots (A_n, s_n)) = G^{E,\omega}_{A_1,\ldots,A_n}(s_1,\ldots,s_n). \tag{1.6}$$

Proposition 1. *Let $(\mathcal{A}, \alpha_t, \omega, \beta)$ be a $C^* - \beta$ - KMS system and let $G^{E,\omega}$ denote the corresponding Euclidean Green functional on the Euclidean algebra $E_\beta(\mathcal{A})$. Then:*

EG(1) : the Euclidean Green functional $G^{E,\omega}$ is continuous on $\mathcal{A}^{\times n}$ for any n as a multilinear functional and in the Euclidean multi-time $(s_1, \ldots, s_n)$,

EG(2) : the functional $G^{E,\omega}$ is reflection positive on $E_\beta(\mathcal{A})$:

$$G^{E,\omega}((RP) \cdot P) \geq 0$$

for every $P \in E_\beta^+(\mathcal{A})$ and is reflection invariant :

$$G^{E,\omega}(P) = \overline{G^{E,\omega}(RP)}$$

EG(3) : $G^{E,\omega}$ is (locally) shift invariant,

EG(4) : for every $P \in E_\beta^+(\mathcal{A})$ and every $A \in \mathcal{A}$

$$\mid G^{E,\omega}((RP) \cdot (A,0) \cdot P \cdot (A,0)) \mid \leq \parallel A \parallel^2 \, G^{E,\omega}((RP) \cdot P)$$

Proof :
We pass to the corresponding GNS quadruplet and apply the Araki theorem. In particular, EG(4) follows from the Cauchy – Schwartz inequality.

It appears that any linear functional on $E_\beta(\mathcal{A})$ fulfilling EG(1) – EG(4) can be analytically continued to the real time algebra $R(\mathcal{A})$.

Proposition 2. *Let G^E be the functional on $E_\beta(\mathcal{A})$ satisfying conditions EG(1) – EG(4). Then there exists a functional G on $R(\mathcal{A})$ such that :*

G(1) : *the Green functional G is continuous on $\mathcal{A}^{\times n}$ and on $\mathbf{R}^n$ separately for every $n \in \mathbf{N}$,*

G(2) : *G is positive in the following sense :*

$$G(P^* \cdot P) \geq 0$$

for every $P \in R(\mathcal{A})$ and $G(P^*) = \overline{G(P)}$,

G(3) : *G is translation invariant.*

The proof is based on the reflection positivity of G^E and the application of the local symmetric semigroup concept [4] (see also Section 2). The following statement explains the main ideas of the proof.

Statement 1. *Let G^E be a functional on $E_\beta(\mathcal{A})$ satisfying EG(1) – EG(4). Then there exists a Hilbert space $\mathcal{H}(G)$, $*$ - representation π_E of $\mathcal{A}$ in $L(\mathcal{H}(G))$, a vector $\Omega_E \in \mathcal{H}(G)$ and a unitary representation U_t of $\mathbf{R}$ such that analytic continuation of G^E is given by :*

$$G_{A_1,\ldots,A_n}(t_1,\ldots,t_n) =$$
$$< \Omega_E, U_{t_n}\pi_E(A_n)U_{-t_n}\cdots U_{t_1}\pi_E(A_1)U_{-t_1}\Omega_E > . \tag{1.7}$$

Moreover, the vector Ω_E is invariant under the action of U_t.

One of the main tasks of our investigation is to find the conditions on G^E which allow the reconstruction of C^* (or W^*) - β - KMS system such that the Euclidean multi-time Green functions of this system coincide with G^E. This problem will be discussed in the next section (see also [5]).

It is well known [6] that the local quantum field theory restricted to the Euclidean points gives a new interesting probabilistic structures. The so-called stochastically positive KMS systems studied by Klein and Landau on the abstract level [7] have the similar property. However, the stochastic positivity seems to be a very restrictive condition [1] and only a few systems are known to satisfy this condition [2].

In Section 3 we obtain a new functional integral representation of the multi-time Green functions for a large class of systems which are not necessary stochastically positive. However, the corresponding measures are not positive. This leads to a very interesting problem of extending the notions of Gibbs states and DLR equations to the case of complex spin distributions. Results in this subject with applications to QSM will be reported elsewhere.

2. RECONSTRUCTIONS FROM THE ONE TIME EUCLIDEAN FUNCTIONALS

Two different (a priori) constructions of a real time quantum structures from a given one time Euclidean Green functional will be presented here. We first discuss a method based on the application of the Poisson kernel.

The Poisson kernel P^β for the strip region $T_\beta^1 = \{z = t + is : 0 < s < \beta\}$ is given as $P^\beta = P_T^\beta + P_\perp^\beta$, where

$$P_{\perp(T)}^\beta(t+is \mid \rho) = \frac{1}{2\beta} \sin\frac{\pi s}{\beta} (\cosh\frac{\pi(t-\rho)}{\beta} \pm \cos\frac{\pi s}{\beta})^{-1} \qquad (2.1)$$

as follows by standard calculation. For a pair $(\varphi_0, \varphi_1) \in C_p(\mathbf{R})$ of continuous and polynomially bounded functions on $\mathbf{R}$ the formula

$$F(\varphi_0, \varphi_1)(t+is) = \int_{-\infty}^{\infty} (P_\perp^\beta(t+is \mid \rho)\varphi_0(\rho) + P_T^\beta(t+is \mid \rho)\varphi_1(\rho))\, d\rho \qquad (2.2)$$

gives a harmonic function in T_β^1 such that

$$\lim_{s \downarrow 0 (s \uparrow \beta)} F(\varphi_0, \varphi_1)(t+is) = \varphi_0\,(\varphi_1)(t)$$

pointwise and for any $t \in \mathbf{R}$. A given harmonic function F is holomorphic iff zF is again harmonic in T_β^1. This remark together with the uniqueness of the solution of the Dirichlet problem in T_β^1 yields the following integral identity :

$$\int_{-\infty}^{\infty} \frac{t-\rho+i\beta/2}{\cosh\frac{\pi(t-\rho)}{\beta}} \varphi_0(\rho)\, d\rho = - \int_{-\infty}^{\infty} \frac{t-\rho-i\beta/2}{\cosh\frac{\pi(t-\rho)}{\beta}} \varphi_1(\rho)\, d\rho \qquad (2.3)$$

as a necessary condition for $F(\varphi_0, \varphi_1)$ to be holomorphic in T_β^1. Taking the Fourier transform of (2.3), we obtain the following identity

$$\widehat{\varphi_1}(k) = e^{-\beta k}\widehat{\varphi_0}(k). \qquad (2.4)$$

---

[1] Examples of stochastic positive systems are listed in [8]. The continuous systems (quantum gases) [9] also satisfy stochastic positivity. It seems to be interesting problem to associate to them a corresponging Markov process.

[2] In [10] the Euclidean formulation of the Bisognano - Wichmann theorem [11] leads to the new application of stochastic positive KMS systems to study the Unruch effect.

In fact it can be proved that (2.4) is also sufficient for $F(\varphi_0, \varphi_1)$ to be holomorphic in T_β^1. The following version of the Paley - Wiener like theorem was proved in [5]:

Theorem 1. *Let $(\varphi_0, \varphi_1) \in C_p(\mathbf{R})$. Then the harmonic extension $F(\varphi_0, \varphi_1)$ is holomorphic in T_β^1 and continuous on the boundary ∂T_β^1 iff (2.4) holds in the sense of $\mathcal{S}'$. If (2.4) is fulfilled then there exists a unique $\widehat{\varphi}(k) \in \mathcal{S}'$ such that $e^{-sk}\widehat{\varphi}(k) \in \mathcal{S}'$ for any $0 \leq s \leq \beta$ and moreover*

$$F(\varphi_0, \varphi_1)(t+is) = \int_{-\infty}^{\infty} e^{-sk} e^{itk} \widehat{\varphi}(k)\, dk. \tag{2.5}$$

Let us put $\varphi_0(t) = \omega(A\alpha_t(B))$ and $\varphi_1(t) = \omega(\alpha_t(B)A)$. Applying Theorem 2.1 we obtain an alternative proof of characterization of KMS state [1], without referring to GNS construction (it may be useful for the study of super - KMS functionals arising in the non - commutative differential geometry (NCG)).[3] It follows from Proposition 1.2 that one time Euclidean Green functions $G_{A,B}^E(s)$ are reflection positive in the following sense :
for any $A_1, \ldots, A_n \in \mathcal{A}$, $s_1, \ldots, s_n \in [0, \beta/2]$, $c_1, \ldots, c_n \in \mathbf{C}$

$$\sum_{i,j=1}^{n} \overline{c_i} c_j G_{A_i^*, A_j}^E(s_i + s_j) \geq 0. \tag{2.6}$$

By the inspection of the Poisson kernel representation (2.2), it follows that

$$G_{A,B}^E(s) = G_{B,A}^E(\beta - s). \tag{2.7}$$

It appears that (2.6) and (2.7) are crucial for the construction of real time functional G on $\mathcal{A} \times \mathcal{A} \times \mathbf{R}$ fulfilling conditions 1 – 7 of Bratelli – Robinson Theorems 6.3.27 and 6.3.28 [1].

Theorem 2. *Let G^E be a functional on $\mathcal{A} \times \mathcal{A} \times [0, \beta]$ such that :*

$EG_2(1)$: *for every $s \in [0, \beta]$, $G^E(s)$ is bilinear on $\mathcal{A} \times \mathcal{A}$ and such that :*

1. $G_{1,1}^E(0) = 1$,
2. *for all $A, B, C \in \mathcal{A}$, $G_{AC,B}^E(0) = G_{A,CB}^E(0)$.*

$EG_2(2)$: *for any pair $A, B \in \mathcal{A}$, $G_{A,B}^E(s)$ is continuous in s and for every $A \in \mathcal{A}$, $G_{A^*, A}^E(s)$ is reflection positive on $[0, \beta]$.*

$EG_2(3)$: *for every $A \in \mathcal{A}$,*

$$G_{A^*, A}^E(s) = G_{A, A^*}^E(\beta - s).$$

Then there exists a unique functional G on $\mathcal{A} \times \mathcal{A} \times \mathbf{R}$ such that :

$G_2(1)$: *for every $t \in \mathbf{R}$, $G(t)$ is bilinear on $\mathcal{A} \times \mathcal{A}$ and such that :*

1. $G_{1,1}(0) = 1$,
2. *for every $A, B, C \in \mathcal{A}$, $G_{AC,B}(0) = G_{A,CB}(0)$.*

---

[3] In fact a hyper-distribution theory formulation of Theorem 2.1 seems to be necessary for NCG applications.

$G_2(2)$: for every $A, B \in \mathcal{A}$, $G_{A,B}(t)$ is continuous in t and

$$\sum_{i,j=1}^{n} G_{A_i^*, A_j}(t_i - t_j) \geq 0$$

for all finite sequences $(A_i)_{i=1,\ldots,n}$ in $\mathcal{A}$ and $(t_i)_{i=1,\ldots,n}$ in $\mathbf{R}$.

$G_2(3)$: for all f with $\hat{f} \in \mathcal{D}$ and all $A, B \in \mathcal{A}$,

$$\int f(t) G_{A,B}(t) \, dt = \int f(t + i\beta) G_{B,A}(-t) \, dt. \tag{2.8}$$

Outline of the proof : (for all details see [5])
From the Widder theorem [12] and $EG_2(2)$ it follows that there exists a unique positive Borel measure $\nu_{A^*,A}$ on $\mathbf{R}$ such that

$$G_{A^*,A}^E(s) = \int_{-\infty}^{\infty} e^{-sp} \, d\nu_{A^*,A}(p). \tag{2.9}$$

Using $EG_2(1)$ and polarization formula, we obtain that for every $A, B \in \mathcal{A}$ there exists a complex bounded measure $\nu_{A,B}$ that

$$G_{A,B}^E(s) = \int_{-\infty}^{\infty} e^{-sp} \, d\nu_{A,B}(p)$$

Then we define

$$G_{A,B}(t) = \int_{-\infty}^{\infty} e^{itp} \, d\nu_{A,B}(p). \tag{2.10}$$

The positive definiteness of (2.10) $G_2(1)$ follows from the observation : for any sequence $A_1, \ldots, A_n \in \mathcal{A}$, $\varphi \in \mathcal{S}$, $\varphi \geq 0$

$$\sum_{i,j=1}^{n} \nu_{A_i^*, A_j}(\varphi) = \sum_{i,j=1}^{n} G_{A_i^*, A_j}^E(\varphi) =$$

$$G_{(\sum_{i=1}^{n} A_i^*, \sum_{i=1}^{n} A_i)}^E(\varphi) \geq 0$$

Therefore,

$$\sum_{i,j=1}^{n} G_{A_i^*, A_j}(t_i - t_j) = \sum_{i,j=1}^{n} \int_{-\infty}^{\infty} e^{i(t_i - t_j)p} \, d\nu_{A_i^*, A_j}(p) \geq 0$$

by the application of the Schur Lemma. The condition $EG_2(3)$ is used to verify the weak KMS condition $G_2(3)$.

Another construction can be realized by using the technique of local semigroups in the sense of Klein and Landau work [7]. For this goal, let us assume that G^E satisfies $EG_2(1)$ and $EG_2(2)$. Let V^β be a complex linear space generated by symbols (A, s), $A \in \mathcal{A}$, $-\beta/2 \leq s \leq \beta/2$ (note the change of parametrization of s), and let V_+^β be the subspace of V^β generated by (A, s) for $0 \leq s \leq \beta/2$. On V_+^β we define positive - definite sesquilinear form $<, >^E$ by

$$< \sum_{k} c_k(A_k, s_k), \sum_{l} d_l(B_l, s_l) >^E = \sum_{k,l} \overline{c_k} d_l G_{A_k^*, B_l}^E(s_k + s_l). \tag{2.11}$$

If $\mathcal{N}$ is the kernel of (2.11), we define (Euclidean) Hilbert space $\mathcal{H}^E$ as the Hilbert space completion of the quotient $V_+^\beta/\mathcal{N}$. The local shift transformations τ_α defined on V_+^β can be lifted to $\mathcal{H}^E$ forming a local symmetric semigroup T_s [7]. The local semigroup generates a unique unitary group U_t acting on $\mathcal{H}^E$ which can be used for the construction of real time functional G.

Proposition 3. Let G^E be a functional on $\mathcal{A} \times \mathcal{A} \times [0, \beta]$ satisfying conditions $EG_2(1) - EG_2(3)$. Then the functional

$$\widetilde{G}_{A,B}(t) = <[(A,0)], U_t[(B,0)]>^E$$

satisfies $G_2(1) - G_2(3)$. Moreover, $\widetilde{G} = G$.

If we have a functional G with properties $G_2(1) - G_2(3)$, we can try to reconstruct some W^* - KMS system for which the one time Green functions are given by G. It appears that the problem of identification of the unitary group U_t constructed above with the modular group of $(\pi(\{(A,0)\})'', \Omega)$ in $\mathcal{H}^E$ can be formulated also in the Euclidean language (for some results in this direction see [5]).

3. FUNCTIONAL INTEGRAL REPRESENTATIONS

Let $\mathcal{A}_{ab}$ be an abelian C^* - subalgebra of $\mathcal{A}$ such that the orbit of $\mathcal{A}_{ab}$ under the dynamics α_t is norm dense in $\mathcal{A}$. Then any α_t - invariant state ω on $\mathcal{A}$ is uniquely determined by its restriction to the algebra generated by $\alpha_t(\mathcal{A}_{ab})$. Let us denote by $\sigma(\mathcal{A}_{ab})$ the spectrum of $\mathcal{A}_{ab}$. From the Gelfand - Naimark Theorem, $\mathcal{A}_{ab}$ is isomorphic to the C^* - algebra $C(\sigma(\mathcal{A}_{ab}))$. The corresponding isomorphism will be denoted by $A \to \widehat{A}$. Let $\Omega_\beta(\sigma(\mathcal{A}_{ab}))$ be the space of all maps from $[-\beta/2, \beta/2]$ to $\sigma(\mathcal{A}_{ab})$. Applying the Araki theorem and using standard probabilistic constructions, we can prove the following result :

Theorem 3. Let $(\mathcal{A}, \alpha_t, \omega)$ be a β - KMS system and let $\mathcal{A}_{ab}$ be an abelian subalgebra as above. Then there exists a unique, complex σ - additive Baire measure μ^β on the path space $\Omega_\beta(\sigma(\mathcal{A}_{ab}))$ such that :

$$G^E_{A_1,\ldots,A_n}(s_1,\ldots,s_n) = \int_{\Omega_\beta(\sigma(\mathcal{A}_{ab}))} d\mu^\beta(q) \widehat{A}_1(q(s_1))\cdots \widehat{A}_n(q(s_n)). \qquad (3.1)$$

The measure μ^β has total variation 1 and is reflection invariant, reflection positive and locally shift invariant. Moreover, for every $A \in \mathcal{A}_{ab}$,

$$\int_{\Omega_\beta(\sigma(\mathcal{A}_{ab}))} d\mu^\beta(q) \mid \widehat{A}(q(\beta/2)) - \widehat{A}(q(-\beta/2)) \mid^2 = 0. \qquad (3.2)$$

The interesting aspect of this theorem is that it can be applied to KMS systems which are not necessary stochastic positive. It appears that Quantum Statistical Mechanics of KMS states satisfying assumption of Theorem 3.1, can be formulated as Classical Statistical Mechanics on the space of paths of the corresponding Quantum Mechanical system, where the Gibbs measures are replaced by complex measures. This leads to many interesting problems of extending the framework of Classical Statistical Mechanics to cover the case of not necessarily probabilistic measures. This topic will be discussed elsewhere.

REFERENCES

1. O. Bratelli and D.W. Robinson, "Operator Algebras and Quantum Statistical Mechanics. II", Springer, New York, (1981).
2. H. Araki, *Publ.RIMS*, **4**, 361(1968).
3. M. Winnink, Some general properties of thermodynamic states in an algebraic approach, *in*: "Statistical Mechanics and Field Theory", Jerusalem (1971).
4. A. Klein and L.J. Landau, *J.Funct.Anal.*, **42**, 368 (1981).
5. R. Gielerak, L. Jakóbczyk and R. Olkiewicz, Reconstruction of KMS structure from Euclidean Green functions, BiBoS preprint 581/6/1993.
6. B. Simon, "The $P(\varphi)_2$ - Euclidean (Quantum) Field Theory", Princeton, (1974).
7. A. Klein and L.J. Landau, *J.Funct.Anal.*, **44**, 121 (1981).
8. W. Driessler, L. Landau and J.F. Perez, *J.Stat.Phys.*, **20**, 123 (1979).
9. J. Ginibre, Some applications of functional integration in statistical mechanics, *in*: "Statistical Mechanics and Quantum Field Theory", Gordon Breach, New-York (1971).
10. R. Gielerak and L. Jakóbczyk, Canonical Markov processes connected to Bisognano - Wichmann theorem., in preparation.
11. J.J. Bisognano and E.H. Wichmann, *J.Math.Phys.*, **16**, 985 (1975).
12. D. Widder, *Bull.Amer.Math.Soc.*, **40**, 321 (1934).

LARGE DEVIATIONS IN THE SPHERICAL MODEL

A.E. Patrick

Institute of Theoretical Physics
Celestijnenlaan 200D, 3001 Heverlee
Belgium

Recently considerable progress was achieved in the investigation of the large deviation probabilities for the total spin (magnetization)

$$\text{Prob}\left[\frac{1}{N}\sum_{i=1}^{N} s_i \in [a, b]\right]$$

in the Ising model [1, 2, 3]. Deep understanding of the large deviation properties of discrete-spin lattice systems was gained, in spite of the fact that most of the results are limited to the 2-D Ising model. It is expected however, that models with continuous spin possess significantly different large deviation properties. Since the investigation of n-vector models (the simplest realistic continuous spin models) is still more difficult then that of the Ising model, it is reasonable to study the spherical model [4] in order to gain insight in the large deviation properties of continuous-spin lattice models. The spherical model is exactly solvable in any dimension and, despite it is often considered as unphysical, it presents a nontrivial example of a random field on $\mathbb{Z}^d$. And so, a detailed investigation of its large deviation properties should not be omitted.

The model is defined as follows. There is a continuous random variable (spin) $\sigma_j \in \mathbb{R}^1$ at each site $j \equiv (j_1, \ldots, j_d)$ of a d-dimensional integer lattice $\mathbb{Z}^d$. The system of spins, confined in the box $\Omega_n \equiv \{j \in \mathbb{Z}^d : \cdot 1 \leq j_k \leq n, \forall k\}$, interacts via the Hamiltonian

$$H_n(\sigma) = -J \sum_{(j;k) \in \mathcal{N}_n} \sigma_j \sigma_k - J_1 \sum_{(j;k) \in \mathcal{B}_n} \sigma_j \sigma_k \qquad (1)$$

where $J > 0$. The first summation in Eq. (1) runs over the set of nearest neighbours

$$\mathcal{N}_n = \left\{(j;k) \in \Omega_n \times \Omega_n : \sum_{\nu=1}^{d} |j_\nu - k_\nu| = 1\right\}.$$

The second summation runs over the pairs of sites bridging the boundary. For instance, the set $\mathcal{B}_n$ is empty in the case of the empty boundary conditions (open edges), and is given by

$$B_n = \bigcup_{\nu=1}^{d} \{(j;k) \in \Omega_n \times \Omega_n : j_\nu = 1, \ k_\nu = n, \ j_l = k_l \text{ for } l \neq \nu\}$$

for periodic boundary conditions in all dimensions (in this case $J_1 = J$).

The choice of boundary conditions is not of vital importance for us, the investigation can be carried out for many of them (e.g., periodic in some dimensions and empty in others, cyclic, antiperiodic, twisted, etc.). Here, we will only consider boundary conditions which are periodic in all dimensions (lattice wrapped in a torus).

The joint probability distribution $P_n(\cdot)$ of the system of spins in the box Ω_n is defined by the density

$$p_n(\sigma_j; j \in \Omega_n) = \frac{1}{\Theta_n} \exp[-\beta H_n(\sigma)] \qquad (2)$$

with respect to the "a priori" measure

$$\mu_n(d\sigma) = \delta\left(\sum_{j \in \Omega_n} \sigma_j^2 - N\right) \prod_{j \in \Omega_n} d\sigma_j, \qquad (3)$$

where $N = n^d = |\Omega_n|$.

Our main objective is the investigation of the large deviation probabilities

$$P_n\left(\frac{1}{|W|} \sum_{j \in W} \sigma_j \in [a;b]\right)$$

for the block-spins $S_W = \sum_{j \in W} \sigma_j$, where $W \subset \Omega_n$ is a parallelepiped which size possibly depends on n. In particular we would like to know the behaviour of the block-spin distribution densities

$$f_{n,W}(\mu) = \frac{1}{\Theta_n} \int_{\mathbb{R}^1} \cdots \int_{\mathbb{R}^1} \delta\left(\tfrac{1}{|W|} S_W - \mu\right) \exp[-\beta H_n(\sigma)] \mu_n(d\sigma), \qquad (4)$$

for large n and $|W|$. Note that the multiple integral in Eq. (4) can be interpreted as the partition function $\Theta_{n,W}(\mu)$ of the spherical model in the microcanonical ensemble with respect to the block-spin S_W.

One can perform the integration in the expressions for $\Theta_{n,W}(\mu)$ and Θ_n using a standard technique of Berlin and Kac [4]. Namely, one first introduces new integration variables y_m, $m \in \Omega_n$: $\sigma_j = \sum_{m \in \Omega_n} V_{j,m} y_m$, where $\{V_{j,m}\}_{j \in \Omega_n}$, $m \in \Omega_n$ are eigenvectors of the matrix $\widehat{C}^{(n)}$ associated with the Hamiltonian (1): $H_n(\sigma) \equiv \sum_{i,j \in \Omega_n} C_{i,j}^{(n)} \sigma_i \sigma_j$. In the case of periodic b.c. the eigenvectors $\mathbf{V}_m$, $m \in \Omega_n$ are given by

$$\mathbf{V}_m = \left\{V_{j,m} = \frac{1}{\sqrt{N}} \prod_{k=1}^{d}\left[\cos\frac{2\pi(j_k-1)(m_k-1)}{n} + \sin\frac{2\pi(j_k-1)(m_k-1)}{n}\right]\right\}_{j \in \Omega_n},$$

and the corresponding eigenvalues λ_m by

$$\lambda_{(m_1,\ldots,m_d)} = \sum_{k=1}^{d} \cos\frac{2\pi(m_k-1)}{n}.$$

Next, one replaces the delta functions by their integral representations

$$\delta(a) = \frac{1}{2\pi} \int_{-\infty}^{+\infty} e^{i\tau a} d\tau,$$

exchanges the integration order, and performs the integration over the variables y_s, $s \in \Omega_n$. One obtains then

$$\Theta_{n,W}(\mu) = \frac{|W|}{4\pi^2 i} \int_{-i\infty+\tau_0}^{i\infty+\tau_0} d\tau\, e^{\tau N} \int_{-\infty}^{\infty} dx\, e^{ix\mu|W|} \prod_{j\in\Omega_n} \sqrt{\frac{\pi}{\tau - \beta J \lambda_j}} \exp\left[-\frac{x^2 \gamma_j^2}{4(\tau - \beta J \lambda_j)}\right], \quad (5)$$

where $\gamma_j \equiv \sum_{l\in W} V_{j,l}$. Integrating over the variable x and introducing a new integration variable z by $\tau = \beta J z$ one arrives at

$$\Theta_{n,W}(\mu) = \frac{\beta J |W|}{2\pi i} \left(\frac{\pi}{\beta J}\right)^{(N-1)/2} \int_{-i\infty+z_0}^{i\infty+z_0} \frac{dz}{\sqrt{\Sigma(z)(z - \lambda_{(1,\ldots,1)})}} \exp\left[N\beta J \Phi_n(z,\mu)\right], \quad (6)$$

where we used the notations

$$\Sigma(z) = \sum_{j\in\Omega_n} \frac{\gamma_j^2}{z - \lambda_j}, \quad (7)$$

$$\Phi_n(z,\mu) = z - \frac{1}{2\beta J N} \sum_{j\in\Omega_n\setminus(1,\ldots,1)} \log(z - \lambda_j) - \frac{|W|^2}{N} \frac{\mu^2}{\Sigma(z)}. \quad (8)$$

The integral in Eq. (6) is evaluated by (a modification of) the saddle point method. In all the cases considered in the present paper one can formulate the result as follows

$$\Theta_{n,W}(\mu) = \left(\frac{\pi}{\beta J}\right)^{N/2} \exp\left[N\beta J \Phi_n(z_n^*,\mu) + O(\log n)\right], \quad (9)$$

where z_n^* is the minimum of the function $\Phi_n(z,\mu)$ on the interval $(s_n^{\max};\infty)$, and $s_n^{\max}$ is the maximal singular point of the integrand in Eq. (6). The singularities of the integrand at the maximal eigenvalue $\lambda_n^{\max} \equiv \lambda_{(1,\ldots,1)}$ of the matrix $\widehat{C}^{(n)}$ cancel each other (if $\gamma_{(1,\ldots,1)} \neq 0$), hence, $s_n^{\max}$ coincides with the maximal zero $\eta_n^{\max}$ of the function $\Sigma(z)$ (if $\gamma_{(2,1,\ldots,1)} \neq 0$). It is easy to see that $\eta_n^{\max}$ is less than $\lambda_n^{\max}$ but greater than the next maximal eigenvalue λ_l (of the matrix $\widehat{C}^{(n)}$) such that $\gamma_l \neq 0$. Formula (9) is too rough for certain problems, e.g., for calculating the correlation functions. However, its precision is quite sufficient for the aims of the present paper.

If we are only interested in the main exponential asymptotics of the function $f_{n,W}(\mu)$, that is, in an asymptotic formula of the form

$$f_{n,W}(\mu) = e^{-|W|g_0(\mu)+o(|W|)}, \quad (10)$$

then it is possible to replace z_n^* in Eq. (9) (and in the corresponding expression for Θ_n) by its limiting value $z_\mu^* \equiv \lim_{n\to\infty} z_n^*$. In the case $|W| \sim \alpha N$, $\alpha \in (0,1]$ one obtains

$$f_{n,W}(\mu) = \exp\left\{N\left[-\beta J(z^* - z_\mu^*) + \tfrac{1}{2}L(z^*) - \tfrac{1}{2}L(z_\mu^*) - \alpha\beta J\mu^2(z_\mu^* - d)\right] + O(n^{d-1})\right\}, \quad (11)$$

where

$$L(z) \equiv \lim_{n\to\infty} L_n(z) = \frac{1}{(2\pi)^d} \int_0^{2\pi}\cdots\int_0^{2\pi} d\omega_1\ldots d\omega_d \log\left(z - \sum_{\nu=1}^{d} \cos\omega_\nu\right), \quad (12)$$

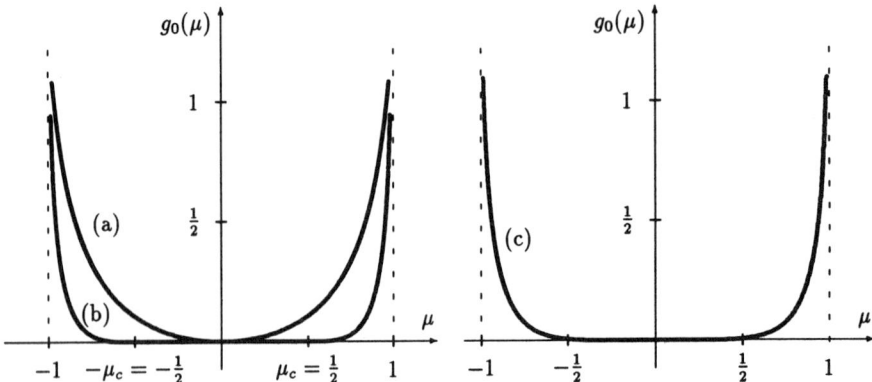

Figure 1. The rate functions $g_0(\mu)$, see (10) and (11) in the case $\alpha = 1$. The curves (a), (b), and (c) correspond to $\beta J = 0.05 < \beta_c J$, $\beta = \frac{4}{3}\beta_c$, and $\beta = \beta_c$, respectively.

z_μ^* is the minimum point of the function

$$\Phi(z, \mu) \equiv \lim_{n \to \infty} \Phi_n(z, \mu) = (1 - \alpha\mu^2)z - \frac{L(z)}{2\beta J}$$

on the interval $[d; \infty)$, and $z^* = z_0^*$. In the case $|W| = o(N)$ one obtains

$$f_{n,W}(\mu) = \exp\left[-|W|(z^* - d)\beta J \mu^2 + O(|W|^2/N)\right]. \tag{13}$$

In the case $|W| \sim \alpha N$, the rate function $g_0(\mu)$, see (10) and (11), defined on the interval $(-\alpha^{-1/2}; \alpha^{-1/2})$ possesses the following properties (see also Fig. 1):

1. $g_0(\mu)$ is a non-negative, even, convex function, and $g_0(0) = 0$;

2. $g_0(\mu) \sim -\frac{1}{2}\log(1 - \alpha\mu^2)$, as $\mu \to \pm\alpha^{-1/2}$;

3. $g_0''(\mu = 0) > 0$, for $\beta < \beta_c$; where $\beta_c = L'(d)/2J$ is the critical temperature of the spherical model found by Berlin and Kac [4];

4. $g_0(\mu) = 0$ for $\mu \in [-\mu_c; \mu_c]$ and $\beta \geq \beta_c$, where $\mu_c = \alpha^{-1/2}\sqrt{1 - \frac{\beta_c}{\beta}}$;

5. $g_0(\mu) \sim \begin{cases} \frac{8\pi^2}{3}(\beta J\alpha)^3(\mu^2 - \mu_c^2)^3, & \text{for } d = 3; \\ -6\pi^2(\beta J\alpha)^2 \frac{(\mu^2 - \mu_c^2)^2}{\log(\mu^2 - \mu_c^2)}, & \text{for } d = 4; \\ -\frac{(\beta J\alpha)^2}{L''(z=d)}(\mu^2 - \mu_c^2)^2, & \text{for } d \geq 5; \end{cases}$ as $\mu \to \pm\mu_c \pm 0$, for $\beta \geq \beta_c$.

Eqs. (11) and (13) provide a satisfactory answer to the problem under investigation for $z_\mu^* > d$ and $z^* > d$, respectively, that is, for $\beta \in [0; \beta_c(\mu))$ and $\beta \in [0; \beta_c)$, where $\beta_c(\mu) \equiv \beta_c/(1 - \alpha\mu^2)$. Indeed, when $z_\mu^* = d$, the rate-function $g_0(\mu) = 0$, and the exact decay rate of the sequence $f_{n,W}(\mu)$ has still to be determined. To gain a higher precision one needs to find how fast the sequence of the saddle points z_n^* converges to the point $z = d$ as $n \to \infty$.

It is possible to show that for any μ and n, the function $\Phi_n(z,\mu)$ is strictly convex on the interval $z \in (s_n^{\max}; \infty)$. Since, $\Phi_n(z,\mu) \to \infty$ as $z \to s_n^{\max}$ or $z \to \infty$, there is a unique minimum point z_n^* of this function on $(s_n^{\max}; \infty)$ which can be identified as the maximal solution of the equation $\partial_z \Phi_n(z,\mu) = 0$. Of course, the exact location of the point(s) z_n^* depends on the choice of W. Here we consider only the (simplest) case

$$W = \{j \in \Omega_n : l+1 \leq j_1 \leq l+k\}, \quad |W| = kn^{d-1}.$$

In this case the coefficients γ_j are given by

$$\gamma_j = n^{d/2-1} \frac{\sin(\pi k(j_1-1)/n)}{\sin(\pi(j_1-1)/n)} \left[\sin \frac{\pi(2l+k)(j_1-1)}{n} + \cos \frac{\pi(2l+k)(j_1-1)}{n} \right] \prod_{\nu=2}^{d} \delta_{1,j_\nu},$$

and the function $\Sigma(z)$ can be calculated exactly [5]

$$\Sigma(z) = \frac{kn^{d-1}}{z-d}\left[1 - \frac{1}{k\sqrt{(z-d)(2+z-d)}} \frac{(1-x_1^k)(x_2^n - x_2^k)}{(x_2^n - 1)}\right], \quad (14)$$

where $x_{1,2} = 1 + z - d \mp \sqrt{(z-d)(2+z-d)}$. To find the minimum of the function $\Phi_n(z,\mu)$ in the case $\beta > \beta_c$ (or $\beta > \beta_c(\mu)$ if $k \sim \alpha n$) it is convenient to make the substitution $z \to d + \zeta n^{-2}$. In the case $k = o(n)$ one obtains

$$\Sigma(d + \zeta n^{-2}) = \frac{k^2 n^d}{h_1(\zeta)} + O(k^3 n^{d-1}),$$

where

$$h_1(\zeta) \equiv \begin{cases} \sqrt{2\zeta} \tanh \sqrt{\zeta/2}, & \text{for } \zeta \geq 0 \\ -\sqrt{-2\zeta} \tan \sqrt{-\zeta/2}, & \text{for } \zeta \leq 0 \end{cases}.$$

The corresponding expression for $\Phi_n(d + \zeta n^{-2}, \mu)$ is given by

$$\Phi_n(d+\zeta n^{-2}; \mu) = d - \frac{L(d)}{2\beta J} + n^{-2}\left[\zeta\left(1 - \frac{\beta_c}{\beta}\right) - \mu^2 h_1(\zeta)\right] + O(kn^{-3}), \quad (15)$$

and hence $z_n^* = d + \zeta^*(\mu)n^{-2} + O(kn^{-3})$, where $\zeta^*(\mu)$ is the maximal solution of

$$1 - \frac{\beta_c}{\beta} = \mu^2 h_1'(\zeta).$$

On substituting the obtained expression for z_n^* in Eq. (9) one arrives at

$$f_{n,W}(\mu) = \exp\left[-n^{d-2}\beta J g_2(\mu) + o(n^{d-2})\right], \quad (16)$$

where

$$g_2(\mu) = -\zeta^*(\mu)\left(1 - \frac{\beta_c}{\beta}\right) + \mu^2 h_1(\zeta^*(\mu)),$$

see Fig. 2. Note that $g_2(\mu)$ is independent of k as long as $k = o(n)$. This is clearly due to the fact that the correlation length of the spherical model in the microcanonical ensemble (with respect to the block spin S_W) is of the order n for $\beta > \beta_c$, and hence, all the lengths $k = o(n)$ are equivalent in many respects.

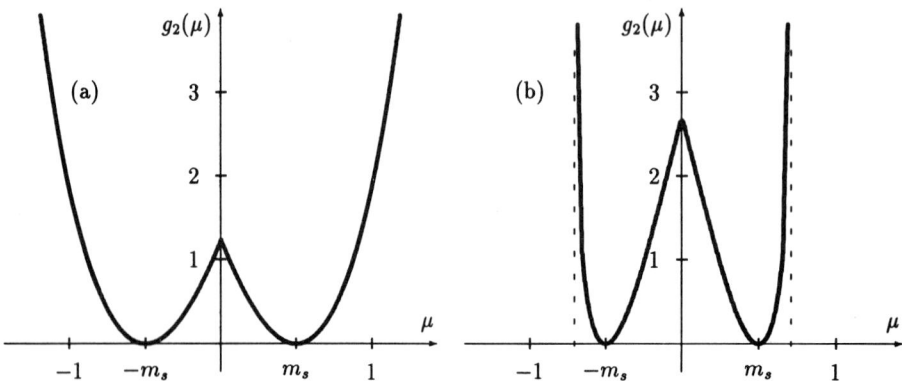

Figure 2. The rate functions (of the order n^{d-2}) $g_2(\mu)$. The curve (a) corresponds to $k = o(n)$ and $\beta = \frac{4}{3}\beta_c$, the curve (b) corresponds to $k = n/2$ and $\beta = \frac{4}{3}\beta_c$. The broken lines are vertical asymptotes for $g_2(\mu)$ at $\pm \alpha^{-1/2} m_s$, $m_s = \sqrt{1 - \frac{\beta_c}{\beta}}$.

In the case $k \sim \alpha n$, $\alpha \in (0;1]$ the substitution $z = d + \zeta n^{-2}$ yields

$$\Sigma(d + \zeta n^{-2}) = n^{d+2} q(\zeta, \alpha) + O(n^{d+1}),$$

where $q(\zeta, \alpha)$ is given by

$$q(\zeta, \alpha) = \frac{\alpha}{\zeta} \times \begin{cases} 1 - \dfrac{\sinh\left[\alpha\sqrt{\zeta/2}\right] \sinh\left[(1-\alpha)\sqrt{\zeta/2}\right]}{\alpha\sqrt{\zeta/2}\,\sinh\sqrt{\zeta/2}} & \text{for } \zeta \geq 0 \\[2ex] 1 - \dfrac{\sin\left[\alpha\sqrt{-\zeta/2}\right] \sin\left[(1-\alpha)\sqrt{-\zeta/2}\right]}{\alpha\sqrt{-\zeta/2}\,\sin\sqrt{-\zeta/2}} & \text{for } \zeta \leq 0 \end{cases},$$

and

$$\Phi_n(d + \zeta n^{-2}, \mu) = d - \frac{L(d)}{2\beta J} + n^{-2}\left[\zeta\left(1 - \frac{\beta_c}{\beta}\right) - \frac{\alpha^2 \mu^2}{q(\zeta, \alpha)}\right] + O(n^{-3}).$$

Hence $f_{n,W}(\mu)$ is given by Eq. (16), where

$$g_2(\mu) = -\zeta^*(\mu)\left(1 - \frac{\beta_c}{\beta}\right) + \frac{\alpha^2 \mu^2}{q(\zeta^*(\mu), \alpha)},$$

see Fig. 2, and $\zeta^*(\mu)$ is the maximal solution of

$$1 - \frac{\beta_c}{\beta} = \mu^2 \frac{\partial}{\partial \zeta} \frac{\alpha^2}{q(\zeta, \alpha)}.$$

The last regime which was not discussed so far is the case of $T = T_c$ (the critical point) and $k = o(n)$. The behaviour of the model in this regime depends significantly on the dimensionality of the lattice. We consider first the case $d = 3$. When $k = \alpha n^\gamma$ and $\frac{1}{2} \leq \gamma < 1$, it is convenient to make substitution $z = d + n^{-2(1-\gamma)}\zeta$ in order to find the location of the saddle point z_n^*. One obtains

$$\Sigma(d+n^{-2(1-\gamma)}\zeta) = \begin{cases} |W|n^{2(1-\gamma)}\zeta^{-1} + O(|W|n^{3-4\gamma}) & \text{for } \frac{1}{2} < \gamma < 1, \\ |W|n\,\sigma(\zeta,\alpha) + O(|W|) & \text{for } \gamma = \frac{1}{2}, \end{cases}$$

$$\Phi_n(d+n^{-2(1-\gamma)}\zeta) = d - \frac{L(d)}{2\beta_c J} + \frac{n^{-3(1-\gamma)}\zeta^{3/2}}{3\sqrt{2\pi}\beta_c J} - \frac{\mu^2|W|^2 N^{-1}}{\Sigma(d+n^{-2(1-\gamma)}\zeta)} + O(k^4 n^{-1}),$$

where

$$\sigma(\zeta,\alpha) = \zeta^{-1}\left[1 - \frac{1-\exp(-\alpha\sqrt{2\zeta})}{\alpha\sqrt{2\zeta}}\right] \tag{17}$$

We used here the small $z - d$ expansion of the function $L(z)$ which can be derived from the results of Barber and Fisher [6]. Hence, $z_n^* = d + n^{-2(1-\gamma)}\zeta^*(\mu) + o(n^{-2(1-\gamma)})$, where $\zeta^*(\mu) = 8\pi^2\alpha^2\mu^4\beta_c^2 J^2$ for $\frac{1}{2} < \gamma < 1$, and where $\zeta^*(\mu)$ is the maximal solution of

$$\frac{1}{4\pi}\sqrt{2\zeta} = \alpha\mu^2\beta_c J \frac{\partial}{\partial \zeta}\frac{1}{\sigma(\zeta,\alpha)},$$

for $\gamma = \frac{1}{2}$. Using Eq. (9) one obtains

$$f_{n,W}(\mu) = \exp\left[-k^3 \frac{8\pi^2}{3}\beta_c^3 J^3 \mu^6 + o(k^3)\right]$$

for $\frac{1}{2} < \gamma < 1$, and

$$f_{n,W}(\mu) = \exp\left[-n^{3\gamma}\left(\frac{\alpha\mu^2\beta_c J}{\sigma(\zeta^*(\mu),\alpha)} - \frac{(\zeta^*(\mu))^{3/2}}{3\sqrt{2\pi}}\right) + o(k^3)\right],$$

for $\gamma = \frac{1}{2}$. Finally, in the case $k = \alpha n^\gamma$, $0 \le \gamma < \frac{1}{2}$ and $\beta = \beta_c$, one should make the substitution $z = d + \zeta n^{-1}$. This yields

$$\Sigma(d+\zeta n^{-1}) = |W|k\sqrt{n}(2\zeta)^{-1/2} + O(k^2|W|),$$

$$\Phi_n(d+\zeta n^{-1};\mu) = d - \frac{L(d)}{2\beta_c J} + n^{3/2}\left[\frac{\zeta^{3/2}}{3\sqrt{2\pi}\beta_c J} - \mu^2\sqrt{2\zeta}\right] + O(kn).$$

Hence, $z_n^* = d + 2\mu^2\beta_c J\pi n^{-1} + o(n^{-1})$ and

$$f_{n,W}(\mu) = \exp\left[-n^{3/2}\tfrac{4}{3}|\mu|^3(\beta_c J)^{3/2}\sqrt{\pi} + o(n^{3/2})\right]. \tag{18}$$

Note that the rate function in Eq. (18) is the same as $g_0(\mu)$ for $k \sim \alpha n$, $\beta = \beta_c$, and $d = 3$. This however, is not surprising since for $k \sim \alpha n^\gamma$, $\frac{1}{2} < \gamma < 1$ and for $k \sim \alpha n$ the corresponding correlation lengths ξ_n (let us remind that $\xi_n \sim c(z_n^* - d)^{-1/2}$) are much less than the minimal size of W, and it is reasonable to expect that the large deviation properties of all such blocks are the same. For $0 \le \gamma < \frac{1}{2}$ the rate function is independent of the actual value of k since in this regime $k \ll \xi_n \sim c\sqrt{n}$, cf. Eq. (16). In the case $k \sim \alpha\sqrt{n}$ the correlation length matches the minimal size of the block W, and hence, as α growth from 0 to ∞ the properties of the system change from those typical for $k \ll \xi_n$ to those typical for $k \gg \xi_n$.

In the case $d \ge 5$, $\beta = \beta_c$ and $k = \alpha n^\gamma$ one has to make the substitutions

$$z = d + \begin{cases} n^{-(1-\gamma)}\zeta & \text{for } \frac{1}{3} < \gamma < 1 \\ n^{-2/3}\zeta & \text{for } 0 \le \zeta \le \frac{1}{3} \end{cases}.$$

Using Eq. (9) one obtains (note that $L''(d) < 0$)

$$f_{n,W}(\mu) = \exp\left[n^{d-2(1-\gamma)}\frac{(\alpha J\beta_c)^2\mu^4}{L''(d)} + o(n^{d-2(1-\gamma)})\right], \quad \text{for} \quad \frac{1}{3} < \gamma < 1,$$

$$f_{n,W}(\mu) = \exp\left[-n^{d-4/3}\mu^{8/3} J\beta_c \left(\frac{J\beta_c}{2|L''(d)|}\right)^{1/3} + o(n^{d-4/3})\right], \quad \text{for} \quad 0 \le \gamma < \frac{1}{3},$$

and

$$f_{n,W}(\mu) = \exp\left[-n^{d-4/3}\left(\tfrac{1}{4}L''(d)(\zeta^*(\mu))^2 + \frac{\beta_c J\alpha\mu^2}{\sigma(\zeta^*(\mu);\alpha)}\right) + o(n^{d-4/3})\right], \quad \text{for} \quad \gamma = \frac{1}{3},$$

where $\sigma(\zeta;\alpha)$ is given by Eq. (17), and $\zeta^*(\mu)$ is the maximal solution of $L''(d)\zeta = -2\beta_c J\alpha\mu^2 \partial_\zeta[\sigma(\zeta;\alpha)]^{-1}$.

Similar expressions can be obtained in the case $d = 4$, they differ from those in the case $d \ge 5$ mainly by "log" corrections.

REFERENCES

1. C.E. Pfister, Large deviations and phase separation in the two-dimensional Ising model, *Helv. Phys. Acta*, **64**, 953 (1991).
2. S. Shlosman, The droplet in the tube: a case of phase transition in the canonical ensemble, *Commun. Math. Phys.*, **125**, 91 (1989).
3. R.H. Shonmann, Second order large deviation estimates for ferromagnetic systems in the phase coexistence region, *Commun. Math. Phys.*, **112**, 409 (1987).
4. T.H. Berlin and M. Kac, The spherical model of a ferromagnet, *Phys. Rev.*, **86**, 821 (1952).
5. A.E. Patrick, The influence of external boundary conditions on the spherical model of a ferromagnet I: Magnetization profiles, Preprint DIAS, (1992).
6. M.N. Barber and M.E. Fisher, Critical phenomena in systems of finite thickness I. The spherical model, *Ann. Phys.*, **77**, 1 (1973).

ENTROPY DENSITY AND THE SPLIT PROPERTY

H. Narnhofer

Institut für Theoretische Physik
Universität Wien, Boltzmanngasse 5
A-1090 Wien, Austria

INTRODUCTION

If we talk of mesomechanics, then we have in mind fairly large systems that are parts of very large systems. These subsystems should be described by a few number of densities, among them evidently the particle density, the energy density, and, since we want to do thermodynamics, the entropy density.

Different from classical systems where the entropy for a reduced system is always defined and smaller than for the larger system in quantum mechanics problems arise. Especially for relativistic quantum fields the entropy of a local subalgebra defined in the usual way is infinite. This can be circumvented by a change of the definition that takes the imbedding of the smaller system into the larger system into account. But even so we need some additional structure of the imbedding, namely the split property introduced by Doplicher and Longo.[1]

DEFINITION OF THE ENTROPY AND POSSIBLE ESTIMATES

Definition (2.1): Consider an algebra $\mathcal{A}$ imbedded in $\mathcal{B}$ together with a state ω over $\mathcal{B}$. Then,[2-4]

$$H_\mathcal{B}(\mathcal{A},\omega) = \sum_{\omega=\sum \omega_i, \omega_i \text{ state over } \mathcal{B}} \omega_i(1) S\left(\omega \Big| \frac{\omega_i}{\omega_i(1)}\right)_\mathcal{A}.$$

Remarks: If $\mathcal{A} = \mathcal{B}$ and ω corresponds to a density matrix ρ, then the definition coincides with the usual one because we can take as $\omega_i = \lambda_i \langle \varphi_i | \cdot | \varphi_i \rangle$ for $\rho = \sum_j \lambda_j |\varphi_j\rangle\langle\varphi_j|$.

Further $H_\mathcal{B}(\mathcal{A},\omega)$ is monotonically increasing in $\mathcal{A}$ (the relative entropy is monotonically increasing[5,6]) and monotonically decreasing in $\mathcal{B}$ (not every decomposition into states over a smaller algebra can be extended into a decomposition over a large algebra).

Definition (2.2): $\mathcal{A} \subset \mathcal{B}$ is split if there exists a type I algebra $\mathcal{N}$ with $\mathcal{A} \subset \mathcal{N} \subset \mathcal{B}$.[1]

Lemma (2.3): Let $\mathcal{A} \subset \mathcal{B}$ are split. Then there exists a state ω over $\mathcal{B}$ such that
$$H_\mathcal{B}(\mathcal{A}, \omega) = 0.$$

Proof: If $\mathcal{A} \subset \mathcal{B}$ are split, then the imbedding can be realized in a representation over $\mathcal{H} \otimes \mathcal{H}$ with $\mathcal{A} \approx \mathcal{A} \otimes 1 \subset \mathcal{B}(\mathcal{H}) \otimes 1 \subset \mathcal{B}(\mathcal{H}) \otimes \mathcal{B}_0 \approx \mathcal{B}$. Notice that for the commutants we have
$$\mathcal{A}' \approx \mathcal{A}' \otimes \mathcal{B}(\mathcal{H}) \supset 1 \otimes \mathcal{B}(\mathcal{H}) \supset 1 \otimes \mathcal{B}'_0 = \mathcal{B}'$$
such that $\mathcal{A} \vee \mathcal{B}' \approx \mathcal{A} \otimes \mathcal{B}'_0 \approx \mathcal{A} \otimes \mathcal{B}'$. This isomorphism characterizes the split property. For our state ω in (2.3) we can choose $\langle \Omega | \cdot | \Omega \rangle \langle \Phi | \cdot | \Phi \rangle$, Ω and Φ arbitrary. For this state $0 = S(\mathcal{N}, \omega) = H_\mathcal{N}(\mathcal{N}, \omega) \geq H_\mathcal{B}(\mathcal{A}, \omega)$.

Theorem (2.4): Assume $\mathcal{A} \subset \mathcal{B}$ is split and the state ω over $\mathcal{A} \subset \mathcal{B}$ is nuclear in the following sense:
$$\omega_{\mathcal{A} \otimes \mathcal{B}'}(AB') = \sum \omega_i(A) \varphi_i(B') - \sum \bar{\omega}_i(A) \bar{\varphi}_i(B')$$
with $\|\varphi_i\| = \|\bar\varphi_i\| = 1$ and $\sum \|\bar\omega_i\| < \varepsilon$.
$$H_\mathcal{B}(\mathcal{A}, \omega) \leq \frac{-1}{1-\varepsilon} \sum \|\omega_i\| \ln \|\omega_i\|.$$

Proof: The state $\bar\omega = \sum \omega_i \cdot \varphi_i$ approximates ω and its entropy dominates that of ω. For $\bar\omega$ the decomposition into $\sum \omega_i \varphi_i$ cannot be improved.[7]

Remark: It is an open problem whether $\mathcal{A} \subset \mathcal{B}$ split is necessary for $H_\mathcal{B}(\mathcal{A}, \omega) < \infty$. At least no counterexample is known.

THE NUCLEARITY CONDITION OF BUCHHOLZ AND WICHMANN

When Buchholz and Wichmann[8] wanted to construct temperature states over relativistic quantum field they could not consider a sequence of Gibbs states: The local algebras in relativistic quantum field theory are type III_1[9] and do not allow density matrices. Similarly, it is not clear how to approach the Hamiltonian by local Hamiltonians with direct spectrum. The replacement is the physical idea that the finiteness of the local phase space corresponds to the fact that more and more energy is needed to produce new states by local operators. In mathematical language this is formulated by

Definition: Consider the map $\Theta_{\beta, \mathcal{A}} : \mathcal{A} \to \mathcal{B}(\mathcal{H})$ given by

$$\begin{aligned} \Theta_{\beta,\mathcal{A}}(A) &= e^{-\gamma H} A |\Omega\rangle & A &\in \mathcal{A}, \; H \text{ Hamiltonian} \\ &= \tau_{i\beta} A |\Omega\rangle & \tau &\text{ time automorphism resp. modular automorphism.} \end{aligned}$$

Let
$$\Theta_{\beta,\mathcal{A}}(A) = \sum \varphi_i(A) \Phi_i, \qquad \|\Phi_i\| = 1. \tag{3.1}$$

Then Θ is nuclear if
$$\|\Theta_{\beta,\mathcal{A}}\|_1 = \inf \sum \|\varphi_i\| < \infty. \tag{3.2}$$

Define also
$$\|\Theta_{\beta,\mathcal{A}}\|_s = \inf \sum -\|\varphi_i\| \ln \|\varphi_i\|. \tag{3.3}$$

Buchholz, D'Antoni and Fredenhagen[10] proved:

Theorem (3.4): Let for the groundstate in relativistic quantum field theories $\|\Theta_{\beta,\mathcal{A}_\Lambda}\|_1$ be bounded. Then $\mathcal{A}_\Lambda \subset \mathcal{A}_{\bar{\Lambda}}$, dist $(\Lambda, \bar{\Lambda}) > 0$, is split.

Theorem: Let for the groundstate in relativistic quantum field theories $\|\Theta_{\beta,\mathcal{A}_\Lambda}\|_s$ be bounded. Then
$$H_{\bar{\mathcal{A}}}(\omega, \mathcal{A}_\Lambda) < \infty. \tag{3.5}$$

Let $\|\Phi_{\beta,\mathcal{A}_\Lambda}\|_s$ tend to zero for $\beta \to \infty$. Then
$$\lim_{\bar{\Lambda} \nearrow \mathbb{R}^3} H_{\bar{\mathcal{A}}}(\omega, \mathcal{A}_\Lambda) = 0. \tag{3.6}$$

Proof: Following Ref. 10 we can write
$$\langle \Omega | AB' | \Omega \rangle = \langle \Omega | A f(H) B' | \Omega \rangle \quad \text{for } A \in \mathcal{A}_\Lambda, \; B' \in \mathcal{A}'_{\bar{\Lambda}}$$

for some function $f(H)$ that sufficiently can be compared with $e^{-\gamma H}$ and therefore also satisfies a nuclearity condition that gives us the possibility to apply (2.3).[7]

Remark: For the groundstate we get immediately
$$H_{B(\mathcal{H})}(\mathcal{A}_\Lambda, \omega) = 0.$$

But this does not give any detailed information on the imbedding of $\mathcal{A}_\Lambda$ into the whole system. In order to construct mesomechanics we need control over
$$\lim_{\Lambda \to \infty} \frac{1}{|\Lambda|} H_{\mathcal{A}_{\bar{\Lambda}}}(\mathcal{A}_\Lambda, \omega)$$

where $\bar{\Lambda} \supset \Lambda$ in a fairly well defined way. For the groundstate in relativistic free quantum field theory this control is given: For the massive case we need dist $(\bar{\Lambda}, \Lambda) \to \infty$ but only in such a way that $\lim |\Lambda|/|\bar{\Lambda}| = 1$.[7,8] For $m = 0$ the available estimates[11] only suffice for (3.6).

THE NUCLEARITY CONDITION IN TEMPERATURE STATES

For temperature states at temperature β we cannot expect the nuclearity condition (3.1) for arbitrary γ for the following reason:

Example (4.1): Consider a KMS state of type I. Then it can be written in the GNS construction

$$|\Omega\rangle = \left|\sum \exp\left[-\frac{\lambda_i}{2}\beta\right]\varphi_i \otimes \psi_i\right\rangle \frac{1}{(\sum e^{-\lambda_i\beta})^{1/2}}$$

$$\pi(A) = \sum a_{\ell m} |\varphi_\ell\rangle\langle\varphi_m| \otimes 1.$$

Therefore

$$\pi(\tau_{i\gamma}A)|\Omega\rangle = \sum \exp\left[-\frac{\lambda_\ell \gamma}{2} - \lambda_m\left(\frac{\beta}{2}-\gamma\right)\right] a_{\ell m}|\varphi_\ell \otimes \psi_m\rangle \frac{1}{(\sum e^{-\lambda_i\beta})^{1/2}}.$$

Thus we have control over the nuclearity bound only for $0 < \gamma < \beta/2$ with the optimal bound for $\gamma = \beta/4$. For massive free relativistic quantum field theory where the temperature states are explicitly given the nuclearity bound can be estimated and is in fact finite for $\gamma < \beta/2$.

Theorem (4.2): Assume that the nuclearity condition holds, i.e.

$$\tau_{i\gamma}A|\Omega\rangle = \sum \varphi_i(A)\Phi_i, \qquad \|\Phi_i\| = 1, \qquad \gamma < \beta/2$$

and $-\sum \|\varphi_i\| \ln \|\varphi_i\| < \infty$. Then

$$H_M(\mathcal{A}_\Lambda, \omega) < \infty.$$

Proof:

$$\langle\Omega|M'A|\Omega\rangle = \langle\Omega|JM'Je^{-\beta H/2}A\Omega\rangle = \sum \langle\Omega|JM'J\exp\left[-\left(\frac{\beta}{2}-\gamma\right)H\right]|\Phi_i\rangle\varphi_i(A)$$

and this decomposition can be used in (2.3).

Theorem (4.3): For relativistic quantum field theories that satisfy the nuclearity condition (4.2)

$$\lim_{\bar{\Lambda} \nearrow \mathbf{R}^3} H_{\mathcal{A}_{\bar{\Lambda}}}(\mathcal{A}_\Lambda, \omega) = H_M(\mathcal{A}_\Lambda, \omega).$$

Proof: The main task is to prove that $H_{\mathcal{A}_{\bar{\Lambda}}}(\mathcal{A}_\Lambda, \omega)$ is finite. This can be done with the same arguments as for (3.5), (3.6). Convergence can then be controlled by the following

Lemma (4.4): Let $\mathcal{A}_\Lambda \nearrow \mathcal{M}$. Define H_Λ to be the modular operator corresponding to $\mathcal{A}''_\Lambda, \omega$. Then

$$\text{st-res}\lim_{\Lambda \nearrow \mathbf{R}^3} e^{-\gamma H_\Lambda} = e^{-H}.$$

This is a consequence of Ref. 12 on convergence of operators that correspond to quadratic forms with increasing domains.

Again for massive relativistic free quantum field theories where the temperature states are explicitly given the nuclearity bound for $\|\Theta_{-\beta/4}\|_s$ is finite. Even more it shows that

$$\overline{\lim_{\Lambda \nearrow \mathbf{R}^3}}, \lim_{\bar{\Lambda} \nearrow \mathbf{R}^3} \frac{1}{|\Lambda|} H_{\mathcal{A}_{\bar{\Lambda}}}(\mathcal{A}_\Lambda, \omega) < \infty$$

i.e. the entropy is an extensive quantity and the entropy density is finite.

REFERENCES

1. S. Doplicher and R. Longo, *Invest. Math.*, **73**, 493 (1984).
2. A. Connes, *C.R. Acad. Sci. Paris*, **301**, 1 (1985).
3. H. Narnhofer and W. Thirring, *Fizika*, **17**, 257 (1985).
4. A. Connes, H. Narnhofer and W. Thirring, *Commun. Math. Phys.* **112**, 691 (1987).
5. H. Kosaki, *J. Operator Theory*, **16**, 335 (1986).
6. M. Ohya and D. Petz, "Quantum Entropy and its Use," Springer, Berlin (1993).
7. H. Narnhofer, Entropy density for relativistic quantum field theory, Vienna preprint UWThPh-1993-11.
8. D. Buchholz and E.H. Wichmann, *Commun. Math. Phys.*, **106**, 321 (1986).
9. K. Fredenhagen, *Commun. Math. Phys.*, **97**, 79 (1985).
10. D. Buchholz, C. D'Antoni and K. Fredenhagen, *Commun. Math. Phys.*, **111**, 123 (1987).
11. D. Buchholz and P. Jacobi, *Lett. Math. Phys.*, **13**, 313 (1987).
12. J. Weidmann, *Math. Scand.*, **54**, 51 (1984).

PERTURBATIONS OF QUANTUM CANONICAL RELATIONS AND Q-INDEPENDENCE

W.A. Majewski[1] and M. Marciniak[2]

1. Institute of Theoretical Physics and Astrophysics
   Gdańsk University
   Wita Stwosza 57, 80-952 Gdańsk, Poland

2. Institute of Mathematics
   Gdańsk University
   Wita Stwosza 57, 80-952 Gdańsk, Poland

INTRODUCTION

The aim of this note is to review some of the recent results on real and complex perturbations of quantum canonical relations. We also give a brief description of the relation between q-perturbations and q-probability calculus.

The detailed proofs will appear in a forthcoming publication.

REAL PARAMETRIZATION AND Q-INDEPENDENCE

q-Canonical Commutation Relations

Let H be a Hilbert space with the inner product $<.,.>$ and $q \in [-1,+1)$. We will consider C-antilinear maps into a C^*-algebra with identity $a : H \to \mathcal{A}$ such that

$$a(f)a^*(g) - qa^*(g)a(f) = <f,g> \mathbf{1}. \tag{2.1}$$

The subalgebra generated by such $a(f)$, $a^*(g)$ will be denoted by $\mathcal{A}_q(H)$.

Remarks. *i*) The two cases $q = 1$ and $q = -1$ are the Bosonic and Fermionic relations respectively. $\mathcal{A}_q(H)$, for $q \neq \pm 1$, is a one parameter family of interpolating algebras. $\mathcal{A}_0(H)$ is exactly the Cuntz algebra ([1], [2]).
ii) For a satisfying Equation (2.1) we have ([3], [4])

$$\|a(f)\| = c(q)\|f\| \tag{2.2}$$

where

$$c(q) = \begin{cases} (1-q)^{-\frac{1}{2}} & \text{for } 0 \le q < 1 \\ 1 & \text{for } -1 < q \le 0 \end{cases} \tag{2.3}$$

iii)

$$\|a^*(f)a^*(f)\| = \|a(f)a(f)\| = \varphi(q)\|f\|^2 \tag{2.4}$$

while

$$\varphi^2(q) = \begin{cases} c^2(q)[qc^2(q)+1] & q \in [-\frac{1}{2},1) \\ \psi(\lambda_0) & q \in (-1,-\frac{1}{2}) \end{cases} \tag{2.5}$$

and $\psi(\lambda_0) = \sup_{\lambda \in \sigma(a(f)a^*(f))}\{q\lambda^2 + \|f\|^2\lambda\} \ne 0$, $\sigma(b)$ denotes the spectrum of b (see [5]). From (2.4) we have that for $q \ne -1$ $\|a^*(f)a^*(f)\| = \|a(f)a(f)\| \ne 0$, i.e. there is a violation of the Pauli principle.

iv) In general, for $q \ne \pm 1$, it is not possible to supplement (2.1) by the relations

$$a(f)a(g) - qa(g)a(f) = 0. \tag{2.6}$$

The following example describes a physically interesting representation of $\mathcal{A}_q(H)$.

Example 2.1. Let $\mathcal{F}(H)$ denote the full Fock space $\bigoplus_{r=0}^{\infty}(\otimes^r H)$ over an infinite dimensional separable Hilbert space H. Define linear maps

$$a^*(f)f_1 \otimes f_2 \otimes ... \otimes f_n = f \otimes f_1 \otimes f_2 \otimes ... \otimes f_n \tag{2.7}$$

$$P_q^{(n)} f_1 \otimes f_2 \otimes ... \otimes f_n = \sum_{\rho \in S_n} q^{|\rho|} f_{\rho(1)} \otimes ... \otimes f_{\rho(n)} \tag{2.8}$$

where $q \in [-1,+1]$, S_n is the symmetric group and $|\rho|$ is the number of inversions of ρ. Let $P_q = \bigoplus_n P_q^{(n)}$. Then ([3]) $P_q > 0$ for $q \in (-1,+1)$. Define

$$a_q(f) = P_q^{-\frac{1}{2}} a(f) P_q^{\frac{1}{2}}, \qquad a_q^*(f) = P_q^{\frac{1}{2}} a^*(f) P_q^{-\frac{1}{2}}. \tag{2.9}$$

Then, i) $a_q^*(f)$, $a_q(g)$ are bounded operators on $\mathcal{F}(H)$ for $q \in (-1,+1)$ and $f, g \in H$ ($q = +1$ is the "singular" point), and
ii)

$$a_q(f)a_q^*(g) - qa_q^*(g)a_q(f) = <f,g> \mathbf{1}.$$

q-Independence

Now we will consider a generalization of some probabilistic notions, especially the notion of independence:

Definition 2.1. A pair $(\mathcal{A}, \phi)$ is called a probability system if $\mathcal{A}$ is a unital C^*-algebra and ϕ is a state on $\mathcal{A}$, i.e. a positive normalized functional on $\mathcal{A}$.

Definition 2.2. ([5]) Let $(\mathcal{A}, \phi)$ be a probability system and $(\mathcal{A}_i)_{i \in I}$ be a family of subalgebras of $\mathcal{A}$ and $q \in [-1,+1]$. We say that the family $(\mathcal{A}_i)_{i \in I}$ is q-independent if for an arbitrary n and a n-tuple $(a_1,...,a_n)$ with two properties: i) $i_1 \ne i_2 \ne ... \ne i_n$ where $a_j \in \mathcal{A}_{i_j}$ and ii) $\phi(a_j) = 0$ for $j = 1,...,n$, one has

$$\phi(a_1...a_n) = \alpha_V(q)\phi(V_1)...\phi(V_p), \tag{2.10}$$

where $V = \{V_1,...,V_p\}$ is the partition of $\{1,...,n\}$ associated with $(a_1,...,a_n)$, $\phi(V_k) = \phi(a_{j_1}...a_{j_s})$ if $V_k = \{j_1,...,j_s\}$ ($j_1 < ... < j_s$) and $\alpha_V(q)$ is such a function of q that $\alpha_V(0) = 0$.

Remarks. *i)* The case $q = 0$ corresponds to the Voiculescu definition ([6]).
ii) As an example let us consider a probability system $(\mathcal{L}(\mathcal{F}(H)), \omega)$ where $\mathcal{L}(\mathcal{F}(H))$ is the algebra of bounded operators on the full Fock space from Example 2.1 and $\omega(A) = <\Omega, A\Omega>$ where $\Omega \in \otimes^0 H$, $\|\Omega\| = 1$. Let $(e_n)_{n=1}^\infty$ be an orthonormal basis of H and $\mathcal{A}_n$ the algebra generated by the operators $a_q(e_n)$ and $a_q^*(e_n)$. Then the family $(\mathcal{A}_n)_{n=1}^\infty$ is q-independent with respect to $(\mathcal{L}(\mathcal{F}(H)), \omega)$.
iii) Having this type of independence one can prove "q-version" of central limit theorems with q-deformed Gaussian distribution (see [5]). In particular, we have

Theorem 2.1. Let $(a_i)_{i=1}^\infty$ be a sequence in $\mathcal{A}$ such that:
i) $\phi(a_i) = 0$, $i = 1, 2, ...$,
ii) $(a_i)_{i=1}^\infty$ generates q-independent family of subalgebras in $\mathcal{A}$,
iii) $\phi(a_i^2) = \beta^2$.
Define $S_N = N^{-\frac{1}{2}}(a_1 + ... + a_N)$. Then, for $r \in \mathbf{N}$ one has

$$\lim_{N\to\infty} \phi(S_N^r) = \begin{cases} 0 & \text{if } r \text{ is odd,} \\ \sum_V \alpha_V(q)\beta^2 & \text{if } r \text{ is even,} \end{cases}$$

where the summation is over all partitions $V = \{V_1, ..., V_{\frac{r}{2}}\}$ such that $cardV_i = 2$.

COMPLEX PARAMETRIZATION

θ-Symmetrizer

Now we will describe another attempt to deform the commutation relations. Remark *iv)* (preceding Example 2.1) and "exotic" statistics can serve as a motivation for this approach.

Let us start with a remark that for models based on 2-dimensional configuration space one can define an analog of symmetrizer in such a way that the symmetry group is replaced by the braid group B_n. The main difficulty in carrying out this construction is, that for the usual Fock space $\mathcal{F}(L^2(\mathbf{R}^d))$ there is no natural faithful representation of B_n on n-th component of the Fock space, *i.e.* on $L^2(\mathbf{R}^{dn})$. So, we have to replace $\mathbf{R}^d$ by a more complicated configuration space.

Let $\Delta_{d,n} = \{(x_1, ..., x_n) : x_j \in \mathbf{R}^d, x_j \neq x_k, j \neq k\}$. There is a natural action of S_n on this space. Let $M_{d,n}$ denote the orbit space of this action. Recall (*cf.* [7], [8]) that if $\pi_1(X)$ denotes the fundamental group of a space X then

$$\pi_1(M_{d,n}) = \begin{cases} S_n & \text{for } d > 2 \\ B_n & \text{for } d = 2 \end{cases} \qquad (3.1)$$

Relation (3.1) tells us that we should restrict our consideration to 2-dimensional configuration space. This is not a surprise since the similar conclusion follows from DHR and BF analysis: models based on a lower dimensional configuration space can exhibit "unusual" statistics (see [9], [10]).

Let $p_n : X_n \to \Delta_{2,n}$ denote the universal covering space of $\Delta_{2,n}$. Obviously, Equation (3.1) implies that there is a natural action of B_n on the space X_n. Hence, one can define a representation of B_n on a space $C_c(X_n)$ of continuous functions on X_n with compact supports

$$(\pi(\sigma)\varphi)(x) = \varphi(\sigma^{-1}x), \qquad \sigma \in B_n, \quad x \in X_n,$$

where $\sigma^{-1}x$ is the action of σ^{-1} on x.

Let $\chi_\theta : B_n \to \mathbf{C}$ be a homomorphism given by
$$\chi_\theta(\sigma_j) = e^{i\theta}, \quad j = 1, ..., n-1,$$
where σ_j are the generators of B_n.

Definition 3.1. Let $\theta \in [0, 2\pi)$. For $\varphi \in C_c(X_n)$ we define
$$P_\theta^{(n)}\varphi = \sum_{\sigma \in B_n} \chi_\theta(\sigma^{-1})\pi(\sigma)\varphi. \tag{3.2}$$

Remarks. i) $P_\theta^{(n)}\varphi$ is a well defined function on X_n, however this function is not in $C_c(X_n)$.
ii)
$$\sigma P_\theta^{(n)} = \chi_\theta(\sigma)P_\theta^{(n)}, \quad \sigma \in B_n. \tag{3.3}$$

So, one can consider functions in the range of $P_\theta^{(n)}$ as "θ-symmetrized" ones. For $\theta = 0$ this is an analog of the symmetric Fock space, for $\theta = \pi$ - of the antisymmetric one.
iii) It is possible to induce from $\Delta_{2,n}$ a canonical structure of a pre-Hilbert space on $C_c(X_n)$. Its complexification contains θ-symmetrized functions and $P_\theta^{(n)}$ occurs as a positive definite operator in this structure.

Example 3.1. Let us describe the case $n = 2$ in detail.

The universal covering space X_2 of $\Delta_{2,2} = \{(x_1, x_2) : x_1, x_2 \in \mathbf{R}^2, x_1 \neq x_2\}$ is the product of $\mathbf{R}^2$ and the helical covering space $\mathcal{H}$ of $\mathbf{R}^2 \setminus \{(0, 0)\}$ (cf. [7]). Moreover, one can write an element of $\mathcal{H}$ in polar coordinates by (r, λ) where $r \in (0, \infty)$ and $\lambda \in \mathbf{R}$. Hence, the action of an element n of $B_2 \cong \mathbf{Z}$ on X_2 is of the form
$$n(x, r, \lambda) = (x, r, \lambda + n\pi), \quad x \in \mathbf{R}^2, r > 0, \lambda \in \mathbf{R}. \tag{3.4}$$

Then, our θ-symmetrizer is defined by
$$(P_\theta^{(2)}\varphi)(x, r, \lambda) = \sum_{n \in \mathbf{Z}} e^{-in\theta} \varphi(x, r, \lambda - n\pi), \tag{3.5}$$
where $\varphi \in C_c(X_2)$. Therefore, (3.3) implies that a function on X_n is a θ-symmetrized function if and only if
$$\varphi(x, r, \lambda - \pi) = e^{i\theta}\varphi(x, r, \lambda). \tag{3.6}$$

Equation (3.6) suggests that the space of functions on $\mathbf{R}^2 \times (0, \infty) \times [0, \pi]$ satisfying $\varphi(x, r, 0) = e^{i\theta}\varphi(x, r, \pi)$ and equipped with a weighted scalar product can be considered as a θ-analog of the subspace $L^2(\mathbf{R}^2 \times \mathbf{R}^2)$ in the usual Fock space (see [7]). Consequently, this concrete example makes more clearer the above remark iii).

Laughlin's functions

For simplicity we will write our θ-symmetrized functions in the Laughlin's-type form (see [11], [12])
$$\Phi_n(x_1, ..., x_n) = \prod_{i<j} (z_i - z_j)^\theta G_n(x_1, ..., x_n), \tag{3.7}$$
where $G_n(x_1, ..., x_n)$ is a single valued symmetric function on $(\mathbf{R}^2)^n$, z_j is the complex number $x_j^{(1)} + ix_j^{(2)}$ corresponding to $x_j = (x_j^{(1)}, x_j^{(2)})$.

Define

$$\left(a^+_\theta(f)\Phi_n\right)_{n+1}(x_1,...,x_{n+1}) = \prod_{1\leq i<j\leq n+1}(z_i-z_j)^\theta \left(a^+_S(f)G_n\right)(x_1,...,x_{n+1}), \qquad (3.8)$$

where $f \in \mathcal{S}(\mathbf{R}^2)$, $a^+_S(f)$ denotes the usual Bosonic creation operator. Moreover

$$\mathcal{D}(a^+_\theta)|_{n\text{-point fcts}} = \{\prod_{i<j}(z_i-z_j)^\theta G^S_n : G^S_n \in \mathcal{S}(\mathbf{R}^{2n})\},$$

where $\mathcal{S}$ is the Schwartz space. Then the adjoint operator $a_\theta(f)$ is of the form

$$\left(a_\theta(f)\Phi_{n+1}\right)_n(x_1,...,x_n) =$$

$$= \frac{1}{\sqrt{n+1}} \sum_{k=1}^{n+1} \int \prod_{1\leq i<k}\overline{(z_i-z)^\theta} \prod_{k\leq i\leq n}(z-z_i)^\theta f(x) \Phi_{n+1}(x_1,...,x_{k-1},x,x_k,...,x_n)dx.$$

Consequently, our θ-perturbed creation and annihilation operators fulfill the following relations:

$$\left(a_\theta(f)a^+_\theta(g) - a^+_\theta(g)a_\theta(f)\right)\Phi_n(x_1,...,x_n) = \qquad (3.9)$$

$$= \Phi_n(x_1,...,x_n) \int \prod_{1\leq i\leq n}|z-z_i|^{2\theta}\overline{f(x)}g(x)dx+$$

$$+ \prod_{i<j}(z_i-z_j)^\theta \sum_{k=1}^n g(x_k) \int (|z-z_k|^{2\theta}-1)\prod_{i\neq k}|z-z_i|^{2\theta}\overline{f(x)}G_n(x_1,...,x_{k-1},x,x_{k+1},...,x_n)dx,$$

where G_n is a symmetric part of Φ_n. Relations (3.9) can be considered as a representation of the complex perturbations of quantum canonical relations if the function (3.7) is modified as in [12], p. 2678 (see also [5]).

Acknowledgments

We are very greatful to K.-H. Rehren for his valuable remarks.

The first author (W.A.M.) wishes to express his thanks to the organizers of the Leuven NATO AR Workshop for a partial support while the work of the second author (M.M.) has been partially supported by the grant KBN PB 1432-2-91.

REFERENCES

1. J. Cuntz, Simple algebra generated by isometries, *Comm. Math. Phys.*, **57**, 173 (1977).
2. D.E. Evans, On O_n, *Publ. RIMS, Kyoto Univ.*, **16**, 915 (1980).
3. M. Bożejko and R. Speicher, An example of a generalized Brownian motion, *Comm. Math. Phys.*, **137**, 519 (1991).
4. P.T.E. Jørgensen, L.M. Schmitt and R.F. Werner, q-canonical commutation relations and stability of the Cuntz algebra, *Pacific J. Math* (to appear).
5. W.A. Majewski, M. Marciniak, On q-perturbations of commutation relations and q-independence, *Acta Phys. Polon. B*, **24**(4), 815 (1993).
6. D. Voiculescu, Symmetries of some reduced free product C^*-algebras, in: "Operator Algebras and their Connection with Topology and Ergodic Theory", Lect. Notes Math. **1132**, Springer-Verlag, Heidelberg (1985).
7. J.S. Birman, "Braids, Links, and Mapping Class Groups", Princeton University Press, Princeton, New Jersey (1974).

8. A. Lerda, "Anyons", Springer-Verlag, Heidelberg (1992).
9. R. Haag, "Local Quantum Physics", Springer-Verlag, Berlin, Heidelberg (1992).
10. G.A. Goldin, R. Menikoff and D.H. Sharp, Representations of a local current algebra in non-simply connected space and the Aharonov-Bohm effect, *J. Math. Phys.*, **22**(8), 1664 (1981).
11. F. Wilczek, "Fractional Statistics and Anyon Superconductivity", World Scientific, Singapore (1990).
12. R.B. Laughlin, Superconducting ground state of noninteracting particles obeying fractional statistics, *Phys. Rev. Lett.*, **60**, 2677 (1988).

THE SPECTRUM OF THE SPIN-BOSON MODEL

M. Hübner and H. Spohn

LMU, Theoretische Physik
Theresienstraße 37
80333 München, Germany

Many physical problems can be described as a single (or a few) degrees of freedom interacting with a free field, regarded as a bath or reservoir. It is commonly assumed that a nontrivial coupling to the field serves as a mechanism for the small system to dissipate its energy. This somewhat vague formulation should be made precise by mathematical physics and we take here the point of view that it is reflected in the spectrum of the coupled Hamiltonian.

Mathematical theorems assure that quantum states $|\psi\rangle$ in the absolutely continuous (a.c.) spectral subspace $\mathcal{H}_{ac}$ of a Hamiltonian H decay weakly to zero under the time evolution induced by H,

$$\langle f|e^{-iHt}|\psi\rangle = \langle f|\psi(t)\rangle \to 0 \quad \text{as} \quad t \to \infty \quad \forall\, |f\rangle \in \mathcal{H}. \tag{1}$$

Of course the Hilbert space norm of $|\psi(t)\rangle$ is preserved and we interpret the weak convergence as a decay of any excitation measurable in a bounded region of the configuration space R^d. In the models which we shall consider, we expect this decay to hold for any physical state orthogonal to the ground state. The problem is then to prove that for interacting models the Hamiltonian has a unique ground state and that the remainder of the spectrum is purely a.c.. Here we would like to outline how such a property is proved for the spin-boson Hamiltonian.

In the spin-boson model the bath is a free Bose field and the small system is a localized degree of freedom, which corresponds to a particle in a double well potential. If the well is reflection symmetric, then the particle has a symmetric ground state and an antisymmetric first excited state. In the spin-boson Hamiltonian we take only these two lowest states into account and model the interaction with the bath by a linear

coupling,
$$H = \frac{\mu}{2}\sigma_z \otimes I + I \otimes H_B + \sigma_x \otimes (a^*(\lambda) + a(\lambda)). \tag{2}$$

The first term is the energy of the localized degree of freedom with (uncoupled) energy separation $\mu > 0$. σ_x, σ_z are Pauli spin matrices and H_B is the selfadjoint Hamiltonian of noninteracting bosons. In the momentum representation, it is given by

$$H_B = \int d^d k \, \omega(k) a^*(k) a(k) = d\Gamma(\omega). \tag{3}$$

with domain $D(H_B)$ in the standard Fock space $\mathcal{F}$. The annihilation and creation operators $a(k), a^*(k)$ are defined over R^d and satisfy canonical commutation relations $[a(k), a^*(k')] = \delta(k - k')$. The second quantization functor is denoted by $d\Gamma(\cdot)$. For example, the boson number N_B equals $d\Gamma(I)$. The Hilbert space $\mathcal{H}$ for the model is the tensor product $C^2 \otimes \mathcal{F}$ of the spin 1/2 representation space and Fock space. We also use the shorthand $a^*(\lambda) = \int d^d k \, \lambda(k) a^*(k)$.

In general, the dispersion relation ω and the coupling λ are fixed by the physics of the problem. Here we consider them as free parameters and would like to have a spectral information for a whole class of ω, λ's. We state our technical conditions on ω. The dispersion is a real function on momentum space which induces by pointwise multiplication a 1-particle operator on the Hilbert space $L^2(R^d)$, whose second quantization is just H_B. We require

i) $\omega : R^d \to R$ is spherically symmetric, i.e. only a function of $|k|$, and everywhere positive with the possible exception of the origin $k = 0$. ω is a.c. as a real function, with positive derivative away from the origin

$$|\nabla_k \omega| = \omega'(|k|) > 0 \quad \text{for} \quad k \neq 0, \quad \text{and} \quad \lim_{k \to \infty} \omega(k) = \infty. \tag{4}$$

These properties are obviously satisfied by the dispersion $w(k) = \sqrt{k^2 + m^2}$ and its limiting cases $w(k) = |k|$ or $w(k) = k^2/2m$. Condition (4) implies that ω is not constant on sets of positive measure and is in fact an operator with purely a.c. spectrum on the 1-particle Hilbert space $L^2(R^d)$. Its second quantization $d\Gamma(\omega)$ is then purely a.c. as well, apart from the Fock vacuum. Clearly, in order to prove a.c. spectrum for the coupled H, we better assume absolute continuity of H_B.

We turn to the technical conditions on λ.

ii) λ should belong to the 1-particle Hilbert space, i.e. $(\lambda, \lambda) = \int d^d k |\lambda|^2 < \infty$. If $\inf \omega > 0$, this is enough to ensure selfadjointness of H on the natural domain $C^2 \otimes D(H_B)$ by the Kato-Rellich theorem, since $a(\lambda), a^*(\lambda)$ are then H_B-bounded with H_B-bound zero. By a unitary multiplication on momentum space, we can transform an arbitrary complex coupling function to a real nonnegative one. Thus it is no further restriction to assume $\lambda \geq 0$.

iii) In order that the Hamiltonian be bounded from below, we impose $\int d^d k \lambda^2/\omega < \infty$. If $\inf \omega = 0$, this is also the natural condition for selfadjointness of H on $C^2 \otimes D(H_B)$, which can again be established using the Kato-Rellich theorem.

iv) To ensure the existence and uniqueness of the ground state in Fock space, we need the even stronger condition $\int d^d k (\lambda/\omega)^2 < \infty$. The ground state of the spin-boson model was thoroughly investigated in [2-5]. Notice that ii) implies iii),iv) if $\inf \omega > 0$.

Our technical tools to prove the desired spectral properties are *Mourre estimates*, which are discussed in the book [6], following the original papers [7,8]. The idea is to find a so-called conjugate operator $-iA$ which has a positive commutator with the Hamiltonian, up to a compact perturbation,

$$[H, iA] \geq \alpha I - C, \quad \alpha > 0. \tag{5}$$

We say then that a selfadjoint operator obeys a (global) Mourre estimate. *Mourre estimates*, which are discussed in the book [6], following the even differentiably to higher energies. Intuitively, this is possible only in the a.c. spectral subspace. Mourre proved his theorems for *selfadjoint* conjugate operators A. The conjugate operators we will use have no selfadjoint extensions, but generate strongly continuous semigroups of isometries. We first give a generalization of Mourre's virial theorem.

Lemma 1: Let H be a selfadjoint operator with domain $D(H)$ and let iA be the closed generator of a semigroup of isometries. We assume:

a) $D(A) \cap D(H)$ is a core for H.

b) $e^{iA\alpha}$ leaves $D(H)$ invariant and for each $\psi \in D(H)$

$$\sup_{0 < \alpha < 1} \|H e^{iA\alpha} \psi\| < \infty. \tag{6}$$

c) The quadratic form $[H, iA] = i(HA - AH)$ defined on $D(A) \cap D(H)$ is bounded below and closable and the selfadjoint operator associated to its closure is H-bounded.

Then for every eigenvector $\psi \in \mathcal{H}_{pp} = P_{pp}\mathcal{H}$ (the closed subspace generated by the eigenvectors)

$$\langle \psi | [H, A] | \psi \rangle = 0. \tag{7}$$

The proof of the virial theorem follows Mourre's proof with minor modifications. Whereever Mourre assumes the resolvent set of his selfadjoint A to contain the complement of the real line, in our case only the lower complex halfplane belongs to the resolvent set of our semigroup generator A.

Mourre [6,7] proves that a global estimate like (5) implies that H has only finitely many eigenvalues, and each eigenvalue has finite multiplicity. In particular, this shows that eigenvalues cannot accumulate. Furthermore, he proved absence of singular continuous spectrum. We need here an explicit bound on the number of eigenstates.

Lemma 2: Under the assumptions of Lemma 1, let us suppose that H obeys a global Mourre estimate (5), with $\alpha > 0$ and C a positive operator of trace class. Then the following bound on the number of eigenvalues, counted with their multiplicity, holds

$$\dim P_{pp} \leq \alpha^{-1} \operatorname{tr} C. \tag{8}$$

Proof: We use the virial theorem which states that

$$\langle \psi | [H, iA] | \psi \rangle = 0 \tag{9}$$

for every eigenvector of H. Then

$$0 = \operatorname{tr} P_{pp}[H, iA] \geq \operatorname{tr} P_{pp}(\alpha I - C) \geq \alpha \dim P_{pp} - \operatorname{tr} C, \tag{10}$$

which proves the lemma.

We turn to find a suitable conjugate operator for H. On the 1-particle level, a conjugate operator for ω fulfilling (4) is the radial derivation on momentum space, properly multiplied with the group velocity, and symmetrized,

$$D = \frac{1}{2}(\frac{1}{|\nabla_k\omega|^2}\nabla_k\omega \cdot \nabla_k + \nabla_k \cdot \nabla_k\omega\frac{1}{|\nabla_k\omega|^2}). \tag{11}$$

It is easy to see that $[D,\omega] = 1$. The obvious guess for a suitable operator conjugate to H is then

$$-iA = d\Gamma(D). \tag{12}$$

D and $-iA$ generate semigroups of shifts directed outward to increasing $|k|$ on 1-particle momentum space and Fock space, respectively. These semigroups of isometries have no extensions to unitary groups. Thus our conjugate operators have no selfadjoint extensions. Because of this, Mourre's theorems don't apply directly to our choice (12) of A. Nevertheless it should be possible to generalize Mourre's theorems to generators of isometric semigroups.

It is not hard to verify the conditions of the lemmata for H of (2) provided $\inf \omega > 0$. If $\inf \omega = 0$, we compress H to a subspace $P_N \mathcal{H}$ of maximal boson number N. This means that we replace H with $P_N H P_N$ and P_N is the projection onto the closed subspace with at most N bosons. Then the conditions in Lemma 1 and 2 are satisfied and also the technical conditions d),e) in Mourre's theorem [7,p.392], which we did not mention explicitly.

We now observe that, because of reflection symmetry, the Hamiltonian commutes with the parity operator

$$P = -\sigma_z \otimes (-1)^{N_B} \quad \text{and} \quad [P, H_{SB}] = 0. \tag{13}$$

The Hilbert space $\mathcal{H}$ decomposes then into the two eigenspaces $\mathcal{H}_+, \mathcal{H}_-$ of the operator P corresponding to the eigenvalues ± 1. To obtain H restricted to $\mathcal{H}_\pm$ we first apply the unitary transformation

$$U = \exp(i\pi\frac{1-\sigma_x}{2} \otimes N_B) = U^* = U^{-1} \tag{14}$$

with the result

$$\begin{aligned} U\sigma_z \otimes (-1)^{N_B} U &= \sigma_z, \\ UH_{SB}U &= \frac{\mu}{2}\sigma_z \otimes (-1)^{N_B} + I \otimes H_B + I \otimes (a^*(\lambda) + a(\lambda)). \end{aligned} \tag{15}$$

Note that the interaction term contains no explicit spin term anymore. The conserved parity has changed to $-\sigma_z$ with the eigenvalues ± 1. Thus $\mathcal{H}_\pm$ is isomorphic to $\mathcal{F}$ and the Hamiltonians $H_\pm$ on the eigenspaces reducing H_{SB} are unitarily equivalent to

$$H_\pm = \mp\frac{\mu}{2}(-1)^{N_B} + \int d^d k \omega(k) a^*(k) a(k) + a^*(\lambda) + a(\lambda). \tag{16}$$

Our result on the spectral structure on the even subspace is

Theorem 1: Let $\lambda \in D(D)$ and $(D\lambda, D\lambda) < \frac{1}{2}$. The compression $P_N H_+ P_N$ of H_+ to a maximal boson number N has a unique ground state and the rest of the spectrum is purely a.c.. If $\inf \omega > 0$, this holds also for H_+.

Sketch of proof: Let us compute the commutator

$$[H_+, iA] = N_B + a^*(D\lambda) + a(D\lambda). \tag{17}$$

We displace the harmonic oscillator in the mode $D\lambda$ and obtain a new boson number operator with the old eigenvalues $0, 1, 2, \ldots$, however. This will be bounded from below as a form as

$$\begin{aligned}[H, iA] &= N'_B - (D\lambda, D\lambda)I \\ &\geq I - (D\lambda, D\lambda) - |vac'\rangle\langle vac'| \\ &\geq (1 - (D\lambda, D\lambda))I - |vac'\rangle\langle vac'|.\end{aligned} \tag{18}$$

Here $|vac'\rangle$ denotes a shifted Fock vacuum due to completing the square. Application of Lemma 2 leads then to the estimate for the number of eigenvalues,

$$\dim P_{pp} \leq \frac{1}{1 - (D\lambda, D\lambda)}. \tag{19}$$

The $\mathcal{H}_-$ sector is more difficult, since we have to make sure that the number of eigenvalues is less than one. Here we try a conjugate operator with an undetermined function f,

$$-iA = d\Gamma + a(f) - a^*(f). \tag{20}$$

f will have to be optimized at the end. Of course the commutator is now somewhat more complicated

$$\begin{aligned}[H_-, iA] &= N_B + a(D\lambda + \omega f) + a^*(D\lambda + \omega f) \\ &\quad + (f, \lambda) + (\lambda, f) - (-1)^{N_B} a(\mu f) - a^*(\mu f)(-1)^{N_B}.\end{aligned} \tag{21}$$

We remark that $(f, \lambda), (\lambda, f)$ are the good terms which offer a chance that the commutator will be positive at all. For the formulation of the next theorem, we replace λ by $\alpha\lambda$ with λ considered to be fixed, and introduce thereby a real coupling constant.

Theorem 2: For every λ in the domains of D and ω there exists a coupling constant α_c depending on μ and λ such that for $0 < \alpha < \alpha_c$ the spectrum of $P_N H_- P_N$ is purely a.c.. If $\inf \omega > 0$, no cutoff in the boson number is needed.

The odd sector is much harder to treat than the even sector. The strategy is to partition the commutator into two summands and to estimate them separately from below. Our results in this direction are not completely satisfory yet. More details will appear in a future publication [9].

REFERENCES

1. A.J. Leggett et al., *Rev. Mod. Phys.*, **59**, 1 (1987).
2. H. Spohn and R. Dümcke, *J. Stat. Phys.*, **41**, 389 (1985).
3. H. Spohn, *Comm. Math. Phys.*, **123**, 277 (1989).

4. M. Fannes, B. Nachtergaele and A. Verbeure, *J. Phys. A*, **21**, 1759 (1988).
5. M. Fannes, B. Nachtergaele and A. Verbeure, *Comm. Math. Phys.*, **114**, 537 (1988).
6. H.L. Cycon, R.G. Froese, W. Kirsch and B. Simon, "Schrödinger Operators", Springer, Berlin (1987).
7. E. Mourre, *Comm. Math. Phys.*, **78**, 391 (1981).
8. R. Froese and I. Herbst, *Duke Math. J.*, **49**, 1075 (1982).
9. M. Hübner and H. Spohn, to be published.

SOME RESULTS ON THE PROJECTED TWO-DIMENSIONAL ISING MODEL

J. Lőrinczi

University of Groningen, Institute for Theoretical
Physics Nijenborgh 4, 9747 AG Groningen, the Netherlands

1. INTRODUCTION

In this note our aim is to review an example of interacting spin systems on the lattice where non Gibbsian states occur. In the theory of Gibbs measures it is important to study the properties a transformation and a Gibbs measure should have in order that the Gibbsian character shall be preserved under the transformation. In particular this will give some insight in distinguishing Gibbsian from non-Gibbsian states.

In the next section we give some notation, collect some basic results, and give a brief description of the problem we want to look at. In the final section we present our ideas concerning the attempt of reestablishing the Gibbsian property of the projection by excluding some configurations. We will refer to this as the *Dobrushin-Shlosman alternative*. Some of the statements below hold in greater generality, but here we want to discuss only a particular example. Proofs and details will be published elsewhere.

2. BASIC NOTATIONS AND RESULTS

We consider here a system of interacting spins on a lattice. The architecture of the traditional theory is adapted to Polish spaces having an underlying product structure. This is a convenience making it possible to display a number of nice features, some of which we summarize as follows.

The spins live on a countable index set $\mathcal{L}$. From the point of view of the forthcoming set-up no further specification of its geometry is needed, except that one requires that the action of the amenable group $\mathbb{Z}^\nu$ on $\mathcal{L}$ be given. The so called amenable sequences of the lattice play an important role because the averages of observables converge along them. To every spin we assign a set of values, attaching to each site $k \in \mathcal{L}$ the same set S, called the *single spin space*. Throughout this paper S will be metrizable, separable and complete with respect to its metric (i.e., a Polish space). The configuration space is $\Omega = S^\mathcal{L}$. It is endowed with the product topology which is metrizable, since S is metrizable. Ω is separable and complete whenever S is. Thus Ω is a Polish space. If S is compact then by Tychonoff's theorem Ω is also compact. We further equip S and Ω with their associated Borel $\sigma-$ fields $\mathcal{F}_0$ and $\mathcal{F}$, respectively. Because of separability both of them are countably generated, moreover $\mathcal{F}$ coincides with the product σ-field $\mathcal{F}_0^{\mathcal{L}}$. $(\Omega, \mathcal{F})$ is thus a so called standard Borel-space. Let us denote by $\mathcal{P}_f(\mathcal{L})$ the finite subsets of the lattice, and take a $\Lambda \in \mathcal{P}_f(\mathcal{L})$. Consider the finite products $\Omega_\Lambda = S^\Lambda$, $\mathcal{F}_\Lambda = \mathcal{F}_0^\Lambda$. Suppose there is given a finite positive measure λ on the one-site Borel-field. Construct the product measures $\lambda_\Lambda = \prod_{k \in \Lambda} \lambda$ on the measurable spaces $(\Omega_\Lambda, \mathcal{F}_\Lambda)$, for each $\Lambda \in \mathcal{P}_f(\mathcal{L})$. Then, according to Kolmogorov's extension theorem, since $\lambda_{\Lambda'}$ is the marginal distribution of λ_Λ whenever $\Lambda' \subset \Lambda \in \mathcal{P}_f(\mathcal{L})$, there exists a unique finite positive measure $\lambda^\otimes$ on $(\Omega, \mathcal{F})$, whose projections to finite Λ sets coincide with λ_Λ. In case S is finite the measure λ usually is chosen to be counting measure. We will call $\lambda^\otimes$ the *reference measure*, and sometimes we will also call it *free measure*, as it describes a noninteracting spin system. As Ω is Polish, every measure on it is tight and regular. Moreover $(\Omega, \mathcal{F})$ is a perfect measure space (this remains true even if separability and completeness are dropped). These features are useful when deriving central limit theorems and large deviation principles for measures. An *interaction* (also called *potential*) is a family of functions $\Phi_\Lambda: \Omega_\Lambda \to \mathbb{R}$, with $\Lambda \in \mathcal{P}_f(\mathcal{L})$, such that Φ_Λ is $\mathcal{F}_\Lambda$-measurable and $\Phi_\emptyset = 0$. The interaction is quite often supposed to fulfill a summability condition like

$$\sup_{k \in \mathcal{L}} \sum_{\Lambda \ni k} |\Lambda|^{r-1} \sup_{\omega \in \Omega_\Lambda} |\Phi_\Lambda(\omega)| < \infty$$

for some $r \geq 0$. Here $|\Lambda| = \text{card}\Lambda$. (This relation becomes yet simpler if also translation invariance is required on Ω.) Note that the left hand side defines a norm on the set of interactions that renders it into a Banach-space denoted by $\mathcal{B}_r$.

The theory of Gibbs measures is formulated in terms of specifications. This is a way to avoid the difficulties arising in the definition of the *Hamiltonian* when one considers thermodynamic rather than finite systems. For finite volumes the Hamiltonian is

$$\mathcal{H}_\Lambda^\Phi(\omega) = \sum_{\substack{X \cap \Lambda \neq \emptyset \\ X \in \mathcal{P}_f(\mathcal{L})}} \Phi_X(\omega)$$

but for infinite volumes the sum almost surely diverges. A *specification* is a family of probability kernels $\Pi = \{\pi_\Lambda\}_{\Lambda \in \mathcal{P}_f(\mathcal{L})}$ on $(\Omega, \mathcal{F})$ satisfying the following requirements for all $\omega \in \Omega$ and $E \in \mathcal{F}$:
1. $\pi_\Lambda(\cdot, E)$ is an $\mathcal{F}_{\Lambda^c}-$ measurable real function
2. π_Λ are $\mathcal{F}_{\Lambda^c}-$ proper, i.e. for any $F \in \mathcal{F}_{\Lambda^c}$, $\pi_\Lambda(\omega, F) = 1_F(\omega)$
3. for any $\Lambda' \subset \Lambda$, the conditional probabilities are compatible, i.e. $(\pi_\Lambda \pi_{\Lambda'})(\omega, E) := \int \pi_\Lambda(\omega, d\tau) \pi_{\Lambda'}(\tau, E) = \pi_\Lambda(\omega, E)$.

If Π is given in the form

$$\pi_\Lambda(\omega, E) = Z_\Lambda(\omega_{\Lambda^c})^{-1} \int e^{-\mathcal{H}_\Lambda^\Phi(\omega)} 1_E(\omega) d\lambda_\Lambda(\omega_\Lambda)$$

then it is called a *Gibbs specification*. A probability measure consistent with a Gibbs specification (that is, its conditional probabilities coinciding with π_Λ) is called a *Gibbs measure* for the interaction Φ and reference measure $\lambda^\otimes$. A specification is called *uniformly nonnull* if all open sets have positive measure with respect to it, and it is called *quasilocal* or *almost-Markovian* if

$$\lim_{\Lambda \to \mathcal{L}} \sup_{\substack{\omega, \tau \in \Omega \\ \omega_\Lambda = \tau_\Lambda}} |(\pi_\Lambda f)(\omega) - (\pi_\Lambda f)(\tau)| = 0$$

for all bounded measurable functions f. For more details see [1,2]. In recent years there have been displayed a number of interesting examples some of which show that certain operations performed on Gibbs measures fail to yield Gibbs measures again. Such examples come from:
- real-space renormalization theory (decimation transformation in zero or small field and majority rule applied to certain models; block spin averaging for dimensions equal to or higher than two; Kadanoff transformations)
- projected massless Gaussian models
- Schonmann's example (projection of a two-dimensional ferromagnetic nearest- neighbour Ising model onto the line)
- the invariant measures for various stochastic dynamics (Martinelli-Scoppola model, voter model).

Whereas for most of these examples one may not have any intuition in advance about the Gibbsianness of the relevant measure, for renormalization transformations it is generally expected that Gibbs measures should be transformed into Gibbs measures. On the other hand these examples point to the fragility of the traditionally accepted picture, indicating that not all interesting measures are Gibbs for some interaction. We further on consider Schonmann's example, but we will give an a posteriori justification of why this can be interesting. The failure of Gibbsianness here can be narrowed down to the lack of such regularity properties like $\mathcal{B}_1$-summability of the associated interaction. Let us first restate Schonmann's theorem and sketch the main steps of its original proof [3]. Consider the two dimensional nearest neighbour ferromagnetic Ising model, in the subcritical temperature range, with interaction J.

Theorem 1 *(Schonmann):* Suppose $J > J_c$. Then there is no translation invariant $\mathcal{B}_1$-summable interaction on $\mathbb{Z} \otimes \{0\}$ such that $\nu_{+,J}$, the marginal of the + phase taken onto this line, is Gibbsian for it.

The proof is achieved through the following observations: Suppose $\nu_{+,J}$ were Gibbs for some interaction $\Phi \in \mathcal{B}_1$. Then Schonmann shows that $\nu_{-,J}$ would also be Gibbsian for the same interaction Φ. On the other hand, for $\nu_{+,J}$ the following large deviation property is known to hold: Suppose $J > J_c$. Then there exists a convex function $I_J : [-1, +1] \to [0, \infty]$, called *rate function*, such that I_J has but one zero, namely at $m = \int \omega \mu_{+,J}(d\omega)$, and

$$\lim_{n \to \infty} |\Lambda_n|^{-1} \log \nu_{+,J}\{\omega \in \Omega : |\Lambda_n|^{-1} \sum_{k \in \Lambda_n} \omega_k \in (a,b)\} = - \inf_{a < m < b} I_J(m)$$

for all $-1 \leq a < b \leq +1$, and $\Lambda_n = \{-n, ..., +n\}, n = 1, 2, ...$. However, the zeroes of $I_J(m)$ can be unique if and only if there is no translation invariant Gibbs measure for Φ other than $\nu_{+,J}$. As $\nu_{-,J}$ is another translation invariant Gibbs measure for Φ, it follows that the large deviation property cannot hold with the required rate function. Therefore neither $\nu_{+,J}$, nor $\nu_{-,J}$ can be a Gibbs measure.

3. RESTORING GIBBSIANNESS: THE DOBRUSHIN-SHLOSMAN ALTERNATIVE

In order to study the causes which may lead to the failure of Gibbsianness, we make use of the following characterization theorem:

Theorem 2 [1]: *Let Π be a specification, $\lambda^\otimes$ a reference measure, and suppose that S is finite. Then the following are equivalent:*
i. There exists an interaction $\Phi \in \mathcal{B}_1$ such that Π is a Gibbs specification for Φ and the reference measure $\lambda^\otimes$.
ii. Π is almost Markovian, and uniformly nonnull with respect to $\lambda^\otimes$.

Thus at least one of the next properties can be lost under projection: 1. $\mathcal{B}_1$-summability; 2. uniform nonnullness; 3. almost-Markovianness.
ad 1: In principle some summability of the interaction can be restored provided one chooses a suitable space. However, there is no 'natural' Banach space for which one could opt. $\mathcal{B}_0$, which would be reasonable from the point of view of the variational principle, carries already many undesirable features. A slightly better approach could be the renouncing of the Banach-space structure altogether. Indeed, it is shown in [4] that the norm involved makes the definition of Gibbs-measures considerably easier, but a normed structure is not compelling.
ad 2: The projected state in the quantum picture has the Tomita structure. There is a natural embedding of the set of measures as gauge invariant (classical) states into the dual of the quantum algebra $\mathcal{A}$, i.e. of the C*-algebra of quasi-local observables obtained as the operator-norm completion of the union over the whole lattice of all local algebras spanned by $2^{|\Lambda|} \times 2^{|\Lambda|}$ matrices ($\Lambda \in \mathcal{P}_f(\mathcal{L})$) [5,6]. From this we can infer that uniform nonnullness is preserved under the projection, once we prove the following

Lemma 3: *Let us assume S is finite, and $\pi_\varrho(\mathcal{A})$ is the representation algebra of the projected two dimensional system in the quantum picture. Let ϱ be the unique extension of a classical state to the von Neumann algebra $\pi_\varrho(\mathcal{A})''$. If ϱ has the Tomita structure on $\pi_\varrho(\mathcal{A})''$ then its restriction to the classical algebra is a measure on Ω consistent with a uniformly nonnull specification.*

The Tomita structure is a weakened version of the KMS condition (the latter is known to be the quantum counterpart of the Gibbsian property), for more details see [4] and references therein.
ad 3: The only serious reason for which Gibbsianness is bound to be destroyed under the projection is the violation of quasilocality (or almost-Markovianness). This could have been even 'worse' if also the Tomita structure had been lost; so it has at least some regularity properties. In [1] it is actually shown, by using a contour argument, that the projected state is indeed not almost-Markovian. A more recent result obtained by R. Fernández and Ch. Pfister [7], is that the formation of a large droplet of + spins which eventually leads to the propagation of information at large distances is due to a complete wetting phenomenon. Whether in the temperature range between the complete wetting transition point and the critical point of the two-dimensional lattice the underlying mechanism remains the same, has not been proven. Schonmann's proof does not rule out the possibility that the projected measure could be Gibbsian for *some* interaction (which is certainly not in $\mathcal{B}_1$, except for possibly negligible sets of configurations). A possible remedy is proposed by an idea of Dobrushin and Shlosman, suggesting to exclude a 'small' (viz., preferably of measure zero) subset of configurations and recuperate some summability on the restriction. A way of conceiving this is to have infinitely large

interaction on the removed set of configurations, and some regularity in the complement. If we adopt the method of removing certain configurations then it becomes natural to look first at those properties of the configuration space $\Omega = \{-1, +1\}^{\mathbb{Z}}$ (equipped with a Polish topology), and measures thereupon, which are stable under the restriction:

1. In general the restricted configuration space (say, Ω^+ for $\nu_{+,J}$) is not any longer a product space, but it is still a projective limit. We will denote by p_Λ the canonical map $\Omega^+ \to \Omega_\Lambda^+$, which can be indexed by Λ, once Ω_Λ^+ does not contain those configurations which are ruled out also from Ω itself.

2. It is important that on the restriction specifications still can be defined. The idea of the proof is the following: Suppose $\Omega^+ \subset \Omega$, $\Omega \neq \Omega^+ \in \mathcal{F}$, and denote by $\mathcal{F}_{\Omega^+}$ the σ-field of all subsets of Ω^+ of the form $E \cap \Omega^+$ with $E \in \mathcal{F}$. Then $(\Omega^+, \mathcal{F}_{\Omega^+})$ is a standard Borel space. Consequently, $\mathcal{F}_{\Omega^+}$ is countably generated. Moreover, the restricted measure $\nu_{+,J}|_{\mathcal{F}_{\Omega^+}}$ can be shown to be perfect. By Jiřina's theorem then there exists a family of regular conditional probabilities for $\nu_{+,J}|_{\mathcal{F}_{\Omega^+}}$, given any sub σ-field $\mathcal{G} \subset \mathcal{F}_{\Omega^+}$ [8]. Finally, if $\mathcal{G}$ is countably generated, then these regular conditional probabilities are $\nu_{+,J}|_{\mathcal{G}}$–a.e. proper. Compatibility follows immediately by [9]. Note that this idea can be extended to renormalized measures. From the 1st Fundamental Theorem of [1] we learn that a (deterministic) renormalization transformation is a continuous surjection of the configuration space. Such an image of a Polish space is called a Suslin space. Suslin spaces, on the other hand, are perfect measure spaces in the sense that every probability measure on them is perfect. Thus Jiřina's theorem applies, and the existence of specifications for renormalized measures can be concluded. Moreover, a continuous image of a Suslin space is a Suslin space [10], thus the iteration of the renormalization map, as far as the existence of specifications is concerned, does not give rise to difficulties.

3. Compactness of the allowed configuration space is in general lost. Our first statement will enable us to establish a positivity property of the restricted measure. We stick to $\nu_{+,J}$, but the same is true for $\nu_{-,J}$.

Lemma 4: *Suppose $\Pi = \{\pi_\Lambda\}_{\Lambda \in \mathcal{P}_f(\mathcal{L})}$ is a uniformly nonnull specification on Ω with respect to the free measure. Then ϱ, a measure consistent with Π, is a finite regular Borel measure on Ω such that $\mathrm{supp}\varrho = \Omega$. Conversely, to any finite regular measure ϱ supported on Ω, there can be found a uniformly nonnull specification consistent with it.*

Proposition 5: *Any nonzero restriction of the measure $\nu_{+,J}$ is strictly positive on all open sets. Therefore there is no compact proper subset of Ω carrying full $\nu_{+,J}$ (or $\nu_{-,J}$) measure.*

4. As a subspace of a complete metric space is complete if and only if it is closed, one may start worrying whether the Polish space structure is lost, as we know that the restriction is noncompact, hence not closed. It can be shown, however, that Ω^+ continues to be Polish, though completeness holds with respect to a metric equivalent to, and which can be reconstructed in terms of, that given on Ω. In other words Ω^+ is topologically complete.

5. Regularity and tightness of the measures on Ω^+ survive. It follows that the support of the measure can be approximated by compact sets. It then follows also that any sequence of tight measures $\{\mu_n\}_{n \in \mathbb{N}}$ has a weakly convergent subsequence, that is, the set $\{\mu_n\}_{n \in \mathbb{N}}$ is relatively compact.

6. The projective limit of compact spaces is compact. Here, however, the sets Ω_Λ^+ are not compact. The theorem of Kolmogorov has to be replaced by a more general

theorem of Prokhorov, which says that the projective limit $\lim_{\leftarrow} \lambda_\Lambda$ exists, provided $\mathcal{L}$ is countable (which is indeed the case here). If the canonical map separates the points of Ω^+, then the projective limit of the measure is unique. We shall call $\lim_{\leftarrow} \lambda_\Lambda = \tilde{\lambda}$ *quasiproduct measure*. It is natural to ask what the relation between $\lambda^\otimes$ and $\tilde{\lambda}$ is, apart from absolute continuity. The latter is no longer a free measure, the Radon-Nikodym derivative can be interpreted as an *induced potential*, which is compensating the averages taken with respect to the two different measures. This observation can be useful in improving the results of [11] by introducing correction terms coming from the Radon-Nikodym derivative, in the cluster expansion. The induced potential contains the information about the configurations to be excluded. Then, by optimizing over the convergence condition of the relative energies, aiming to get the largest possible set of configurations on which it holds, the correction terms could be determined recursively. Thus the induced potential could be reconstructed and we could learn how to suppress more 'gently' the bad behaviour than in [11], allowing many more configurations.

7. It is still a question whether the large deviation properties and the decomposition properties following from the Choquet-simplex structure in general also survive under the restriction. Some minor modifications may be required, e.g. the level sets of the relative entropy in the large deviation property may in general be noncompact. In fact we expect, however, that the collective behaviour of spins (as clusters, for example) is not essentially affected under the restriction. Indeed, the mechanism responsible for generating ill-behaving configurations seems to be the too probable occurrence of large islands of identically valued spins. Such droplets can form if many-body correlations are not sufficiently well decaying.

Suppose $\mathcal{F}_1 \subset \mathcal{F}_2 \subset \mathcal{F}$ are sub $\sigma-$ fields, and $\pi_{+,J}(\omega, \mathcal{F}_1)$ resp. $\pi_{-,J}(\omega, \mathcal{F}_1)$, $\omega \in \Omega$, are specifications for $\nu_{+,J}$ resp. $\nu_{-,J}$, given $\mathcal{F}_1$. Then the relative entropy governing the large deviation principle on the restriction satisfies [12]

$$I_{\mathcal{F}_2}(\nu_{+,J}|\nu_{-,J}) = I_{\mathcal{F}_1}(\nu_{+,J}|\nu_{-,J}) + \int I_{\mathcal{F}_2}(\pi_{+,J}(\omega, \mathcal{F}_1)|\pi_{-,J}(\omega, \mathcal{F}_1))\nu_{+,J}(d\omega)$$

As a conclusion of this paragraph we note that the existence of specifications as well as the conservation of many of the good properties are partly due to the special situation of a restriction (e.g. the uniform nonnullness of the state on Ω^+, inherited from the state on Ω). Some of them, as we have seen, will be preserved also under suitable mappings of Ω. We are going now to discuss the consequences of the observation that the restrictions of the form $\nu_{+,J}|_{\mathcal{F}^+}$ resp. $\nu_{-,J}|_{\mathcal{F}^-}$ obey essentially the same large deviation principle as the one in Theorem 1 (modulo some small alterations as we pointed out earlier). The aforementioned property of the relative entropy will entail this. Then by reversing Schonmann's argument, we can prove our next result.

Proposition 6 : *If there exists at all an interaction on some subfield $\mathcal{F}^+$ such that the measure $\nu_{+,J}$ restricted to $\mathcal{F}^+$ is Gibbs, then this measure is the unique Gibbs-measure for it. The same holds for $\nu_{-,J}$ with some $\mathcal{F}^-$. These two restricted measures cannot coincide, because at least the magnetization in the origin separates them. Therefore also the two interactions differ.*

An explicit construction of the subfields $\mathcal{F}^+$ and $\mathcal{F}^-$, and the respective interactions on them is , of course, the heart of the matter. For the moment we have only partial results on this problem. In general the interaction for the restricted measure, if there is any, will contain some magnetic field-term, and two-body, three-body etc., terms. It can be shown that the ferromagnetic and antiferromagnetic Ising pair-interactions, with spins $+1$ and -1, however long range they may have, are not satisfactory in the description of the restricted spins' interaction, thus many-body terms cannot be ignored as unimportant perturbative corrections, nor the unbounded-like character of the interaction can be neglected. The intuition might develop along the following remarks: The decay of the interaction is not slower than the decay of correlations, at any temperature. Correlations, however, decay exponentially. Then, by applying Dyson's results [13] (with later improvements by Aizenman and others [14]), one can conclude that the admissible configurations are roughly those on which the ferromagnetic interaction would have a power law decay with exponent at least as large as 2. The antiferromagnetic case can be treated in a similar way, by noticing the fact that all correlation functions of the ferromagnetic interaction majorize the correlation functions of the antiferromagnetic interaction. This description is far from being optimal but at least indicates the kind of criterion we want to establish for the construction of the allowed configurations and the interaction. Next we want to say something about how much to exclude from the whole configuration space. Suppose $\{\pi_\Lambda(\omega|\mathcal{F}_{\Lambda_n\setminus\Lambda})\}$ is a sequence of (trivially) Feller specifications for $\nu_{+,J}$ converging pointwise to some $\pi_\Lambda(\omega|\mathcal{F}_{\Lambda^c})$, where the conditioning is made on the annuli $\Lambda_n \setminus \Lambda$, $\{\Lambda_n\}$ being an amenable sequence. In [4] it is then shown that there exists a dense set of Ω, say Ω_{rel}, on which the limit $\pi_\Lambda(\omega|\mathcal{F}_{\Lambda^c})$ is Feller, therefore this set could be a possible candidate for collecting up the 'good' configurations. Ω_{rel} is measurable with respect to the tail field of the whole space. Moreover, by Proposition 5, $\nu_{+,J}|_{\Omega_{rel}}$ takes strictly positive values on all relatively open sets. We then have the next

Corollary 7 : $\nu_{+,J}(\Omega_{rel}) = 1$.

This is the only case we know of where a suggested restriction provably carries full measure. We want to point out, further, that unlike in Schonmann's example, there may exist situations where the restriction of a Gibbs measure can be given in terms of the restriction of the interaction. (In Schonmann's example the projected nearest neighbour Ising system actually can not be written even as a long-range Ising interaction.) This remark to some extent also counterpoints the objection why should one at all be surprised that in Schonmann's case this is so far from being true. We do not claim that the next example is anywhere near to being 'physical', but it would at least hint repeatedly to why our expectation is that the unbounded-like character of some sort of collective modes of the projected spins could indeed play a role in the emerging non-Gibbsian behaviour. The key is to perform a 'master-stroke' (like the *coup de Jarnac*), just the right cut in the configuration space. Consider $\Omega = \{-1,+1\}^\mathbb{Z}$, i.e. the entire configuration space, the standard Borel-space $(\Omega, \mathcal{F}, \mu)$ and the canonical map $p_\Lambda : \Omega \to \Omega_\Lambda$. (Here μ could be for instance $\nu_{+,J}$.) By a theorem of Godement-Bourbaki-Schwartz [10] there exists a 'μ-concassage' of Ω, i.e., a μ-null set N and a countable family of disjoint compact sets $\{K_\alpha\}_{\alpha \in \mathcal{I}}$ such that $\Omega = N \cup \bigcup_{\{\alpha \in \mathcal{I}\}} K_\alpha$ and $\mathrm{supp}\mu \circ i_\alpha = K_\alpha$, where $i_\alpha : \Omega \to K_\alpha$ is the restriction map $i_\alpha = 1_{K_\alpha} \cdot id_{K_\alpha}$. Suppose on the compact pieces there can be defined the interactions $\Phi_\Lambda^{(\alpha)} : p_\Lambda K_\alpha \to \mathbb{R}$, for all $\alpha \in \mathcal{I}$, as well as specifications $\{\pi_\Lambda^{\Phi^{(\alpha)}}\}_{\Lambda \in \mathcal{P}_f(\mathcal{L})}$ corresponding to them. Then the next proposition can be proven:

Proposition 8 : *Let* $\{\pi_\Lambda^{\Phi^{(\alpha)}}\}_{\Lambda \in \mathcal{P}_f(\mathcal{L})}$ *be a Gibbs specification given on* $(\Omega, \mathcal{F})$ *for the interaction* $\Phi = \{\Phi_\Lambda^{(\alpha)}\}_{\Lambda \in \mathcal{P}_f(\mathcal{L}), \alpha \in \mathcal{I}}$. *Then the family* $\{\pi_\Lambda^{(\alpha)} \circ p_\Lambda \circ i_\alpha\}_{\Lambda \in \mathcal{P}_f(\mathcal{L}), \beta \neq \alpha}$ *is a Gibbs-specification for the interaction* $\{\Phi_\Lambda^{(\alpha)}\}_{\Lambda \in \mathcal{P}_f(\mathcal{L}), \beta \neq \alpha \in \mathcal{I}}$, *for any* $\beta \in \mathcal{I}$.

Finally let us summarize why Schonmann's example is interesting to us. First of all, it is an example in some sense intermediate between the so called 'pathologies' of renormalization group transformations and the steady states of some nonequilibrium dynamics. Secondly, it may help in fixing at least a sense of a 'non-Gibbsian' measure. Thirdly, it may serve to assess the discrepancy between a KMS state and a state having the Tomita structure. Indeed, it is an example of a state having the Tomita structure, but not being KMS.

Acknowledgments

I am grateful to A.C.D. van Enter, R. Fernández and M. Winnink for comments on this work. The careful reading of the manuscript by M. Winnink is also gratefully acknowledged.

REFERENCES

1. A.C.D. van Enter, R. Fernández and A. Sokal, Regularity properties and pathologies of position-space renormalization-group transformations: Scope and limitations of Gibbsian theory, *J. Stat. Phys.*, **72**, 879 (1993).
2. H.-O. Georgii, "Gibbs Measures and Phase Transitions", Walter de Gruyter (de Gruyter Studies in Mathematics, 9), Berlin, New-York (1988).
3. R.H. Schonmann, Projections of Gibbs measures may be non-Gibbsian, *Commun. Math. Phys.*, **124**, 1 (1989).
4. J. Lőrinczi and M. Winnink, Some remarks on almost Gibbs-states, *in*: Proceedings of the NATO Advanced Studies Institute Workshop on Cellular Automata and Cooperative Systems (Les Houches 1992), Kluwer, Dordrecht (1993).
5. R.B. Israel, "Convexity in the Theory of Lattice Gases", Princeton University Press, Princeton, New Jersey (1979).
6. M. Winnink, Some remarks on the Gibbs phase rule, *Delft Progress Report*, **9**, 115 (1984).
7. R. Fernández, private communication
8. M. Jiřina, Conditional probabilities on σ–algebras with countable basis, *Selected Translations in Mathematical Statistics and Probability*, **2**, AMS (1962).
9. A.D. Sokal, Existence of compatible families of proper regular conditional probabilities, *Z. Wahrscheinlichkeitstheorie verw. Geb.*, **56**, 537 (1981).
10. L. Schwartz, "Radon Measures on Arbitrary Topological Spaces and Cylindrical Measures", Oxford University Press, Oxford (1973).
11. C. Maes and K. Vande Velde, Defining relative energies for the projected Ising-measure, *Helv. Phys. Acta*, **65**, 1055 (1992).
12. V.M. Donsker and S.R.S. Varadhan, Asymptotic evaluation of certain Markov process expectations for large time IV, *Commun. Pure Appl. Math.*, **36**, 183 (1983).
13. F. Dyson, Existence of phase transition in a one-dimensional Ising ferromagnet, *Commun. Math. Phys.*, **12**, 91 (1969).
14. M. Aizenman et al., Discontinuity of the magnetization in one-dimensional $1/|x-y|^2$ Ising and Potts models, *J. Stat. Phys.*, **50**, 1 (1988).

THE SECOND VIRIAL COEFFICIENT FOR QUANTUM-MECHANICAL STICKY SPHERES

M.D. Penrose[1] and O. Penrose[2]

1. Department of Mathematical Sciences
   University of Durham
   South Road, Durham DH1 3LE, England

2. Department of Mathematics
   Heriot-Watt University
   Riccarton, Edinburgh EH14 4AS, Scotland

This paper gives a non-rigorous demonstration of a result announced in poster form at the 'On three levels' conference in July 1993; it may be useful since the rigorous proof given in [1] uses probabilistic techniques which are not familiar to all physicists. The result concerns the second virial coefficient $B_{D,a}$ for a quantum-mechanical system of hard spheres (a true mathematician would perhaps call them 'hard balls') with an attractive two-body interaction, typically a square well of width a. More precisely, the interaction between any pair of molecules with relative displacement $\mathbf{r}$ is taken to be

$$\phi(\mathbf{r}) := \begin{cases} +\infty & \text{if} \quad |\mathbf{r}| \leq D, \\ \mathcal{E} u\left(\frac{|\mathbf{r}|-D}{a}\right) & \text{if} \quad |\mathbf{r}| > D, \end{cases} \quad (1)$$

where the colon indicates a definition, $|\mathbf{r}|$ denotes the magnitude of the vector $\mathbf{r}$, $\mathcal{E}$ denotes the depth of the well, a its width, D the diameter of the hard spheres, and u is the 'unit square well' potential defined by

$$u(x) := -\mathbf{1}_{(0,1]}(x) = \begin{cases} -1 & \text{if} \quad 0 < x \leq 1, \\ 0 & \text{if} \quad 1 < x. \end{cases} \quad (2)$$

The potential (1) has been much used as a model of real fluids [2,3], colloids [4] and some biological systems [5].

The result proved in [1] is that in the "sticky limit", where $a \to 0, \mathcal{E} \to \infty$ in such a way that the quantity

$$\alpha := a^2 \mathcal{E} m/(2\hbar^2), \quad (3)$$

(m = mass of particles, $2\pi\hbar$ = Planck's constant) is held fixed at a value less than $\pi^2/8$, the second virial coefficient $B_{D,a}$ has the behaviour

$$B_{D,a} = B_D - a\left(\frac{\tan\sqrt{(2\alpha)}}{\sqrt{(2\alpha)}} - 1\right)\frac{d}{dD}B_D + o(a) \quad (\alpha < \pi^2/8) \quad (4)$$

where B_D is the second virial coefficient for hard spheres of diameter D without the potential well, i.e. for the case $\mathcal{E} = 0$.

Like the proof in [1], our demonstration is based on the general formula for the second virial coefficient given in [6], or section 14.3 of [7], which is

$$B = \frac{1}{2}\int_{R^3} \{1 - (4\pi\hbar^2\beta/m)^{3/2}[G(\beta;\mathbf{r},\mathbf{r}) + CG(\beta;\mathbf{r},-\mathbf{r})]\}d^3\mathbf{r} \quad (5)$$

where β means $1/(kT)$ as usual and C is a constant which depends on the spin and statistics of the particles, being zero for Boltzmann statistics, $1/n$ for Bose particles with n internal spin states (so that $n = 2S+1$ for particles of total spin S) and $-1/n$ for Fermi particles. The function $G(\beta;\mathbf{r},\mathbf{r}')$ is the fundamental solution (Green's function) of the Bloch equation for relative motion in a two-particle system with the interaction potential $\phi(\mathbf{r})$. That is to say, $G(\beta;\mathbf{r},\mathbf{r}')$ is the solution of the partial differential equation

$$\partial G/\partial \beta = [(\hbar^2/m)\nabla^2 - \phi(\mathbf{r})]G(\beta;\mathbf{r},\mathbf{r}') \quad (\beta > 0) \quad (6)$$

(where ∇^2 acts on functions of $\mathbf{r}$ at fixed $\mathbf{r}'$ and β) with the initial condition

$$G(0;\mathbf{r},\mathbf{r}') = \delta(\mathbf{r}-\mathbf{r}') \quad (7)$$

and with the interpretation that $G(\beta;\mathbf{r},\mathbf{r}') = 0$ if, with $\beta > 0$, $\phi(\mathbf{r})$ or $\phi(\mathbf{r}')$ is equal to $+\infty$.

Equation (6) for $G(\beta;\mathbf{r},\mathbf{r}')$, written in terms of spherical polar co-ordinates $\mathbf{r} = (r,\theta,\phi)$, implies that

$$\frac{\partial G}{\partial \beta} = \frac{\hbar^2}{m}\left(\frac{\partial^2 G}{\partial r^2} + \frac{2}{r}\frac{\partial G}{\partial r} + \frac{1}{r^2\sin\theta}\frac{\partial}{\partial \theta}(G\sin\theta) + \frac{1}{r^2\sin^2\theta}\frac{\partial^2 G}{\partial \phi^2}\right) + \mathcal{E}G$$
$$(D < r < D+a) \quad (8)$$

Making the substitution $r = D + ax$ and considering the limit $a \to 0$ we find that for small a the dominant terms (they are asymptotically proportional to $1/a^2$ in this limit) are

$$0 \simeq \frac{\hbar^2}{m}\frac{\partial^2 G}{\partial r^2} + \mathcal{E}G \quad (D < r < D+a) \quad (9)$$

with solution (since $G = 0$ when $r = D$)

$$G(\beta;\mathbf{r},\mathbf{r}') \simeq f(\theta,\phi,\mathbf{r}',\beta)\sin\sqrt{\frac{\mathcal{E}m}{\hbar^2}}(r-D) \quad (10)$$

where the function f (which we do not need to know) is determined by the conditions on the sphere $\{\mathbf{r}:r=D+a\}$. At the inner surface of this sphere we have (using (3))

$$\left.\begin{array}{rl}G(\beta;\mathbf{r},\mathbf{r}') & \simeq f(\theta,\phi,\mathbf{r}',\beta)\sin\sqrt{2\alpha} \\ \frac{\partial G(\beta;\mathbf{r},\mathbf{r}')}{\partial r} & \simeq f(\theta,\phi,\mathbf{r}',\beta)\frac{\sqrt{2\alpha}}{a}\cos\sqrt{2\alpha}\end{array}\right\} \quad (r = D+a-0) \quad (11)$$

so that

$$G = a\frac{\tan\sqrt{2\alpha}}{\sqrt{2\alpha}}\frac{\partial G}{\partial r} \quad (r = D+a-0) \quad (12)$$

Now consider the 'outer' solution of (6), i.e. the solution in the region where $r > D+a$. In this region (6) is simply the heat equation, and since the first derivative of G is continuous, the relevant solution obeys (by (12)) the Robin boundary condition

$$G = a\frac{\tan\sqrt{2\alpha}}{\sqrt{2\alpha}}\frac{\partial G}{\partial r} \quad (r = D+a+0). \quad (13)$$

This boundary condition has the interpretation that the tangent to the graph of G against r (with θ, ϕ and $\mathbf{r}'$ held fixed) cuts the horizontal axis at the point where

$$r = D + a - a\frac{\tan\sqrt{2\alpha}}{\sqrt{2\alpha}}. \tag{14}$$

For sufficiently small a, the tangent should be a good approximation to the graph itself over the relevant range of values of r, and hence we expect the 'outer' solution to extrapolate to a value close to zero on the sphere $\{\mathbf{r} : r = D'\}$, where by definition

$$D' := D + a - a\frac{\tan\sqrt{2\alpha}}{\sqrt{2\alpha}}. \tag{15}$$

This suggests approximating G, in the region where $r > D + a$, by the fundamental solution of the heat equation over the slightly larger region $\{\mathbf{r} : r > D'\}$ with the boundary condition $G = 0$ on the boundary of this larger region. This latter function is the Green's function for hard spheres of diameter D'.

Now compare the integral in (5) for the second virial coefficient with the formula for the second virial coefficient for hard spheres of diameter D'. Except in a thin shell of thickness $O(a)$ the integrands are approximately the same. They differ only inside the thin shell; but provided that $\alpha < \pi^2/8$ (so that there are no bound states) the difference there is $O(a)$ and therefore the contribution of the shell to the difference between the two integrals is $O(a^2)$. So, at the level of approximation we are working to (the $O(a)$ level), the two integrals are the same, and we have

$$\begin{aligned} B_{D,a} &\simeq B_{D'} \\ &= B_D + (D' - D)\frac{\partial}{\partial D}B_D + O(D' - D)^2 \end{aligned} \tag{16}$$

By (15), this is a non-rigorous version of the desired result (4).

The above analysis is easily generalized to dimension numbers other than three, and to the case where the function u in (1) is replaced by a more general function on the interval $(0, 1)$. The resulting formulas are the same as in [1].

Acknowledgments

OP thanks the organizers for an excellent conference and the Royal Society of London for a grant towards his expenses in attending it.

REFERENCES

1. M D Penrose, O Penrose and G Stell, Sticky spheres in quantum mechanics, submitted to *Rev. Math. Phys.*.
2. J.A. Barker and D. Henderson, What is "liquid"? Understanding the states of matter, *Rev. Mod. Phys.*, **48** 587 (1976).
3. H. van Beijeren, Kinetic theory of dense gases and liquids, *in:* "Fundamental Problems in Statistical Mechanics VII", 357, North Holland (1990).
4. P.W. Rouw, A. Vrij and G. de Kruif, *Prog. Colloid Polymer Sci.*, **76**, 1 (1988); Adhesive hard-sphere colloidal dispersions III. Stickiness in n-dodecane and benzene. *Colloid Surf.*, **31**, 299 (1988).
5. R. Saunders, R.J. Kryscio and G.M. Funk, Poisson limits for a hard-core clustering model, *Stochastic Processes Appl.*, **12**, 97 (1982).

6. E.H. Lieb, Calculation of exchange second virial coefficient of a hard sphere gas by path integrals, *J. Math. Phys.*, **8**, 43 (1967).
7. K. Huang, "Statistical Mechanics", Second edition, Wiley, New York (1987).

BETHE–ANSATZ SOLUTION
OF A MODIFIED SU(3)-XXZ-MODEL

H. Grosse and E. Raschhofer

Institut für Theoretische Physik
Universität Wien
A-1090 Wien, Boltzmanngasse 5
Austria

Supported in part by the 'Fonds zur Förderung der wissenschaftlichen Forschung in Österreich' under project No. P8916-PHY.

ABSTRACT

We investigate a solvable SU(3)-XXZ-model modified by elements of the Cartan-subalgebra. The nested Bethe-Ansatz is discussed in detail and the corresponding quantum spin model is stated. The coupled Bethe-equations are solved in the thermodynamic limit. Effects of the modification on the finite size scaling are derived.

INTRODUCTION

Low dimensional integrable models, both in the continous formulation and on the lattice play a dominant role in field theory and solid state physics. They provide rigorous results for physical quantities.

The Yang-Baxter equation (YBE) guarantees the existence of an infinite number of conserved currents for the transfer matrix. Commutation relations for the elements of the so called monodromy matrix implied by the YBE allow the construction of eigenstates of the transfer matrix and the calculation of eigenvalues provided that a number of consistency equations are solved.

Expanding the logarithm of the transfer matrix gives one dimensional quantum mechanical models with local interactions. Conformal field theory predicts the behaviour of integrable models in the critical regime for large but finite volume. In particular, the central charge of the model determines how the groundstate energy approaches the thermodynamic limit.

One way of generalization consists of turning on 'electrical fields' to the vertex model (for the quantum model this gives diffusion terms).

The poster is organized as follows: in section 2 we describe the model and employ the YBE for its generalization. Section 3 is devoted to the nested Bethe-Ansatz solution. In section 4 we review the thermodynamic limit properties and in section 5 finite size effects are discussed and the central charge is computed.

THE MODEL

We start with the statistical model presented in[1]. The R-matrix for this model reads

$$R_{ab,ij} = \sinh \gamma \, \delta_{ia}\delta_{jb} \, e^{\lambda \, \text{sign}(a-b)} + \sinh \lambda \, \delta_{ib}\delta_{ja} \quad i \neq j \qquad (1)$$
$$R_{aa,ii} = \sinh(\gamma + \lambda) \, \delta_{ia} \, .$$

R acts in $\mathbf{C}^Q \otimes \mathbf{C}^Q$. λ is called the spectral parameter and γ determines the anisotropy. We shall call this model the SU(Q)-XXZ-model, because for $\gamma = 0$ it enjoys SU(Q) invariance[1]. There are several choices for the range of the parameters γ, λ. In the ferroelectric region we take $\lambda > 0$, $\gamma > 0$ in eq.(1). Another regime is obtained by allowing imaginary parameters $\gamma \to i\gamma$, $\lambda \to i\lambda$. The weights then are complex due to the exponential in eq.(1). The model in the trigonometric region is of interest because it is critical. An antiferroelectric region is also present.

The vertex model has $Q(2Q-1)$ nonvanishing statistical weights determined by the local operator $l_{0n} := P_{0n} R_{0n}$, where P_{0n} denotes the permutation operator. In the following, we use the abbreviation $a := \sinh(\gamma + \lambda)$, $b := \sinh \lambda$, $c := \sinh \gamma \, e^{\lambda}$, $d := \sinh \gamma \, e^{-\lambda}$ for the weights of the allowed vertex configurations. With this conventions, the Yang-Baxter equation

$$R_{12}(\lambda - \lambda', \gamma) \, l_{13}(\lambda, \gamma) \, l_{23}(\lambda', \gamma) = l_{13}(\lambda', \gamma) \, l_{23}(\lambda, \gamma) \, R_{12}(\lambda - \lambda', \gamma) \qquad (2)$$

is fulfilled. We are interested in solutions of the YBE for models that are generated from the above ones by multiplication with elements of the Cartan subalgebra of SU(Q). For ease of computation, we restrict ourself to the case of SU(3), but every step could in principle be made for SU(Q) as well. For our discussion let us define:

$$K(\vec{v}) := e^{(v_3 \lambda^3 + v_8 \lambda^8)/N} \qquad I(\vec{h}) := e^{(h_3 \lambda^3 + h_8 \lambda^8)/N} \qquad L_{0n} := I_0 \, K_n \, l_{0n}. \qquad (3)$$

In this context we use the notation $\vec{v} := (v_3, v_8)$, $\vec{h} := (h_3, h_8)$. λ^3 and λ^8 are the usual diagonal Gell-Mann generators of SU(3). $\vec{v}$ and $\vec{h}$ will be called 'horizontal' and 'vertical' fields respectivly. The crucial point is, that with the given conventions the local operator $L_{0n}(\lambda, \vec{v}, \vec{h}_n)$ solves the YBE and thereby generates an integrable model:

$$\tilde{R}_{12}(\lambda'', \vec{v}, \vec{v}') \, L_{13}(\lambda, \vec{v}) \, L_{23}(\lambda', \vec{v}') = L_{13}(\lambda', \vec{v}') \, L_{23}(\lambda, \vec{v}) \tilde{R}_{12}(\lambda'', \vec{v}, \vec{v}') \, , \qquad (4)$$

where the new R-matrix $\tilde{R}$ reads

$$\tilde{R}_{12}(\lambda - \lambda', \gamma, \vec{v}, \vec{v}') = K_1(\vec{v}) \, R_{12}(\lambda - \lambda', \gamma) \, K_1^{-1}(\vec{v}') \, . \qquad (5)$$

Notice that the vertical fields have the properties of spectral parameters, whereas the horizontal fields have to be kept fixed in eq.(4). In order to prove our claim we recall from[1] the symmetries of the original R-matrix R_{12}. Although there is no SU(3) invariance for $\gamma \neq 0$, R_{12} is still invariant under the abelian subgroup generated by λ^3, λ^8 in the sense that

$$[R_{12}, \lambda^i \otimes \lambda^i] = 0 , \qquad (6)$$

for $i = 3, 8$. Whenever eq.(6) holds, there is a general strategy for adding horizontal fields[1]. Moreover the structure of the equations allow space dependent horizontal fields. Remarkably this step does not affect the R-matrix. Inclusion of the vertical fields is achieved in the following manner: write the original YBE in terms of the modified model as

$$R_{12}(\lambda'') K_3^{-1}(\vec{v}) L_{13} K_3^{-1}(\vec{v}') L'_{23} = K_3^{-1}(\vec{v}') L'_{13} K_3^{-1}(\vec{v}) L_{23}(\lambda) R_{12}(\lambda'') , \qquad (7)$$

(with $\lambda'' = \lambda - \lambda'$) and multiply both sides with $K_3(\vec{v}) K_3(\vec{v}') K_1(\vec{v})$ from the left and with $K_1^{-1}(\vec{v})$ from the right. Using the commutation relation eq.(6), the desired result is immediatly obtained. Since the constructions for horizontal and vertical fields are independent of each other this proves the YBE for our model.

ALGEBRAIC BETHE-ANSATZ

As it is well known, the YBE for the elements of the monodromy matrix give commutation relations for the elements of the transfer matrix T. Together with the existence of a reference eigenvector ('local vacuum' $|\omega\rangle$), these relations enable the construction of eigenvectors for T in many cases. For our models with $Q \geq 3$, one has to proceed in steps, called nested algebraic Bethe-Ansatz. Explicitly, the YBE for

$$M = \begin{pmatrix} A & B_1 & B_2 \\ C_1 & D_{11} & D_{12} \\ C_2 & D_{21} & D_{22} \end{pmatrix} \qquad (8)$$

gives 81 relations for the elements of M. 15 of these relations are of special interest for the Bethe-Ansatz and can be summarized as:

$$\begin{aligned} A A' &= A' A \\ A B'_1 &= \frac{\beta}{\alpha} \frac{a(\lambda' - \lambda)}{b(\lambda' - \lambda)} B'_1 A - \frac{\beta}{\alpha} \frac{d(\lambda' - \lambda)}{b(\lambda' - \lambda)} B_1 A' \\ A B'_2 &= \frac{\gamma}{\alpha} \frac{a(\lambda' - \lambda)}{b(\lambda' - \lambda)} B'_2 A - \frac{\gamma}{\alpha} \frac{d(\lambda' - \lambda)}{b(\lambda' - \lambda)} B_2 A' \\ B \otimes B' &= (B' \otimes B) R^{(2)}(\lambda - \lambda') = R^{(2)T}(\lambda - \lambda')(B' \otimes B) \\ D \otimes B' &= \frac{1}{b(\lambda - \lambda')} S^{(2)} (B' \otimes D) R^{(2)}(\lambda - \lambda') - \frac{c(\lambda - \lambda')}{b(\lambda - \lambda')} S^{(2)} (B \otimes D'), \end{aligned} \qquad (9)$$

where we have used the tensor notation $(B \otimes B')_{ij} = B_i B'_j$, $(D \otimes B')_{i,jk} = D_{ij} B'_k$, $(B \otimes D')_{i,jk} = B_j D'_{ik}$. Due to our modifications, the operator B_1 and B_2 give different expressions when commuted with A. The matrices $R^{(2)}$ and $S^{(2)}$ in eqs.(9) are given by:

$$R^{(2)}(\lambda) = \begin{pmatrix} 1 & 0 & 0 & 0 \\ 0 & c/a & b_3/a & 0 \\ 0 & b_6/a & d/a & 0 \\ 0 & 0 & 0 & 1 \end{pmatrix} \qquad S^{(2)} = \begin{pmatrix} \beta/\alpha & 0 \\ 0 & \gamma/\alpha \end{pmatrix} . \qquad (10)$$

Here $\alpha = e^{(v_3+v_8)/N}$, $\beta = e^{(-v_3+v_8)/N}$, $\gamma = e^{-2v_8/N}$, and $b_3 = b\beta/\gamma$, $b_6 = b\gamma/\beta$. Now we turn to the action of L and M on the reference vectors $|\omega_i\rangle := (1,0,0) \in \mathbf{C}^3$ and $|\Omega\rangle := \otimes^N |\omega_i\rangle$. From

$$L_{0i}(\lambda) |\omega_i\rangle = \alpha \begin{pmatrix} ax|\omega_i\rangle & \star & \star \\ 0 & by|\omega_i\rangle & 0 \\ 0 & 0 & bz|\omega_i\rangle \end{pmatrix}, \qquad (11)$$

where $\star$ means some nonzero vector (nonproportional to $|\omega_i\rangle$). We seek for eigenvectors $|\psi, \vec{\mu}; r, s\rangle$ of T which are of the form

$$|\psi, \vec{\mu}; r, s\rangle = \sum_{\{i_n=2,3\}} X_{i_1,\ldots,i_s} B_{i_1}(\mu_1) \cdots B_{i_s}(\mu_s) |\Omega\rangle . \qquad (12)$$

In this linear combination, B_1 is supposed to appear r-times and therefore, B_2 appears $(r-s)$-times. The parameters $\vec{\mu} = \{\mu_1, \ldots, \mu_s\}$ are determined by coupled Bethe equations as we shall see later on. Note that $X_{i_1,\ldots,i_s}$ can be regarded as a vector in $\otimes \mathbf{C}^2$!

Applying the operator $A(\lambda)$ onto $|\psi\rangle$ and commuting it with the B's gives 2^s terms according to eq.(9). The 2^s terms can be collected into $2s$ terms, one of them is of special interest. Taking into account the action of A on $|\Omega\rangle$ gives

$$A(\lambda)|\psi, \vec{\mu}; r, s\rangle = \beta^r \gamma^{s-r} \alpha^{N-s} (\prod x_i) a^N(\lambda) \prod_{i=1}^s \frac{a(\mu_i - \lambda)}{b(\mu_i - \lambda)} |\psi\rangle \qquad (13)$$

+nondiagonal terms .

Computing the contribution of $D_{\alpha\alpha} = D_{11} + D_{22}$ to the eigenvalue is less easy, because the commutation relations are more involved. It can then be written as

$$D_{\alpha\alpha}|\psi\rangle = \alpha^{N-s} b^N(\lambda) \prod_{i=1}^s \frac{a(\lambda - \mu_i)}{b(\lambda - \mu_i)} (\tilde{T}X)_{i_1 \cdots i_s} B_{i_1} \cdots B_{i_s} |\Omega\rangle . \qquad (14)$$

+nondiagonal terms ,

where $\tilde{T}$ is the transfer matrix of a modified six vertex model. This ends the discussion of the summands that suit the eigenvalue equation.

Non diagonal terms are generated if the second terms of the eqs.(9) are used at least once. Then one of the μ_i's is interchanged with the spectral parameter λ. To obtain a typical term, we first make a cyclic permutation of the B-operators, using the expression in the third line of eqs.(9). Forcing the nondiagonal terms to cancel each other implies that X has to be an eigenvector of the transfer matrix $\tilde{T}$ with eigenvalue

$$\tilde{T}(\mu_k, \vec{\mu})|X\rangle = \beta^r \gamma^{s-r} (\prod x_i) \frac{a^N(\mu_k)}{b^N(\mu_k)} \prod_{1=i \neq k}^s (-1) \frac{a(\mu_i - \mu_k)}{a(\mu_k - \mu_i)} |X\rangle . \qquad (15)$$

On the other hand, our prime object of interest, the eigenvalue of the transfer matrix T of our model is the sum of eq.(13) and eq.(14) and reads:

$$\beta^r \gamma^{r-s} \alpha^{N-s} (\prod x_i) a^N(\lambda) \prod_{i=1}^s \frac{a(\mu_i - \lambda)}{b(\mu_i - \lambda)} + \alpha^{N-s} b^N(\lambda) \prod_{i=1}^s \frac{a(\lambda - \mu_i)}{b(\lambda - \mu_i)} (\tilde{T}X) . \qquad (16)$$

This finishes the first step of the nested Bethe-Ansatz. We are left with the determination of the eigenvectors X of $\tilde{T}$, which is much easier to perform, because the dimension of the horizontal space is reduced by 1. Together with the consistency equations for the parameters ν_i (obtained from the second Bethe-Ansatz) this gives s coupled Bethe equations for the set of parameters μ_i, ν_i:

$$\prod_{i=1}^{s} \frac{b(\nu_k - \mu_i)}{a(\nu_k - \mu_i)} = (\prod \frac{y_i}{z_i}) \prod_{1=i \neq k}^{p} (-1) \frac{a(\nu_i - \nu_k)}{a(\nu_k - \nu_i)} \qquad (17)$$

$$\prod_{i=1}^{p} \frac{a(\nu_i - \mu_k)}{b(\nu_i - \mu_k)} = (\prod \frac{x_i}{y_i}) \frac{a^N(\mu_k)}{b^N(\mu_k)} \prod_{1=i \neq k}^{s} (-1) \frac{a(\mu_i - \mu_k)}{a(\mu_k - \mu_i)}. \qquad (18)$$

We conclude that the eigenvalue of the transfer matrix T acquires the form:

$$\beta^r \gamma^{s-r} \alpha^{N-s} a^N(\lambda) \left[(\prod x_i) \prod_{i=1}^{s} \frac{a(\mu_i - \lambda)}{b(\mu_i - \lambda)} + \frac{b^N(\lambda)}{a^N(\lambda)} \prod_{i=1}^{s} \frac{a(\lambda - \mu_i)}{b(\lambda - \mu_i)} \tilde{\tau}(\lambda) \right] \qquad (19)$$

$$\tilde{\tau}(\lambda, \vec{\mu}, \vec{\nu}) = (\prod y_i) \prod_{i=1}^{p} \frac{a(\nu_i - \lambda)}{b(\nu_i - \lambda)} + (\prod z_i) \prod_{i=1}^{s} \frac{b(\lambda - \mu_i)}{a(\lambda - \mu_i)} \prod_{i=1}^{p} \frac{a(\lambda - \nu_i)}{b(\lambda - \nu_i)}.$$

We see, that the vertical fields do not show up in the Bethe equations eqs.(17,18) but amount to overall factors in the eigenvalue eq.(19). The detailed structure of the horizontal fields is not important, only their sum contributes. As far as the Bethe equations are concerned, the horizontal fields amount to a constant shift of the quantum numbers attached to the solutions.

THERMODYNAMIC LIMIT

We shall be interested in the way the horizontal fields influence the behaviour at large but finite sample sizes. Since there will be an effect in the gapless regime, we use complex parametrization $\gamma \to i\gamma$, $\lambda \to i\lambda$, $\vec{h} \to i\vec{h}$ from now on. The solution of the Bethe equations (17,18) in the infinite chain limit is obtained by the standard method of deriving an integral equation for the density of roots. We shall concentrate on the ground state behaviour. Details can be found in[1].

In the critical (trigonometric) parameter region of the model it is convenient to define an auxiliary function $\Phi(z, \gamma) = i \log [\sinh(z + i\gamma)/\sinh(z - i\gamma)]$. We take the logarithm of the Bethe equations and shift μ_k, ν_k to a new set of variables σ_k, ρ_k defined by $\mu_k = i\sigma_k - \gamma$, $\nu_k = i\rho_k - \frac{\gamma}{2}$, and set

$$Z_N^\sigma(\sigma) = \frac{1}{2\pi} \left[\Phi(\sigma, \gamma/2) - \frac{1}{N} \sum_{i=1}^{s} \Phi(\sigma - \sigma_i, \gamma) + \frac{1}{N} \sum_{i=1}^{p} \Phi(\sigma - \rho_i, \gamma/2) - \frac{2h_3}{N} \right]$$

$$Z_N^\rho(\rho) = \frac{1}{2\pi} \left[-\frac{1}{N} \sum_{i=1}^{p} \Phi(\rho - \rho_i, \gamma) + \frac{1}{N} \sum_{i=1}^{s} \Phi(\rho - \sigma_i, \gamma/2) + \frac{(h_3 - 3h_8)}{N} \right] \qquad (20)$$

The ground state densities $\eta_\infty^{\sigma,\rho}$, are given by $\eta_\infty^{\sigma,\rho} = \lim_{N \to \infty} \frac{dZ_N^{\sigma,\rho}}{d\sigma}$. In the limit of large N we can replace the sums in eq.(20) by an integral weighted with the corresponding density function and we get a matrix integral equation for the density of roots. The solutions are given as:

$$\eta_\infty^\sigma(\sigma) = \frac{1}{\sqrt{3}\gamma} \frac{\cosh\frac{\pi}{3\gamma}\sigma}{\cosh\frac{\pi}{\gamma}\sigma}$$

$$Z_\infty^\sigma(\sigma) = -\frac{1}{\pi}\arctan\frac{\coth\frac{\pi}{3\gamma}\sigma}{\sqrt{3}} + \frac{5[-1]}{6}$$

$$\eta_\infty^\rho(\rho) = \frac{1}{\sqrt{3}\gamma} \frac{\sinh\frac{\pi}{3\gamma}\rho}{\sinh\frac{\pi}{\gamma}\rho}$$

$$Z_\infty^\rho(\rho) = -\frac{1}{\pi}\arctan(\sqrt{3}\coth\frac{\pi}{3\gamma}\rho) + \frac{2[-1]}{3}.$$

(21)

For negative values of the arguments one has to take the numbers in brackets in the formulae for $Z_\infty^{\sigma,\rho}$. With this convention, $Z_\infty^{\sigma,\rho}$ is continous at zero and the distinction arises from a branch cut that has to be introduced when computing the integral for $Z_\infty^{\sigma,\rho}$.

FINITE SIZE EFFECTS

The Density

In the last section we analysed some properties of the system in the thermodynamic limit and found that they are independent on the fields $\vec{h}$, $\vec{v}$. Studying the model for large but finite N it is expected, that there will be some dependence (see[1] for the XXZ–case).

To study the deviation of the densities and the eigenvalue from their limit values, we shall consider the differences $Z_N^{\sigma,\rho}(x) - Z_\infty^{\sigma,\rho}(x)$. Differentiating gives an integral equation which can be recasted into the form:

$$(\vec{\eta}_N - \vec{\eta}_\infty)(x) = \int dx'\, \mathbf{F}(x-x')\, \vec{S}_N(x'),$$

(22)

where $\vec{\eta} = (\eta^\sigma, \eta^\rho)$. $\mathbf{F}(x-x')$ is some matrix resolvent connected to eqs.(20). $\vec{S}_N(x') = (S_N^\sigma(x'), S_N^\rho(x'))$ is given by $S_N^{\sigma,\rho}(x') = \frac{1}{N}\sum_{i=1}^{s}\delta(x'-[\sigma_i,\rho_i]) - \eta_N^{\sigma,\rho}(x')$. Since the unknown functions $\eta_N^{(\sigma,\rho)}$ appear at the right hand side of eq.(22), this equation is no solution to the problem but eq.(22) is a suitable expression for a systematic approximation with the help of the Euler–Maclaurin formula[3]. According to this formula $\int dy f(y) S_N(y)$ can be approximated by $-(\int_{-\infty}^{-\Lambda}+\int_\Lambda^\infty)\eta(y)f(y) + \frac{1}{2N}[f(-\Lambda)+f(\Lambda)] + \frac{1}{12N^2\eta(\Lambda)}[f'(-\Lambda)-f'(\Lambda)]$, where Λ is the largest Bethe root.

Define the 'half Fourier transforms' $X_\pm^{\sigma,\rho}(\omega)$ as:

$$X_\pm^{\sigma,\rho}(\omega) = \int_{-\infty}^{\infty} dt\, e^{i\omega t}\Theta(\pm t)\eta_N^{\sigma,\rho}(t+\Lambda^{\sigma,\rho}).$$

(23)

This are analytic functions on the upper resp. lower half of the complex plane. The integral equation eq.(22) then turns into a Riemann-Hilbert problem of the form:

$$|X_-(\omega)\rangle + \hat{\mathbf{R}}(\omega)|X_+(\omega)\rangle = |\eta(\omega)\rangle + (\hat{\mathbf{R}}-\mathbf{1})(\omega)|A(\omega)\rangle.$$

(24)

where $|A(\omega)\rangle$ is a two component vector with components $(\frac{1}{2N}+\frac{i\omega}{12N^2\eta^{\sigma,\rho}(\Lambda^{\sigma,\rho})})$. $\hat{R}=F+1$ can be factorized into $\hat{R}=G_+^{-1}G_-^{-1}$.

The solution of the inhomogenous equation then reads:

$$|X_+(\omega)\rangle = G_+(\omega)\,[\,|P(\omega)\rangle + |Q_+(\omega)\rangle\,] + |A(\omega)\rangle, \qquad (25)$$

together with $|X_+(0)\rangle = Z_N(\infty) - Z_N(\Lambda)$ and $\lim_{\omega\to\infty} \omega|X_+(\omega)\rangle = iN|\eta(\Lambda)\rangle$. $|P(\omega)\rangle$ and $|Q_+(\omega)\rangle$ are given by:

$$|P(\omega)\rangle = -\frac{1}{2N}\begin{pmatrix}1-\frac{ig_1-i\omega}{6N\eta_N^\sigma}\\ 1-\frac{ig_2-i\omega}{6N\eta_N^\rho}\end{pmatrix}. \qquad |Q_+(\omega)\rangle = \frac{i}{\sqrt{3}\gamma}\frac{1}{i\alpha+\omega}G_+(i\alpha)|V\rangle. \quad (26)$$

up to leading order in $e^{-\alpha\Lambda}$. $g_{1,2}$ are expansion coefficients of $G_+(\omega)$ at $\omega=\infty$ and $|V\rangle$ is defined as $|V\rangle = (e^{-\alpha\Lambda^\sigma}, e^{-\alpha\Lambda^\rho})$.

The Eigenvalue

The expression for the eigenvalue, eq.(19), is a sum of two termes. For $\lambda < \gamma/2$, we have $b^N(\lambda)/a^N(\lambda) = [\sin(\lambda)/\sin(\gamma-\lambda)]^N < 1$ and one concludes, that the second summand is suppressed exponentially for large N. Since in our analysis of the finite size corrections we shall only keep the leading terms, we consider only the first term of the transfer matrix eigenvalue.

As in the discussion of the density, one considers the correction L_N of the 'free energy' to the thermodynamic limit value:

$$L_N^h(\lambda) = -\frac{1}{N}\log \Lambda_N(\lambda) + \lim_{N\to\infty}\frac{1}{N}\log \Lambda_N(\lambda). \qquad (27)$$

Performing elementary manipulations leads to

$$L_N^h(\lambda) = -i\frac{h_3+h_8}{N} - \int dy\, 2\pi i Z_N^\sigma(y+i\lambda)S_N^\sigma(y) - \int dy\, 2\pi i Z_N^\rho(y+i\lambda)S_N^\rho(y) \quad (28)$$

Here we used the fact, that the groundstate belongs to $s=\frac{2N}{3}$, $p=\frac{1N}{3}$. Using the Euler–Maclaurin formula and the solution for the densities eqs.(25) one finds:

$$L_N^h(\lambda) = \frac{\pi}{N^2}\left[\frac{2}{6} - \frac{(h_3^2+3h_8^2)}{\pi(\pi-\gamma)}\right]\sin\frac{2\pi}{3\gamma}\lambda. \qquad (29)$$

For conformal invariance one must properly this expression in order to get a relativistic energy momentum relation (see Refs.[4],[1]). Comparing to the prediction of conformal field theory for the finite size corrections to the eigenvalue, $L_N^h = \frac{-\pi}{6N^2}c$, we find, that the vertical fields modify the central charge c to

$$c = 2 - \frac{6}{\pi(\pi-\gamma)}(h_3^2+3h_8^2) \qquad (30)$$

CONCLUSION

In this paper we found, that certain diffusion terms implemented into the system do not spoil integrabilty of multistate vertex models. The nested Bethe-Ansatz procedure is done and we showed that the thermodynamic limit is independent of the fields. In order to discuss finite size properties, we computed the central charge of the model in the critical region and found a result which generalizes that of Ref.[1].

REFERENCES

1. H.J. de Vega, *Int. J. Mod. Phys. A*, **A**, 2371 (1989)
2. B. Sutherland, *Phys. Rev. B*, **12**, 3795 (1975)
3. M. Abramowitz and I. Stegun, "Handbook of Mathematical Functions", 806, Dover, New York (1970)
4. F. Woynarovich and H.P. Eckle, *J. Phys. A*, **20**, L97 (1987)

THE QUANTUM MEAN FIELD STATE AS A LIMIT OF CANONICAL STATES: MAXWELL-BOLTZMANN STATISTICS

N. Angelescu

Institute for Atomic Physics
P.O.Box MG-6, Bucharest, Romania

INTRODUCTION

This is a report on joint work with M.Pulvirenti and A.Teta[1]. Stimulated by a recent paper[2] devoted to the mathematics of the quantum mean-field description of a one-component plasma (existence, uniqueness and classical limit of solutions to the Schrodinger-Poisson problem), we proposed to understand what kind of many-body system is that theory relevant to, more precisely to obtain it as a limit of N-particle quantum theories with suitably scaled interactions. The result, which I shall present below in a slightly more general form, is that the canonical k-particle reduced density matrices at fixed temperature for N particles obeying Maxwell- Boltzmann statistics, with charges of the order $N^{-1/2}$, and in a fixed confining potential (e.g. in a finite box) converge, as $N \to \infty$, to the tensor product of k one-particle density matrices, which are solutions of the self-consistency equations considered in ref. 2. Another possible choice would be to scale temperature like $N^{1/3}$ and both mass and charge like $N^{-1/3}$, what would correspond to a hot dense plasma. Though this convergence relates the Hartree theory of ref. 2 to interacting quantum Coulomb systems, not much insight is gained in this way concerning the range of applicability of the former. It turned out however that the techniques used in deriving it are simple enough to provide an alternate more transparent treatment and a better understanding of the Hartree theory itself (in particular, an easier control of the classical limit), and to apply hopefully to more subtle cases of mean field limits, e.g. for quantum statistics and not purely repulsive forces. This is why I shall include also a short outline of the proof.

STATEMENT OF THE RESULT

Consider N quantum particles obeying Maxwell-Boltzmann statistics, confined by some fixed external potential V_e and interacting via a two-body potential V with coupling constant N^{-1}:

$$H_N = -\frac{\epsilon}{2} \sum_{1 \leq i \leq N} \Delta_i + N^{-1} \sum_{1 \leq i < j \leq N} V(x_i, x_j) + \sum_{1 \leq i \leq N} V_e(x_i), \quad (1)$$

where $\epsilon = \hbar^2/2m$ and Δ_i is the Laplacian for the ith particle. We assume that the interaction is repulsive: $V(x_1, x_2) = V(x_2, x_1) \geq 0$, continuous except possibly an integrable diagonal singularity, and positive definite. The Coulomb case $V(x_1, x_2) = 1/4\pi|x_1 - x_2|$ is thus allowed. The external potential V_e is supposed continuous, bounded below and such that $\exp(-V_e) \in L_1(\mathbf{R}^3)$.

Let ρ_j^N be the j-particle canonical reduced density matrices corresponding to the Hamiltonian (1) at temperature $\beta^{-1} = 1$:

$$\rho_j^N(X^j, Y^j) = \Xi^{-1} \int dZ^{N-j} e^{-H_N}(X^j \cup Z^{N-j}, Y^j \cup Z^{N-j}), \quad (2)$$

where $X^j := (x_1, \ldots, x_j) \in \mathbf{R}^{3j}$, $e^{-H_N}(X^N, Y^N)$ is the integral kernel of e^{-H_N}, and

$$\Xi_N = Tr e^{-H_N} = \int dZ^N e^{-H_N}(Z^N, Z^N). \quad (3)$$

The main result is the following:

Theorem. *Under the assumptions above, for all $j \geq 1$ and $X^j, Y^j \in \mathbf{R}^{3j}$, $\rho_j^N(X^j, Y^j)$ converge as $N \to \infty$, and*

$$\lim_{N \to \infty} \rho_j^N(X^j, Y^j) = \prod_{1 \leq i \leq j} \rho(x_i, y_i), \quad (4)$$

where $\rho(x, y)$ is the kernel of a positive operator of unit trace $\hat{\rho}$ on $L_2(\mathbf{R}^3)$; $\hat{\rho}$ is the unique solution of the equation:

$$\hat{\rho} = \exp(\frac{\epsilon}{2}\Delta - V \star n - V_e)/Tr \exp(\frac{\epsilon}{2}\Delta - V \star n - V_e). \quad (5)$$

In (5), the following notations have been used:

$$n(x) := \rho(x, x); \qquad V \star n(x) := \int V(x, y) n(y) dy. \quad (6)$$

Remarks. (i) The convergence of ρ_j^N can be proved in the same way for particles confined in some fixed bounded domain $\Lambda \subset \mathbf{R}^3$ with smooth boundary and V_e continuous on $\bar{\Lambda}$. In this case, the Laplace operators in (1) and (5) are supplemented with boundary conditions making them positive self-adjoint operators in $L_2(\Lambda)$. Also, it may be necessary, from physical reasons, to alter the interaction too; e.g. in the Coulomb case and for Λ with 'perfectly conducting' boundary, it is appropriate to take the interaction to be the fundamental solution of the Poisson equation in Λ with Dirichlet boundary conditions, which fits equally well in our approach.

(ii) In all cases, Eq.(5) is the self-consistency equation of a Hartree approximation. The existence and uniqueness of solutions to (5) in the Coulomb case are discussed in Markowich[2]. Our result settles the existence under weaker conditions, by providing a solution via Eq.(4). Uniqueness follows from a variational principle for $\hat{\rho}$, which enters naturally in the proof; for V positive definite, the variational 'free-energy' is strictly convex, hence it has one minimum point. In the case of several minima, ρ_j^N converges along subsequences to a superposition of product states.

(iii) The corresponding problem in the case of classical particles has been considered by Messer and Spohn[3] for bounded interactions and by Cagliotti et al[4] for Coulomb interactions. In fact, the proof in the quantum case consists in going to a 'classical' problem in which the 'particles' are closed Brownian paths appearing in the Ginibre representation of Ξ_N; hence, the techniques in refs. 3 and 4 apply. Let me mention however that, for classical particles, attractive interactions have been managed too, in which case new interesting features are encountered (cf.[3,4] and refs therein). The extension of such results to quantum particles is not easy, due to the fact that Brownian paths are extended objects, what makes the estimates more complicated. Also, the relation with a classical problem is much more involved if quantum statistics is used. Such problems are left for further consideration.

OUTLINE OF PROOF

The main ingredients of the proof are the following:
1. The Ginibre representation[5] in terms of Wiener integrals for the reduced density matrices:

$$\rho_j^N(X^j, Y^j) = \int P_{X^j,Y^j}^\epsilon(d\omega^j) \rho_j^N(\omega^j), \tag{7}$$

where $P_{X^j,Y^j}^\epsilon(d\omega^j) = \prod_{1 \leq i \leq j} P_{x_i,y_i}^\epsilon(d\omega_i)$, with P_{xy}^ϵ the Wiener measure over the continuous paths in $\mathbf{R}^3$ of time interval ϵ conditioned to start at x and finish at y, and

$$\rho_j^N(\omega^j) = \int dX^{N-j} \int P_{X^{N-j},Y^{N-j}}^\epsilon(d\eta^{N-j}) \Xi^{-1} e^{-\bar{U}_N^\epsilon(\omega^j \cup \eta^{N-j})}, \tag{8}$$

$$\bar{U}_N^\epsilon(\omega^N) = \epsilon^{-1} \int_0^\epsilon dt [N^{-1} \sum_{1 \leq i < j \leq N} V(\omega_i(t), \omega_j(t)) + \sum_{1 \leq i \leq N} V_e(\omega_i(t))]. \tag{9}$$

We remark that in the bounded domain case, P_{xy}^ϵ has to be replaced by the appropriate Wiener type measure[6], which has sufficiently nice properties for the argument to go through.

The representation (7) allows to formulate the problem in terms of 'classical' objects: Let Ω_ϵ be the space of all continuous closed paths of time ϵ, which is a Polish space when endowed with the sup-distance, and the probabilities on Ω_ϵ^j:

$$\mu_j^N(d\omega^j) = \rho_j^N(\omega^j) \bar{d}\omega^j, \qquad \bar{d}\omega^j := \prod_{1 \leq i \leq j} dx_i P_{x_i x_i}^\epsilon(d\omega_i). \tag{10}$$

For $j = N$, μ_N^N is a classical Gibbs measure on Ω_ϵ^N, hence a variational principle is valid: ρ_N^N minimizes the free-energy functional defined on probability densities ρ on $(\Omega_\epsilon^N, \bar{d}\omega^N)$ by:

$$F_N(\rho) = N^{-1} \int \rho(\omega^N) \log \rho(\omega^N) \bar{d}\omega^N + N^{-1} \int \rho(\omega^N) \bar{U}_N^\epsilon(\omega^N) \bar{d}\omega^N. \tag{11}$$

2. The compactness bound: There exists $K > 0$ such that for all N:

$$\rho_j^N(\omega^j) \leq K^j \epsilon^{3j/2} \exp[-\sum_{1 \leq i \leq j} \bar{V}_e^\epsilon(\omega_i)]. \tag{12}$$

This is here an easy consequence of the positivity of V. It implies the existence of cluster points μ_j of μ_j^N for every j; a diagonal trick ensures the existence of a subsequence of N along which $\mu_j^N \to \mu_j$ for all j, where $\rho_j(\omega^j) = \frac{d\mu_j}{d\omega^j}(\omega^j)$ is symmetric and fulfills also the bound (12). The Hewitt-Savage theorem[7] ensures that μ_j are the marginals of a certain superposition of 'one-particle' (i.e. product) measures on Ω_ϵ^N: there exists a Borel probability measure ν_0 on $M_{+,1}(\Omega_\epsilon)$ such that:

$$\mu_j = \int \nu_0(d\rho)\rho^j. \tag{13}$$

The bound (12) shows that ν_0 is concentrated on absolutely continuous measures such that

$$\rho(\omega) := \frac{d\rho}{d\omega}(\omega) \leq K\epsilon^{3/2}\exp[-\bar{V}_e^\epsilon(\omega)]$$

3. The variational principle for the mixture measure ν: For every probability ν on $M_{+,1}(\Omega_\epsilon)$, one can define $f(\nu) = s(\nu) + e(\nu)$, where:

$$s(\nu) := \lim_{j\to\infty} j^{-1}\int d\mu_j \log \frac{d\mu_j}{d\omega^j} = \int \nu(d\rho)\int d\rho \log \frac{d\rho}{d\omega}, \tag{14}$$

$$e(\nu) := \int \nu(d\rho)[\frac{1}{2}\int \bar{V}^\epsilon(\omega_1,\omega_2)d\rho(\omega_1)d\rho(\omega_2) + \int \bar{V}_e^\epsilon(\omega)d\rho(\omega)], \tag{15},$$

and μ_j is defined in terms of ν by Eq.(13). The variational principle for ρ_N^N (eq.(11)) and the convergence of μ_j^N imply that $F_N(\rho_N^N) \to f(\nu_0)$, and ν_0 minimizes f. Hence, ν_0 is concentrated on the minima of the functional $\hat{f}(\rho) = f(\delta_\rho)$, given on probability densities on Ω_ϵ by:

$$\hat{f}(\rho) = \int \bar{d}\omega\rho(\omega)\log\rho(\omega) + \frac{1}{2}\int \bar{d}\omega_1\bar{d}\omega_2\bar{V}^\epsilon(\omega_1,\omega_2)\rho(\omega_1)\rho(\omega_2) + \int \bar{d}\omega\bar{V}_e^\epsilon(\omega)\rho(\omega). \tag{16}$$

4. Uniqueness of the minimum point of $\hat{f}$. This follows here from the strict convexity of $\hat{f}$, which is implied in turn by V being positive-definite. We have $\nu_0 = \delta_\rho$ independently of the subsequence chosen in 2., hence the convergence of the sequence μ_j^N. In particular, $\rho_j^N(\omega^j) \to \prod_{1\leq i\leq j}\rho(\omega_i)$ for $\omega^j \in \Omega_\epsilon^j$. One can extend ρ to all Brownian paths (not necessarily closed) in a natural way using the minimum condition for (16), such that the latter convergence is preserved (apply a dominated convergence theorem in Eq.(8)). The theorem follows hence by making proper use of Eq.(7).

Note added: During the Workshop, Professors A. Verbeure and R. F. Werner pointed out to me an alternative approach in terms of operator algebras to this kind of mean-field limits, they developed in e.g. Refs. 8,9 (of which we were not aware). The role of the Hewett-Savage theorem in obtaining the one-particle variational principle is taken up there by Størmer's theorem on symmetric states over product algebras. The domain and regularity problems involved when applying this to unbounded interactions are solved in our approach by the compactness bound on the path-space measures, etc. I am grateful for their valuable comments.

REFERENCES

1. N. Angelescu, M. Pulvirenti, and A. Teta, Derivation and classical limit of the mean-field equation for a quantum Coulomb system; Maxwell-Boltzmann statistics, to appear.
2. P.A. Markowitch, Boltzmann distributed quantum steady states and their classical limit, *Forum Math.* (1993) (to appear).
3. J. Messer and H. Spohn, Statistical mechanics of the isothermal Lane-Emden equation, *J. Stat. Phys.*, **29**, 561 (1982).
4. E. Cagliotti, P.L. Lions, C. Marchioro, and M. Pulvirenti, A special class of stationary flows for two-dimensional Euler equations: a statistical mechanics description, *Commun. Math. Phys.*, **143**, 501 (1992).
5. J. Ginibre, Some applications of functional integration in statistical mechanics, *in*: "Statistical Mechanics and Quantum Field Theory", Gordon and Breach, New-York (1971).
6. N. Angelescu and G. Nenciu, On the independence of the thermodynamic limit of the pressure on boundary conditions in quantum statistical mechanics, *Commun. Math. Phys.*, **29**, 15 (1973).
7. E. Hewitt and Savage, Symmetric measures on Cartesian products, *Trans. Amer. Math. Soc.*, **80**, 470 (1955).
8. M. Fannes, H. Spohn, and A. Verbeure, Equilibrium states for mean-field models, *J. Math. Phys.*, **21**, 355 (1980).
9. R.F. Werner, Large deviations and mean-field quantum systems, *in*: "Quantum Probability and Related Topics", **7**, World Scientific, Singapore, (1992).

MULTIFRACTAL PROPERTIES OF DISCRETE STOCHASTIC MAPPINGS

U. Behn[1], J.L. van Hemmen[2], R. Kühn[3], A. Lange[1], and V.A. Zagrebnov[4]

1. Institut für Theoretische Physik, Universität Leipzig
   D-04109 Leipzig, Germany

2. Physik-Department, TU München
   D-85747 Garching, Germany

3. Institut für Theoretische Physik, Universität Heidelberg
   D-69120 Heidelberg, Germany

4. Instituut voor Theoretische Fysica, Katholieke Universiteit Leuven
   B-3001 Leuven, Belgium

In different fields of statistical physics, such as 1d random field Ising models (RFIM) and neural networks, there appear discrete stochastic mappings of the form

$$x_n = f_n(x_{n-1}) \tag{1}$$

where $f_n(x)$ is chosen with equal probability from two functions $f_\sigma(x)$, $\sigma = +$ or $-$, and $0 < f'_\sigma(x) \leq 1$. The dynamics is nonchaotic and, in a region of physical parameters, converges to a *strange* (fractal) attractor as may be visualized in Fig. 1. The mapping generates an invariant measure which undergoes, as parameters are changed, qualitative transitions, for instance, a transition from thin to fat multifractal.

To be specific, for the 1d RFIM[1-7] with random field h_n on lattice site n taking the values $h_n = h_0 \pm h$, the effective local random field is generated by (1) choosing $f_n(x) = h_n + A(x)$, where $A(x) = (2\beta)^{-1} \ln\left[\cosh\beta(x+J)/\cosh\beta(x-J)\right]$. β and J are inverse temperature and exchange, respectively. In neural networks (1) is the learning rule for a forgetful memory[7,8] choosing $f_n(x) = \tanh(\varepsilon_n + x)$. Here x_{n-1} is the synaptic coupling between two neurons, say i and j, and $\varepsilon_n = \pm\varepsilon$ is the weight added through the learning of the nth random pattern. The analytic formalism for both models is the same. In presenting numerical results we restrict ourselfes here to the case of the learning rule which has only *one* parameter ε. For details the reader is referred to Refs. 6 and 8.

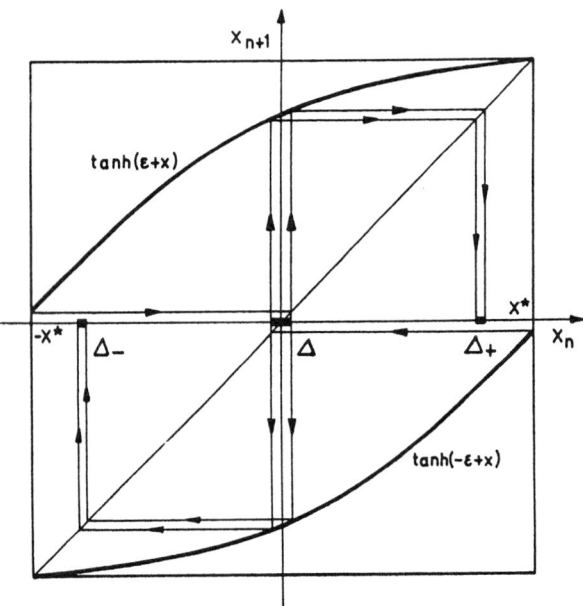

Figure 1. Discrete stochastic mapping in the case of nonoverlapping bands. Only gaps of first (Δ) and second (Δ_σ) generations are indicated.

The Frobenius-Perron equation corresponding to (1) reads

$$p_n(x) = \int dy\, p_{n-1}(y) \frac{1}{2} \sum_{\sigma=\pm} \delta(x - f_\sigma(y)). \qquad (2)$$

The fixed point of (2) gives the invariant measure which is a multifractal. To evaluate (2) it is advantageous to introduce a symbolic dynamics which encodes the result of the mapping (1) by the sequence of signs characterizing the history of the dynamical system. We denote the result of the nth iteration of (1) by

$$x_{\{\sigma\}_n, y} = f_{\sigma_n}(f_{\sigma_{n-1}}(\ldots f_{\sigma_1}(y) \ldots)), \qquad (3)$$

where $\{\sigma\}_n$ is the sequence of n signs $+$ or $-$ corresponding to the given realization of the random sequences $\{h_n\}$ or $\{\varepsilon_n\}$. The result of infinitely many iterations is denoted by $x_{\{\sigma\}}$ where $\{\sigma\}$ symbolizes an infinite sequence of signs. Infinitely many iterations with the *same* function, say f_+, leads to a fixed point denoted by $x_{\{+\}}$. It is easy to see that $x_{\{+\}}$ and $x_{\{-\}}$ are the boundaries of the support of the invariant measure. For the learning rule and the RFIM in the case $h_0 = 0$ the measure is symmetric so that $x_{\{+\}} = -x_{\{-\}}$. The iteration of (2) starting with the initial measure $p_0(x)$ generates a probability measure which may be also encoded by symbolic dynamics. In the first step we obtain two bands $p_\sigma(x)$ living on the intervals $I_\sigma = [x_{\sigma\{-\}}, x_{\sigma\{+\}}]$

$$p_\sigma(x) = p_0(y_{\sigma,x})\,[2 f'_\sigma(y_{\sigma,x})]^{-1} \quad \text{for } x \in I_\sigma, \qquad \sigma = \pm \qquad (4)$$

where $y_{\sigma,x} = f_\sigma^{-1}(x)$ denotes the preimage of x for the mapping f_σ. Each band generates in the next step two new bands so that in the nth generation the measure consists of

2^n bands,

$$p_{\{\sigma\}_n}(x) = p_0(y_{\{\sigma\}_n,x}) \prod_{\nu=1}^{n} \left[2f'_{\sigma_\nu}(y_{\{\sigma\}_\nu,x})\right]^{-1} \qquad \text{for } x \in I_{\{\sigma\}_n} \qquad (5)$$

labeled by the 2^n possible configurations $\{\sigma\}_n$. The notations $y_{\{\sigma\}_n,x}$ and $I_{\{\sigma\}_n}$ are obvious. Thus the measure in the nth iteration may be represented as $p_n(x) = \sum_{\{\sigma\}_n} p_{\{\sigma\}_n}(x)$. This explicit representation is helpful to investigate the qualitative behaviour of the measure and to calculate generalized scaling exponents.

The two bands in the first step may overlap or not and correspondingly the *support* of the measure is the whole interval $I = [x_{\{-\}}, x_{\{+\}}]$ or a fractal with the topology of a Cantor set. The *measure* constitutes in the former case a *fat* multifractal (Fig. 2a-c) and in the latter case a *thin* multifractal (Fig. 2d). In the case of the learning rule the condition of zero overlap defines the critical parameter $\varepsilon_c^{(1)} \simeq 0.957$.

The histograms generated by numerical simulation of (1) (or alternatively by numerical solution of (2)) reveal that, depending on physical parameters, there are qualitative changes in the behaviour of the measure at the boundaries of the support, cf. Fig. 2. The invariant measure for the membrane potential in a single-neuron model[9] shows a similar behaviour. These changes can be analyzed exploiting that the preimage of the fixed point is the fixed point itself. For example we consider the right boundary $x_{\{+\}} = x^*$. In the nth iteration we obtain from (5) for the rightmost band

$$p_{\{+\}_n}(x^*) \sim \left[2f'_+(x^*)\right]^{-n} \qquad (6)$$

which goes in the limit $n \to \infty$ to ∞ or 0, unless $f'_+(x^*) = 1/2$. The latter condition separates two parameter regions in which the invariant measure at its right boundary diverges (Fig. 2c, d) or goes to zero (Fig. 2a,b), respectively.

To investigate the scaling behaviour of the coarse grained measure at the boundaries of the support we consider

$$P(l) = \lim_{n \to \infty} P_n(l) = \lim_{n \to \infty} \int_{x^*-l}^{x^*} dx\, p_n(x) \qquad (7)$$

which is expected to behave like $P(l) \sim l^\alpha$ as $l \to 0$ with the scaling exponent α to be determined. In the *scaling limit* we chose $l = l_n \sim \left[f'_+(x^*)\right]^n \to 0$ as $n \to \infty$ which ensures that only the righmost band $p_{\{+\}_n}(x)$ contributes to (7). Application of (5) and expansion near the fixed point leads to the scaling law[8]

$$P_n(l) = \int_{x^*-l}^{x^*} dx\, p_{\{+\}_n}(x) \sim l^\alpha, \qquad \alpha = -\frac{\ln 2}{\ln f'_+(x^*)}. \qquad (8)$$

The coarse grained density at the boundaries of the support $\tilde{p}(l) = P(l)/l$ scales as $l^{\alpha-1}$, i.e., it goes to ∞ or 0 unless $f'_+(x^*) = 1/2$ as already found from (6). The left derivative of the coarse grained density at x^* scales as $l \to 0$ like

$$-\partial_l \tilde{p}(l) \sim -(\alpha-1) l^{\alpha-2}. \qquad (9)$$

Correspondingly, there are parameter regions in which the left derivative of the density goes to zero (for $\alpha > 2$, i.e., $f'_+(x^*) > 1/\sqrt{2}$), to $-\infty$ (for $2 > \alpha > 1$, i.e., $1/\sqrt{2} > f'_+(x^*) > 1/2$), or to ∞ (for $1 > \alpha$, i.e., $1/2 > f'_+(x^*)$). This explains analytically the qualitative changes found in the numerical simulations, cf. Fig. 2a-c. In the case of the learning rule the above conditions for qualitative changes define the critical parameters $\varepsilon_c^{(2)} \simeq 0.174$ and $\varepsilon_c^{(3)} \simeq 0.064$.

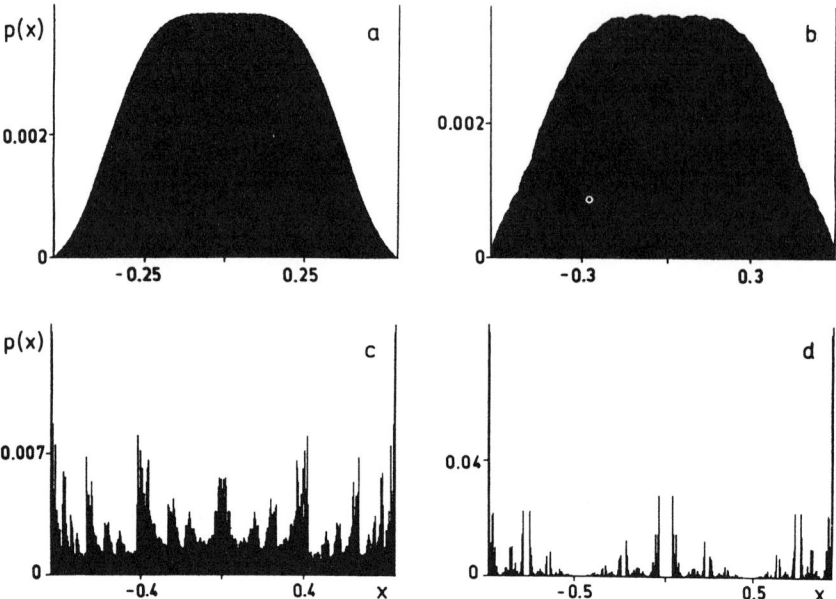

Figure 2. Qualitatively different shapes of invariant measures. In (a)-(d) we have $\varepsilon = 0.05$, 0.1, 0.4, and 1.0, respectively. The measure at the boundaries of the support is either zero, (a) and (b), or infinite, (c) and (d). The former case is further distinguished by the derivative of the outmost bands at the boundaries of the support, which either vanishes (a) or diverges (b). The multifractal measure in (a)-(c) covers the whole interval (fat fractal), whereas in (d) it is a thin fractal. The histograms are calculated by a digital simulation of a trajectory of length 10^9. Note the different scales on both axes.

After examining the behaviour of the invariant measure at the boundaries of its support we turn to characterize the scaling behaviour of the invariant measure on the *whole* support by generalized fractal dimensions,

$$D_q = \frac{1}{q-1} \lim_{l \to 0} \frac{\ln \sum_{i=1}^{N(l)} P_i^q}{\ln l}. \qquad (10)$$

Here the sum runs over all $N(l)$ nonoverlapping cells of length l used to cover the support, and P_i denotes the total weight of the measure on cell i. For $q = 0, 1, 2$ (10) gives the Hausdorff, information, and correlation dimension, respectively. The limiting values D_∞ and $D_{-\infty}$ characterize the most dominating and most rare parts of the multifractal, respectively.

In cases, where the cells at the boundary of the support correspond to the most rare or most dominant events, our previous scaling analysis leads to $D_{-\infty} = \alpha$ or $D_\infty = \alpha$ with α given in (8).

To compute the fractal dimensions in general, it is advantageous to use instead of the equipartion of the support the *natural* partition generated by the mapping itself. In the thermodynamic formalism[10], the partition function $\Gamma_n(q,\tau) = \sum_i P_i^q / l_i^\tau$ calculated

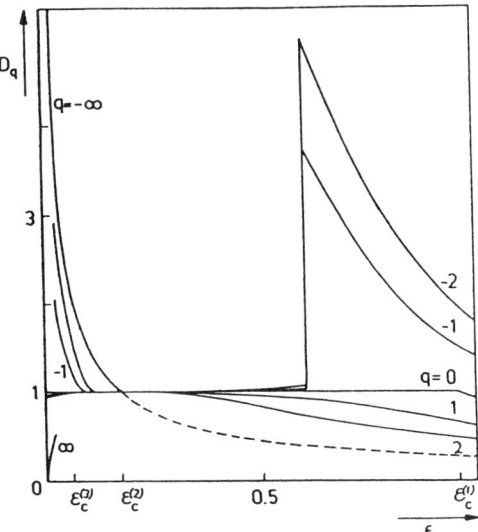

Figure 3. Generalized fractal dimensions D_q of the invariant measure corresponding to the learning rule, for $q = 0, \pm 1, \pm 2$. The most remarkable feature is that the width of the multifractal spectrum becomes narrow near $\varepsilon_c^{(2)}$ where the measure at the boundaries of the support jumps from zero to infinity. For $\varepsilon < \varepsilon_c^{(2)}$, $D_{-\infty}(\varepsilon)$ is also displayed. For $\varepsilon > \varepsilon_c^{(2)}$, the dashed line shows the scaling exponent α of the right(left)most band, which approaches $D_\infty(\varepsilon)$ as ε becomes large. For $\varepsilon \to 0$, we have drawn $D_\infty(\varepsilon) \to 0$ only schematically.

with the natural partition in a given generation of the hierarchy goes to zero or to infinity as $n \to \infty$ unless $\tau = (q-1)D_q$. This can be used to determine the D_q which can be shown[11] to agree with those determined from (10).

The natural partition is for the case of nonoverlapping bands given by the 2^n images $I_{\{\sigma\}_n}$ of the initial interval I, each carrying the same total measure 2^{-n}. For overlapping bands $I_{\{\sigma\}_n}$ where the support of the measure is the whole interval we use the 2^{n+1} endpoints of the $I_{\{\sigma\}_n}$ for repartitioning into $2^{n+1} - 1$ nonoverlapping new intervals[6,8]. Each of these new intervals in general carries weight from several bands. We determined the D_q by solving $\Gamma_n(\tau) - \Gamma_{n-1}(\tau) = 0$ ($n = 10, \ldots, 15$) as an eigenvalue equation[12] for τ.

Qualitatively, the behaviour of the generalized dimensions shown in Fig. 3 can be understood as follows. The tails of the measure at the boundaries of the support determine $D_{-\infty}(\varepsilon)$ for $\varepsilon < \varepsilon_c^{(2)}$. In particular, $D_{-\infty}(\varepsilon)$ diverges, as $\varepsilon \to 0$. In the same limit, the mass of the distribution concentrates at the center of the support in a δ-function like fashion, so that $D_\infty(\varepsilon)$ decreases to zero as $\varepsilon \to 0$. In the opposite limit, $\varepsilon \to \infty$, the invariant measure is a *thin* fractal, collapsing into two δ-functions at the endpoints of the support. This means that all fractal dimensions approach zero, implying that the width of the spectrum of fractal dimensions approaches zero, too.

Now we look at the fate of the multifractal as ε decreases from large to small values. For still large ε, the fractal is thin and most bins are empty at any resolution. With decreasing ε, the fractal becomes more dense, and all fractal dimensions increase.

If the gaps close, the fractal becomes fat, and $D_0 = 1$ for all $\varepsilon < \varepsilon_c^{(1)}$. However there are still very deep valleys and high peaks on all scales that cause D_q for $q > 0$ or $q < 0$ to be smaller or larger than D_0, respectively. Upon further decrease of ε the D_q for $q < 0$ decrease drastically and reach values close to 1, which announces that a dense nonfractal "background" on the whole support has emerged. Still, superimposed on this background, there are peaks on all scales, so that $D_q < 1$ for $q > 0$. At $\varepsilon_c^{(2)}$, the behaviour of the measure at the boundaries of the support changes qualitatively, $\tilde{p}(\pm x^*)$ jumps from ∞ to zero as ε is decreased through this critical value, and $D_{-\infty} = 1$. For $\varepsilon_c^{(3)} < \varepsilon < \varepsilon_c^{(2)}$, there are tails decreasing to zero with infinite slope as $x \to \pm x^*$. At $\varepsilon_c^{(3)}$ there is yet another qualitative change, the slope jumps from ∞ to zero, i.e., $D_{-\infty} = 2$. The scaling behaviour of this slope determines $D_{-\infty}$ as discussed earlier.

The 1d RFIM shows a similar behaviour of the invariant measure varying h for suitably choosen nonzero temperature and homogeneous field. Since there the parameter space is three dimensional the behaviour of fractal dimensions exhibits a richer phenomenology including discontinuous transitions[6].

Transitions in the behaviour of invariant measures (or their projections) generated by iterated function systems have been recently discussed in a different context[13].

REFERENCES

1. G. Györgyi and P. Ruján, Strange attractors in disordered systems, *J. Phys. C*, **17**, 4207 (1984).
2. U. Behn and V.A. Zagrebnov, One-dimensional random field Ising model and discrete stochastic mappings, *J. Stat. Phys.*, **47**, 939 (1987); One-dimensional Markovian field Ising model: Physical properties and characteristics of the discrete stochastic mapping, *J. Phys. A*, **21**, 2151 (1988); Comment on "Random-field Ising model as a dynamical system", *Phys. Rev. B*, **38**, 7115 (1988); U. Behn, V.B. Priezzhev, and V.A. Zagrebnov, One dimensional random field Ising model: Residual entropy, magnetization, and the "perestroyka" of the ground state, *Physica A*, **167**, 457 (1990).
3. P. Szépfalusy and U. Behn, Calculation of a characteristic fractal dimension in the one-dimensional random field Ising model, *Z. Phys. B*, **65**, 337 (1987).
4. J. Bene and P. Szépfalusy, Multifractal properties in the one-dimensional random field Ising model, *Phys. Rev. A*, **37**, 1702 (1988); J. Bene, Multifractal properties of a class of non-natural measures as an eigenvalue problem, *Phys. Rev. A*, **39**, 2090 (1988).
5. T. Tanaka, H. Fujiska, and M. Inoue, Free-energy fluctuations in a one-dimensional random Ising model, *Phys. Rev. A*, **39**, 3170 (1989); Scaling structures of free-energy fluctuations in a one-dimensional dilute Ising model, *Progr. Theor. Phys.*, **84**, 584 (1990).
6. U. Behn and A. Lange, 1D random field Ising model and nonlinear dynamics, in: "From Phase Transition to Chaos," G. Györgyi, I. Kondor, L. Sasvári, and T. Tél, eds., World Scientific, Singapore (1992).
7. J.L. van Hemmen, G. Keller, and R. Kühn, Forgetful memories, *Europhys. Lett.*, **5**, 663 (1988).
8. U. Behn, J.L. van Hemmen, R. Kühn, A. Lange, and V.A. Zagrebnov, Multifractality in forgetful memories, *Physica D*, **68**, (1993).
9. P.C. Bressloff, Analysis of quantal synaptic noise in neural networks using iterated function systems, *Phys. Rev. A*, **45**, 7549 (1992).
10. T.C. Halsey, M.H. Jensen, I. Procaccia, and B.I. Shraiman, Fractal measures and their singularities: The characterization of strange sets, *Phys. Rev. A*, **33**, 1141 (1989).
11. H.B. Lin, "Elementary Symbolic Dynamics and Chaos in Dissipative Systems," World Scientific, Singapore (1989).
12. M.J. Feigenbaum, I. Procaccia, and T. Tél, Scaling properties of multifractals as eigenvalue problem, *Phys. Rev. A*, **39**, 5359 (1989).
13. G. Radons, H.G. Schuster, and D. Werner, Fractal measures and diffusion as results of learning in neural networks, *Phys. Lett. A*, **174**, 293 (1993); G. Radons, A new transition for projections of multifractal measures and random maps, *J. Stat. Phys.*, **72**, 227 (1993).

GIBBS STATES OF THE CHERN-SIMONS CHARGED PARTICLE SYSTEM IN THE MEAN-FIELD TYPE LIMIT

W.I. Skrypnik

Institute of Mathematics
Tereshchenkivskya St.3, Kiev-4, Ukraine

Topological electrodynamics is a theory describing an interaction of a $U(1)$ gauge field $A(x,t)$, a vector-valued function on three-dimensional space, with a charged matter field, characterized by a current $j(x,t)$, a vector-valued measure with a discrete support. The Lagrangian is given by

$$L(A,j) = \frac{k}{2}\int \varepsilon^{\alpha\beta\nu} A_\nu(x,t)\partial_\alpha A_\beta(x,t)d^2x - \frac{1}{c}\int A_\alpha(x,t)j^\alpha(x,t)d^2x + \frac{1}{2m_0}\sum_{s=1}^{n}||v_s||^2.$$

where greek indices run over the set $(0,1,2)$, repeated indices are summed over and $\varepsilon^{\alpha\beta\nu}$ is the antisymmetric tensor. The integral is taken over 2-d space, $||v_j||$ is the Euclidean norm of the two-dimensional velocity vector of a particle with two-dimensional position vector x_j and charge σ_j; t is the time ($x^\circ = ct$), c is the velocity of light,

$$j^\alpha(x,t) = \sum_{k=1}^{n} v_k^\alpha \sigma_k \delta(x - x_k(t)), \alpha = 1,2, j^\circ(x,t) = c\rho(x,t) = c\sum_{k=1}^{n}\sigma_k\delta(x - x_k(t)).$$

The quadratic form in A in the Lagrangian is the integral of the Chern-Simons form for an abelian principal fibre bundle. It is remarkable that the space components A_i satisfying the equation of motion are equal to

$$A_i(x,t) = k^{-1}\varepsilon^{is}\partial_s \sum_{l=1}^{n}\sigma_l \ln ||x - x_l(t)||$$

As a result the particle variables satisfy the equation of motion of a system with the Hamiltonian,

$$H(P_n, X_n) = (2m_0)^{-1}\sum_{l=1}^{n}||p_l - \sigma_l A(x_l)||^2$$

A system of particles with this Hamiltonian we shall call the Chern-Simons (CS) charged particle system. The quantized CS-particle system is an interesting object from the point of view of the theory of high-temperature superconductivity [1]–[4] and integrable systems [1]. The CS-Hamiltonian is remarkable because of its singularity which generates noninteger spin statistics and might be at the origin of a high–temperature superconductivity leaving its trace on the classical level.

At the same time this singularity creates difficulties in treating the system in the limit where the number of particles tends to infinity. There are no difficulties when the ultraviolet and infrared cut-offs are introduced, i.e. when the space components of the electromagnetic potential is changed into a regular exponentially decreasing function at infinity. The resulting system is integrable [5] if the electromagnetic potential is a pure gauge and the Gibbs (grand canonical) correlation functions are easily computed in the thermodynamic limit.

The problem of removing the cut-offs can be tackled in different ways.

We solve the problem for the Gibbs correlation functions in the mean-field limit.

Let us recall that the usual procedure of the mean-field limit for the Newtonian system of particles interacting via the potential

$$\Phi(x) = \varphi(x) + \varepsilon^d C(\varepsilon x)$$

— ϕ is a positive singular short range potential and C is a regular positive-definite potential — is fulfilled by taking the limit of vanishing ε (d is the dimension)[6]-[7]. This method is an analog of the Bogoliubov method of an approximating Hamiltonian [8]. The generalization of both is given in [9]. The probabilistic approach to the mean-field limit or large deviation method is developed in a series of papers [10]–[12]. When $\varphi = 0$, rescaling the variables in the expression for the Gibbs correlation functions results in rescaling the temperature β and the activity z by multiplying the former by ε^d and the latter by ε^{-d}. The size of the compact domain containing the particles is also changed. If this domain prior to the change of the position vectors is the ball of the radius R centered at the origin, then the resulting domain is the ball with radius εR. If it is assumed that $R = R_o \varepsilon^{-1}$ then the thermodynamic limit and the mean-field limit can be performed independently. The mean-field limit then corresponds to the high temperature dense gas. The correlation functions satisfy the Kirkwood-Salzburg equation in which the temperature and the activity are rescaled by ε in the aforementioned fashion. If the potential is integrable then the limiting equation is easily solved for small values of $z\varepsilon$ and the correlation functions are factorized into the product of one particle correlation functions[13],[14]. For charged particle system the limiting correlation functions are found with the help of functional integrals (Sine-Gordon transformation) [15]. The Debye screaning is proved this way for neutral systems [16]–[18].

Let us now return to the CS–particle system and rescale A_i

$$A_i(x) \Rightarrow \varepsilon^2 A^i(\varepsilon x) = \varepsilon A_i(x).$$

This means that we change k into $\varepsilon^{-1} k$. And now the dependence of R on ε has to be chosen. We prove that if

$$\varepsilon^2 \ln R(\varepsilon) \to 0$$

then the limiting correlation functions coincide with the correlation functions of the free gas (the ultraviolet cut-off is taken off in the limit). We prove also that if $R = \exp r_o \varepsilon^{-2}$ then the limiting correlation functions define a Gaussian distribution in (momentum variables) which does not factorize. The "faster" divergence of R does not produce a sensible result (the limit does not exist). That is, the behavior of the CS–particle system differs radically in the mean field limit from the usual charged particle system. We demonstrate this also assuming that R does not depend on ε, rescaling only the temperature and the activity.

REFERENCES

1. R.Jackiw, So-Young Pi, *Phys. Rev. D*, **15**, 3500 (1990); Classical and Quantum non-relativistic Chern-Simons theory, BU-HEP-90-11,Preprint.
2. J. D. Lykken, J. Sonnenschein, and N. Wess, The theory of anyonic superconductivity. A review, TAUP-1858-91, Preprint.
3. E. Fradkin, "Field theories of condensed matter systems". Addison-Wesley Publishing Company.
4. F. Wilczek (ed), "Fractional statistics and anyon superconductivity," World Scientific, (1990).
5. W. I. Skrypnik, Infinite particle Hamiltonian dynamics of Chern-Simons type, DIAS-STP-91-11, Preprint.
6. J. L. Lebowitz, and O. Penrose, *J. Math. Phys.*, **6**, 98 (1966).
7. P. C. Hemmer, and J. L. Lebowitz, Systems with weak long-range potentials, *in* "Phase transitions and critical phenomena", M. S. Green ed., C.Domb-N.Y., Academic Press, (1973).
8. N.N.Bogoliubov, "Collected papers", Naukova Dumka, Kiev, (1970); N. N. Bogoliubov (jr), I. B. Brankov, V. A. Zagrebnov, and A. M. Kurbatov, "Method of approximating Hamiltonian in Statistical Physics," Sofia, Bulgarian Academy of Sciences, (1981).
9. M. V. Shcherbina, "Some asymptotic problems of Statistical Mechanics", Candidate Thesis, Phys. Tech. Inst. Low Temp., Harkiv, (1985).
10. J. T. Lewis, Why do bosons condense ? *in* : "Statistical Mechanics and Field Theory," *Lecture notes in Physics*, **257**, Groningen (1985).
11. M. van den Berg, and J. T. Lewis, *Comm. Math. Phys.*, **81**, 475 (1981).
12. M. van den Berg, J. T. Lewis, and P. de Smedt, *J. Stat. Phys.*, **37**, 697 (1984).
13. H. Spohn, *Rev. Mod. Phys*, **52**, 569 (1980).
14. N. Grewe, and W. Klein, *J. Math. Phys.*, **18**, 1729 (1977).
15. J. Frohlich, and Y. M.Park, *Comm. Math. Phys.*, **59**, 235 (1978).
16. P. Brydges, *Comm. Math. Phys.*, **73**, 197 (1980).
17. T. Kennedy, *Comm. Math. Phys.*, **92**, 269 (1983).
18. W. W. Gorunovich, W. I. Skrypnik, *Teor. Mat. Fiz.*, **86**, 257 (1991).

RIGOROUS BETHE ANSATZ FOR THE NONLINEAR SCHROEDINGER MODEL

T.C. Dorlas

University College Swansea
Singleton Park, Swansea SA2 8PP, U.K.

INTRODUCTION

The author's recent proof of the completeness and orthogonality of the Bethe Ansatz eigenstates of the nonlinear Schroedinger model is outlined. The completeness follows from monotonicity in the coupling parameter of the model combined with orthogonality of the Bethe Ansatz eigenfunctions. The latter follows from the orthogonality of the corresponding eigenstates of a lattice approximation to the nonlinear Schroedinger model introduced by Korepin and Izergin. This, in turn follows from the fact that these states are simultaneous eigenstates of a complete set of commuting "transfer operators", i.e. the fact that the model is completely integrable.

1. THE BETHE ANSATZ FOR THE NONLINEAR SCHROEDINGER MODEL

The nonlinear Schroedinger model is a model of a boson gas in one dimension with a delta-function interaction. In the N-particle sector the Hamiltonian is given by

$$H_\kappa^N = -\sum_{j=1}^N \frac{\partial^2}{\partial x_j^2} + 2\kappa \sum_{1\leq i<j\leq N} \delta(x_i - x_j), \tag{1.1}$$

where $\kappa \geq 0$ is the coupling constant. The operator H_κ^N operates on the Hilbert space of symmetric L^2-functions of N variables: $\mathcal{H}_N = L^2_{sym}\left([0,L]^N\right)$, and we impose periodic boundary conditions.

In [1], Lieb and Liniger showed that one can obtain an infinite set of eigenfunctions for this Hamiltonian using the so-called Bethe Ansatz:

$$\psi^{BA}_{\{m_j\}}(x_1,\ldots,x_N) = \sum_{\sigma\in\mathcal{S}_N} A_\sigma \exp\left[i\sum_{j=1}^N k_{\sigma(j)}x_j\right], \tag{1.2}$$

for $0 \leq x_1 \leq \ldots \leq x_N \leq L$. Inserting this hypothetical form for the wavefunction into the Schroedinger equation one finds that the latter can be satisfied provided that the wavenumbers $k_1, \ldots, k_N$ satisfy the BA equations:

$$k_j = k_j^0 - \frac{1}{L} \sum_{i=1}^{N} \theta_\kappa(k_j - k_i), \tag{1.3}$$

where

$$k_j^0 = \begin{cases} \frac{2\pi}{L} m_j & \text{if } N \text{ is odd,} \\ \frac{2\pi}{L}(m_j + \frac{1}{2}) & \text{if } N \text{ is even,} \end{cases} \tag{1.4}$$

and

$$\theta_\kappa(k) = 2\tan^{-1}(k/\kappa) \tag{1.5}$$

and $m_1 < \ldots < m_N$ are integers, and provided the coefficients A_σ satisfy

$$\frac{A_\tau}{A_\sigma} = -\exp[-i\theta_\kappa(k_{\tau(j)} - k_{\sigma(j)})] \tag{1.6}$$

if the permutations σ and τ differ by a tranposition at $(j, j+1)$. It was shown by Yang and Yang in [2] that, for any given set of integers $m_1 < \ldots < m_N$, the set of nonlinear equations (1.3) has a unique solution $k_1 < \ldots < k_N$. However, it remained an open problem whether one obtains a complete set of eigenfuntions this way for every N.

2. YANG-YANG THERMODYNAMICS

Before outlining a proof of the completeness of the BA eigenfunctions let us briefly consider the thermodynamics of the model. The model describes a one-dimensional gas of bosons, interacting via a delta-function potential. One can therefore define a grand-canonical pressure at inverse temperature β and chemical potential μ by

$$p(\beta, \mu) = \lim_{L \to \infty} \frac{1}{\beta L} \ln \sum_{N=0}^{\infty} e^{\beta \mu N} \text{Trace } e^{-\beta H_\kappa^N}. \tag{2.1}$$

Assuming that the BA eigenfunctions are complete, one can write this as:

$$p(\beta, \mu) = \frac{1}{\beta L} \ln \sum_{N=0}^{\infty} e^{\beta \mu N} \sum_{m_1 < \ldots < m_N} \exp\left[-\beta \sum_{j=1}^{N} k_j(\underline{m})^2\right]. \tag{2.2}$$

Using heuristic arguments analogous to those used in the case of the free gas, Yang and Yang [2] succeeded in evaluating this limit. Their result was made rigorous in [3] using Large Deviation methods. (For a short explanation see [4]) It can be formulated as follows:

$$p(\beta, \mu) = \sup_{m \in E} \{\mu ||m|| - f[m]\}, \tag{2.3}$$

where E is the space of positive bounded measures on the real line and

$$f[m] = \int_{\mathbb{R}} f_m(k)^2 \, m(dk) - \beta^{-1} s[m], \tag{2.4}$$

where $s[m]$ is the *fermionic* entropy function and $f_m(k)$ satisfies the analogue of the BA equations:

$$f_m(k) = k - \int_{\mathbb{R}} \theta_\kappa(f_m(k) - f_m(k')) \, m(dk'). \tag{2.5}$$

3. COMPLETENESS

Completeness of the BA eigenstates (1.2) was proved in [5] by means of the following strategy:

I. A modification of the continuity argument proposed by Yang and Yang in [2]. Yang and Yang noticed that the Hamiltonian (1.1) tends to the Girardeau model Hamiltonian [6] as κ tends to ∞. (The pressure of the corresponding gas is just the free-fermion pressure, as can be seen from (2.3)) The BA eigenfunctions, in this case, are simply symmetrised running waves, and are easily seen to be complete. Yang and Yang argued that it should follow by continuity that, for finite κ, the BA eigenfunctions are still complete. As explained below, we replace in this argument continuity at $\kappa = \infty$ by monotonicity at $\kappa = 0$.

II. This monotonicity argument uses the fact that all BA eigenfunctions are independent for all κ. This fact follows from the orthogonality of the BA eigenfunctions, which can be proved using a lattice approximation introduced by Izergin and Korepin [7]. This is explained in Sections 4,5 and 6.

Monotonicity of $\kappa \mapsto H_\kappa^N$ implies (by the Min-Max Theorem):

Lemma 1. *If $\mu_1(\kappa) \leq \mu_2(\kappa) \leq \ldots$ are the eigenvalues of H_κ^N, counting multiplicity, then*

$$\mu_n(\kappa_0) = \inf_{\kappa > \kappa_0} \mu_n(\kappa). \quad (\kappa_0 \geq 0) \tag{3.1}$$

Assuming orthogonality of the BA eigenstates, the following lemma is easy:

Lemma 2. *Let $\lambda_1(\kappa) \leq \lambda_2(\kappa) \leq \ldots$ be the BA eigenvalues, ordered according to their value. Assume that, for some n, $\mu_n(\kappa) < \lambda_n(\kappa)$. Then $\mu_{n+k}(\kappa) \leq \lambda_{n+k-1}(\kappa)$ for all $k \geq 1$.*

The completeness proof now runs as follows:
Assume that for some κ and for some n_0, $\mu_{n_0}(\kappa) < \lambda_{n_0}(\kappa)$. Define

$$\kappa_1 = \inf\{\kappa > 0 | \mu_{n_0}(\kappa) < \lambda_{n_0}(\kappa)\}. \tag{3.2}$$

Then $\mu_n(\kappa_1) \leq \lambda_{n-1}(\kappa_1)$ for $n > n_0$ by Lemma 2 and Lemma 1 and continuity of $\lambda_n(\kappa)$. If n_1 is the smallest integer such that $n_1 > n_0$ and $\lambda_{n_1}(\kappa_1) > \lambda_{n_0}(\kappa_1)$ then

$$\mu_{n_1}(\kappa_1) \leq \lambda_{n_1-1}(\kappa_1) = \lambda_{n_0}(\kappa_1) < \lambda_{n_1}(\kappa_1). \tag{3.3}$$

Thus we can define κ_i and n_i ($i = 1, 2, \ldots$) inductively. But n_i cannot tend to ∞ because $\mu_{n_i}(\kappa) \leq \mu_{n_0}(\kappa_0) \leq \lambda_p(0)$ for some $p \in \mathbb{N}$. The sequence must therefore terminate and $\kappa_i = 0$ for some i.

4. THE LATTICE MODEL AND THE ALGEBRAIC BETHE ANSATZ

The lattice model introduced by Izergin and Korepin is formulated in second quantised form. We map $\mathcal{H}_N$ onto the N-particle Fock space $\mathcal{F}_N$ according to

$$\psi(x_1, \ldots, x_N) \mapsto \int \ldots \int dx_1 \ldots dx_N \, \psi(x_1, \ldots, x_N) \, a(x_1)^* \ldots a(x_N)^* \Omega, \tag{4.1}$$

where Ω is the ground state and the creation and annihilation operators $a(x)^*$ and $a(x)$ satisfy the usual commutation relations, $[a(x), a(y)^*] = \delta(x - y)$. We subdivide the interval $[0, L]$ into M intervals $[x_{i-1}, x_i]$ of equal length $\Delta = x_i - x_{i-1} = L/M$. If we put $a_n^\sharp = \int_{x_{n-1}}^{x_n} a(x)^\sharp dx$, then

$$[a_n, a_m^*] = \Delta \delta_{n,m}. \tag{4.2}$$

Let $\mathcal{H}_1^\Delta$ be the 1-particle Hilbert space spanned by $a_n^* \Omega$ ($n = 1, 2, \ldots, M$). Then $\mathcal{F}_N^\Delta = \left(\bigotimes_{k=1}^N \mathcal{H}_1^\Delta\right)_{sym}$ is a subspace of $\mathcal{F}_N$. We define local operators on $\mathcal{F}_N^\Delta$ by

$$\hat{\alpha}_n(\lambda) = \alpha(\lambda)\mathbf{1} + \frac{1}{2}\kappa a_n^* a_n \tag{4.3}$$

and

$$\hat{\beta}_n = i\sqrt{\kappa}\,\hat{\rho}_n\, a_n, \tag{4.4}$$

where λ is a real number called the spectral parameter,

$$\alpha(\lambda) = 1 - i\lambda\Delta/2, \tag{4.5}$$

and $\hat{\rho}_n = \left(1 + \frac{1}{4}\kappa a_n^* a_n\right)^{1/2}$. Next we define 2×2 matrices of these operators, the so-called monodromy matrices:

$$L_n(\lambda) = \begin{pmatrix} \hat{\alpha}_n(\lambda) & \hat{\beta}_n^* \\ \hat{\beta}_n & \hat{\alpha}_n(\lambda) \end{pmatrix}. \tag{4.6}$$

These operator matrices contain the local scattering information about the lattice model. Finally, we define transfer matrices $\tau(\lambda)$ as follows: Put

$$T_M(\lambda) = L_1(\lambda) \ldots L_M(\lambda) = \begin{pmatrix} A_M(\lambda) & B_M^*(\lambda) \\ B_M(\lambda) & A_M^*(\lambda) \end{pmatrix}. \tag{4.7}$$

Then

$$\tau_M(\lambda) = \text{Trace}_2 T_M(\lambda) = A_M(\lambda) + A_M^*(\lambda). \tag{4.8}$$

Formally one can define a Hamiltonian for this model as a certain logarithmic derivative of $\tau_M(\lambda)$. However, we shall not use this Hamiltonian but work instead with the transfer matrices $\tau_M(\lambda)$. The essential feature of this lattice model is that the operators $\tau_M(\lambda)$ have a *common* set of eigenfunctions $\Psi_\Delta(m_1, \ldots, m_N)$. This can be shown using the algebraic Bethe Ansatz method developed by Faddeev et.al. [8]. The precise result reads:

Theorem 1. *Suppose that* $\lambda_1 < \ldots < \lambda_N$ *are real numbers satisfying*

$$\left(\frac{\alpha(\lambda_k)}{\bar{\alpha}(\lambda_k)}\right)^M = \prod_{l \neq k} \frac{\lambda_k - \lambda_l - i\kappa}{\lambda_k - \lambda_l + i\kappa} \tag{4.9}$$

or equivalently,

$$2M \tan^{-1}(\lambda_j \Delta/2) = k_j^0 L - \sum_{i=1}^N \theta_\kappa(\lambda_j - \lambda_i), \tag{4.10}$$

where k_j^0 is defined by (1.4) for some set of integers $m_1 < \ldots < m_N$. Then the wave function

$$\Psi_\Delta(\lambda_1, \ldots, \lambda_N) = B_M^*(\lambda_1) \ldots B_M^*(\lambda_N)\Omega \tag{4.11}$$

is an eigenfunction of $\tau_M(\lambda)$ for all $\lambda \in \mathbb{R}$ with eigenvalues given by

$$E_M(\lambda; \lambda_1, \ldots, \lambda_N) = \prod_{j=1}^N \frac{\lambda_j - \lambda - i\kappa}{\lambda_j - \lambda}\alpha(\lambda)^M + \prod_{j=1}^N \frac{\lambda_j - \lambda + i\kappa}{\lambda_j - \lambda}\bar{\alpha}(\lambda)^M. \tag{4.12}$$

5. ORTHOGONALITY

Orthogonality of the BA eigenfunctions for the nonlinear Schroedinger model now follows from:

- 1. The operators $\tau_M(\lambda)$ have a common set of eigenfunctions $\Psi_\Delta(m_1,\ldots,m_N)$.

- 2. If $m_1 < \ldots < m_N$ and $m'_1 < \ldots m'_N$ are two different sets of integers then the corresponding eigenfunctions $\Psi_\Delta(m_1,\ldots,m_N)$ and $\Psi_\Delta(m'_1,\ldots,m'_N)$ are orthogonal.

- 3. As $\Delta \to 0$, $\Psi_\Delta(m_1,\ldots,m_N) \to \psi^{BA}_{\{m_j\}}$.

Statement 1 is given by Theorem 1. Statement 2 follows from:

Lemma 3. *Suppose that $\lambda_1 < \ldots < \lambda_N$ and $\lambda'_1 < \ldots < \lambda'_N$ are two solutions of the above system of coupled nonlinear equations (4.9). Assume that $E_M(\lambda;\lambda_1,\ldots,\lambda_N) = E_M(\lambda;\lambda'_1,\ldots,\lambda'_N)$ for all real λ. Then $\lambda_i = \lambda'_i$ for $i=1,\ldots,N$.*

Proof. Let $f(\mu) = \prod_{j=1}^{N}(\nu - \lambda_j)$ and $g(\mu)$ the same with λ'_j instead of λ_j. Let $W(\mu)$ be the Wronskian $W(\mu) = f(\mu+i\kappa)g(\mu) - f(\mu)g(\mu+i\kappa)$. Then it follows from (4.12) that $W(\lambda - i\kappa) = \left(\frac{\alpha(\lambda)}{\bar\alpha(\lambda)}\right)^M W(\lambda)$. On the other hand, a direct computation shows that $W(\lambda - i\kappa) = W(\lambda)$. Hence $W(\lambda) = 0$ for all but a finite number of λ's and by continuity, for all $\lambda \in \mathbb{R}$. QED

The proof of Statement 3 is technical. Some explanation is given in Section 6. The precise result is:

Theorem 2. *For arbitrary N, the eigenstates $\Psi_\Delta(m_1,\ldots,m_N)$ of the lattice model converge to the corresponding BA eigenstates $\psi^{BA}_{m_1,\ldots,m_N}$ up to normalisation as $\Delta \to 0$:*

$$\lim_{\Delta \to 0} \Psi_\Delta(m_1,\ldots,m_N) = (-i\sqrt{\kappa})^N e^{-i(k_1+\ldots+k_N)L/2} \psi^{BA}_{m_1,\ldots,N} \text{ in } \mathcal{F}_N-norm. \quad (5.1)$$

Hence, the Bethe Ansatz eigenstates for different sets $m_1 < \ldots < m_N$ and $m'_1 < \ldots < m'_N$ are orthogonal, and the set of all Bethe Ansatz eigenstates is complete in $\mathcal{F}_N$.

6. CONVERGENCE OF THE LATTICE APPROXIMATION

The first part of the proof is to show that the factors $\hat\rho(\lambda)$ are irrelevant. The wave functions can then be written as follows: (We omit hats on β-operators)

Lemma 4.

$$\Psi_\Delta(\lambda_1,\ldots,\lambda_N) \approx \left(\prod_{k=1}^{N}(B_0^*(\lambda_k) + \ldots + B_{k-1}^*(\lambda_k))\right)\Omega + O(\Delta), \quad (6.1)$$

where

$$B_j^*(\lambda) = \sum_{1 \le n_1 < \ldots < n_j \le M} \beta^*_{M,n_j+1}(\lambda)\beta_{n_j}\beta^*_{n_j-1,n_{j-1}+1}(\lambda)\ldots\beta^*_{n_1-1,1}(\lambda) \quad (6.2)$$

and

$$\beta^*_{m,n}(\lambda) = \sum_{r=n}^{m} \alpha(\lambda)^{m-r}\bar\alpha(\lambda)^{r-n}\beta^*_r. \quad (6.3)$$

This is proved by writing $L_n(\lambda)$ in the form $L_n(\lambda) = L_n^0(\lambda) + \tilde{L}_n$ with
$$L_n^0(\lambda) = \begin{pmatrix} \alpha(\lambda)\mathbf{1} & \beta_n^* \\ 0 & \bar{\alpha}(\lambda)\mathbf{1} \end{pmatrix} \text{ and } \tilde{L}_n \approx \begin{pmatrix} 0 & 0 \\ \beta_n & 0 \end{pmatrix}.$$
One then proceeds by induction on N. For $N = 1$,

$$\Psi_\Delta(\lambda_1) = \beta_{M,1}^*(\lambda_1)\Omega = \sum_{r=1}^M \alpha(\lambda_1)^{M-r}\bar{\alpha}(\lambda_1)^{r-1}(-i\sqrt{\kappa})a_n \tag{6.4}$$

which tends to

$$(-i\sqrt{\kappa})\int_0^L dx\, e^{-ik_1(L-x)/2}e^{ik_1x/2}a(x)\Omega. \tag{6.5}$$

(Notice that the equations (4.10) tend to (1.3) so that their solutions also converge.) We shall not discuss the general case but only the case $N = 2$. We write the BA eigenstates as follows:

$$\psi_{\{m_1,m_2\}}^{BA} = A\exp[i(k_1x_1 + k_2x_2)] + B\exp[i(k_1x_2 + k_2x_1)] \tag{6.6}$$

for $x_1 < x_2$, where

$$A = 1 - \frac{\kappa}{i(k_1 - k_2)} \tag{6.7a}$$

and

$$B = 1 + \frac{\kappa}{i(k_1 - k_2)}. \tag{6.7b}$$

(Formula (6.6) can be generalised to general N; see Gaudin [9]. It is with this normalisation that Theorem 2 holds.) Now, according to Lemma 4,

$$\Psi(\lambda_1, \lambda_2) \approx B_0^*(\lambda_1)(B_0^*(\lambda_2) + B_1^*(\lambda_2))\Omega = B_0^*(\lambda_1)B_0^*(\lambda_2)\Omega + \Psi_1(\lambda_1, \lambda_2) \tag{6.8}$$

with

$$\Psi_1 = \sum_{n=1}^M \sum_{r_1=1}^{n-1} \sum_{r_2=n+1}^M \sum_{r_3=1}^M \alpha(\lambda_2)^{M-r_2+n-1-r_1}\bar{\alpha}(\lambda_2)^{r_2-n+r_1-2}$$
$$\times \alpha(\lambda_1)^{M-r_3}\bar{\alpha}(\lambda_1)^{r_3-1}\beta_{r_2}^*\beta_n\beta_{r_1}^*\beta_{r_3}^*\Omega. \tag{6.9}$$

It is clear that the first term in the RHS of (6.8) corresponds to the 1's in A and B inserted in (6.6). (One distinguishes $r_1 < r_2$ and $r_1 > r_2$ in the B_0^* operators.) To see that the term Ψ_1 corresponds to the factors $\kappa/i(k_1 - k_2)$ notice first that $\beta_r^* \approx -i\sqrt{\kappa}a_r^*$ and that $[a_n, a_{r_3}^*] = \Delta\delta_{n,r_3}$ in formula (6.9). Moreover, single factors of $\alpha(\lambda_i)$ are negligible. Hence,

$$\Psi_1 \approx -\kappa^2\Delta \sum_{1\leq r_1 < r_2 \leq M} \sum_{n=r_1+1}^{r_2-1} \left(\frac{\alpha(\lambda_2)\bar{\alpha}(\lambda_1)}{\alpha(\lambda_1)\bar{\alpha}(\lambda_2)}\right)^n$$
$$\times \alpha(\lambda_1)^M\alpha(\lambda_2)^{M-r_1-r_2}\bar{\alpha}(\lambda_2)^{r_1+r_2}a_{r_1}^*a_{r_2}^*\Omega. \tag{6.10}$$

The crucial step is now that,

$$\kappa\Delta\sum_{n=r_1+1}^{r_2-1}\zeta^n = \kappa\Delta\frac{\zeta^{r_2} - \zeta^{r_1+1}}{\zeta - 1} \approx \frac{\kappa}{i(\lambda_1 - \lambda_2)}(\zeta^{r_2} - \zeta^{r_1+1}), \tag{6.11}$$

where
$$\zeta = \frac{\alpha(\lambda_2)\bar{\alpha}(\lambda_1)}{\alpha(\lambda_1)\bar{\alpha}(\lambda_2)} \approx 1 + i(\lambda_1 - \lambda_2)\Delta + O(\Delta^2). \qquad (6.12)$$

Inserting this into (6.10) we find

$$\Psi_1 \approx -\frac{\kappa^2}{i(\lambda_1 - \lambda_2)} \sum_{1 \leq r_1 < r_2 \leq M} \left(\alpha(\lambda_1)^{M-r_2}\bar{\alpha}(\lambda_1)^{r_2}\alpha(\lambda_2)^{M-r_1}\bar{\alpha}(\lambda_2)^{r_1} - \right.$$
$$\left. \alpha(\lambda_1)^{M-r_1}\bar{\alpha}(\lambda_1)^{r_1}\alpha(\lambda_2)^{M-r_2}\bar{\alpha}(\lambda_2)^{r_2}\right) a^*_{r_1} a^*_{r_2} \Omega. \qquad (6.13)$$

As $\Delta \to 0$, this tends to

$$-\frac{\kappa^2}{i(k_1 - k_2)} e^{-i(k_1+k_2)L/2}$$
$$\times \int_0^L dx_1 \int_{x_1}^L dx_2 \left(e^{-i(k_1 x_1 + k_2 x_1)} - e^{-i(k_1 x_1 + k_2 x_2)} \right) a(x_1)^* a(x_2)^* \Omega. \qquad (6.14)$$

The first exponential in the integrand corresponds to the second part of the B-term; the second exponential to the second part of the A-term in (6.6).

REFERENCES

1. E.H. Lieb and W. Liniger, *Phys. Rev.*, **130**, 1605–1624 (1963).
2. C.N. Yang and C.P. Yang, *J. Math. Phys.*, **10**, 1115–1122 (1969).
3. T.C. Dorlas, J.T. Lewis and J.V. Pulé, *Commun. Math. Phys.*, **124**, 365–402 (1989).
4. T.C. Dorlas, *in:* "Proceedings of the IX-th I.A.M.P. Congress", Swansea, 17-27 July 1988, World Scientific, Singapore 1992.
5. T.C. Dorlas, *Commun. Math. Phys.*, **154**, 347–376 (1993).
6. M. Girardeau, *J. Math. Phys.*, **1**, 516–523 (1960).
7. A.G. Izergin and V.E. Korepin, *Sov. Phys. Dokl.*, **26**, 653–654 (1981).
8. L.D. Faddeev, *Sov. Sc. Rev.*, **C1**, 107–155 (1980).
9. M. Gaudin, "La Fonction d'Onde de Bethe", Paris, Masson, 1983.

FLAT NON-REGULAR STATES ON WEYL ALGEBRAS

F. Acerbi

International School for Advanced Studies and GNFM
via Beirut 2-4, I-34013, Trieste, Italy

1. MOTIVATIONS

Some models of Many Body Theory and Quantum Statistical Mechanics are formulated and sometimes solved in terms of variables or fields which formally satisfy the Canonical Commutation Relations (CCR), but which actually cannot be represented as operators in a Hilbert space because of their bad infrared behaviour[1,2]. Typical examples are:

1. the infinite quantum harmonic lattice in thermal equilibrium and the Free Bose gas, both in space dimensions $d \leq 2$ [3,4];

2. the electrons in a periodic potential (Bloch electrons)[3];

3. the massless scalar field in two space–time dimensions[5].

The strategies usually adopted to discuss the above models fall essentially in two categories: to represent the singular fields as operators on an indefinite metric space[5]; or to use a restricted set of field variables and describe the remaining degrees of freedom as morphisms of the regular field algebra[6].

We introduced [1,2] an alternative approach based on the representation of the Weyl exponentials of the singular variables in a positive metric Hilbert space. This allowed to keep the canonical structure (in Weyl form) also for the singular variables and clarify the relation between infrared singularities and the so arising *nonregular representations* of the CCR.

2. NONREGULAR REPRESENTATIONS OF CCR ALGEBRAS

Let be given a symplectic space (V, σ), i.e., V is a real linear space and σ is a symplectic form on it (σ is *not* supposed to be nondegenerate).

A CCR algebra[7,8,9] $\mathcal{A}(V,\sigma)$ is the *-algebra generated by elements $\delta(F)$, $F \in V$; the product and the involution in the algebra are defined by

$$\delta(F)\delta(G) = \delta(F+G)e^{-\frac{i}{2}\sigma(F,G)}$$
$$\delta(F)^* = \delta(-F) \qquad F, G \in V.$$

These equations imply $\delta(0) = \mathbb{1}$ and $\delta(F)^{-1} = \delta(F)^*$.
$\mathcal{A}(V,\sigma)$ can always be completed to a C^*-algebra. A state ω on $\mathcal{A}(V,\sigma)$ is a positive and normalized linear functional on it.

A representation π of $\mathcal{A}(V,\sigma)$ is said to be *regular* if the map $\lambda \in \mathbb{R} \longmapsto \pi(\delta(\lambda F))$ is strongly continuous for every $F \in V$. A state ω on $\mathcal{A}(V,\sigma)$ is said to be *regular* if the associated GNS representation π_ω of $\mathcal{A}(V,\sigma)$ is regular. The regularity of ω is equivalent to the fact that $\lambda \in \mathbb{R} \longmapsto \omega(\delta(\lambda F))$ is a continuous mapping for every $F \in V$.

If the representation π is regular then there are selfadjoint operators $\Phi(F)$ on $\mathcal{H}_\omega$ such that $\pi(\delta(\lambda F)) = e^{i\lambda \Phi(F)}$ for every $\lambda \in \mathbb{R}$, $F \in V$. The operators $\Phi(F)$ satisfy the (unbounded form of) the Canonical Commutation Relations.

It is easily seen that the set $\mathcal{R}_V^\pi$ of $F \in V$ such that $\lambda \in \mathbb{R} \longmapsto \pi(\delta(\lambda F))$ is strongly continuous is a linear subspace of V. Hence the $\delta(\cdot)$ indexed by the elements of this set give rise in a natural way to a CCR subalgebra of $\mathcal{A}(V,\sigma)$.
In particular, if π is the GNS representation induced by a (possibly nonregular) state Ω, $\pi \equiv \pi_\Omega$, the set $\mathcal{R}_V^\pi$ coincides with the set

$$\mathcal{R}_V^\Omega := \{F \in V : \lambda \in \mathbb{R} \longmapsto \Omega(\delta(\lambda F)) \text{ is continuous}\}.$$

The CCR algebra $\mathcal{A}_\Omega \equiv \mathcal{A}(\mathcal{R}_V^\Omega, \sigma)$ will be referred to as the *regular subalgebra* of $\mathcal{A}(V,\sigma)$ in representation π_Ω.

We now introduce a particular class of nonregular representations of Weyl algebras. We suppose to this end to be given a CCR algebra $\mathcal{A}(V_0, \sigma_0)$ and a state ω on it. Suppose then to have a symplectic space (V, σ) such that V_0 is a linear subspace of V, and that $\sigma_{|V_0 \times V_0} = \sigma_0$. In this case we say that (V_0, σ_0) is a *symplectic subspace* of (V, σ). $\mathcal{A}(V_0, \sigma_0)$ is then a *-subalgebra of $\mathcal{A}(V, \sigma)$.

Prop. 1 below says that ω can *always* be extended to a *nonregular state* Ω on $\mathcal{A}(V,\sigma)$. In the proofs, the fact that (V_0, σ_0) is a symplectic subspace of (V, σ) is crucial. Indeed, one can give two different proofs of this proposition: we sketch here their main ideas. The first proof generalizes an argument by Buchholz and Fredenhagen[10] and displays the "charged sectors" structure which will be further analyzed in Prop. 4. The second proof uses the fact that one can always construct an abelian (and hence amenable) group of automorphisms of $\mathcal{A}(V, \sigma)$, whose action leaves $\mathcal{A}(V_0, \sigma_0)$ pointwise invariant. The existence of this group follows *solely* from the fact that (V_0, σ_0) is a symplectic subspace of (V, σ). The extended nonregular states are then obtained by taking the mean of any extension of ω to $\mathcal{A}(V, \sigma)$ over the group.

Proposition 1

Let $\mathcal{A}(V,\sigma)$ be a CCR algebra. Let $\mathcal{A}(V_0, \sigma_0)$ be a CCR subalgebra of $\mathcal{A}(V, \sigma)$ and ω be a state on $\mathcal{A}(V_0, \sigma_0)$. Then the linear functional on $\mathcal{A}(V, \sigma)$ defined by

$$\Omega(\delta(F)) = \begin{cases} \omega(\delta(F)) & F \in V_0 \\ 0 & F \in V \backslash V_0. \end{cases} \qquad (1)$$

is positive and hence is a state (flat nonregular state) on $\mathcal{A}(V, \sigma)$.

We then give

Definition 2

Let $\mathcal{A}(V_0, \sigma_0)$ be a CCR subalgebra of $\mathcal{A}(V, \sigma)$. Then a regular state ω on $\mathcal{A}(V_0, \sigma_0)$ has $\mathcal{A}(V_0, \sigma_0)$ as its *maximal domain of regularity* in $\mathcal{A}(V, \sigma)$ if there is *no regular extension* of ω to a larger CCR subalgebra $\mathcal{A}(V_1, \sigma_1)$ of $\mathcal{A}(V, \sigma)$, $V_0 \not\subseteq V_1 \subset V$. A nonregular state Ω on a CCR algebra $\mathcal{A}(V, \sigma)$ is *minimally nonregular* if the regular subalgebra $\mathcal{A}_\Omega$ of $\mathcal{A}(V, \sigma)$ is its maximal domain of regularity.

- This definition makes sense by Prop. 1, saying that *there are nonregular extensions of ω to $\mathcal{A}(V, \sigma)$*.
- A nonregular state on a CCR algebra always exists: take $V_0 = \{0\}$ in Prop. 1. This state induces the type II_1 representation constructed by Slawny[7].

Our goal is to characterize minimally nonregular *flat* states. To this end we do the following, crucial, assumption:

Condition A

There is a sequence $\{F_n\}_{n \in \mathbb{N}} \in V_0$, such that $\lim_{n \to +\infty} \sigma(F_n, \cdot) \in V_{\mathbb{R}}^*$, and

$$\lim_{n \to +\infty} \omega(\delta(F_n)) = 1.$$

Remark. There are cases in which $\{F_n\}_{n \in \mathbb{N}}$ in Condition A can be taken a constant sequence: $\exists F \in V_0$, $F \neq 0$, such that $\omega(\delta(F)) = 1$. One can show that, if Ω extends ω to $\mathcal{A}(V, \sigma)$,

$$\pi_\Omega(\delta(F)) \in \mathcal{Z}_\Omega(\mathcal{A}(\mathcal{R}_V^\Omega, \sigma)).$$

A situation of this type occurs, for instance, in the quantization of systems with constraints (like gauge theories). First class constraints are identified precisely by the above condition (see the works by Grundling and Hurst[11,12]), the corresponding states are referred to as "Dirac states".

The existence of *nontrivial* (i.e., $\exists G \in V : \lim_{n \to +\infty} \sigma(F_n, G) \neq 0$) sequences in Condition A forces any state Ω which extends ω to $\mathcal{A}(V, \sigma)$ to be nonregular; moreover one has

Proposition 3

Let $\mathcal{A}(V, \sigma)$ be a CCR algebra. Let $\mathcal{A}(V_0, \sigma_0)$ be a CCR subalgebra of $\mathcal{A}(V, \sigma)$ and ω be a regular state on $\mathcal{A}(V_0, \sigma_0)$. Consider the two statements

i. $\forall G \in V \setminus V_0$, $\exists \{F_n\}_{n \in \mathbb{N}} \in V_0$ satisfying Condition A such that

$$\lim_{n \to +\infty} e^{i\sigma(F_n, G)} = e^{i\alpha} \neq 1.$$

ii. ω admits a unique extension Ω to $\mathcal{A}(V, \sigma)$, namely the flat nonregular state given by eq. (1).

It follows that i. $\Rightarrow$ ii. and in particular Ω is minimally nonregular. Moreover, if ω is pure so is Ω.

In the following proposition we describe the GNS representation π_Ω of $\mathcal{A}(V, \sigma)$ in terms of representations of $\mathcal{A}(V_0, \sigma_0)$, in the hypotesis of the above Prop. 3 (uniqueness of extension). Similar results hold under less restrictive assumptions; they are outlined in Section 3 on the basis of a very relevant example.

Proposition 4

Let Ω be a nonregular state on $\mathcal{A}(V,\sigma)$ and let ω be its restriction to $\mathcal{A}(V_0,\sigma_0)$. Let i. in Prop. 3 hold. Then π_Ω decomposes into the direct sum of inequivalent representations of $\mathcal{A}(V_0,\sigma_0)$, labelled by the equivalence classes $F \in V/V_0$:

$$\pi_\Omega = \bigoplus_{F \in V/V_0} \pi_{\omega_F}, \qquad \mathcal{H}_\Omega = \bigoplus_{F \in V/V_0} \mathcal{H}_{\omega_F}$$

where $\omega_F := \omega \circ \rho_F$, $\rho_F \in Aut\mathcal{A}(V_0,\sigma_0): \delta(G) \mapsto e^{i\sigma(F,G)}\delta(G) \quad \forall G \in V$.
Moreover every π_{ω_F} is irreducible (factorial) iff π_ω is.

3. BLOCH ELECTRONS AND θ ANGLE STRUCTURE

The framework discussed above allows to clarify the occurrence of a structure of sectors labelled by an angle θ in the case of electrons in a periodic potential $V(x) = V(x+a)$, and its relation with the θ vacua structure of QCD.

The analogy has been stressed by Jackiw[13] and it is discussed here in a mathematically rigorous version which does not rely on the semiclassical approximation: the non-normalizable Bloch wave functions are treated as non regular states over the CCR algebra generated by the Weyl operators.

The CCR algebra $\mathcal{A}(V,\sigma)$ is generated by the Weyl operators $W(\alpha,\beta)$, with $\alpha, \beta \in \mathbb{R}$, and the Bloch states Ω_k^n are defined by the Bloch wave functions[1] (we set $a = 1$ in the following)

$$\psi_k^n(x) = e^{ikx}v_k^n(x), \quad v_k^n(x) = v_k^n(x+1), \quad k \in [0,2\pi).$$

The ground state $(n = k = 0)$ is thus given by

$$\Omega_0^0(W(\alpha,\beta)) = 0 \quad \text{if} \quad \alpha \neq 2\pi n, \; n \in \mathbb{N}$$

$$\Omega_0^0(W(2\pi n,\beta)) = e^{i\pi n\beta} \int_0^1 \bar{v}_0^0(x) v_0^0(x+\beta) e^{i2\pi nx}\,dx.$$

The Weyl operators $W(\alpha,0)$, which generate the boosts on $\mathcal{A}(V,\sigma)$, are represented non regularly by Ω_0^0; all the states Ω_k^n defined by the Bloch wave functions are represented by vectors in the GNS representation space defined by Ω_0^0. The nonregularity of the representation is forced by the invariance of Ω_0^0 under the non compact discrete group of lattice translations. The regular subalgebra is generated by the Weyl operators $W(0,\beta) = \exp i\beta p$.

The group of discrete (lattice) translations $T_n = \exp inp$, $n \in \mathbb{N}$ plays the rôle of an unbroken gauge group $\mathcal{G}$. The subalgebra of $\mathcal{A}$ which is left pointwise invariant under such (gauge) group may be identified with the observable algebra $\mathcal{A}_{obs}$; it is generated by $\exp i\beta p$ and $\exp i2\pi mx$ with $\beta \in \mathbb{R}$, $m \in \mathbb{N}$.

The GNS representation space $\mathcal{H}$ given by the state Ω_0^0 over $\mathcal{A}$ decomposes into disjoint irreducible representations of $\mathcal{A}_{obs}$, labelled by an angle $\theta \in [0,2\pi)$

$$\mathcal{H} = \oplus_{\theta \in [0,2\pi)} \mathcal{H}_\theta, \quad \mathcal{H}_\theta \equiv W(\theta,0)\mathcal{H}_0$$

$$\mathcal{H}_0 = \overline{\mathcal{A}_{obs}\Psi_0},$$

where Ψ_0 is the vector in $\mathcal{H}$ which corresponds to the state Ω_0^0.

$\mathcal{A}_{obs}$ has a non trivial center $\mathcal{Z}_{obs}$, generated by the group $\mathcal{G}$ of lattice translations and Ω_0^0 is a pure state over $\mathcal{A}_{obs}$. In fact, $\mathcal{G}$ is unbroken, since $\Omega_0^0(T_n) = 1$ implies $T_n \Psi_0 = \Psi_0$; therefore

$$\Omega_\theta(T_m) \equiv \Omega_0^0(W(-\theta,0)T_m W(\theta,0)) = e^{im\theta},$$

which implies
$$T_m \Psi_\theta = e^{im\theta} \Psi_\theta \quad (\Psi_\theta \equiv W(\theta,0)\Psi_0).$$

Hence, $\mathcal{Z}_{obs}$ is non trivial and the representations of $\mathcal{A}_{obs}$ given by the states Ω_θ are inequivalent; they are irreducible since all the states Ω_θ are pure as a consequence of purity of Ω_0^0.

The group τ of charged automorphisms of $\mathcal{A}_{obs}$ is generated by $\exp i\alpha x$, $\alpha \in [0, 2\pi)$, and the following equations are satisfied:

$$e^{i\alpha x} \Psi_\theta = \Psi_{\theta+\alpha}$$

$$T_n e^{i\alpha x} T_n^{-1} = a^{i\alpha(x+n)}.$$

They are necessarily broken in each irreducible representation of $\mathcal{A}_{obs}$ given by the vectors Ψ_θ. The crucial point is that the algebra of observables has a non trivial center which is not left pointwise invariant under the action of τ (no use has been made of the semiclassical approximation and of the tunnelling mechanism).

As far as the analog of the n-vacua is concerned, one may investigate in general the representations of $\mathcal{A}_{obs}$ defined by wave functions $\psi(x) \in L^2$, i.e., by states with some localization in x, e.g., wave functions $\psi_n(x)$ localized around the minima x_n of the periodic potential. In this case one necessarily gets reducible representations of $\mathcal{A}_{obs}$. In fact, the operators $T_n \in \mathcal{Z}_{obs}$ cannot be represented by c-numbers t_n (a necessary condition for irreducibility), since, by the unitarity of T_n one must then have $|t_n| = 1$, and, on the other side, $T_n \psi(x) = \psi(x+n)$ implies $|(\psi, T_n \psi)| < 1$. For the Wannier wave functions ψ_W one actually gets $(\psi_W, T_n \psi_W) = 0$.

The above argument applies to any representation of $\mathcal{A}_{obs}$ in which the transformations $\exp i\alpha x$ are implemented by strongly continuous unitary operators, so that the associated charge is well defined; in fact, in this case the wave functions are in L^2 and one necessarily gets reducibility.

In conclusion, one is led to use the representations defined by the vectors Ψ_θ (or by the Bloch states) quite generally in order to have irreducible representations of $\mathcal{A}_{obs}$ and this automatically implies the breaking of the automorphisms in τ. The so emerging θ angle structure is merely a consequence of the structure of $\mathcal{A}_{obs}$ (in particular its non trivial center).

Acknowledgments

The author is grateful to the Gruppo Nazionale di Fisica Matematica of CNR and to the local organizers of the conference for financial support.

REFERENCES

1. F. Acerbi, G. Morchio and F. Strocchi, Infrared singular fields and nonregular representations of CCR algebras, J. Math. Phys., **34**, 899 (1993).
2. F. Acerbi, G. Morchio and F. Strocchi, Theta vacua, charge confinement and charged sectors from nonregular representations of CCR algebras, Lett. Math. Phys., **27**, 1 (1993).

3. N.W. Ashcroft and N.D. Mermin, "Solid State Physics," Holt-Saunders, Tokyo, Phila- delphia (1981).
4. O. Bratteli and D.W. Robinson, "Operator Algebras and Quantum Statistical Mechanics," vol. I and II. Springer-Verlag, Berlin, Heidelberg, New York (1981).
5. A. S. Wightman, Introduction to some aspects of the relativistic dynamics of quantized fields, *in*: "High Energy Electromagnetic Interactions and Field Theory," M.Lévy ed., Cargèse 1964, 171, Gordon and Breach, New York, London, Paris (1967).
6. R. F. Streater and I. F. Wilde, Fermion states of a boson field, *Nucl. Phys. B*, **24**, 561 (1970).
7. J. Slawny, On factor representations and the C^*-algebra of canonical commutation relations, *Commun. Math. Phys.* **24**, 151 (1972).
8. J. Manuceau, M. Sirugue, D. Testard and A. Verbeure, The smallest C^*-algebra for canonical commutation relations, *Commun. Math. Phys.*, **32**, 231 (1973).
9. D. Petz. "An Invitation to the Algebra of Canonical Commutation Relations," Leuven Notes in Mathematical and Theoretical Physics **2**, Series A: Mathematical Physics, Leuven University Press (1990).
10. D. Buchholz and K. Fredenhagen, Locality and the structure of particle states in gauge field theory, *in*: "Mathematical Problems in Theoretical Physics," Berlin 1981 Lecture notes in Physics **153**, 368, Springer-Verlag, Berlin, Heidelberg, New York (1982).
11. H.B.G.S. Grundling and C.A. Hurst, Algebraic quantization of systems with a gauge degeneracy, *Commun. Math. Phys.*, **98**, 369 (1985).
12. Hurst C.A.: Dirac's theory of constraints, *in*: "Recent Developments in Mathematical Physics," Schladming 1987, 18, Springer-Verlag, Berlin, Heidelberg, New York (1987).
13. R. Jackiw, Topological investigations of quantized gauge theories, *in*: " Current Algebra and Anomalies," 263, World Scientific, Singapore (1985).

SECOND VIRIAL COEFFICIENT FOR ONE-DIMENSIONAL SYSTEMS

P. Kurasov[1,2], V. Kurasov[1] and B. Pavlov[1]

1. Dept. of Math. and Comp. Physics
   St Petersburg Univ., 198904, Russia

2. Stockholm Univ.
   Frescativagen 24, 10405 Stockholm, Sweden

This work is dedicated to the memory of S.P. Merkuriev

INTRODUCTION

Exactly solvable models are very important in modern statistical physics. Such models can give essential information about investigating physical objects. Unfortunately these models can not posses all properties of real systems. But exact analytical solutions permit us to calculate very important quantities, which can not be done for the real ones, or can be done only numerically. One of the most popular models in statistical mechanics is the model of hard core spheres. This model has been used to investigate properties of the second ([8,9]) and third ([10]) virial coefficients. Even phase transitions can be observed in this model [7].

Another method which allows us to construct exactly solvable quantum mechanical problems is the method of zero-range potentials [5] which is widespread in mathematical, nuclear and atomic physics [1,4]. However, constructions in the framework of the method of zero-range potentials are rather poor to give a realiable description of many real physical processes. Moreover, direct application of the zero-range potentials method to some physical problems can lead to paradoxes (see [4]). This method is based on the theory of selfadjoint extensions of symmetric operators [2]. Standard Fermi delta potential can be generalized considering selfadjoint extensions with some additional degrees of freedom. This method of zero range potentials with internal structure was developed first in [14,15]. Rigorous mathematical treatment of the two- and three-body scattering problems in one dimension can be found in [12,13]. The first exactly solvable three-body model with Fermi delta potentials was considered in [3].

The present paper is devoted to the second virial coefficient which is investigated for the case of one-dimensional particles. We chose this system as one of the simplest models in statistical physics which gives nontrivial results. We discuss the method of zero-range potentials applied to this problem. It is shown that the standard Fermi zero-range potential leads to a "nonphysical" behaviour of the cluster integral. Considering an additional space of interactions allows us to introduce a realistic model of the problem. Exact calculation of the second cluster integral for the model operator is carried out. Important high and low temperature limits are investigated.

TWO-BODY HAMILTONIAN

In this section, we shall consider a model for the scattering of two one-dimensional particles. The interaction between the particles will be introduced with the help of the extension theory of symmetric operators. The spectrum and the eigenfunctions for this model operator can be calculated exactly and expressed in terms of elementary functions.

We start from the orthogonal sum of the unperturbed kinetic operator for two one-dimensional particles $\mathcal{A} = -d^2/dx^2$ and some finite dimensional operator $\mathcal{A}_{in}$ acting in the othogonal sum of the Hilbert spaces $L_2(\mathbf{R}) \oplus H_{in}$. Restriction of the differential operator to the set of functions vanishing together with their first derivatives in a neighbourhood of the origin gives us a symmetric operator $\mathcal{A}_0 \oplus \mathcal{A}_{in}, Dom(\mathcal{A}_0) = \{u \in W_2^2(\mathbf{R}), u(0) = 0, u'(0) = 0\}$ (after a closure). The adjoint operator is given by $\mathcal{A}_0^* \oplus \mathcal{A}_{in}$ on $Dom(\mathcal{A}_0^*) = \{u \in W_2^2(\mathbf{R} \setminus \{0\})\}$. The operator with interaction is defined as follows:

$$\mathcal{L}\begin{pmatrix} u \\ u_{in} \end{pmatrix} = \begin{pmatrix} \mathcal{A}_0 u \\ \mathcal{A}_{in} u_{in} + (a[\frac{du}{dx}]|_{x=0} + b\{u\}|_{x=0}) < u_{in}, \theta > \end{pmatrix},$$

where $\theta \in H_{in}, a$ and b are real numbers, $< *, * >$ denotes the scalar product in the internal space H_{in}. Vector θ and numbers a, b are parameters of the interaction. The bracket and the brace denote the jump and the average value of the function:

$$[f]|_x = \lim_{y \to x+0} f(y) - \lim_{y \to x-0} f(y),$$

$$2\{f\}|_x = \lim_{y \to x+0} f(y) + \lim_{y \to x-0} f(y).$$

The operator $\mathcal{L}$ is selfadjoint on the domain of functions in $Dom(\mathcal{A}_0^*) \oplus H_{in}$ satisfying the boundary conditions

$$\left(c[\frac{du}{dx}] + d\{u\}\right)|_{x=0} = < u_{in}, \theta >,$$

$$[u]|_{x=0} = 0, \tag{1}$$

with real parameters c and d, such that

$$\det \begin{vmatrix} a & b \\ c & d \end{vmatrix} = -1. \tag{2}$$

A proof of this result can be carried out by following the ideas in the papers [12,13].

The operator constructed as above can be investigated in details. The equation for the eigenfunction $\Psi(\lambda) = (\psi(\lambda, x), \psi_{in}(\lambda))$ corresponding to the energy λ can be reduced to the Helmholtz equation for the first component in $L_2(\mathbf{R})$,

$$-\frac{d^2}{dx^2}\psi(\lambda, x) = \lambda\psi(\lambda, x), \tag{3}$$

with the energy dependent boundary conditions at the origin:

$$[\frac{d\psi}{dx}]\,|_{x=0} = -\mathcal{D}(\lambda)\{\psi\}\,|_{x=0} = 0,$$

$$[\psi]\,|_{x=0} = 0. \qquad (4)$$

The function $\mathcal{D}(\lambda)$ is defined by the internal operator and the parameters θ, a, b, c, d:

$$\mathcal{D}(\lambda) = \frac{b\mathbf{R}(\lambda) + d}{a\mathbf{R}(\lambda) + c}, \quad \mathbf{R}(\lambda) = <(\mathcal{A}_{in} - \lambda)^{-1}\theta, \theta>. \qquad (5)$$

It is a function with positive imaginary part in the upper half plane of the spectral parameter λ. We note that the standard method of zero-range potentials [1,4] gives the same boundary conditions but with some real constant $\mathcal{D}$.

The two-body scattering matrix can be introduced using the following ansatz for the continuous spectrum eigenfunctions:

$$\psi(\lambda, x) = \frac{1}{\sqrt{2\pi k}} \begin{cases} e^{ikx} + R(k)e^{-ikx}, & x < 0; \\ T(k)e^{ikx}, & x > 0. \end{cases} \qquad (6)$$

The transition and reflection coefficients can be calculated from the boundary conditions

$$T(k) = \frac{2ik}{\mathcal{D}(k^2) + 2ik} \quad \text{and} \quad R(k) = \frac{-\mathcal{D}(k^2)}{\mathcal{D}(k^2) + 2ik}. \qquad (7)$$

The dispersion equation for the eigenvalues $-\xi_s^2$,

$$2\xi_s = \mathcal{D}(-\xi_s^2), \qquad (8)$$

has a finite number of negative solutions.

THE SECOND CLUSTER INTEGRAL

The second virial coefficient can be calculated using the second cluster integral B. Contributions of the discrete and continuous spectrums to the cluster integral can be separated:

$$B = B_d + B_c. \qquad (9)$$

The discrete spectrum contribution B_d is given by the following expression

$$B_d = \sum_{\sigma_d} e^{\gamma \xi_s^2} \qquad (10)$$

where the sum is taken over all bound states and $-\xi_s^2$ are the energies of the bound states; $\gamma = (k_B T)^{-1}$, k_B is Boltzman's constant and T is the absolute temperature. This sum can be calculated using the dispersion relation (8).

To calculate the contribution of the continuous spectrum the corresponding integral should be regularized because the continuous spectrum eigenfunctions are not square integrable on the real axis:

$$B_c = \int_{\sigma_c} dE \int_{-\infty}^{+\infty} dx(|\psi(E,x)|^2 - |\psi_0(E,x)|^2)e^{-\gamma E} \qquad (11)$$

425

Here the first integral is the integral over the continuous spectrum with the measure $dE = 2k\,dk$. The function $\psi_0(E, x)$ is the eigenfunction of the unperturbed operator. The eigenfunctions of the perturbed and original operators should be normalized by the standard condition:

$$\int_{-\infty}^{+\infty} \psi(\lambda, x)\overline{\psi}(\lambda', x)dx = \delta(\lambda - \lambda').$$

The probability density of the perturbed state characterized by the function $|\psi|^2$ is reduced by the probability density of the unperturbed (free) state (the function $|\psi_0|^2$). The regularisation mentioned above leads to the exclusion of the free state probability. The integral in (11) does not diverge contrary to the ordinary form of the second cluster integral [11]. In order to calculate B_c we shall use the orthogonal decomposition of $L_2(\mathbf{R})$ into the subspaces of symmetric and antisymmetric functions:

$$\psi(\lambda, x) = \psi_s(\lambda, x) + \psi_a(\lambda, x),$$

where

$$\psi_s(\lambda, x) = \frac{1}{2}(\psi(\lambda, x) + \psi(\lambda, -x)),$$

$$\psi_a(\lambda, x) = \frac{1}{2}(\psi(\lambda, x) - \psi(\lambda, -x)).$$

According to (4) the antisymmetric wave ψ_a is not perturbed by the boundary condition and consequently coincides with the free one. The continuous spectrum contribution can be calculated by integrating the symmetric waves only

$$\psi_s(E, x) = \frac{1}{2\sqrt{2\pi k}}(e^{-ik|x|} + S(k)e^{ik|x|}) \tag{12}$$

where

$$S(k) = T(k) + R(k) = \frac{2ik - D(k^2)}{2ik + D(k^2)} \tag{13}$$

and

$$\psi_{0s}(E, x) = \frac{1}{2\sqrt{2\pi k}}(e^{-ik|x|} + e^{ik|x|}) \tag{14}$$

As a result B_c can be expressed by the following formula:

$$B_c = \frac{1}{2\pi}\int_0^{+\infty} dk \int_0^{+\infty} dx((S(k)-1)e^{2ikx} + (\overline{S(k)}-1)e^{-2ikx})e^{-\gamma k^2}. \tag{15}$$

This integral can be transformed into an integral over the real line with respect to the spectral parameter k. In this way a delta-functional singularity should be extracted:

$$B_c = \frac{S(0)-1}{4} + \frac{1}{2\pi}\int_{-\infty}^{+\infty} dk \int_0^{+\infty} dx(S(k)-S(0))e^{2ikx}e^{-\gamma k^2} \tag{16}$$

and we have the following expression for the continuous spectrum contribution to the cluster integral

$$B_c = \frac{S(0)-1}{4} + \frac{i}{4\pi}\int_{-\infty}^{+\infty} dk\,\frac{S(k)-S(0)}{k}e^{-\gamma k^2}. \tag{17}$$

LOW AND HIGH TEMPERATURE LIMITS

In this section the limits of the cluster integral for high and low values of the absolute temperature T will be investigated. The limits of the contribution of the bound states B_d can be easily calculated and we restrict our consideration to the continuous spectrum part.

The low temperature asymptotics of B_c can be calculated with the help of the steepest descent method

$$B_c \sim_{T \to 0} \frac{S(0) - 1}{4} + \frac{i}{4\pi} \sqrt{\frac{\pi}{\gamma} \frac{dS}{dk}} |_{k=0} . \tag{18}$$

The derivative of the scattering matrix at the origin is purely imaginary

$$\frac{dS}{dk}|_{k=0} = \Im(\frac{dS}{dk}|_{k=0})$$

and the asymptotic expansion for B_c is real. Then the asymptotics of B_c at low temperatures is given by the following expression:

$$B_c(T) \sim_{T \to 0} \frac{S(0) - 1}{4} + \sqrt{T} \frac{i}{4} \sqrt{\frac{k_B}{\pi} \frac{dS}{dk}} |_{k=0} . $$

The zero order term vanishes if there is no zero energy resonance or bound state present. In this case $S(0) = 1$ and B_c vanishes like $\sqrt{T}$. We want to emphasize that this condition coincides with the condition obtained for the usual Schroedinger equation with the potentials from the Faddeev class $\int_0^\infty |xV(x)|dx < \infty$ (see [6]). The standard zero-range potential gives a scattering matrix $S(k)$ with $S(0) = -1$ and $B_c|_{T=0}$ is not trivial in this case. Using a zero-range potential with internal structure one can consider this condition as some additional restriction on the model. It leads to the following restriction on the function $\mathcal{D}(\lambda)$

$$\mathcal{D}(0) = 0. \tag{19}$$

This condition can be easily satisfied for a special choice of the parameters.

For large values of the absolute temperature, $T \to \infty$, formula (17) can be simplified

$$B_c = \frac{i}{4\pi} \int_{-\infty}^{+\infty} dk \frac{S(k) - 1}{k}. \tag{20}$$

We notice that this integral exists iff

$$S(k) \to_{|k| \to \infty} S(0). \tag{21}$$

This condition is not satisfied for the standard zero-range potential $\mathcal{D} = const$, and it gives another restriction in our model. Combining conditions (19) and (21) we get the following restriction on the function $\mathcal{D}(\lambda)$

$$\mathcal{D}(\lambda) \to_{\lambda \to \infty} 0. \tag{22}$$

Conditions (19) and (21) define a natural vanishing of the quantum corrections after the transition to the classical case $S(k) \to_{|k| \to \infty} 1$, which takes place at high temperatures (large values of k). In this case the function $S(k)$ can be represented by a finite Blaschke product as a ratio of two rational functions

$$S(k) = \prod_{j=1}^{n} \frac{k + k_j}{k - k_j}. \tag{23}$$

Then the continuous part of the cluster integral for large temperatures is given by the following expression

$$B_c \sim_{T\to\infty} -\sum_{\Im k_j > 0} \prod_{l \neq j} \frac{k_l + k_j}{k_l - k_j}. \tag{24}$$

The formula for the low temperature asymptotics can also be simplified. Using the representation (23) the function $dS/dk\,|_{k=0}$ can be calculated explicitely

$$\frac{dS}{dk}\bigg|_{k=0} = 2\sum_{l=1}^{n} \frac{1}{k_l}.$$

Then the asymptotic expansion takes the form

$$B_c \sim_{T\to 0} \sqrt{T}\frac{\imath}{2}\sqrt{\frac{k_B}{\pi}}\sum_{l=1}^{n} \frac{1}{k_l}.$$

We have introduced two restrictions on the model during the calculation of the limits of the continuous spectrum contribution to the cluster integral. These two conditions, (19) and (22), can be satisfied only in the case of zero-range potentials with internal structure. We note that for the constructed model one can calculate the third cluster integral (see [13]).

REFERENCES

1. S. Albeverio, F. Gesztesy, R. Hoegh-Krohn and H.Holden, "Solvable Models in Quantum Mechanics," Springer-Verlag, Berlin, New-York, 1988.
2. F.A. Berezin and L.D. Faddeev, A remark on Schroedinger's equation with a singular potential, *Soviet Math. Dokl.* ,**2**, 372 (1961).
3. V.S. Buslaev, S.P. Merkuriev and S.P. Salikov, On the diffraction character of the scattering in the quantum system of three one-dimensional particles, *in:* "Problems of Mathematical Physics", **9**, Leningrad Univ. Press (in Russian) 1979.
4. Yu.N. Demkov and V.N. Ostrovskii, "Zero-Range Potentials and their Applications in Atomic Physics," Plenum Press, New-York, London, 1988.
5. E. Fermi, Sul moto dei neutroni nelle sostanze idrogenate, *Ricerca Scientifica*, **7**, 13 (1936).
6. L.D. Faddeev, Inverse problem in the quantum theory of scattering, *Uspekhi Mat. Nauk*, **14**, 57 (1959); English Translation: *J. Math. Phys.*, **4**, 73 (1963).
7. M. Gorzelanczyk, Phase transition in a gas of hard core spheres, *Commun. Math. Phys.*, **136**, 43 (1991).
8. B. Jancovici, Quantum-mechanical equation of state of a hard-sphere gas at high temperature, *Phys. Rev*, **178**, 195 (1969).
9. B. Jancovici, Quantum-mechanical equation of state of a hard-sphere gas at high temperature II, *Phys. Rev*, **184**, 119 (1969).
10. B. Jancovici and S.P. Merkuriev, Quantum-Mechanical third virial coefficient of a hard-sphere gas at high temperatures, Preprint LPTHE 75/14 (1975).
11. B. Kahn and G.E. Ulenbeck, *Physica*, **5**, 399 (1938).
12. P. Kurasov, Inverse scattering problem and zero-range potentials with internal structures, *Lett. Math. Phys.*, **25**, 287 (1992).
13. P. Kurasov, Scattering theory for three one dimensional particles, Preprint MSI 92-10, ISSN-1100-214X (1992).
14. B.S. Pavlov, A model of zero-radius potential with internal structure, *Teor. Mat. Fiz.*, **59**, 345 (1984).
15. B.S. Pavlov, The theory of extensions and explicitly-solvable models, *Russ. Math. Surv.*, **42**, 127 (1987).

SELF-ADJOINTNESS AND THE EXISTENCE OF DETERMINISTIC TRAJECTORIES IN QUANTUM THEORY

K. Berndl[1], D. Dürr[1], S. Goldstein[2] and N. Zanghì[3]

1. Mathematisches Institut der Universität München, Theresienstraße 39 80333 München, Germany

2. Department of Mathematics, Rutgers University, New Brunswick, NJ 08903, USA

3. Dipartimento di Fisica, Università di Genova, Sezione INFN Genova Via Dodecaneso 33, 16146 Genova, Italy

ABSTRACT

We show that the particle motion in Bohmian mechanics as the solution of an ordinary differential equation exists globally, i.e., the singularities of the velocity field and infinity will not be reached in finite time for typical initial configurations and a large class of potentials, including the physically most relevant potential of N-particle Coulomb interaction with arbitrary charges and masses. The analysis is based on the probabilistic significance of the quantum flux. We point to the connection between the global existence of Bohmian mechanics and the self-adjointness of the Schrödinger Hamiltonian.

INTRODUCTION

The title alludes to two of the tenets of orthodox quantum theory: 1) the self-adjointness of observables, especially of the Hamiltonian

$$H = -\sum_{k=1}^{N} \frac{\hbar^2}{2m_k}\Delta_k + V, \tag{1}$$

and 2) the impossibility of understanding quantum phenomena on the basis of an underlying theory with deterministic trajectories, with quantum randomness arising "classically" from randomness in the initial conditions.

David Bohm [1] showed several decades ago that the second tenet is false: His theory, *Bohmian mechanics*, does precisely what has been held impossible. Bohmian mechanics is a Galilean and time-reversal invariant theory for the motion of point particles. The *state* of an N-particle system is given by the configuration $Q = (\mathbf{Q}_1, \ldots, \mathbf{Q}_N) \in \mathbb{R}^{3N}$ and the wave function ψ on configuration space $\mathbb{R}^{3N}$. $\mathbf{Q}_k \in \mathbb{R}^3$ is the position of the k-th particle. On the subset of $\mathbb{R}^{3N}$ where $\psi \neq 0$ and is differentiable, ψ generates a velocity field $v^\psi = (\mathbf{v}_1^\psi, \ldots, \mathbf{v}_N^\psi)$

$$\mathbf{v}_k^\psi = \frac{\hbar}{m_k} \mathrm{Im} \frac{\nabla_k \psi}{\psi} \qquad (2)$$

determining the motion of particles with masses $m_1, \ldots, m_N$. The *time evolution* of the state (Q_t, ψ_t) is given by a first-order ordinary differential equation for the configuration Q_t

$$\frac{dQ_t}{dt} = v^{\psi_t}(Q_t) \qquad (3)$$

and Schrödinger's equation for the wave function ψ_t

$$i\hbar \frac{\partial \psi_t(q)}{\partial t} = \left(-\sum_{k=1}^N \frac{\hbar^2}{2m_k} \Delta_k + V(q) \right) \psi_t(q). \qquad (4)$$

We shall assume that the potential V is a C^∞-function on an open set $\Omega \subset \mathbb{R}^{3N}$, and the set of singularities $\mathcal{S}(= \mathbb{R}^{3N} \setminus \Omega)$ of the potential is a set of Lebesgue measure zero.

Bohmian mechanics can be regarded as a (completion of) nonrelativistic quantum theory. It resolves all problems associated with the measurement problem in nonrelativistic quantum mechanics [1, 2]. It accounts for the "collapse" of the wave function, for the quantum randomness as expressed by Born's law $\rho = |\psi|^2$, and familiar (macroscopic) reality. Moreover, the usual quantum measurement formalism involving self-adjoint operators as observables emerges from Bohmian mechanics as a phenomenological description. (See [2, 3], and the contribution of M. Daumer et al. in this volume.)

Here we are concerned with the mathematical problem of existence and uniqueness of the motion in Bohmian mechanics, i.e., with establishing that for given Q_0 and ψ_0 at some "initial" time t_0 ($t_0 = 0$), solutions (Q_t, ψ_t) of (3, 4) with $Q_{t_0} = Q_0$ and $\psi_{t_0} = \psi_0$ exist uniquely and globally in time. This problem initiates further analysis on the status of the self-adjointness of the Hamiltonian.

GLOBAL EXISTENCE OF BOHMIAN MECHANICS

The solution of the problem of the existence of dynamics for Schrödinger's equation (4) (which is independent of the solution of equation (3), i.e., the actual motion of the configuration) is, of course, well known: if the Hamiltonian is self-adjoint (on a domain $\mathcal{D}(H)$), there exists a unitary one parameter group $U_t = e^{-itH/\hbar}$, and, for all $\psi_0 \in \mathcal{D}(H)$, $\psi_t := U_t \psi_0$ is a solution of Schrödinger's equation in the L^2-sense.

Thus we assume, as usual, the Hamiltonian to be a self-adjoint extension of $H|_{C_0^\infty(\Omega)}$. (Under the above assumptions on the potential V, the set $C_0^\infty(\Omega)$ is dense in $L^2(\mathbb{R}^{3N})$, and the Hamiltonian as defined by (1) is symmetric on this set. Since the Hamiltonian commutes with complex conjugation, there always exist self-adjoint extensions of $H|_{C_0^\infty(\Omega)}$.)

In order for equation (3) to be well-defined we need the wave function to be smooth. A particular suitable set of wave functions is the set $C^\infty(H)$ of C^∞-vectors of H.[1] This set of wave functions is dense in $L^2(\mathbb{R}^{3N})$ and invariant under the time evolution $e^{-itH/\hbar}$ (and is hence a core, i.e., a domain of essential self-adjointness for H).

We shall therefore assume that the initial wave function $\psi_0 \in C^\infty(H)$. Then $\psi_t = e^{-itH/\hbar}\psi_0$ may be regarded as an element of $C^\infty(\Omega \times \mathbb{R})$ and thus also as a classical solution of Schrödinger's equation [4]. The velocity field v^ψ is thus C^∞ on the complement of the set $\mathcal{N} := \{(q,t) \in \Omega \times \mathbb{R} : \psi_t(q) = 0\}$ of nodes of ψ, and the (space-time) set of singularities $\mathcal{S} \times \mathbb{R}$, i.e., on the set of "good" points $\mathcal{G} := (\Omega \times \mathbb{R}) \setminus \mathcal{N}$, which is an open subset of configuration-space-time $\mathbb{R}^{3N} \times \mathbb{R}$. Let $\mathcal{G}_t$ denote the slice of $\mathcal{G}$ at a fixed time t: $\mathcal{G}_t := \Omega \setminus \mathcal{N}_t$, where $\mathcal{N}_t := \{q \in \Omega : \psi_t(q) = 0\}$. Then by a standard theorem of existence and uniqueness of ordinary differential equations, for any initial value q_0 in $\mathcal{G}_0$ there exists a unique right maximal (non-extendible in positive time direction[2]) solution Q_t of (3) on a time interval $0 \leq t < \tau(q_0)$. The problem is to exclude that $\tau(q_0) < \infty$, in which case the solution Q_t as $t \nearrow \tau$ reaches infinity or points in the boundary of $\mathcal{G}$, i.e., singularities of the velocity field v^ψ.

For large classes of potentials, we shall show that for typical q_0 we have that $\tau(q_0) = \infty$, i.e., that the solution exists globally in time P^{ψ_0}-almost surely,

$$P^{\psi_0}(\tau < \infty) = 0,$$

where P^{ψ_0} denotes the probability measure on configuration space $\mathbb{R}^{3N}$ (supported on $\mathcal{G}_0$) given by the density $|\psi_0|^2$. (We assume the initial wave function ψ_0 to be normalized.) This measure is the natural measure associated with the dynamical system defined by "Bohmian mechanics:" it plays the role of the "equilibrium measure" and defines our notion of "typicality" [2]. Moreover, *given* the existence of the dynamics for configurations Q_t—the result we report on here—the notion of typicality is time independent by "equivariance" [2]:

$$\rho_0 = |\psi_0|^2 \Longrightarrow \rho_t = |\psi_t|^2 \text{ for all } t \in \mathbb{R},$$

where ρ_t denotes the probability density on configuration space $\mathbb{R}^{3N}$ at time t—the image density of ρ_0 under the process Q_t.

We now state the **Theorem** we have established in [4]: *If i) the potential V is a C^∞-function on an open set $\Omega \subset \mathbb{R}^{3N}$, and the set of singularities $\mathcal{S}(= \mathbb{R}^{3N} \setminus \Omega)$ of the potential is contained in a finite union of $(3N-3)$-dimensional hyperplanes S_l, $\mathcal{S} \subset \bigcup_{l=1}^m S_l$, ii) the Hamiltonian H is a self-adjoint extension of $H|_{C_0^\infty(\Omega)}$, iii) the initial wave function ψ_0 is a C^∞-vector of H and normalized, and iv) for all $T > 0$ $\int_0^T \|\nabla \psi_t\|^2 dt < \infty$, then $P^{\psi_0}(\tau < \infty) = 0$, i.e., Bohmian mechanics exists uniquely and globally in time P^{ψ_0}-almost surely.*

We comment on the conditions in the theorem: The condition i) on the shape of $\mathcal{S}$ is very natural from a physical point of view, as it includes pair potentials and central potentials. The condition iv) of "finite integrated kinetic energy" is automatically

---

[1] $C^\infty(H) = \bigcap_{n=1}^\infty \mathcal{D}(H^n)$, where $\mathcal{D}(H^n)$ is the domain of H^n. Special C^∞-vectors are eigenfunctions and "wave packets" $\psi \in \text{Ran}(P_{[a,b]})$, where $P_{[a,b]}$ denotes the spectral projection of H to the finite energy interval $[a,b]$.

[2] From time reversal invariance, $P^{\psi_0}(\tau < \infty) = 0$ implies $P^{\psi_0}(\tau^- < \infty) = 0$, when (τ^-, τ) denotes the existence interval of the maximal solution. Therefore it is sufficient to consider the positive time direction.

satisfied for all $\psi \in \mathcal{Q}(H)(\supset C^\infty(H))$, the form domain of H, provided the quadratic form $(\nabla \psi_t, \nabla \psi_t)(\leq M(\psi_t, H_0\psi_t))$ with $M = (2/\hbar^2)\max(m_1,\ldots,m_N))$ can be bounded in terms of the form $(\psi_t, H\psi_t)$, which is finite and independent of t. Such a bound follows from various bounds on the potential: (a) Potentials which are bounded below, a class which includes, for example, harmonic and anharmonic potentials, and arbitrarily strong positive repulsive potentials. (b) H_0-form bounded potentials with relative bound $a < 1$. This class, with arbitrarily small relative bound a, includes for example $R + L^\infty$ or $L^{3/2} + L^\infty$ on $\mathbb{R}^3$, where R is the Rollnik class. (For details, see for example [5].) Therefore H_0-form bounded potentials include power law interaction $1/r^\alpha$ with $\alpha < 2$ and thus the physically most relevant potential of N-particle Coulomb interaction with arbitrary charges and masses. (The class of H_0-form bounded potentials contains the more familiar class of H_0-operator bounded potentials, which already includes the N-particle Coulomb interaction.)

We remark that also from the point of view of establishing only self-adjointness of the Hamiltonian, these classes of potentials are particularly well understood, in the sense that much is known about cores, lower bounds, etc. [5].

At this point we shall emphasize that on the one hand, this result of global existence of Bohmian mechanics should not be too surprising because the "bad sets" are either very small—the set of nodes has "generically" codimension 2, and the set of singularities has codimension at least 3—or infinitely far away. But on the other hand, especially when compared with the N-body problem in Newtonian mechanics, where general results for systems of more than 4 particles are still missing, the generality of the result of existence of dynamics for Bohmian mechanics and also the essential simplicity of the proof should be quite surprising. Taking now this result for Bohmian mechanics and also the general emergence of the quantum formalism [2, 3] into account, we maintain that the failure of Newtonian mechanics both to describe the physics in the nonrelativistic domain correctly and to be mathematically well understood on the most basic level of existence and uniqueness is not due to its having point particles as fundamental elements; rather it is due to its having the "wrong" dynamics.

BASIC IDEA OF THE PROOF: FLUX ESTIMATES

Consider the random trajectory $(\mathcal{G}_0, P = P^{\psi_0}, \tilde{Q}_t)$, where for $t \geq 0, q_0 \in \mathcal{G}_0$, the process $\tilde{Q}_t(q_0) = Q_t(q_0)$ for $t < \tau(q_0)$ and $\tilde{Q}_t(q_0) = \dagger$ for $t \geq \tau(q_0)$. The image density of $\tilde{Q}_t$ on $\mathcal{B}_t := \mathrm{Ran}\tilde{Q}_t \cap \mathcal{G}_t$ will be denoted by ρ_t. ρ_t is bounded by $|\psi_t|^2$ on $\mathcal{G}_t, t \geq 0$ [4].

Consider now a smooth surface Σ in $\mathcal{G}$. Reflecting on the probabilistic significance of the flux $J_t(q) := (\rho_t(q)v^{\psi_t}(q), \rho_t(q))$, we obtain that the expected number of crossings of Σ by the random trajectory $\tilde{Q}_t$ is given by

$$\int_\Sigma |J_t(q) \cdot U| d\sigma,$$

where U denotes the local unit normal vector. ($\int_\Sigma J \cdot U\, d\sigma$ is the expected number of *signed* crossings of Σ.) The probability of crossing Σ (at least once) is hence bounded by $\int_\Sigma |J_t(q) \cdot U| d\sigma$, which in turn from

$$|J_t(q) \cdot U| \leq |(|\psi_t(q)|^2 v^{\psi_t}(q), |\psi_t(q)|^2) \cdot U| = |(j^{\psi_t}, |\psi_t|^2) \cdot U| = |J^{\psi_t} \cdot U|$$

is bounded by

$$P^{\psi_0}(\tilde{Q}_t \text{ crosses } \Sigma) \leq \int_\Sigma |J^{\psi_t}(q) \cdot U| d\sigma.$$

This insight can be applied to prove (almost sure) global existence by choosing a sequence of surfaces around the "bad points:" consider the surface of $\mathcal{G}^{\epsilon\delta n}$, the set of "$\epsilon$-$\delta$-$n$-good" points in configuration-space-time:

$$\mathcal{G}^{\epsilon\delta n} := (\mathcal{K}^n \times \mathbb{R}) \setminus (\mathcal{N}^\epsilon \cup (\mathcal{S}^\delta \times \mathbb{R})),$$

where $(\mathcal{K}^n)_{n\in\mathbb{N}}$ is a sequence of balls with radius n in configuration space (that serves as cutoff at infinity), $\mathcal{N}^\epsilon$ and $\mathcal{S}^\delta$ are neighborhoods of $\mathcal{N}$ (in configuration-space-time) resp. $\mathcal{S}$ (in configuration space) (that take care of the nodes resp. the singularities). We further denote $\mathcal{G}_0^{\epsilon\delta n} := \mathcal{K}^n \setminus (\mathcal{N}_0^\epsilon \cup \mathcal{S}^\delta)$.

Then we arrive at: for all $T > 0, \epsilon > 0, \delta > 0, n < \infty$

$$\begin{aligned} P^{\psi_0}(\tau < T) &\leq P^{\psi_0}(\mathcal{G}_0 \setminus \mathcal{G}_0^{\epsilon\delta n}) + P^{\psi_0}(\tilde{Q}_t \text{ crosses } \partial \mathcal{G}^{\epsilon\delta n} \cap (\mathbb{R}^{3N} \times (0,T))) \\ &\leq P^{\psi_0}(\mathcal{G}_0 \setminus \mathcal{G}_0^{\epsilon\delta n}) + \int_{\partial\mathcal{N}^\epsilon\cap((\mathcal{K}^n\setminus\mathcal{S}^\delta)\times(0,T))} |J^{\psi_t}(q) \cdot U| \, d\sigma + \\ & \quad \int_{\partial\mathcal{S}^\delta\times(0,T)} |J^{\psi_t}(q) \cdot U| \, d\sigma + \int_{(\partial\mathcal{K}^n\cap\Omega)\times(0,T)} |J^{\psi_t}(q) \cdot U| \, d\sigma \\ &= P^{\psi_0}(\mathcal{G}_0 \setminus \mathcal{G}_0^{\epsilon\delta n}) + \mathbf{N} + \mathbf{S} + \mathbf{I}. \end{aligned}$$

Moreover, (almost sure) global existence follows if the right hand side goes to 0. It is heuristically now rather clear that all the flux integrals should vanish as the limit $\epsilon \to 0, \delta \to 0$, and $n \to \infty$ is suitably approached: For the "nodal integral" $\mathbf{N}$ it seems fairly obvious that this term vanishes as $\epsilon \to 0$ because J^{ψ_t} is zero at the nodes, and, moreover, one expects that $\mathcal{N}$ has codimension 2, so $\partial\mathcal{N}^\epsilon$ should have small area. The "singularity integral" $\mathbf{S}$ should vanish in the limit $\delta \to 0$ since the set $\mathcal{S}$ of singular points of the potential has codimension greater than 1 for potentials that are normally considered. Finally, the "infinity integral" $\mathbf{I}$ should tend to zero as $n \to \infty$ since $\psi_t(q)$ and hence $J^{\psi_t}(q)$ should rapidly go to zero as $|q| \to \infty$.

The main difficulty in making these considerations rigorous lies in controlling the area of the surface of the nodal set and the behavior of the flux at the singularities and at infinity. This is done in [4].

SUMMARY AND PERSPECTIVE: BOHMIAN MECHANICS AND SELF-ADJOINTNESS OF THE HAMILTONIAN

In this final section we shall comment on the connection between the global existence of Bohmian mechanics and the self-adjointness of the Hamiltonian, as mediated by the quantum flux $J = (j, |\psi|^2)$:

Bohmian mechanics $\overset{1}{\longleftrightarrow}$ J—the quantum flux $\overset{2}{\longleftrightarrow}$ self-adjointness of H

"$\overset{1}{\longrightarrow}$" Bohmian mechanics gives meaning to the quantum flux as a flux of particles. Indeed, the quantum current j is a current of particles moving along deterministic trajectories with a velocity given by a functional of the wave function ψ.

"$\overset{2}{\longrightarrow}$" In standard quantum theory, the unitarity of the time evolution U_t of ψ or equivalently the self-adjointness of its generator H is taken as one of the *axioms*. For a concrete physical problem, the generator is given by (1). There is, however, no general rule yielding the *domain* on which this operator should be considered. One first has to find a dense set of vectors where the Hamiltonian is definable by (1) and is symmetric. $C_0^\infty(\mathbb{R}^{3N})$, $C_0^\infty(\Omega)$, or Schwartz space are good and usual candidates. Then one has to analyse whether the Hamiltonian is essentially self-adjoint on that domain, so that the unitary time evolution is uniquely determined. If this is not the case, there are (infinitely many) different self-adjoint extensions, giving rise to different unitary evolutions. It is now a matter of the physics of the system being described to choose the right one. As an example, consider a free particle on $(0,\infty)$: there is a one-parameter family of self-adjoint extensions H^a of $H_0|_{C_0^\infty(0,\infty)}$, characterized by the boundary condition $\psi'(0) = a\psi(0)$ with $a \in \mathbb{R}$ or $\psi(0) = 0$ ("$a = \infty$") defining the respective domain. a determines the law of reflection of the ψ-function at 0 [5].

In Bohmian mechanics, there is *a priori* no reason to demand self-adjointness of the Hamiltonian: Any solution ψ of Schrödinger's equation (4) for which global trajectories Q_t solving (3) exist is fine. Rather, the axiom of self-adjointness experiences an a posteriori justification in Bohmian mechanics. Consider the above mentioned example of a free particle on $(0,\infty)$: To have the Bohmian particle motion well defined on $(0,\infty)$ it seems inevitable to demand $j_t(0) = 0$, i.e., $\psi_t(0) = 0$ or $v^{\psi_t}(0) = 0$ which in view of (2) yields $\frac{\nabla\psi_t(0)}{\psi_t(0)} \in \mathbb{R}$. In this way, by immediately suggesting $j(0) = 0$, Bohmian mechanics leads directly to the necessary boundary condition for self-adjointness in terms of the current. For a more detailed discussion see [4].

"$\xleftarrow{2}$" We have proven that self-adjointness of the Hamiltonian (together with "finite integrated kinetic energy") guarantees the right behavior of the quantum flux at the bad points—"No flux into the bad points"—and thereby

"$\xleftarrow{1}$" global existence of Bohmian mechanics.

Acknowledgments

This work was partially supported by DFG, by NSF Grant No. DMS-9105661 and No. DMS-9305930, and by INFN.

REFERENCES

1. D. Bohm, A suggested interpretation of the quantum theory in terms of "hidden" variables, I and II, *Phys. Rev.* **85**, 166 and 180 (1952).
2. D. Dürr, S. Goldstein, and N. Zanghì, Quantum equilibrium and the origin of absolute uncertainty, *J. Stat. Phys.* **67**, 843 (1992).
3. M. Daumer, D. Dürr, S. Goldstein, and N. Zanghì, On the role of operators in quantum theory, in preparation.
4. K. Berndl, D. Dürr, S. Goldstein, G. Peruzzi, and N. Zanghì, On the global existence of Bohmian mechanics, in preparation.
5. M. Reed, B. Simon, "Methods of Modern Mathematical Physics II," Academic Press, San Diego (1975).

AN APPLICATION OF THE MAES–SHLOSMAN CONSTRUCTIVE CRITERIA

H. de Jong

Institute for Theoretical Physics, University of Groningen
Nijenborgh 4, 9747 AG Groningen, The Netherlands

INTRODUCTION

The Maes–Shlosman constructive criteria for probabilistic cellular automata can be used to improve on the $M < \epsilon$ criterion for ergodicity of continuous time interacting particle systems. We illustrate this for the asymmetric contact process which is computer–assistedly proven to be ergodic for $M < 1.5\epsilon$.

NOTATION

We restrict ourselves to the one–dimensional lattice Z. Attach Ising spins $\sigma(x)$ to all sites $x \in Z$. A configuration is a complete characterization of all these spins. We equip the configuration space Ω with its usual product topology.

$C(\Omega)$ is the set of real–valued continuous functions f on Ω. The set of probability measures μ on Ω is denoted by $E(\Omega)$.

σ^x is the configuration that differs from σ only at x. $\delta_x f$ is the maximal variation of f due to a spin–flip at x:

$$\delta_x f = \sup_{\sigma \in \Omega} |f(\sigma^x) - f(\sigma)|. \tag{1}$$

The total oscillation $|||f|||$ of f is

$$|||f||| = \sum_{x \in Z} \delta_x f. \tag{2}$$

We define $D(\Omega)$ to be the set of continuous functions with finite oscillation. Since $D(\Omega)$ contains the local functions, it is dense in $C(\Omega)$.

WHAT IS A PROBABILISTIC CELLULAR AUTOMATON?

A probabilistic cellular automaton (PCA, from now on) is a discrete–time Markov process $\{\sigma_n\}_{n=0,1,\ldots}$ on Ω. It is characterized by transition probabilities $\{p_x(\pm|\cdot)\}_{x\in Z}$. $p_x(\pm|\tau)$ is local in the configuration τ and usually invariant under equal translations of x and τ. $p_x(+|\tau)$ has to be interpreted as the probability that the spin at site x will be $+$ at the next time, given that the configuration is τ:

$$\text{Prob}[\sigma_{n+1}(x) = +|\sigma_n = \tau] = p_x(+|\tau). \tag{3}$$

All spins are simultaneously and independently updated.

In this way, the transition probabilities $\{p_x(\pm|\cdot)\}_{x\in Z}$ define a map from the configurations at time n to the probability measures at time $n+1$ which we denote by $P(d\sigma|\tau)$. This kernel induces a map from $C(\Omega)$ to $C(\Omega)$:

$$(Pf)(\cdot) = \int f(\sigma) P(d\sigma|\cdot), \tag{4}$$

and another one from $E(\Omega)$ to $E(\Omega)$:

$$(\mu P)(f) = \mu(Pf). \tag{5}$$

WHAT IS ERGODICITY?

We call a PCA ergodic iff there is a probability measure ν such that we have for all $\mu \in E(\Omega)$

$$\mu P^N \longrightarrow \nu \text{ weakly, as } N \longrightarrow \infty. \tag{6}$$

Hence, in the limit of large times, the state of an ergodic PCA is independent of the initial configuration.

THE MAES–SHLOSMAN CONSTRUCTIVE CRITERIA FOR PCA

The idea of the Maes–Shlosman constructive criteria (see [1, 2, 3]) is as follows: if for some $n \in N$ the spin $\sigma_n(0)$ depends weakly enough on the initial configuration σ_0 under the PCA evolution, then the PCA is ergodic.

We thus need a way to compare evolutions started with different initial configurations.

To this end, we introduce the basic coupling (see [4] for an extensive discussion). This is a PCA $\boldsymbol{P}$ on the space $\Omega \times \Omega$ with properties

- The action of $\boldsymbol{P}$ reduces to that of P (the original PCA) on both copies of the space Ω.

- $\boldsymbol{P}(d\sigma, d\bar{\sigma}|\tau, \bar{\tau})$ tries to make σ and $\bar{\sigma}$ as equal as possible.

The above is achieved by the following conditions

- $\sum_{\sigma(x)=\pm} \boldsymbol{P}_x(\sigma(x), \bar{\sigma}(x)|\tau, \bar{\tau}) = p_x(\bar{\sigma}(x)|\bar{\tau})$. The other summation goes analogous.

- Maximize $p_x(+,+|\tau,\bar\tau)$ and $p_x(-,-|\tau,\bar\tau)$ under the above condition. This can be done by choosing $p_x(+,+|\tau,\bar\tau) = \min_{\tau,\bar\tau}\{p_x(+|\tau), p_x(+|\bar\tau)\}$ and analogously for $p_x(-,-|\tau,\bar\tau)$.

We now have a good way of estimating the influence of $\sigma_0(x)$ at $\sigma_N(0)$ under evolution of the basic coupling:

$$\kappa_x^N := \sup_{\sigma \in \Omega} \mathrm{Prob}[\sigma_N(0) \neq \bar\sigma_N(0) | (\sigma_0, \bar\sigma_0) = (\sigma, \sigma^x)], \tag{7}$$

The probability that $\sigma_N(0) \neq \bar\sigma_N(0)$ given any pair of initial configurations is thus at most

$$\gamma_N := \sum_{x \in Z} \kappa_x^N. \tag{8}$$

We now come to a statement of the Maes–Shlosman constructive criteria for PCA:

Theorem 1 *If there is an $N = 0, 1, \ldots$ such that $\gamma_N < 1$, then the PCA is uniformly exponentially ergodic: there is a $\nu \in E(\Omega)$ such that for all $f \in D(\Omega)$:*

$$\|P^n f - \nu(f)\| \leq c_N \gamma_N^{n/N} \|\|f\|\|, \tag{9}$$

for all positive integers n, with c_N a positive constant.

THE DISCRETE–TIME ASYMMETRIC CONTACT PROCESS

The discrete–time asymmetric contact process has transition probabilities

$$p_x(+|\sigma(x-1), \sigma(x)) = \begin{cases} 1 & \text{if } \sigma(x-1) = \sigma(x) = +, \\ 1 - \delta M & \text{if } \sigma(x-1) = - \text{ and } \sigma(x) = +, \\ \delta\epsilon & \text{otherwise.} \end{cases} \tag{10}$$

δ is small and we always take $1 - \delta M \geq \delta\epsilon$. The all $+$ configuration is invariant under application of the PCA–rule. If $M < \epsilon$, then it is easy to see that the measure on the configurations tends to the δ-measure on the all $+$ configuration after infinitely many applications of the PCA–rule, independently of the initial configuration. On the other hand, you can use Toom's argument to show that if $1 - \delta M$ is small enough, a non–zero–density of '$-$' will survive forever, if you start with the all '$-$' configuration. It is easy to show that for any δ and ϵ there is an M separating ergodic and non–ergodic behavior.

It is well–known that for attractive PCA ($\sigma \leq \tau \Rightarrow p_x(+|\sigma) \leq p_x(+|\tau)$) the expression for κ_x^N simplifies to

$$\kappa_x^N = \sup_{\sigma \in \Omega} |\mathrm{Prob}[\sigma_N(0) = +|\sigma_0 = \sigma] - \mathrm{Prob}[\bar\sigma_N(0) = +|\bar\sigma_0 = \sigma^x]|. \tag{11}$$

For the asymmetric contact process the sup in the expression for κ_x^N is reached for $\sigma \equiv +$ (see [5]).

WHAT IS A CONTINUOUS–TIME INTERACTING PARTICLE SYSTEM?

A continuous–time interacting particle system (from now on abbreviated to IPS) is a continuous–time Markov process $\{\sigma_t\}_{t>0}$ on Ω. It is characterized by transition rates $\{c_x(\sigma)\}_{x \in Z}$ which are translation invariant, local, bounded and non–negative. $c_x(\sigma)$ is interpreted via the following formula:

$$\text{Prob}[\sigma_t(x) \neq \sigma_0(x) | \sigma_0 = \sigma] = tc_x(\sigma) + o(t). \tag{12}$$

Ergodicity is defined analogously as for PCA in 6. It is well–known that an IPS is ergodic if $M < \epsilon$ (see [6]), with M and ϵ defined by

$$M = \sum_{x \neq 0} \sup_{\sigma \in \Omega} |c_0(\sigma^x) - c_0(\sigma)|, \tag{13}$$

$$\epsilon = \inf_{\sigma \in \Omega} |c_0(\sigma^0) + c_0(\sigma)|. \tag{14}$$

In the spirit of the Maes–Shlosman constructive criteria for PCA one would like to calculate quantities like κ_x^t and γ_t defined analogously to the discrete–time quantities in (7) and (8), and check whether $\gamma_t < 1$ for some $t > 0$. A complication is that in continuous time a spin can flip infinitely often in any time interval. One can get around this problem by Steif's idea (see [7]) to approximate an IPS with spin–flip rates $\{c_x(\cdot)\}_{x \in Z}$ by a PCA with rates

$$p_x^{(\delta)}(+|\sigma) = \begin{cases} \delta c_x(\sigma) & \text{if } \sigma(x) = -, \\ 1 - \delta c_x(\sigma) & \text{if } \sigma(x) = +. \end{cases} \tag{15}$$

δ is the step size of the approximating PCA.

The relation between IPS and their approximating PCA has been investigated in [2] and [3]. Some results are:

- An approximation of κ_x^t of the IPS by $\kappa_x^{(\delta)N}$ of the approximating PCA.

- The influence of $\sigma_0(x)$ on $\sigma_t(0)$ decays rapidly for x far from 0.

The strategy is now to approximate κ_x^t by its PCA counterpart for x inside some neighborhood U of 0, yielding an error E_1, and to approximate it by 0 for x outside U, yielding an error E_2. See [2] for explicit expressions.

Theorem 2 *Given an IPS, if there exist N, δ and U such that for the approximating PCA we have*

$$\sum_{x \in U} \kappa_x^{(\delta)N} + E_1 + E_2 < 1, \tag{16}$$

then the IPS is uniformly exponentially ergodic.

THE CONTINUOUS–TIME ASYMMETRIC CONTACT PROCESS

The continuous–time asymmetric contact process is defined by

$$c_x(\sigma) = \begin{cases} 0 & \text{if } \sigma(x-1) = \sigma(x) = +, \\ M & \text{if } \sigma(x-1) = - \text{ and } \sigma(x) = +, \\ \epsilon & \text{if } \sigma(x) = -. \end{cases} \tag{17}$$

It is easy to check that M and ϵ have the same meaning as in the $M - \epsilon$ criterion, see (13) and (14). Note that the discrete–time asymmetric contact process, defined in (10) is the approximating PCA.

APPLICATIONS

The $M < \epsilon$ criterion is an easy consequence of Theorem 2. It is easy to check that

$$\gamma_1^{(\delta)} = 1 - \delta(M - \epsilon). \tag{18}$$

Since $E_1 + E_2$ is $o(\delta)$ (see [2]), Theorem 2 directly yields the $M < \epsilon$ criterion if you take δ sufficiently small. This correspondence between the $M < \epsilon$ criterion and the $\gamma_1 < 1$ criterion for the approximating PCA is generic.

For better criteria we have to apply higher order constructive criteria to the approximating PCA. In order to let these be effective, we may not take δ too small because the terms in $\gamma_N^{(\delta)}$ responsible for making the $\gamma_N^{(\delta)} < 1$ criterion better than the $\gamma_1^{(\delta)} < 1$ criterion would vanish then. We thus have to deal with positive error terms E_1 and E_2. It turns out that one has to consider quite high N in order to let $\sum_{x \in U} \kappa_x^{(\delta)N} + E_1 + E_2 < 1$ when $M \geq \epsilon$.

Keeping the simplifications in mind, it is easy to see that for the asymmetric contact process $\sum_{x \in U} \kappa_x^{(\delta)N}$ equals the expectation of the number of '$-$' in $-U$ after applying the approximating PCA N times on the initial configuration that is everywhere $+$ except at 0.

We calculated the above expectation value for various N and U with the aid of a computer; some relevant results are:

ϵ	δ	M	N	U	$1 - (\sum_{x \in U} \kappa_x^{(\delta)N} + E_1 + E_2)$
1	3.5×10^{-3}	1.066	10^2	$\{-3, \ldots, 0\}$	4.6×10^{-4}
1	1.1×10^{-3}	1.28	10^3	$\{-5, \ldots, 0\}$	4.8×10^{-3}
1	1.65×10^{-4}	1.5	10^4	$\{-8, \ldots, 0\}$	3.8×10^{-3}

(19)

(The computation for $N = 10^4$ took a few hours on a Sun SPARC station 1+.) We conclude that the continuous–time asymmetric contact process is ergodic for $M < 1.5\epsilon$. We hope that this example makes clear how to apply the Maes–Shlosman constructive criteria to a wide class of IPS.

Acknowledgments

This article concerns work done in collaboration with Christian Maes, a more detailed report on this work will be published elsewhere (see [5]). I like to thank A.C.D. van Enter, N.W. Schellingerhout and M. Winnink for useful discussions. I am grateful to the Institute for Theoretical Physics in Leuven for warm hospitality and to the ERASMUS-fund for financial support of this visit.

REFERENCES

1. C. Maes and S.B. Shlosman, Ergodicity of probabilistic cellular automata: A constructive criterion, *Commun. Math. Phys.*. **135**, 233 (1991).
2. C. Maes and S.B. Shlosman, When is an interacting particle system ergodic? *Commun. Math. Phys.*, **151**, 447 (1993).

3. C. Maes and S.B. Shlosman, Constructive criteria for the ergodicity of interacting particle systems, *in:* "Cellular Automata and Cooperative Systems", 451 Kluwer Academic Publishers 1993.
4. C. Maes, A note on using the basic coupling in interacting particle systems, Preprint KUL–TF–92/9.
5. H. de Jong and C. Maes, Extended application of constructive criteria for interacting particle systems, to be published.
6. T.M. Liggett, "Interacting Particle Systems", Springer–Verlag, New York 1985.
7. J.E. Steif, "The Ergodic Structure of Interacting Particle Systems", Stanford Mathematics Department Ph.D. Thesis (1988).

BROWNIAN TRAPPING WITH GROUPED TRAPS

L.V. Bogachev[1], A.M. Berezhkovskii[2] and Yu.A. Makhnovskii[3]

[1] Faculty of Mechanics and Mathematics
  Moscow State University
  119 899 Moscow, Russia

[2] Karpov Institute of Physical Chemistry
  103 064 Moscow, Russia

[3] Institute of Petrochemical Synthesis
  117 912 Moscow, Russia

The problem of Brownian particle trapping by randomly distributed static traps is well known in the theory of disordered media. Refs. 1-3 provide us with a commonly accepted framework for the description of a wide range of processes in physics and physical chemistry.[4] The conventional theory dated back to Smoluchowski is essentially based on the assumption that traps are spatially non-correlated. However, there are widespread situations where trap correlations are *a priori* known to be present, either due to trap interaction or being generated in the course of the medium formation. A few attempts to take correlation effects into account were made in the recent years. Refs. 5-13 indicate that trap correlations may affect the trapping rate substantially. In particular, it was argued that correlations due to trap attraction (repulsion) induce the process slowdown (respectively, acceleration) as compared to the case of non-correlated traps.[8–12]

In the present communication, we consider the trapping problem in a random medium that is microscopically inhomogeneous, in the sense that traps are correlated in a "selective" fashion manifested in the formation of trap "clouds". Correlations of such a type may arise, for instance, if traps are attached to a certain supporting object (e.g., a polymer chain) or as a result of the trap generation (e.g., via radiation damage). The model under study, proposed in ref. 14 in a simplified case of clustered traps, is specified by the assumptions that the clouds are statistically identical, independent of each other, and distributed in space in a uniform, non-correlated fashion. We show that such a trap ensemble is positively correlated (which indicates an attractive character of the grouping correlations) and establish the effect of the trapping slowdown at all times. We also derive the long-time asymptotics of Donsker–Varadhan type [15,16].

Denote by $B = (B_t,\ t \geq 0)$ a standard Brownian motion in $\mathbb{R}^d$, $d \geq 1$, and let $\langle \cdot \rangle$ stand for the expectation with respect to the corresponding Wiener measure. A random medium is formed of spherical traps of radius b centered at the points $\{T_\alpha\}$ of a random point process governed by a probability law $\mathbb{P}$ independent of B. The quantity of primary interest in the trapping problem is the survival probability $P(t)$, that is, the probability for the Brownian particle to be untrapped up to time t:

$$P(t) = \left\langle \mathbb{P}\left(\min_{s\leq t} \min_\alpha \|B_s - T_\alpha\| > b\right)\right\rangle = \langle h(S_b(t))\rangle, \qquad (1)$$

where $S_b(t)$ is the Wiener sausage (i.e., the b-neighborhood of the Brownian path) and $h(S)$ is the emptiness probability, $h(S) = \mathbb{P}\,(S \cap \{T_\alpha\} = \emptyset)$. If traps are non-correlated, that is, $\mathbb{P}$ is a Poisson law (with intensity c), the emptiness probability is explicitly given by

$$h_{nc}(S;c) = \exp\left(-c|S|\right), \qquad (2)$$

where $|S|$ is the volume of S, and hence from Eq. (1) one has

$$P_{nc}(t;c) = \langle \exp\left(-c|S_b(t)|\right)\rangle \qquad (3)$$

(the subscript nc refers to the case of non-correlated traps).

The model with grouped traps is set via a *cluster* Poisson law $\mathbb{P}$.[17] More specifically, suppose that each trap is ascribed to a certain group called "cloud", so that the trap ensemble $\{T_\alpha\}$ can be presented in the form $\{X_i + Y_j^{(i)},\ j = 1,\ldots,\nu_i\}$, where the points $\{X_i\}$ interpreted as cloud "centers" are generated by a Poisson law with intensity c_0, and the aggregates of random vectors $\{Y_1^{(i)},\ldots,Y_{\nu_i}^{(i)}\}$ determining the structure of the clouds are assumed to be independent of each other and of $\{X_i\}$ and identically distributed. Note that the total trap concentration for such a trap ensemble is $c = c_0\,\bar{\nu}$, where $\bar{\nu} = \mathbb{E}\nu < \infty$ is the average number of traps in each cloud. It is also worth noting that in the particular case $\nu \equiv 1$ our model obviously reduces to the non-correlated case since the Poisson ensemble is invariant under independent shifts.

To find the emptiness probability $h(S)$, it is convenient to introduce an auxiliary box Ω and calculate first a "cut-off" emptiness probability taking into account only the clouds with $X_i \in \Omega$. Passing then to the limit $\Omega \uparrow \mathbb{R}^d$, we obtain[14]

$$h(S) = \exp\left\{-c_0 \int \mathbb{P}\left(\bigcup_{j=1}^{\nu}\{x + Y_j \in S\}\right) dx\right\}. \qquad (4)$$

It can be verified that the integral (4) is in fact equal to the volume of the region $S^* = S^*(Y_1,\ldots,Y_\nu) = \bigcup_{j=1}^{\nu}(S - Y_j)$ averaged over $\{Y_j\}$. Hence

$$h(S) = \exp\left(-c_0\,\mathbb{E}\,|S^*|\right). \qquad (5)$$

So, from Eqs. (1), (4) and (5) we obtain a representation for the survival probability similar to Eq. (3):

$$P(t) = \langle \exp\left(-c_0\,\mathbb{E}\,|S_b^*(t)|\right)\rangle, \qquad (6)$$

where $S_b^*(t) = \bigcup_{j=1}^{\nu}(S_b(t) - Y_j)$ is the "bunch" of ν copies of the Wiener sausage $S_b(t)$ determined by the intracloud trap configuration $\{Y_j\}$. The presence in Eq. (6) of the concentration c_0 in place of the total concentration c is obviously related to the fact that in our model we have the "ideal gas" of clouds instead of the ideal gas of traps implied in Eq. (3).

From the definition of S^*, it is clear that $|S| \leq |S^*| \leq \nu|S|$, whence

$$h_{nc}(S;c_0) \geq h(S) \geq h_{nc}(S;c) \qquad (7)$$

and, in view of Eqs. (1)–(3),

$$P_{nc}(t;c_0) \geq P(t) \geq P_{nc}(t;c). \qquad (8)$$

In fact, the right-hand side inequality is strict for all $t>0$ because of possible intersections of the Wiener sausages in the bunch $S_b^*(t)$. Thus, we arrive at a conclusion that, regardless of the cloud structure, trap grouping leads to the slowdown of the trapping process in comparison with the case of non-correlated traps.

It is worth mentioning that a cluster Poisson point process can be shown to be positively correlated which implies the presence of effective attraction. One of the most perceptible forms of this property is as follows. Along with the emptiness probability $h(S)$, consider the function $h(S|x_0)$ with $x_0 \in S$, defined as the conditional emptiness probability with a "probe" trap at x_0. Then, similar to the derivation of Eq. (4), one can find $h(S|x_0)$ and show that $h(S) > h(S|x_0)$ for each set S such that $|S| > 0$ (note that in the non-correlated case, $h(S) \equiv h(S|x_0)$). In that sense, the slowdown effect is in agreement with a general conclusion of our previous papers[10-12] about the influence of trap attraction on the trapping rate.

Now let us discuss the limiting behavior of the survival probability (6) as $t \to \infty$. If traps are non-correlated, the celebrated theorem by Donsker and Varadhan[15] (see also Ref. 16) asserts that

$$\lim_{t \to \infty} t^{-d/(d+2)} \log P_{nc}(t;c) = -\gamma_d c^{2/(d+2)}, \tag{9}$$

where $\gamma_d > 0$ is a numerical constant depending on the space dimension, specified as

$$\gamma_d = \omega_d^{2/(d+2)} (1 + d/2) (2\lambda_d/d)^{d/(d+2)}, \tag{10}$$

ω_d being the volume of the unit ball $V_1 \subset \mathbb{R}^d$ and $\lambda_d > 0$ the principal eigenvalue of the operator $-\frac{1}{2}\Delta$ in such a ball with the Dirichlet boundary conditions. We claim that in the case of grouped traps the result (9)–(10) holds up to the replacement of c with c_0:

$$\lim_{t \to \infty} t^{-d/(d+2)} \log P(t) = \lim_{t \to \infty} t^{-d/(d+2)} \log P_{nc}(t;c_0) = -\gamma_d c_0^{2/(d+2)}. \tag{11}$$

In order to prove (11), we use advantage of the fact that, due to a proper *upper* estimate on $P(t)$ given by Eq. (8), it suffices to obtain a consistent *lower* bound. To this end, we employ a standard idea[3,6,13,15,16] assuming that a ball V_R of large radius R is free from traps and, moreover, that the Brownian particle spends all the time in that ball. In the Poissonian case, the emptiness probability $h(V_R)$ is of specific form (2); however, for the asymptotics (9)–(10) it is actually important only that (cf. Refs. 3,6,13)

$$\lim_{R \to \infty} R^{-d} \log h(V_R) = -c\omega_d. \tag{12}$$

So, the usual lower estimate will apply to our case provided that for the emptiness probability (4) the limit (12) is valid with c_0 in place of c. By the change of variables $x = Rx'$, we have

$$-\log h(V_R) = c_0 R^d \int \mathbb{P}\left(\bigcup_{j=1}^{\nu} \{x' + R^{-1}Y_j \in V_1\}\right) dx'. \tag{13}$$

Note that as $R \to \infty$, the probability in Eq. (13) tends to 1 if $\|x'\| < 1$ and 0 if $\|x'\| > 1$ (the points $\|x'\| = 1$ can be neglected). Hence, at least formally, expression (13) is equivalent to $c_0 R^d \omega_d$. Furthermore, passing to the limit can be justified via the dominated convergence theorem, and this proves our claim (11).

We would like to remark that if the cloud structure is non-random, or if the clouds are bounded, the asymptotics (11) easily follows from the results of Refs. 15, 16.

However, the case of *random unbounded* clouds does not seem to yield to an immediate application of those results.

In the context of a general fact of the trapping slowdown argued above, Eq. (11) shows that as $t \to \infty$, the effect is manifested strongly (whenever $\bar{\nu} > 1$). Our final remark is that the behavior of the process at "normal" (large, but not asymptotically large) times is discussed on physical grounds in Ref. 14. Such a preliminary analysis allows one to obtain some information on how the cloud structure may affect the process kinetics.

Acknowledgments

The research of the first author (L.V.B.) was partially supported by the Soros Foundation. His participation in the NATO Advanced Research Workshop "On Three Levels" supported in part by the Volkswagen Foundation through the GUS-Project. The work of Yu.A.M. was partially supported by the Russian Foundation for Fundamental Researches under Grant 93-03-5328.

REFERENCES

1. J.W. Haus and K.W. Kehr, Diffusion in regular and disordered lattices, *Phys. Rep.*, **150**, 263 (1987).
2. S. Havlin and D. Ben-Avraham, Diffusion in disordered media, *Adv. Phys.*, **36**, 695 (1987).
3. F. den Hollander and G.H. Weiss, "Aspects of Trapping in Transport Processes", preprint No.716, University Utrecht (1992).
4. S.A. Rice, "Diffusion-Limited Reactions", Elsevier, Amsterdam (1985).
5. G.H. Weiss and S. Havlin, Trapping of random walks on a line, *J. Stat. Phys.*, **37**, 17 (1984).
6. R.F. Kayser and J.B. Hubbard, Reaction diffusion in a medium containing a random distribution of non-overlapping traps, *J. Chem. Phys.*, **80**, 1127 (1984).
7. A.R. Kerstein, Diffusion in a medium with connected traps, *Phys. Rev. B*, **32**, 3361 (1985).
8. P.M. Richards, Diffusion to non-overlapping or spatially correlated traps, *Phys. Rev. B*, **35**, 248 (1987).
9. G.S. Oshanin and S.F. Burlatsky, Reaction kinetics in polymer systems, *J. Stat. Phys.*, **65**, 1109 (1991).
10. A.M. Berezhkovskii, Yu.A. Makhnovskii, R.A. Suris, L.V. Bogachev, and S.A. Molchanov, Trap correlation influence on Brownian particle death. One-dimensional case, *Phys. Lett. A*, **161**, 114 (1991).
11. A.M. Berezhkovskii, Yu.A. Makhnovskii, R.A. Suris, L.V. Bogachev, and S.A. Molchanov, Trap correlation influence on diffusion-limited process rate, *Phys. Rev. A*, **45**, 6119 (1992).
12. A.M. Berezhkovskii, Yu.A. Makhnovskii, R.A. Suris, L.V. Bogachev, and S.A. Molchanov, Diffusion-limited reactions with correlated traps, *Chem. Phys. Lett.*, **193**, 211 (1992).
13. A.S. Sznitman, Brownian survival among Gibbsian traps, *Ann. Probab.*, **21**, 490 (1993).
14. A.M. Berezhkovskii, Yu.A. Makhnovskii, L.V. Bogachev, and S.A. Molchanov, Brownian-particle trapping by clusters of traps, *Phys. Rev. E*, **47**, 4564 (1993).
15. M.D. Donsker and S.R.S. Varadhan, Asymptotics for the Wiener sausage, *Comm. Pure Appl. Math.*, **28**, 525 (1975).
16. A.S. Sznitman, Lifschitz tail and Wiener sausage I, *J. Funct. Anal.*, **94**, 223 (1990).
17. D.J. Dadley and D. Vere-Jones, "An Introduction to the Theory of Point Processes", Springer, New York (1988).

LOCAL THERMODYNAMIC EQUILIBRIUM AND CONTINUOUS MEDIA: A PROGRAMME

H. Roos and R. N. Sen[1]

Institut für Theoretische Physik, Universität Göttingen
Bunsenstraße 9, D-37073 Göttingen
Federal Republic of Germany

MASS IN NONRELATIVISTIC CLASSICAL MECHANICS

The aim of this programme is to study the possible *states of matter* in nonrelativistic classical physics in relation to the concept of *local thermodynamic equilibrium*. Since the phrase "states of matter" has different meanings in different contexts, we shall begin by defining our terms.

We understand "matter" in nonrelativistic physics as being "that which possesses mass"; states of matter, for us, will therefore be the states in which one finds mass. Now mass, in nonrelativistic *classical* mechanics, may be regarded as a *measure* on $\mathbf{R}^3$ (see for example Roos, Sen and Steinitz 1993). Mathematically, a measure μ on $\mathbf{R}^3$ may be decomposed as

$$\mu = \mu_{pp} + \mu_{ac} + \mu_{sc}, \tag{1}$$

where μ_{pp} is a *pure point* measure, μ_{ac} is *absolutely continuous* with respect to Lebesgue measure on $\mathbf{R}^3$, and μ_{sc} is *singular continuous*. Physically, one should distinguish between

$$\mu_{pp} = \mu_{pp}^{(f)},$$

the superscript (f) meaning that the support of μ_{pp} is a finite set of points in $\mathbf{R}^3$, and

$$\mu_{pp} = \mu_{pp}^{(\infty)},$$

when the support is not a finite set of points. The measure

$$\mu^{(f)} = \mu_{pp}^{(f)}$$

---

[1]On Sabbatical leave from the Department of Mathematics and Computer Science, Ben Gurion University, P. O. Box 653, 84150 Beer Sheva, Israel

corresponds to finitely many *point-particles*, and therefore to finitely many degrees of freedom; we use it as shorthand for a classical system of finitely many particles. The measure

$$\mu^{(\infty)} = \mu_{pp}^{(\infty)} + \mu_{ac} + \mu_{sc}$$

corresponds to infinitely many degrees of freedom. For brevity, we shall call systems described by such measures "infinite systems". Here μ_{ac} stands for continuous media understood in the ordinary sense; $\mu_{pp}^{(\infty)}$ and μ_{sc} stand for other cases which may, or may not, be of physical relevance.

THE DESCRIPTION OF INFINITE SYSTEMS

In any laboratory situation, one cannot obtain more than a finite set of numbers, each with only a finite accuracy. Therefore the physical description of infinite systems might appear to be problematical, requiring as it may an infinite set of real numbers (consider a system consisting of infinitely many point-particles). What is true of particulate systems is not true of *fields*, for the latter can be determined to any given accuracy by a *finite* set of numbers — using, however, certain *mathematical hypotheses*, the simplest of which are continuity and differentiability. Experience shows that certain infinite systems can indeed be described by a set of fields which possess these mathematical properties. Given the mathematical hypotheses, description by a finite set of numbers (to a certain accuracy) — e.g., Fourier or mulitpole expansion coefficients — corresponds to a laboratory situation.

This, however, is not the only possibility. One may think of physical situations which can be adequately characterized by a finite or countable number of expectation values; the state itself need not be defined down to the last atom, e.g., only a finite number of particles may be considered individually, the rest being described by a continuous measure. A similar argument may apply to a μ_{sc} situation. If one accepts this, it would appear that the physical relevance of singular continuous media can only be tested after the corresponding mathematical hypotheses have been uncovered.

The theory of infinite systems, in short, is brought under control by shifting much of the burden to mathematical hypotheses. One is replacing the description in terms of $\mu_{pp}^{(\infty)}$ by one in terms of μ_{ac}. Or rather, one is replacing a description in terms of

$$\lim_{N \to \infty} \mu_{pp}^{(N)} \qquad (2)$$

(in a sense which has to be made precise) by one in terms of μ_{ac}.

THE EMERGENCE OF IRREVERSIBILITY

The limit (2) gives rise to a new feature, *irreversibility*, which is *not* found in systems described by $\mu_{pp}^{(N)}$ for any finite N. What is generally regarded as thermodynamic irreversibility is precisely that which is associated with this limiting process.

A few comments are in order. One is that the limit $\lim_{N\to\infty} \mu_{pp}^{(N)}$ may at times be some $\mu_{pp}^{(\infty)}$ or μ_{sc}; these cases have to be studied separately. However, for the present we are mainly interested in μ_{ac} since we suspect that local thermodynamic equilibrium may not compatible with $\mu_{pp}^{(\infty)}$ or μ_{sc}; this remains to be proven. The second is that μ_{ac} does not necessarily imply irreversibility; consider, for example, the hydrodynamic equations without viscosity, admittedly a rather special situation. Finally, it is important to note that the so-called *arrow of time* is a fairly common property of infinite systems. This, apparently, is not always realized; the confusion which followed Boltzmann's announcement of the H-theorem has not yet been fully dispelled (see, for example, the contributions in the volume edited by Gal-Or (1974); the article by Grad (1958) supplies the background, and provides a clear discussion of the Boltzmann-Grad limit). We are confining ourselves to thermodynamic irreversibility alone, and not considering cosmological or other sources of arrows of time.

THE STUDY OF CONTINUOUS MEDIA

The preceding discussion shows that, even in a purely "atomistic" context, an autonomous investigation of systems described by μ_{ac}, namely *continuous media*, is highly desirable. First of all, $\mu_{pp}^{(\infty)}$ may not correspond directly to a laboratory situation. To arrive at the commonest laboratory situations one has to replace description in terms of $\mu_{pp}^{(\infty)}$ by one in terms of μ_{ac}, plus mathematical hypotheses, which is hardly possible if the latter description is not available. In particular, the limiting processes required for the above may introduce a new feature, namely *irreversibility*. In that case (to avoid circular reasoning) this feature *must somehow be present* in the description of matter by μ_{ac}. This line of argument cannot be reconciled with the idea that, in the final analysis, everything to do with *heat* has to be explained in terms of molecular motions, because, unless B is well-defined, the statement $\lim A = B$ has no meaning.

CONTINUOUS MEDIA AND LOCAL THERMODYNAMIC EQUILIBRIUM

The limit

$$\lim_{N\to\infty} \mu_{pp}^{(N)} \to \mu_{ac}$$

is a combined hydro-thermodynamic limit. To see this, consider the phenomenological theory of hydrodynamics (see Serrin (1959)).

The five equations of hydrodynamics are:
1. The equation of continuity (conservation of mass).
2. The (three) equations of motion (conservation of linear momentum).
3. The energy equation (conservation of energy).

They involve the five *hydrodynamic fields* which are the (mass) density, the velocity fields and the total energy density. Since continuous media may admit body stresses which, in turn, may result in internal work, new fields — generally the stress tensor and the heat-flow vector — have to be introduced. As a result, the system of equations becomes *underdetermined*, and new equations have to be set up.

However, the phenomena which they have to describe are thermodynamic rather than purely mechanical, because of the possibility of gain or loss of internal energy. That is, the mechanics of continuous media has, in some sense, to be "completed" by the incorporation of (irreversible) thermodynamics.

One such programme has been carried out (mainly for viscoelastic media) by Coleman, Noll, Truesdell and others, and the resulting theory has been described by Truesdell as *rational thermodynamics* (Truesdell (1969); for a quick introduction, see Truesdell (1966b); see also Owen (1984); see Levanda (1978) for a discussion, as well as an introduction to generalized thermodynamics). These authors *postulate* the existence of a *temperature field* $T(\vec{x})$ (which we take as the definition of local thermodynamic equilibrium, hereafter abbreviated LTE). For a particulate system, the LTE condition is rather restrictive; most states of such a system are *not* LTE states. Therefore the assumption that a temperature field always exists is a strong physical assumption. It could be that in a certain class of materials the interaction which produces the condensed state is strong enough to induce local thermalization as well, and therefore rational thermodynamics is applicable to them. However, *this needs to be proven.* In any case, rational thermodynamics is not universally applicable (Its founders went to great lengths to stress this point; see Truesdell (1966a), third lecture.). We conclude that it is necessary to develop the mechanics of continuous media in a more general setting, away from local thermodynamic equilibrium, possibly by incorporating something like generalized thermodynamics, or by considering singular continuous media, neither of which has yet been done.

THE PROBLEMS

We may therefore formulate the problems as follows:

1) It is necessary to study the limiting processes which may be symbolically represented as

$$\lim_{N \to \infty} \mu_{pp}^{(N)} \to \mu_{ac}$$

and to elucidate how they differ from other possible limiting processes

$$\lim_{N \to \infty} \mu_{pp}^{(N)} \to \mu_{pp}^{(\infty)} \quad \text{or} \quad \ldots \to \mu_{sc},$$

and how, and when, irreversibility may appear. The right-hand sides of the above may not correspond to physically interesting situations, but have well-defined mathematical meanings.

2) Special attention should be paid to the case in which the limit is μ_{ac}: does the limiting procedure give any clue as to how one singles out the cases in which a local temperature can be defined? That is: if the system is coupled *locally* at a point x to a suitable heat bath at (uniform) temperature $T(x)$, and if the the limit can be performed without energy exchange between the system and the heat bath, the local temperature would be $T(x)$. Under what conditions can this limiting process be performed for all x? And does the time evolution — which is to be defined suitably in the limit — transform LTE states into LTE states?

3) It is necessary to begin an *autonomous* study of μ_{sc}, to gain some insight into the possible limiting processes

$$\lim_{N \to \infty} \mu_{pp}^{(N)} \to \mu_{sc}$$

which may give some sort of a particle interpretation to singular continuous forms of matter. This might also help in answering the previous questions.

Galilei group considerations immediately make it clear that there exists a gap in the theory of continuous media which may at times be filled by thermodynamic inputs. However, they do not provide guiding principles for finding solutions to the above problems (Roos, Sen and Steinitz 1993). A first study of the questions 2) in a quantum version has been done, but only with negative results so far (see Roos 1993).

We agree with Truesdell (op. cit.) that thermodynamics (as distinct from thermostatics) is basically the science of entropy production; but we would like to add that a new physical principle is required, firstly, to *define* the entropy production, and, secondly, to isolate the subclass of situations in which a local temperature, and thus an equation of state, can be defined (which is a basic assumption of rational thermodynamics). *Such a principle cannot be arrived at from considerations of symmetry or dynamics.*

SINGULAR CONTINUOUS MEDIA

Finally, we would like to make a few remarks about singular continuous media. Let C denote the Cantor set (in $\mathbf{R}$). Then the sets $\mathbf{R}^2 \times C$, $\mathbf{R} \times C^2$ and C^3 are all sets of zero Lebesgue measure in $\mathbf{R}^3$. A nonzero measure supported on any of these would be a singular continuous measure. One approach to the study of singular continuous mass distributions would be to define *standard measures* on these and then define fields as Radon-Nikodym derivatives of measures which are a.c. with respect to these. Whether or not this bears any significance for physics is an open question. A singular continouus measure may appear in the limiting process $\lim_{N \to \infty} \mu_{pp}^{(N)}$; it may be needed for non-LTE situations, e.g., stellar atmospheres with strong magnetic fields (Steinitz 1990).

Acknowledgements

R. N. Sen would like to thank the Heraeus-Stiftung for financial support, and the members of the Institut für Theoretische Physik, Universität Göttingen, for their hospitality.

REFERENCES

Gal-Or, B.(ed.), 1974, "Modern Developments in Thermodynamics", Keter Publishing House, Jerusalem.
Grad, H., 1958, Principles of the kinetic theory of gases, *in*: "Encyclopedia of Physics", vol. XII, S. Flügge, ed., Springer-Verlag, Berlin.
Levanda, B., 1978, "Thermodynamics of Irreversible Processes", MacMillan, London.
Owen, D. R., 1984, "A First Course in the Mathematical Foundations of Thermodynamics", Springer-Verlag, Berlin.
Roos, H., 1993, On the Problem of Defining Local Thermodynamic Equilibrium, to appear in Report of the Workshop on "Mathematical Physics Towards the 21st Century", R. N. Sen, A. Gersten, eds., Beer Sheva, 14-19 March 1993.
Roos, H., Sen, R. N., Steinitz, R., 1993, Galilei invariance, continuous media and local thermodynamic equilibrium, *in*: "Anales de Fisica, Monografias", Vol. 2 (Proceedings of the XIX International Colloquium on Group Theoretical Methods in Physics), M. A. del Olmo, M. Santander and J. Mateos, eds., MCIEMAT/RSEF, Madrid.
Serrin, J., 1959, Mathematical principles of classical fluid dynamics, *in*: "Encyclopedia of Physics", vol. VIII/1, S. Flügge, ed., Springer-Verlag, Berlin.
Steinitz, R., 1990, Diamagnetic Heating and Cooling at Stellar Surfaces, *in*: "Nonequilibrium Statistical Mechanics", E. S. Hernandez, ed., World-Scientific, Singapore, pp. 171-175.
Truesdell, C., 1966a, "Six Lectures on Modern Natural Philosophy", Springer-Verlag, Berlin.
Truesdell, C., 1966b, "The Elements of Continuum Mechanics", Springer-Verlag, Berlin.
Truesdell, C., 1969, "Rational Thermodynamics", McGraw-Hill, New York.
Truesdell, C. and Noll, W., 1965, The non-linear field theories of mechanics, *in*: "Encyclopedia of Physics", vol. III/3, S. Flügge, ed., Springer- Verlag, Berlin.
Truesdell, C. and Toupin, R. A., 1960, The classical field theories, *in*: "Encyclopedia of Physics", vol. III/1, S. Flügge, ed., Springer-Verlag, Berlin.

STOCHASTIC MODEL OF A QUANTUM DIFFUSION-REACTION PROCESS

V.P. Belavkin

University of Nottingham, NG7 2RD UK

Supported by Nottingham University Grant NLRG007

We develop a differential treatment of discontinuous unitary evolution for a system of quantum particles which are scattered by quantum "bubbles" at random instants of time. The large number and central limits of the model give macroscopic dynamics of the system in the mean field approximation and its fluctuations, described by a quantum stochastic unitary equation.

A HAMILTONIAN REACTION MODEL

Let $\mathcal{H}$ be a Hilbert space called the state space of a particle and R be a selfadjoint operator in $\mathcal{H}$, considered as a position of the particle $x \in \mathbf{R}$ A quantum apparatus with a pointer scale $\Lambda \subseteq \mathbf{R}$ is described by the Hilbert space $L^2(\Lambda)$, $f : \Lambda \to \mathbf{C}$.

A singular evolution corresponding to a scattering of the particle at the time instant $t = 0$ with its own Hamiltonian H in $\mathcal{H}$ is described in the product space $\mathcal{H}_1 = \mathcal{H} \otimes L^2(\Lambda)$ by a time-dependent Hamiltonian

$$H_1(t) = H_0 + \kappa R \otimes \delta(t) P, \tag{1}$$

given in the interaction picture with respect to to a free evolution of the measurement apparatus in $L^2(\Lambda)$. Here $H_0 = H \otimes \mathbf{1}$, $\mathbf{1}$ is the identity operator in $L^2(\Lambda)$, and P is a selfadjoint operator in $L^2(\Lambda)$, say $P = \hbar \partial / i \partial \lambda$. It generates the scattering operator

$$S_t = \exp\left(-\frac{i}{\hbar} \kappa R \otimes P_t\right) = \begin{cases} S, & t > 0 \\ I, & t \le 0 \end{cases} \tag{2}$$

in $\mathcal{H}_1$, defined by $P_t = 1_t P$, $1_t = 1$ if $t > 0$, otherwise $1_t = 0$.

The singular time dependence of the Hamiltonian (1) does not allow to define the Schrödinger equation $i\hbar d\psi/dt = H_1(t)\psi(t)$ in the usual sense. But one can define a unitary stochastic evolution $U_1(t) : \mathcal{H}_1 \to \mathcal{H}_1$ for a $t_0 < 0$ as the single-jump unitary process

$$U_1(t) = \exp\left(-\frac{i}{\hbar} \int_{t_0}^{t} H_1(s) \mathrm{d}s\right) = e^{iH_0(t_0-t)/\hbar} S_t$$

if $[R, H] = 0$. Usually it is not so, and the unitary evolution corresponding to (1) must be redefined by the solution $\psi(t) = U_1(t)\psi_0$ of a regularized wave equation. It can be done in terms of the forward differentials

$$d\psi(t) = \psi(t + dt) - \psi(t), \quad dl_t = 1_{t+dt} - 1_t,$$

defining the generalized Schrödinger equation as

$$d\psi(t) + \frac{i}{\hbar}H_0\psi(t)dt = (S - I)\psi(t)dl_t, \quad \psi(t_0) = \psi_0. \tag{3}$$

Theorem 1 *The difference equation (3) has a unique solution given for every $\psi_0 \in \mathcal{H}$, corresponding to an initial $t_0 \leq 0$, by the unitary operator $U_1(t) = U_0(t - t_0)S_t(-t_0)$, where $U_0(t) = \exp\{-iH_0t/\hbar\}$ and $S_t(r) = U_0^\dagger(r)S_tU_0(r)$.*

PROOF. To prove this we write the generalized Schrödinger equation as an integral one

$$\psi(t) = e^{-iH_0t/\hbar}\left(e^{iH_0t_0/\hbar}\psi_0 + \int_{t_0}^{t} e^{iH_0r/\hbar}(S - I)\psi(r)dl_r\right) \tag{4}$$

which is equal to the integral of (3) as one can prove by straightforward differentiation. Denoting $\psi(0) = U_0(-t_0)\psi_0$ the solution of (4) for $t = 0$ as $\int_{t_0}^{t} \psi(r)dl_r = 0$ for any $t_0 \leq 0$, it can be rewritten as

$$\psi(t) = U_0(t)(\psi(0) + (S - I)1_t\psi(0)) = U_0(t)S_t\psi(0)$$

for any $t_0 < 0$ as $\int_{t_0}^{t} \psi(r)dl_r = 1_t\psi(0)$ and $(S - I)1_t = S_t - I$ due to the definition of the scattering operator (2). Hence $U(t)$ is the unitary operator $U_0(t)S_tU_0(-t_0) = U_0(t - t_0)S_t(-t_0)$. This gives the solution of the equation (3) also for $t_0 = 0$ as $\psi(t) = U_0(t)S_t\psi(0)$ is equal ψ_0 for $t = 0$. ∎

Now we study a spontaneous process of the scattering interactions (1) of a quantum particle at random time instants $t_n > 0$; $t_1 < t_2 < \ldots$ with a renewable pointer in an apparatus like a cloud chamber with identycal bubbles serving as the meter. We consider the increasing sequences $(t_1, t_2, \ldots)$ as countable subsets $\tau \subset \mathbf{R}_+$ having only finite intersections $\tau_t = \tau \cap [0, t)$ for any $t \geq 0$ in accordance with the finiteness of the numbers of the scattered bubbles on the finite intervals $[0, t)$ of observation. The set of all such infinite τ will be denoted as Γ_∞ while Γ stands for all finite $\tau \subset \mathbf{R}_+$ as the inductive limit $\cup \Gamma_t$ at $t \to \infty$ of $\Gamma_t = \{\tau_t | \tau \in \Gamma_\infty\}$ which is the disjoint union $\Gamma_t = \sum_{n=0}^{\infty} \Gamma_t(n)$ of n-simplexes $\Gamma_t(n) = \{t_1 < \ldots < t_n\} \subset [0, t)^n$. The coordinate q_n of a bubble labeled by the scattering number $n \in \mathbf{N}$ show the position $\lambda_n \in \Lambda$ of the pointer at the time $t_n \in \tau$.

The corresponding Hamiltonian of the moving particle is given as a series

$$H(t, \tau) = H_0 + \kappa R \otimes \sum_{n=1}^{\infty} \delta(t - t_n)P(n) \tag{5}$$

having at most two nonzero summands if $t \in \tau$. Here $H_0 = H \otimes \mathbf{1}$ is the Hamiltonian describing the time evolution during the intervals between the scatterings $t \in \tau$ and $P(n)$ acts on the n-th scattered bubble say as the operator $P(n) = -i\hbar d/d\lambda_n$.

The generalized Schrödinger equation corresponding to the Hamiltonian (1) can be written for a fixed $\tau \in \Gamma_\infty$ similarly to a single-kick case

$$d\psi(t) + \frac{i}{\hbar}H_0\psi(t)dt = (S(n_t) - I)\psi(t)dn_t, \quad \psi(0, \tau) = \psi_0. \tag{6}$$

Here $S(n) = \exp\{-\frac{i}{\hbar}\kappa R \otimes P(n)\}$, $n_t(t) = |\tau_t|$ is the number process, giving the cardinality $|\tau_t| = \sum_{r \in \tau} 1_{t-r}$ of the localized subset $\tau_t = \{t_n < t\}$, such that $dn_t(t)$ is equal 1 if $t \in \tau$, otherwise it is zero.

Theorem 2 *The solution of the equation (6) is uniquely defined for every $\tau \in \Gamma_\infty$ by an initial state ψ_0 of the system as $\psi(t,\tau) = U(t,\tau)\psi_0$, where $U(t,\tau) = U_0(t)V_t^\dagger(\tau)$, and $V_t^\dagger(\tau)$ is the chronological product $\prod_{r\in\tau}^{\leftarrow} S_t(r) = S(t_{n_t})\ldots S(t_1)$,*

$$V_t(\tau) = S_t^\dagger(t_1) S_t^\dagger(t_2) \ldots = \left(\prod_{r\in\tau}^{\leftarrow} S_t(r)\right)^\dagger. \tag{7}$$

Here $S_t(t_n) = U_0^\dagger(t_n) S(n) U_0(t_n)$ for $t_n < t$, where $S(n) = \exp\{-\frac{i}{\hbar}\kappa R \otimes P(n)\}$, and $S_t(t_n) = I$ if $t_n \geq t$ such that the infinite product (7) contains just a finite number $n_t = \sum_{r\in\tau} 1_{t-r}$ of multipliers different from the identical operator I.

PROOF. To prove this we recall that the differential equation (6) is equivalent to the integral equation given by the recurrency

$$\psi(t,\tau) = e^{-iH_0 t/\hbar}\left(\psi_0 + \sum_{r\in\tau}^{r<t} e^{iH_0 r/\hbar}(S(n_r) - I)\psi(r,\tau)\right) \tag{8}$$

for every $\tau \in \Gamma_\infty$. Hence $\psi(t,\tau) = U_0(t) V_t^\dagger(\tau)\psi_0$, where $U_0(t) = e^{-iH_0 t/\hbar}$ and $V_t(\tau)$ is a solution of the operator equation

$$V_t(\tau) = I + \sum_{r\in\tau}^{r<t} V_r(t)(S_t^\dagger(r,\tau) - I), \quad V_0(\tau) = I,$$

with $S^\dagger(t,\tau) = U_0(t)^\dagger S(n_t(\tau))^\dagger U_0(t)$. But this equation has a unique solution (6) written as binomial sum

$$[L_t(t_1) + I][L_t(t_2) + I]\ldots = \sum_{\sigma \subseteq \tau_t} L(s_1,\tau)\ldots L(s_n,\tau)$$

in terms of $\sigma = \{s_1,\ldots,s_n\}$, $s_1 < \ldots < s_n$, $n \leq n_t$, $L_t(r) = S_t^\dagger(r) - I$ $(= 0$ if $r \geq t)$ and $L(r,\tau) = S^\dagger(r,\tau) - I$. Indeed, this sum contains I as the empty product, corresponding to $\sigma = \emptyset$ and the rest is equal to

$$V_t(\tau) - I = \sum_{r\in\tau}^{r<t} \sum_{\sigma \subseteq \tau_r} L(s_1,\tau)\ldots L(s_m,\tau) L(r,\tau) =$$

$$= \sum_{r\in\tau}^{r<t} V_r(\tau) L(r,\tau) = \sum_{r\in\tau}^{r<t} V_r(\tau)(S^\dagger(r,\tau) - I)$$

with $m \leq n_t - 1$. ∎

Note that the differential equation (6) depending on $\tau \in \Gamma_\infty$ as $n_t = n_t(\tau)$ is not a stochastic one until we have not fixed a probability distribution for the instants $\tau = (t_1, t_2, \ldots)$ of the spontaneous interactions. To obtain a continuous at least in the mean dynamics for such instantaneous process one may assume the probability distribution of random number process $n_t(\tau)$ is given by the Poissonian law $\pi_0(d\tau)$ on Γ_∞ as the projective limit at $t \to \infty$ of the probability measures

$$\pi_0(d\tau_t) = e^{-\nu t} \nu^{|\tau_t|} d\tau_t, \quad \nu > 0. \tag{9}$$

Here $\tau_t = \tau$ is a finite time ordered sequence $\tau(n) = (t_1,\ldots,t_n) \in \Gamma_t$ with $n = n_t$, $d\tau_t = \prod_{k=1}^{n_t} dt_k$ is the measure on Γ_t given by the sum of product measures $dt_1,\ldots,dt_n = d\tau(n)$ on the simplexes $\Gamma_t(n)$ with $d\tau(0) = 1$ on $\Gamma_t(0) = \{\emptyset\}$, such that

$$\int_{\Gamma_t} \nu^{|\tau|} d\tau := \sum_{n=0}^\infty \nu^n \int \ldots \int_{0\leq t_1<\ldots<t_n<t} dt_1\ldots dt_n = e^{\nu t}.$$

453

MACROSCOPIC AND DIFFUSION LIMITS

Now we consider the mean field approximation of the measurement apparatus fixing its total effect $\nu\kappa = -\gamma$ given by the mean number ν of the scattered bubbles per second and an interaction constant κ coupling each bubble to a particle in the Hamiltonian (5). We look for the limits of the unitary evolutions (6) under the condition $\nu \to \infty$ and $\kappa \to 0$ such that γ is a real constant. To perform these limits we need the expansions

$$S(n) = I \otimes \mathbf{1} - i\frac{\kappa}{\hbar} R \otimes P(n) - \left(\frac{1}{2}\right)\left(\frac{\kappa}{\hbar}\right)^2 (R \otimes P(n))^2 + \ldots \tag{10}$$

of the scattering operator $S(n) = \exp\{-\frac{i}{\hbar}\kappa R \otimes P(n)\}$ with respect to the coupling constant κ. The first term of the expansion for $S(n)$ is disappearing in the r.h.s. of Eq.(6) while the second and third terms are appearing as the differentials of the operator-valued stochastic integrals

$$\hat{n}_t[P] = \int_0^t P(n_r)\,dn_r, \quad \hat{n}_t[P^2] = \int_0^t P(n_r)^2\,dn_r,$$

defined for each $\tau \in \Gamma_\infty$ as the finite sums

$$\hat{n}_t[L](\tau) = \sum_{n=1}^{n_t(\tau)} L(n). \tag{11}$$

Hence the main term in the r.h.s. of eq.(6) for $\kappa \to 0$ is given by the renormalized stochastic integral

$$\hat{\lambda}(t) = \frac{1}{\nu t}\int_0^t L(n_r)\,dn_r = \frac{1}{\nu t}\hat{n}_t[L] \tag{12}$$

of the operator-valued stochastic functions $L(t,\tau) = L(n_t(\tau))$ respectively to the number-process $n_t(\tau)$ which has the Poissonian probability distribution (9) on Γ_∞.

To perform the large number limit $\nu \to \infty$ in (12) for an arbitrary operator L in $L^2(\Lambda)$ we need to use the quantum stochastic representation of the integral (12) the Fock space $\mathcal{F}$ over $L^2(\mathbf{R}_+ \times \Lambda)$. The space $\mathcal{F}$ is defined as $L^2(\Upsilon)$-space of all square-integrable functions $\varphi : \Upsilon \to \mathbf{C}$, $\|\varphi\|^2 = \int_\Upsilon |\varphi(v)|^2 \lambda(dv) < \infty$ of time ordered finite sequences $v = (y_1,\ldots,y_n)$, $y = (t,\lambda)$ identified with the subsets $v \subset \mathbf{R}_+ \times \Lambda$ of the cardinality $|v| = 0,1,2,\ldots$. The measure $\lambda(dv)$ on the union $\Upsilon = \sum_{n=0}^\infty \Upsilon(n)$ of the disjoint subsets $\Upsilon(n) = \{v \in \Upsilon \mid |v| = n\}$ is given as the sum $\lambda(A) = \sum_{n=0}^\infty \lambda(\Upsilon(n) \cap A)$ of the product $\lambda(dv) = \prod_{y \in v} dy$ of measures $dy = dt\,d\lambda$ on $\mathbf{R}_+ \times \Lambda$ such that

$$\|\varphi\|^2 = \sum_{n=0}^\infty \iint_{0 \leq t_1 < \ldots \leq t_n < \infty} |\varphi(y_1,\ldots,y_n)|^2 \prod_{i=1}^n dy_i.$$

Let us denote by $N_t[L]$ the number integral as an operator

$$(N_t[L]\varphi)(\tau,\boldsymbol{v}) = \sum_{n=1}^{n_t(\tau)} L(n)\varphi(\tau,\boldsymbol{v}) = \hat{n}_t[L](\tau)\varphi(\tau,\boldsymbol{v}) \tag{13}$$

in the Fock space $\varphi \in L^2(\Upsilon)$. It represents the operator-valued stochastic integral (11) by pointwise multiplication of the functions $\varphi(v) = \varphi(t,\boldsymbol{v})$ of $v \in \Upsilon$, by $\hat{n}_t[L](\tau)$ which is considered as the function of a finite sequence $\tau \in \Gamma$ because of its independence of $\tau_{|t} = \{t_n \geq t\}$. In order to obtain the initial probability measure $P_0(dv) = \pi(d\tau)\mu_0^\infty(d\boldsymbol{v})$ on

Υ_∞ induced by an initial Fock vector $\varphi_0 \in L^2(\Upsilon)$, we need an isomorphic transformation of (13)

$$\hat{N}_t[L] = N_t[L] + \sqrt{\nu}(A_t[f_0^\dagger L] + A_t^\dagger[Lf_0]) + \nu t f_0^\dagger L f_0 \tag{14}$$

which can be locally performed by a unitary transformation

$$\hat{N}_t[L] = U_s^\dagger N_t[L] U_s, \quad U_s = \exp\{\sqrt{\nu}(A_s^\dagger[f_0] - (A_s[f_0^\dagger])\}$$

for every $t < s$. Here $A_t^\dagger[f]$ and $A_t[f^\dagger]$ are the creation and annihilation integrals of $f \in L^2(\Lambda)$, $f^\dagger \in L^2(\Lambda)^*$, given by the operators

$$\left(A_t^\dagger[f]\varphi\right)(v) = \sum_{y \in v_t} f(\lambda)\varphi(v \setminus y)$$

$$\left(A_t[f^\dagger]\varphi\right)(v) = \int_{([0,t) \times \Lambda)} f(\lambda)^* \varphi(v \sqcup y) dy$$

in the Fock space $L^2(\Upsilon)$, where $v \setminus y$ means the sequence $v \in \Upsilon$ with a canceled $y = (r, \lambda)$, $r < t$, and $v \sqcup y$ means the seqence $v \in \Upsilon$ with an additional element $y \notin v$. The characterstic functional of the stochastic operators $\hat{n}_t[L]$ respectively to the initial state-vector $f_0^\infty \in \mathcal{E}$ and the Poissonian probability measure (9) is now given simply by the vacuum expectation:

$$\int_{\Gamma_\infty} (f^\infty, e^{i\hat{n}_t[L]} f^\infty) \pi_0(d\tau) = \langle \delta_\phi, e^{i\hat{N}_t[L]} \delta_\phi \rangle,$$

where $\delta_\phi(v) = 1$ if $v = \emptyset$ otherwise $\delta_\phi(v) = 0$.

The corresponding representation $\hat{l}(t) = \frac{1}{\nu t} \hat{N}_t[L]$ for (12) helps to obtain immediately the quantum large number limit

$$\lim_{\nu \to \infty} \frac{1}{\nu t} \hat{N}_t[L] = f_0^\dagger L f_0 \hat{1}$$

as the mean value $l_0 = (f_0, Lf_0) \equiv f_0^\dagger L f_0$ of a single-bubble operator with respect to an initial wave packet $f_0 \in L^2(\Lambda)$. This gives the following:

Theorem 3 *The macroscopic limit*

$$d\psi(t) + \frac{i}{\hbar} H_0 \psi(t) dt = \frac{i}{\hbar} \gamma(R \otimes p_0 \hat{1}) \psi(t) dt$$

of the generalized Schrödinger equation (6) turns out to be a nonsingular one in the initial Hilbert space $\mathcal{H}$ with an additional potential $-\gamma p_0 R$ corresponding to the mean momentum $p_0 = (f_0, P f_0)$ of a bubble in the initial state f_0.

Let us now pay attention to the fluctuations respectively to the obtained large number limits. Such fluctuations might appear for $-\kappa = \gamma/\nu \to 0$ in the large time scale $t \sim 1/\kappa$. We can get these fluctuations without rescaling the time t if we assume that $p_0 = 0$ and $-\kappa = \gamma/\sqrt{\nu}$, such that we have to take into account also the κ^2-terms in (10).

It follows from the Fock space representation (14) that there exists a quantum central limit

$$\lim_{\nu \to \infty} \frac{1}{\sqrt{\nu}} \hat{N}_t[L] = A_t[f_0^\dagger L] + A_t^\dagger[Lf_0]$$

455

for any single-bubble operator L with zero mean value $(f_0, Lf_0) = 0$. We apply this central limit theorem firstly to the r.h.s. of (6) represented in $\mathcal{H} \otimes \mathcal{F}$ as

$$\hat{n}_t[S-I] = i\frac{\gamma}{\hbar}\frac{1}{\sqrt{\nu}}\hat{N}_t[R \otimes P] - \frac{1}{2}\left(\frac{\gamma}{\hbar}\right)\frac{1}{\nu}\hat{N}_t[(R \otimes P)^2] + \ldots .$$

This yelds $\lim \frac{1}{\sqrt{\nu}} \hat{N}_t[P] = \hat{u}_t$, $\lim \frac{1}{\nu} \hat{N}_t[PP] = (Pf_0, Pf_0)t$, and the following:

Theorem 4 *The central limit*

$$d\psi(t) + K_0\psi(t)dt = \frac{i}{\hbar}\gamma(R \otimes d\hat{u}_t)\psi(t). \tag{15}$$

of the generalized Schrödinger equation (6) turns out to be a stochastic Schrödinger-Ito one of the diffusive type driven by a Wiener process u_t with the dispertion $\sigma^2 = (Pf_0, Pf_0)$. Here $K_0 = K \otimes \hat{1}$, $K = \frac{i}{\hbar}H + \frac{1}{2}\left(\frac{\gamma}{\hbar}\right)^2 R\sigma^2 R$,

$$\hat{u}_t = A_t[f_0^\dagger P] + A_t^\dagger[Pf_0] = 2\Re A_t^\dagger[Pf_0]$$

is the Fock space representation of the Wiener process u_t, defined by the multiplication formula

$$d\hat{u}d\hat{u} = dA_t[f_0^\dagger P]dA_t^\dagger[Pf_0] = f_0^\dagger PPf_0 dt.$$

CRYSTALISATION OF ITINERANT ELECTRONS

A. Messager

C.P.T. Marseille

I. INTRODUCTION AND RESULTS

The Hubbard model on a lattice $\mathbb{Z}^\nu$ was introduced to describe itinerating electrons [1]. The problem of electron crystalisation was pointed out by Wigner [2]: it is believed that the Hubbard model exibits both ferromagnetism and antiferromagnetism. This model is also a natural candidate for the description of conductivity, supraconductivity and more recently for high T_c supraconductors.

The Hamiltonian of this model is written in terms of fermionic creation and annihilation operators $C_{x\sigma}^\star$ and $C_{y\sigma}$ respectively where x, y are lattice sites and σ is the spin of the electron. In this paper we will restrict ourself to spin up (denoted by +) and spin down (denoted by -). Given a finite subvolume $V \subset \mathbb{Z}^\nu$ the Hamiltonian can be written as follows:

$$H_V = \sum_{\sigma=+,-} \sum_{x,y \in V} t_{xy}^\sigma \{C_{x\sigma}^\star C_{y\sigma} + C_{y\sigma}^\star C_{x\sigma}\}$$
$$+ U \sum_{x \in V} n_{x+} n_{x-} + h \sum_{x \in V} n_{x+}$$

t_{xy}^σ are numbers which are either all positive or all negative, and $t_{xy}^\sigma = t_{yx}^\sigma$. They are called the hopping parameters. U is called the on site energy and h the magnetic field. $n_{x\sigma}$ is the number operator of electrons of spin σ at site x. Let us define the total number operator in V

$$N_V^\sigma = \sum_{x \in V} n_{x\sigma}$$

N_V is the number of lattice site in V. It is of particular interest to study this model in the canonical ensemble with a constraint on the total number of electrons:

$$(1-\rho)N_V = N_V^+ + N_V^- \qquad \rho < 1$$

The first case of interest is the half-filled band corresponding to $\rho = 0$. The second case is the dopped case $\rho > 0$. We will consider the repulsive case $U > 0$ with $(h = 0)$; the case $U < 0$ can be treated by using the symmetry of the Hamiltonian.

The free energy and the correlation functions are defined as usual:

$$F(\beta t^+, \beta t^-, \beta U) = \lim_{V \to \infty} \frac{1}{\beta V} \ln Tr_{\mathcal{H}_V} e^{-\beta H_V}$$

$$<A>_V^{bc} = \frac{Tr_{\mathcal{H}_V}\{e^{-\beta H_V} A\}}{Tr_{\mathcal{H}_V}\{e^{-\beta H_V}\}}$$

with $t_{xy}^+ = t^+$ $(t_{xy}^- = t^-)$ for nearest neighbors and $t_{xy}^+ = 0$ otherwise.

The trace is taken over the Fock space $\mathcal{H}_V$, A is a product of creation and annihilation operators, the boundary conditions (b.c.) are defined by a vector in Fock space V^c denotes boundary conditions.

To give some insight into this model we want to point out Anderson's paper [14] where he argued that for $t^\sigma = t$ and $h = 0$ the Hubbard model for large U behaves like the antiferromagnetic quantum Heisenberg model with effective Hamiltonian:

$$H_V^{eff} = \sum_{<x,y>\subset V} \frac{t^2}{U} \vec{\sigma_x} \cdot \vec{\sigma_y}$$

where $\vec{\sigma_x}$ is the triple of Pauli matrices $(\sigma_x^1, \sigma_x^2, \sigma_x^3)$ at site x and $<.,.>$ denotes nearest neighbors.

Let us remark that similar arguments suggest that the asymmetric Hubbard model
$(t^+ \neq t^-)$ behaves like the anisotropic quantum Heisenberg model:

$$H_V^{eff} = \sum_{<x,y>\subset V} \left\{ \frac{{t^+}^2 + {t^-}^2}{U} \sigma_x^3 \sigma_y^3 + \frac{2t^+ t^-}{U}(\sigma_x^2 \sigma_y^2 + \sigma_x^1 \sigma_y^1) \right\}$$

This suggests the occurence of a phase transition of Ising type for $t^+ \neq t^-$.

Unfortunately very few rigourous results are known for the Hubbard model. Let us mention the deep analysis of the Falicov-Kimbal model [4] (defined from the Hubbard model by taking $t^- = 0$) containing a proof of a phase transition and a domain of uniqueness by Kennedy and Lieb [3]. Moreover, for the Hubbard model, a domain of uniqueness for the correlation functions is known, provided $\beta t^2 U^{-1}$ is small [6].

The goal of these notes is to describe the results obtained in [7] for the asymmetric Hubbard model in the half-filled band at $h = 0$, and for the Falicov-Kimbal model [8] in a more general case. The results are contained in the following theorem with the following choice of parameters: we will write β for βt^+, U for $\frac{U}{t^+}$ and $\alpha = \frac{t^-}{t^+}$.

Theorem 1: *The asymmetric Hubbard model in the half-filled band for $h = 0$ has the following properties:*
  i) *For $\beta(1 + \alpha^2)U^{-1}$ small there is a unique equilibrium state.*
  ii) *asymmetric case: $U > 8\nu$; $\beta > C_0 U$; $\alpha < \alpha_0$*
      a) *There is a phase transition between the two Néel states defined as follows (provided that there exists a bipartition A and B of the lattice into two Néel sublattices)*

(1) $\qquad <n_{x^+}>^{b.c.1} \geq \frac{1}{2} \quad x \in A \ ; \ <n_{x^-}>^{b.c.1} \geq \frac{1}{2} \quad x \in B$

(2) $\quad <n_{x-}>^{b.c.2}> \dfrac{1}{2} \quad x \in A \; ; \; <n_{x+}>^{b.c.2}> \dfrac{1}{2} \quad x \in B$

b.c.1 and b.c.2 are the boundary conditions associated to the two pure phases. C_0 and α_0 are positive constants.

b) Every translation invariant equilibrium state can be written as an ergodic decomposition of the two previous states.

We can obtain a more complete result in the case of the Falicov-Kimbal model (the proof of phase transition in this model has been obtained also in [12]).

Theorem 2: *The Falicov-Kimball model for small dopping $\rho < CU^{-1}$ and for $|h| < 4U^{-1} - k_\nu U^{-3}$ (k_ν is a positive constant which depends on the space dimension ν, C is a constant) has the following properties:*

i) For βU^{-1} small there is a unique equilibrium state.

ii) under the additionnal conditions: $U > 8\nu$; $\beta > C_0 U$

a) There exist two different Néel states with long range Néel order (provided that there exists a bipartition A and B of the lattice into two Néel sublattices) described as follows:

$$<n_x^e>^{b.c.1}> \dfrac{1}{2} \quad x \in A \; ; \; <n_x^->^{b.c.1}> \dfrac{1}{2} \quad x \in B$$

$$<n_x^i>^{b.c.2}> \dfrac{1}{2} \quad x \in A \; ; \; <n_x^e>^{b.c.2}> \dfrac{1}{2} \quad x \in B$$

b.c.1 and b.c.2 are the boundary conditions associated to the two chessboard vectors. C_0 is a positive constant.

b) Every translation invariant equilibrium state (defined by all the different boundary conditions) can be written as an ergodic decomposition of the two previous states.

We describe the ideas entering the proof of the existence of a phase transitions in the quantum model. We think that they are new in this context. We first notice that for $t^\sigma = 0$ the model becomes classical. The difficulty which arises, contrary for example to the anisotropic quantum Heisenberg model, is that the classical model has infinitely many ground states as we shall prove: patch randomly n lattice sites with + electrons and the $N - n$ remaining sites with − electrons. There is an analogy with classical statistical mechanics which is useful: Let us look at the A.N.N.I. model. For certain values of the coupling constants the A.N.N.I. model has infinitely many ground states. The situation can be handled as follows: we look at the thermal fluctuations of every ground state, then the set of equilibrium states at small positive temperature will correspond to the infinite number of ground states with the largest fluctuations (they are called the dominant ground states). This situation was understood by Fisher and Selke [9] and put into rigorous theory by Sinai and Dinaburg [10] and by Bricmont and Slawny [11].

One has to develop a similar theory for the case of the quantum system: *quantum fluctuations will play the role of thermal fluctuations* in classical statistical mechanics. In the context of the Feynman-Kac representation we will look for the classical ground states with the largest quantum fluctuations. We have to suppose that there is a bipartition of the lattice into two Néel sublattices: A and B. Then the dominant classical ground states will be the following ones: put the + electrons on A sites and the − electrons on B sites and conversly the − electrons on A sites and the + ones on B sites; It is clear that these two classical ground states allow for the maximum quantum fluctuations: each electron can jump onto one of its neighboring sites. We will show that the equilibrium states correspond to these two dominant ground states. We want to emphasize that this *phase transition* is of *quantum nature*: the quantum part of the Hamiltonian produces the phase transition.

FEYNMAN KAC REPRESENTATION FOR THE HUBBARD MODEL: LOOP REPRESENTATION

The configurations are described in terms of trajectories of the electrons on the cylinder $\Lambda \times \{0, \beta\}$ [13], we will call Λ the basis and the time lines fibers. To each trajectory we draw an up arrow when there is a + electron $\omega^+(j)$ and a down arrow when there is a − electron $\omega^-(j)$ the following four situations occur:
  i) There is an up arrow trajectory $\omega^+(j)$.
  ii) There is a down arrow trajectory $\omega^-(j)$.
  iii) Two trajectories with opposite orientations, i.e. there is a pair of a + and − electron: $\omega^+(j) \cap \omega^-(j)$.
  iv) There is no trajectory at all: there is a hole.

We consider now the following new representation:
  i) we draw on the vertical lines a dashed line with an up arrow.
  ii) we draw a dashed line with a down arrow.
  iii) we draw a continuous segment with an up arrow.
  iv) we draw a continuous segment with a down arrow.

We complete this representation with additional horizontal segments: on the horizontal bonds at which a jump occurs we draw a segment of the same type and with the same arrow as the vertical line coming at this bond. We have obtained sets of loops either continuous or dashed ones with a fixed orientation. We denote by $\{\lambda'_1 \ldots \lambda'_n\}$ the continuous loops and by $\{\lambda''_1 \ldots \lambda''_n\}$ the dashed ones. We notice that all the vertical lines are covered with parts of oriented loops, two loops of different kinds always meet along the horizontal segments of the loops. Some of these loops may wind around the cylinder. This certainly occurs to the dashed loops outside the half-filled band. There is a one-to-one correspondence between the quantum configurations and such a family of *oriented loops*. We may also notice that the symmetries described above are easy to see in the loop representation. The canonical partition function is invariant by changing the sign of the hopping constants and by the change of the sign of U which, physically means, to change the repulsion of the electrons of different kinds into an attraction is equivalent to the change of the continuous loops into the dashed one. Finally, the hole particle symmetry corresponds to the change of the orientation of the loops. The construction is described in the picture.

An oriented loop λ is defined by the set of bonds $(x_o, x_1)(x_1, x_2)\ldots(x_n, x_o)$ where the points are ordered according to the orientation of the loop, these bonds correspond to jumps of the electron, the variable σ specify if a jump corresponds to the hop of a + or of a − electron; by the set of times $(t_0, t_1 \ldots t_n)$ where t_0 is the time corresponding to the first jump of an electron and t_i is the interval of time between the two successive jumps occuring on the bonds (x_{i_1}, x_i) and (x_{i+1}, x_i). We denote by $S_+(\lambda')$ (resp.

$S_-(\lambda')$ the number of jumps of + electrons (resp.of − electrons) in the loopλ'. The following density is associated to a continuous loop λ' with 0 winding number:

$$\varphi(\lambda') = \varphi(x_0, x_1...x_n; t_0, t_1...t_n; \sigma_1...\sigma_n)$$
$$= e^{-\frac{U}{2}|t_1|} e^{-\frac{U}{2}|t_2|} ....e^{-\frac{U}{2}|t_n|} \left\{ 1^{s_+(\lambda')} \alpha^{s_-(\lambda')} + 1^{s_+(\lambda')} \alpha^{s_-(\lambda')} \right\}$$

Our goal is to write the partition function as a gas of loops in the grand canonical ensemble. The crucial point is the compatibility condition between the loops and to characterize the interaction between the loops. We first give a classification of continuous loops.

Definition: the *classical loops* $\mathcal{L}_C$ are the loops such that either all its jumps correspond to + electrons or all the jumps correspond to − electrons.

Definition: the *quantum loops* $\mathcal{L}_Q$ are the loops which contain at least one jump of a + electron and one jump of a − electron.

We focus on the following essential properties:
  the classical loops in the grand canonical ensemble are not interacting,
  the partition function written as a gaz of loopsl.

We write the partition function (it depends of the boundary conditions) by summing over all the sets of *compatible loops*: the *non-interacting classical loops* λ', the *interacting quantum loops* μ, and the *dashed loops* λ''. The corresponding densities are integrated over the times, over the number of jumps and over the different possible jumps (corresponding to + or to − spin). Every compatible continuous loop with 0 winding number and arbitrary orientation corresponds to an admissible configuration. This is no more true for winding loops, because the half-filled band condition could be violated. Nevertheless, we will drop this constraint to be able to derive a cluster expansion on th classical loops. The constraint will be restored by using the equivalence of ensembles.

$$Z(\Lambda) = \sum_{\{\lambda'_1...\lambda'_r; \mu_1...\mu_s; \lambda''_1...\lambda''_n\}} \int ... \int \in (\mu_1...\mu_s) \prod_{j=1}^{s} \varphi(\mu_j) d\tau(\mu_j) \times$$
$$\prod_{i=1}^{r} \varphi(\lambda'_i) \epsilon(\lambda'_i) d\tau(\lambda'_i)$$

where $\in (\mu_1...\mu_s)$ is the sign associated to the set of quantum loops.

GROUND STATES AND CONTOURS

Classical ground states

The classical ground states of the Hubbard model are obtained for $t^\sigma = 0$

$$n_{x+} n_{x-} = 0$$

The model exhibits *infinitely many classical ground states*: n + electrons are located randomly, on the lattice the $(N - n)$ − electrons are located on the complementary sites. This means that the classical ground-state configurations are dashed lines without any loop.

Definition: bipartite Néel lattice. Let us choose an arbitrary site of a lattice Λ and call it an A sites (resp. a B sites). The nearest neighbor sites of x will be B site (resp. an A site). If this procedure can be iterated over all the lattice, we will say that the lattice has a bipartite Néel decomposition (A, B). This is supposed to be true in the following.

Definition: Néel ground states $\mathcal{G}_\Lambda^* = \{g_{1\Lambda}^*, g_{2\Lambda}^*\}$
i) $g_{1\Lambda}^*$: put the $+$ electrons on the A sites and the $-$ ones on the B sites.
ii) $g_{2\Lambda}^*$: put the $-$ electrons on the A sites and the $+$ ones on the B sites.

It is clear that these two classical ground states will have the largest quantum fluctuations because each electron can jump to its neighbor site.

Restricted ensembles

Now we define the restricted ensemble that plays the role of quantum ground state.

Definition: Restricted ensemble in Λ compatible with the Néel classical ground states:
$$Res(\Lambda|g_1^*) = \{X_\Lambda \subset \mathcal{L}_C | Ret X_\Lambda = g_1^*\}$$

$Res(\Lambda|g_2^*)$ is defined analogously.

Definition: The restricted partition function in Λ with respect to $Res(\Lambda|g_1^*)$ is given by:
$$Z^R(\Lambda|g_1^*) = \sum_{\{\lambda_1'\ldots\lambda_p'\} \subset Res(\Lambda|g_1^*)} \int \cdots \int \prod_{i=1}^p \varphi(\lambda_i')\epsilon(\lambda_i')d\tau(\lambda_i')$$

where $\epsilon(\lambda_i')$ is defined in lemma 3.2.7.
The free energy associated to this restricted partition function is
$$f(g_1^*) = \lim_{\Lambda \to \infty} \frac{1}{|\Lambda|} \ln Z^R(\Lambda|g_1^*)$$

Lemma: *The above free energy (which is the ground state energy per site of the Hamiltonian) is an analytical function of U provided that $U > 8\nu$:*
$$f(g_1^*) = U^{-1} - k_\nu U^{-3} + 0(U^{-4})$$

where k_ν is a constant depending of the space dimension ν.
Proof: it follows from a cluster expansion valid provided that $U > 8\nu$. We can derive perturbatively a classical representation of the restricted partition function.

Definition of the large scale contours

To every configuration X_Λ we have associated $Ret X_\Lambda$, defined by removing the classical loops $\mathcal{L}_C$.
$Ret X_\Lambda$ is partitionned as follows:
i) Subvolumes of Néel ground states $\mathcal{G}_\Lambda^*$.
ii) Subvolumes of non dominant ground states $\mathcal{G}_\Lambda/\mathcal{G}_\Lambda^*$.
iii) Subvolumes containing quantum loops of $\mathcal{L}_Q$.

A useful characterization of non dominant ground states is the following: we draw vertical planes on the dual lattice between electrons of the same spin. These planes are decomposed into maximal connected components $\Gamma = \{\gamma_1 \ldots \gamma_n\}$. Notice that a section at fixed time of the cylinder of Γ corresponds to contours of the antiferromagnetic Ising model. We will call them "A.F. contours".

There is no usual Peierls condition on the Γ'^s and on the quantum loops, so that we have to define large scale contours; we will need, for U large, only a time scale. We need to define the vertical slabs:

$$S_0 = \{\text{rectangle defined by the 4 points:}[(0,0);0];[(0,0);\tau];[(0,1);0];[(0,1);\tau]\}$$

The two first coordinates are space like and the last one is the time τ which has to be a multiple of β. The set of slabs S is obtained by translating S_0 by $a = \{\tau, 1\}$.

Each configuration X_Λ can be decomposed into three classes of slabs:

$$B^1 = \{\tau \subset S | Ret\ X_\Lambda \subset \mathcal{G}_\Lambda^*\}$$
$$B^2 = \{\tau \subset S | Ret\ X_\Lambda \subset \mathcal{G}_\Lambda/\mathcal{G}_\Lambda^*\}$$
$$B^3 = \{\tau \subset S | \text{ each slab is intersected by, at least, one loop of } \mathcal{L}_Q\}$$

The volume of pure phase are the maximal connected set of τ slabs of type B^1.

Definition: A large scale contour (usually thick) is a pair $\{\theta^j, Y_{\theta^j}\}$ where θ^j is a maximal connected set of slabs of type B^2 and B^3 together with a configuration Y_{θ^j}, defined on θ^j.

Definition: A geometric contour is the set of large scale contours with the same support.

Definition: Let $\Xi_\Lambda(\theta^1 \ldots \theta^n | g^*)$ be the partition function with geometric contours $\theta^1 \ldots \theta^n$ with boundary condition g^* corresponding to one of the Néel ground states and rescaled by the restricted partition function defined above. We need to prove the Peierls condition to derive the existence of a first order phase transition.

Peierls condition

Proposition: (Peierls condition for geometric contours) *Under the following conditions:* $U > 8\nu, \beta > C_1 U, \alpha < \alpha_0$ *there exists* $\tau = C_2 U_0$ *such that*

$$\Xi_\Lambda(\theta^1 \ldots \theta^n | Y^1 \cup Y^2) \leq e^{-\rho |\theta^1|_\tau} \Xi_\Lambda(\theta^2 \ldots \theta^n | Y^1 \cup Y^2)$$

where U_0, C_1, α_0 *and* C_2 *are positive constants.* ρ *depends of the choice of the constants and can be choosen as large as we want.* $|\theta^1|_\tau$ *denotes the number of* τ *slabs contained in* θ^1.

CONCLUSION

In these notes we have sketched a part of the existence proof of a first order phase transition for the asymmetric Hubbard model in the half-filled band case for h small. The method, we have developped is fairly general and can be applied in several cases. Let us recall that the model exhibits infinitely many classical ground states among which the Néel ground states are dominant. The situation will change when h is increasing, then other classical ground states with larger periodicity will become dominant for appropriate values of h. This has been shown for the ground states of the Falicov-Kimbal model in [16]. We will prove that there is a cascade of phase transitions in the Falicov-Kimbal model. We think that our method could prove the Peierls-Frölich instability [17] for an appropriate model. Another situation of interest which has been pointed out for the high T_c supraconductor is the occurence of flux phases. We think that such a situation can be seen in the Falicov-Kimbal model coupled to a magnetic field: flux can appear on all the plaquettes or on some restricted set of plaquettes according to the value of h.

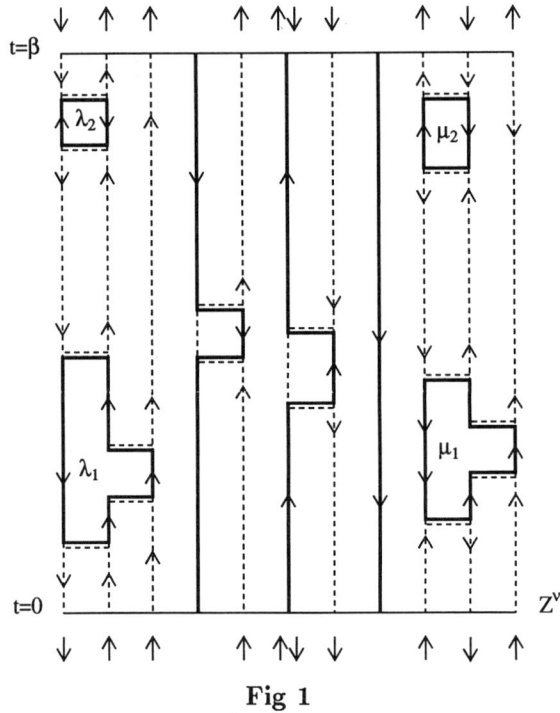

Fig 1

Finally let us mention the "dopped case" in the Falicov-Kimbal model. We conjecture that, for very large U and ρ large, a ferromagnetic phase should appear. Moreover, we have proved that there exists an antiferromagnetic phase for U large and small dopping. We conjecture that between these two cases a cascade of phase transitions could occur for appropriate values of U and of the dopping parameter ρ.

REFERENCES

1. J. Hubbard, *Proc. Roy. Phys. A*, 276 (1963).
2. E. Wigner, *Phys. Rev.*, **46** (1934).
3. T. Kennedy and E. Lieb, *Physica A*, **138**, 320 (1986).
4. L. Falicov and J. Kimbal, *Phys. Rev. Lett.*, **22**, 997 (1969).
5. E. Lieb and F. Wu, *Phys. Rev. Lett.*, **20**, 1445 (1968).

6. A. Messager, S. Pirogov and Y. Suhov, Preprint CPT 1993.
7. A. Messager, Preprint CPT 1993.
8. A. Messager and S. Miracle-Sole, Preprint CPT 1993.
9. N. Fisher and W. Selke, *Phil. Trans. Royal. Soc.*, **302** (1981).
10. E. Dinaburg and Ya. Sinai, *Comm. Math. Phys.*, **98**, 119 (1985).
11. J. Bricmont and J. Slawny, *J. Stat. Phys.*, **54**, 89 (1989).
12. J. Lebowitz and N. Macris, preprint Rutgers (1993).
13. J. Ginibre, *in:* "Statistical Mechanics and Quantum Field Theory", Gordon and Breach, New-York (1971).
14. P. Anderson, *Solid State Phys.*, **14**, 99 (1963).
15. Ya. Sinai, "Theory of Phase Transitions: Rigorous Results", Pergamon Press Gordon (1982).
16. C. Gruber, J. Jedrzejewski and P. Lemberger, *J. Stat. Phys.*, **66**, 913 (1992)
17. J.V. Pulé, A. Verbeure and V.A. Zagrebnov, Peierls-Frohlich instability and Kohn anomaly, preprint-KUL-TF 93/11.

MOLECULAR CHAIN: DYNAMICAL VARIABLES, QUANTIZATION AND STATISTICAL MECHANICS

Ramon F. Alvarez-Estrada

Departamento de Fisica Teorica
Universidad Complutense
E–28040, Madrid, Spain

Standard models for a flexible polymer describe the latter as a chain of N (classical) point particles, such that the relative distances between successive particles equal a constant, d, and, through purely probabilistic (random flight) arguments, they imply that the probability W for the end-to-end distance to be D is approximately gaussian in D. Later, Thermodynamics is introduced by identifying $k_B T ln W$ with the entropy at absolute temperature T (k_B being Boltzmann's constant), etc.[1]. One may try to formulate models for the molecular chain or polymer which be based i) either directly on Quantum Mechanics plus some approximations typical of molecular physics, ii) or in Classical Mechanics from the outset, followed by a suitable quantization. Here, we shall summarize some recent works [2,3,4], in which both formulations i) and ii) as well as their mutual relationships and their consistency with standard probabilistic models[1] have been analyzed.

First, we shall treat the N non-relativistic particles in the three-dimensional molecular chain or polymer quantum-mechanically, and let $m_i, i = 1..N$ be their masses. We shall introduce the center-of-mass (CM) $\vec{X}_{CM}$ and relative position vectors $\vec{y}_s, s = 1..N-1$ successively along the chain. Let T be the total kinetic energy operator: $T = (2M)^{-1}\vec{\Pi}_{CM}^2 + 2^{-1}\sum_{s,j=1}^{N-1} A_{sj}\vec{\pi}_s\vec{\pi}_j$. $\vec{\Pi}_{CM}, \vec{\pi}_s$ are the momentum operators canonically conjugate to $\vec{X}_{CM}, \vec{y}_s$, respectively. M is the total mass and A_{sj} are constants depending only on the m_i. The interaction potential is supposed to be

$$V = 2^{-1} \sum_{s=1}^{N-1} A_{ss}\omega_s^2(y_s - d_s)^2,$$

$\vec{y}_s = y_s\vec{u}_s$, $\vec{u}_s = (\sin\theta_s\cos\varphi_s, \sin\theta_s\sin\varphi_s, \cos\theta_s)$. $\omega_s(> 0)$ and $d_s(> 0)$ are given frequencies and (bond) lengths, respectively. The total hamiltonian is $H_1 = T + V$. We shall study this quantum system when all $\omega_s \to +\infty$, for fixed d_s. Physically, one expects, then, that y_s evolves very rapidly and, on the average, equals d_s, while angular variables evolve slowly: this agrees with the spirit of the Born-Oppenheimer approximation[5]. We shall consider variational wave functions for the chain bearing the form $\phi_0\psi$ where ϕ_0 is the product of suitable (normalized) real gaussians depending on $\omega_s^{1/2}(y_s - d_s)$ for $s = 1..N-1$, and $\psi = \psi(\vec{X}_{CM}; \vec{u}_1..\vec{u}_{N-1})$ is independent on $y_1..y_{N-1}$ but, otherwise, arbitrary. Then, one searches for the new operator H_2 such that $\lim \int_0^{+\infty} [\prod_{s=1}^{N-1} y_s^2 dy_s]\phi_0[H_1(\phi_0\psi)] \equiv H_2\psi$ as $\omega_s \to +\infty$, for fixed $\vec{X}_{CM}, \vec{u}_s, d_s, s = 1..N-1$ and for any ψ. The underlying justification for this calculation is the variational principle.Thus, the ground state energy of H_1 is less or equal than the expectation value of H_2 in any normalized wave function ψ: of course, the latter expectation value is computed by integrating over $\vec{X}_{CM}, \vec{u}_1..\vec{u}_{N-1}$, but not over $y_1..y_{N-1}$. The result is [2] : $H_2 = H + E_0, E_0 = 2^{-1}\sum_{s=1}^{N-1}\hbar\omega_s$:

$$H = (2M)^{-1}\vec{\Pi}_{CM}^2 + 2^{-1}\sum_{s,j=1}^{N-1} A_{sj}(d_s d_j)^{-1}\vec{e}_s\vec{e}_j \qquad (1)$$

$$\vec{e}_s = i\hbar\vec{u}_s - \vec{a}_s, \quad \vec{a}_s = \vec{u}_s \times \vec{l}_s, \qquad (2)$$

$\vec{l}_s$ is the standard orbital angular momentum operator associated to $\vec{y}_s$. E_0 is a constant vibrational energy shift.In agreement with the spirit of the Born-Oppenheimer approximation, H describes, at least approximately, the low energy states of the microscopic chain associated to internal rotations. Eq.(1) differs from Eq.(6) in our first work [2] and from Eq.(1) in a subsequent work [3] just by the additive constant term $-c = -2^{-1}\sum_{s=1}^{N-1} A_{ss}\hbar^2 d_s^{-2}$.The discrepancy can be explained just by taking into account in our preceding work [2] further contributions of (non-leading) order ω_s^0 in Eqs. (9) and (16), as well as terms of order ω_s^{-1} in the normalization of ϕ_0. The addition of all those (non-leading) corrections gives rise (after cancellations of frequency factors) to a new net contribution equal to $+c$, as $\omega_s \to +\infty$, in our previous work[2], as one can check easily. The inclusion of $+c$ and its cancellation with $-c$ leads to the correct result given in Eq. (1) above. The cartesian components $e_{s\alpha}, \alpha = 1, 2, 3$ of $\vec{e}_s$ and H are hermitean and essentially self-adjoint, respectively. An important property is the following: the commutators of $e_{s\alpha}$ and of the cartesian components $l_{s\alpha}$ of $\vec{l}_s$ satisfy the following closed Lie algebra ($\alpha, \beta, \gamma = 1, 2, 3$):

$$[e_{s\alpha}, e_{s\beta}] = -i\hbar\epsilon_{\alpha\beta\gamma}l_{s\gamma}, \quad [l_{s\alpha}, e_{s\beta}] = i\hbar\epsilon_{\alpha\beta\gamma}e_{s\gamma}, \quad [l_{s\alpha}, l_{s\beta}] = i\hbar\epsilon_{\alpha\beta\gamma}l_{s\gamma} \qquad (3)$$

$\epsilon_{\alpha\beta\gamma}$ is the standard antisymmetric tensor with three indices.One can prove [3] that: i) the above Lie algebra is semi-simple and has rank two, ii) the Lie group generated through standard exponentiations of linear combinations of $e_{s\alpha}, l_{s\alpha}$ is non compact. The two Casimir operators of the above Lie algebra have also been constructed [3].

Next, by taking Classical Mechanics as starting point, we shall discuss [4] another class of models for molecular chains formed by N classical(non-relativistic) point particles.Use will be made of the same $\vec{X}_{CM}, \vec{y}_s, \vec{u}_s$, etc. as before. We shall now assume that the distances y_s between successive neighbours are constant: $y_s = d_s > 0, s = 1..N-1$ (that is, they are holonomic constraints). Here, the canonically conjugate variables are $\vec{X}_{CM}, \theta_s, \varphi_s$ and the classical momenta $\vec{\Pi}_{CM,c}, \pi_{\theta_s,c}, \pi_{\varphi_s,c}$. The kinetic energy H_c' of the chain becomes, then, a function of the above angles and depends quadratically of all the above classical momenta. It turns out that H_c' differs from the classical limit H_c of H. One can introduce the classical transverse momenta as $-\vec{a}_{s,c} = -\vec{u}_s \times \vec{l}_{s,c}, \vec{l}_{s,c}$ being the classical orbital angular momentum. The Poisson brackets of the cartesian components of $\vec{a}_{s,c}, \vec{l}_{s,c}$ satisfy a closed algebra which turns out to be the classical ($\hbar \to 0$) limit of the Lie algebra (3). Now, let all particles in the chain be regarded quantum-mechanically [3]. Then, the above H_c' gives rise to a quantum mechanical hamiltonian

H', through the known quantization recipes for curvilinear coordinates[6]. It turns out that H is different from H'. Moreover the scalar product required in the actual quantization procedure differs from the one associated with H. Consequently, neither the standard orbital angular momentum operators $\vec{l}_s$ nor $\vec{a}_s$ given in (2) are now hermitean. However, they can be used to construct new quantum operators $\vec{l'}_s, \vec{e'}_s$ the cartesian components of which are self-adjoint and hermitean, respectively, with respect to the actual scalar product. Notice that $\vec{l}_s, \vec{e}_s$ are different from $\vec{l'}_s, \vec{e'}_s$. In each model, one can regard $\vec{e}_s, \vec{e'}_s$ as quantum versions of the classical transverse momenta. Actually, the closed Lie algebra (3) is also satisfied if the cartesian components of $\vec{e}_s, \vec{l}_s$ are replaced by those of $\vec{e'}_s, \vec{l'}_s$, respectively. One can prove that H' is essentially self-adjoint with respect to the actual scalar product. The preceding developments and results for the model based upon Classical Mechanics as starting point [3] can be generalized directly to the case in which all y_s are assumed to be known positive functions of the angles : $y_s = d_s(\theta_1\varphi_1..\theta_{N-1}\varphi_{N-1}) > 0, s = 1..N-1$.

We shall now turn to Classical Statistical Mechanics for both models [2,4]. Let all particles in the chain be identical (with equal mass m and $d_s = d$ for $s = 1..N-1$) and regarded as classical. We shall now suppose that N is very large and that the chain is in thermodynamical equilibrium, with $k_BT \gg (2md^2)^{-1}\hbar^2$. The statistical behaviour corresponds to the large number of internal degrees of freedom (that is, the angles), but not to the motion of the center of mass .Let Z_c be the classical partition function determined by H_c (the classical limit of H), according to the standard rules of the canonical ensemble. In Z_c, one also includes the potential of external stretching forces applied to both ends of the chain. It can be shown that the classical internal energy of the chain, U_c, satisfies the equipartition principle exactly. Now, let Z'_c be the classical (canonical) partition function associated to H'_c: its associated classical internal energy U'_c also satisfies the equipartition principle exactly. For both Z_c and Z'_c one can obtain approximately the entropies S_c and S'_c respectively and show that both of them coincide and increase when the chain is stretched: the latter is an experimentally known polymer property conventionally referred to as "rubber elasticity". Moreover, both S_c and S'_c yield the same approximate probability for the end-to-end distance in the chain to be D. The latter probability turns out to coincide with the probability W provided by the standard (probabilistic) models for flexible polymers, given succinctly at the beginning of this contribution.

Our main conclusions are the following:

1) The two models, which start directly either from Quantum Mechanics or from Classical Mechanics are different, both at the classical and quantum levels.

2) New dynamical variables, namely, transverse momenta, arise in both models.At the classical level: a) transverse momenta coincide in both models, b) the Poisson brackets of the cartesian components of the transverse momenta and of the orbital angular momenta close an algebra.At the quantum level, transverse momenta turn out to be given by different expressions in both models, and the same occurs for the orbital angular momenta. The commutators of the cartesian components of the transverse momenta and of the orbital angular momenta close a Lie algebra in each model. The structures of those two Lie algebras coincide. Some general properties of that Lie algebra have been established.

3) The exact partition functions, in Classical Statistical Mechanics, for both models are different. However, the equipartition principle holds exactly for both models. To a first approximation, both models yield the same entropy and the same probability for the end-to-end distance in the chain, which, in turn, coincides with the one(W) provided by the standard (probabilistic) models for flexible polymers. The last fact provides, at least, a partial check of consistency for both models.

Acknowledgments

The author is grateful to Profs. A. Galindo and C. Moreno for some discussions about Lie algebras and to Mr. A. Gomez Nicola for constructive help. The partial financial support and kind facilities provided by the Local Organizing Committee of the NATO Advanced Research Workshop "On Three Levels" is acknowledged. The partial financial support of Proyecto AEN90-0034, CICYT, Spain, is also acknowledged.

REFERENCES

1. F.W. Wiegel, "Introduction to Path-Integral Methods in Physics and Polymer Science," World Scientific, Singapore (1986). M. Doi and S.F. Edwards, "The Theory of Polymer Dynamics," Oxford University Press, Oxford (1986).
2. R.F. Alvarez-Estrada, String-like model for linear polyatomic molecules and polymers, *Phys. Lett. A*, **157**, 469 (1991).
3. R.F. Alvarez-Estrada, String-like model for internal rotations in linear polyatomic molecules: New dynamical variables, *Phys. Lett. A*, **159**, 118 (1991).
4. R.F. Alvarez-Estrada, Freely jointed molecular chain: dynamical variables, quantization and statistical mechanics, *Phys. Rev. A*, **46**, 3206 (1992).
5. P.W. Atkins, "Molecular Quantum Mechanics," 2nd ed., Oxford University Press, Oxford (1983).
6. B.S. De Witt, Dynamical theory in curved spaces (I): A review of the classical and quantum action principles, *Rev. Mod. Phys.*, **29**, 377 (1957).

QUASIPARTICLE'S SPIN AND FRACTIONAL STATISTICS IN THE FRACTIONAL QUANTUM HALL EFFECT

Dingping Li

International School for Advanced studies, SISSA
I-34014 Trieste
Italy

INTRODUCTION

The possibility of fractional statistics on two dimensional surfaces was discovered in Ref. [1–3]. When two fractional-statistics particles (anyons) are interchanged, the wave function changes by a phase $\exp(i\theta)$, where θ is neither given by $\theta = 0$ (Bose statistics) nor by $\theta = \pi$ (Fermi statistics). Fractional statistics has attracted a lot of attention after it was found that a gas of fractional statistics objects should be superconducting and could provide a mechanism for high-T_c superconductivity [4]. Furthermore, the quasiparticles in fractional quantum Hall systems (for a review on the fractional quantum Hall effect (FQHE), see Ref. [5]) are anyons [6, 7] and this picture had been used to construct the hierarchical wave function in the FQHE [7] (for a review, see Refs. [8, 9]). On higher dimensional spaces ($D > 2$), only Fermions and Bosons exist and the Fermions' spin is half-integer, the Bosons' spin is integer. It will be very interesting to know what is the spin of anyons and the spin-statistics relation of anyons. The spin-statistics relation of anyons is a generalized one if the spin s satisfies $s = \theta/2\pi$. In various models, for example, in non-linear sigma models, Chern-Simons field theories and relativistic quantum field theories on $2D$ dimensional spaces, anyons indeed satisfy the generalized spin-statistics relation [10, 11]. Naturally, we will ask the question what is the spin-statistics relation of the quasiparticle in the FQHE. Recent discussions of the quasiparticle's spin (QPS) can be found in Refs. [12–15]. Refs. [12, 13] calculated the QPS by analyzing the hierarchical wave function or by calculating the Berry phase of the quasiparticles moving in a closed path on the sphere. Ref. [14] obtained the QPS by analyzing the Ginzburg-Laudau-Chern-Simons (GLCS) theory of the FQHE on the sphere. On the other hand, Ref. [15] calculated the QPS based on the GLCS theory in a disc geometry. However, the results in the Refs. [12, 15] are different from each other. The ambiguity of the QPS in the literature is due to lack of a good definition of the QPS.

We will calculate here the QPS by using braid relations of anyons on Riemann surfaces [16–22]. We will show that the QPS calculated in the following is consistent with physical restrictions and is intrinsic in the sense that the notion of spin is well defined for all Riemann surfaces. The results we obtain here agree with the results of Refs. [12, 13].

BRAID GROUP ON RIEMANN SURFACES

Let us consider N spinning anyons on an oriented compact Riemann surface with g handles [20–22]. To define the anyon's spin, we attach an oriented local frame to every particle. When a particle moves on a curved surface (the torus is a flat surface), the attached frame is parallel transported and a path-dependent frame-rotation is associated with the particle transport. Let us denote the clockwise 2π rotation of the frame attached to the particle by $R_{2\pi}$. The action of the operator $R_{2\pi}$ on the wave function will give a phase $\exp(i2\pi S)$. We define S as the spin of the particle. S is equal to $1/2$ for the electron by this definition. The braiding operators are $\sigma_i, \rho_{n,i}, \tau_{n,i}$, where σ_i interchanges (clockwise) particle i and particle $i+1$ and $\rho_{n,i}, \tau_{n,i}$ take particle i around noncontractable loops on the mth handle. Here we use the same definition of operators $\rho_{n,i}, \tau_{n,i}$ as in Ref. [20].

The braid relations for spinning anyons on the Riemann surface with g handles are

$$\sigma_j \sigma_{j+1} \sigma_j = \sigma_{j+1} \sigma_j \sigma_{j+1}, \tag{1}$$

$$\tau_{m,j+1} = \sigma_j^{-1} \tau_{m,j} \sigma_j, \quad \rho_{m,j+1} = \sigma_j^{-1} \rho_{m,j} \sigma_j, \tag{2}$$

$$\rho_{m,j}^{-1} \tau_{m,j+1} \sigma_j^{-2} \rho_{m,j} \sigma_j^2 \tau_{m,j+1}^{-1} = \sigma_j^2, \tag{3}$$

$$\sigma_1 \sigma_2 \cdots \sigma_{N-1}^2 \cdots \sigma_2 \sigma_1 = R_{2\pi}^{2(g-1)} \prod_n^g \rho_{n,1}^{-1} \tau_{n,1}^{-1} \rho_{n,1} \tau_{n,1}. \tag{4}$$

For the spinless anyons on the sphere, Equation (4) becomes

$$\sigma_1 \sigma_2 \cdots \sigma_{N-1}^2 \cdots \sigma_2 \sigma_1 = 1$$

which had been derived in Ref. [17]. It expresses the fact that a closed (clockwise) loop of particle 1 around all the other particles can be continuously shrunk to a point on the rear side of the sphere. Eq. (4) is the generalization of the case of the spinless anyons on the sphere to the spinning anyons on general Riemann surfaces [20, 21]. When we deform the loop of the left hand side of Eq. (4) to the loop of the right side of Eq. (4), described by $\prod_n^g \rho_{n,1}^{-1} \tau_{n,1}^{-1} \rho_{n,1} \tau_{n,1}$, the attached spin frame is rotated ($4\pi(g-1)$ rotation) and we obtain a phase $R_{2\pi}^{2(g-1)}$. We need to include this phase on the right hand side of the equation. However, for charged anyons in a magnetic field, as is the case for quasiparticles in the FQHE, the braid relation (4) should be changed. We also need to include the Aharonov-Bohm phase $\exp(2\pi i q \Phi)$ on the right hand side of Eq. (4) because the charged anyon interacts with the magnetic field, where q is the anyon's charge and Φ is the magnetic flux out of the surface. Thus instead of Eq. (4), for the charged anyons on the magnetic field, we have

$$\sigma_1 \sigma_2 \cdots \sigma_{N-1}^2 \cdots \sigma_2 \sigma_1 = \exp(2\pi i q \Phi) R_{2\pi}^{2(g-1)}$$

$$\times \prod_n^g \rho_{n,1}^{-1} \tau_{n,1}^{-1} \rho_{n,1} \tau_{n,1}. \tag{5}$$

We will only consider Abelian fractional statistics, which means that the representation of the operator σ_i is given by $\sigma_i = \sigma = \exp i\theta \mathbf{1}_M$, where $\mathbf{1}_M$ is the $M \times M$ identity matrix. Inserting $\sigma_i = \sigma = \exp i\theta \mathbf{1}_M$ in Eq. (2) and Eq. (3), one obtains that $\tau_{m,j} = \tau_m$, $\rho_{m,j} = \rho_m$, $\tau_m \rho_m = \sigma^2 \rho_m \tau_m$. These relations and Eq. (5) yield

$$\exp[2i(N-1)\theta] = \exp[2\pi i q\Phi + 4\pi(g-1)S] \times \exp(-2ig\theta). \qquad (6)$$

If there are several kinds of anyons, we shall introduce mutual statistics [23]. The mutual statistics $\theta_{i,j}$ means that when a particle of the ith kind moves clockwise around a particle of the jth kind, we get a phase $\exp(2i\theta_{i,j})$. Actually, $\theta_{i,i} = \theta_i$ is the fractional statistics parameter of the particle of the ith kind. The left hand side of Eq. (6) is a phase which is obtained by moving one particle (clockwise) around all other particles. If there are other kinds of particles, we have instead of Eq. (6)

$$\exp[2i(N_i - 1)\theta_i + 2i\sum_{j\neq i}^{l} N_j \theta_{i,j}] = \exp(-2ig\theta_i) \times \exp[2\pi i q_i \Phi + 4\pi(g-1)S_i], \qquad (7)$$

where there are l different kinds of particles and the spin of the particle of the ith kind is S_i and its charge is q_i. Eq. (7) gives a constraint on the parameters (numbers, statistics and spin) of anyons, namely

$$[((N_i - 1 + g)\theta_i + \sum_{j\neq i}^{l} N_j \theta_{i,j})/\pi] - q_i \Phi - 2(g-1)S_i \qquad (8)$$

is an integer.

QUASIPARTICLE'S SPIN IN THE LAUGHLIN STATE

Now we consider the FQHE on a surface with g handles. For a FQHE state with filling factor $1/m$, the relation between the electron number and the flux is $m(N + g - 1) = \Phi$ (see Ref. [24] for details). If N_q quasiparticles are created, one has $m(N + g - 1) + N_q = \Phi$. The mutual statistics between the electron and quasiparticle is $2\theta_{mut} = 2\pi$, the charge of the quasiparticle is $1/m$ and the statistics parameter is π/m. We remark that, if $2\theta_{i,j} = 2\pi, i \neq j$ in Eq. (8) or Eq. (7), we can simply omit the term $N_j \theta_{i,j}$. By applying Eq. (8) to those N_q quasiparticles, one can show that

$$[(N_q - 1 + g)/m] + N - [\Phi/m] - 2(g-1)S_q \qquad (9)$$

is an integer. From Eq. (9) and the equation $m(N + g - 1) + N_q = \Phi$, the QPS turns out to be $S_q = [1/2m] + [n/2(g-1)]$ (n is an integer and $g \neq 1$ is assumed). Let us discuss how to fix n. We consider a cluster of particles which contains n_i particles of the ith kind with mutual statistics $\theta_{i,j}$. By using the method developed in Ref. [17], we get the statistics, spin and charge of the cluster, as

$$S_c = \sum_i [n_i(n_i - 1)\theta_i/2\pi] + n_i S_i + \sum_{i\neq j} n_i n_j \theta_{i,j}/2\pi \qquad (10)$$

$$q_c = \sum n_i q_i \quad \text{and} \quad \theta_c = \sum_i n_i^2 \theta_i + \sum_{i\neq j} n_i n_j \theta_{i,j}$$

If the cluster's charge is an odd (even) integer and the cluster satisfies Fermi (Bose) statistics, we suppose that the cluster contains only an odd (even) number of electrons (for example, see Ref. [25], and thus the cluster's spin is a half-integer (integer). If the cluster contains m quasiparticles in the above example, the cluster's charge is 1 and its statistics of fermionic. This cluster shall be a hole and the cluster's spin is a half-integer (see also Ref. [11]). By using Eq. (10) for this cluster, we get a restriction for the QPS, $[mn/2(g-1)] = integer$. If m and $g-1$ are coprime to each other (for example, g is equal to $0, 2, 3$), n must be equal to $2n'(g-1)$ (n' is an integer). Thus S_q is equal to $1/2m$ (up to an integer) and the spin-statistics relation is the standard one. However, when m and $g-1$ are not coprime to each other for some higher genus surfaces, there exist other solutions for the spin than $1/2m$. Writing $m = kp$ and $g - 1 = kq$, where p and q are coprime to each other, we have the solutions $n = 2n'q$. Other solutions of the QPS are $S_q = [1/2m] + [n'/k]$ where $n' = 1, \cdots, k-1$. Therefore S_q can not be completely fixed by using the braid group analysis. However, we shall point out that it is highly unlikely that those **other** solutions are the **true** QPS, as we expect that the value of the spin should be intrinsic and not depend on the surface on which the quasiparticles live. To completely fix the QPS on Riemann surfaces, we can obtain the QPS by analyzing the wave function of the quasiparticles on Riemann surfaces (we plan to do this elsewhere), as it was done for the case on the sphere [12, 13].

QUASIPARTICLE'S SPIN IN THE HIERARCHICAL STATE

Let us calculate the QPS in the standard hierarchical state (see Ref. [7, 12, 13, 26–31]). We remark that the method here can be used to calculate the QPS in other kinds of quantum Hall fluids, for example, the multi-layered FQHE or in Jain states [32] etc.. The hierarchical state is described by a symmetric matrix $\Lambda_{i,j}, i, j = 1, 2, \cdots, l$, where $\Lambda_{i,i+1} = \Lambda_{i+1,i} = \pm 1$, $\Lambda_{1,1}$ is an odd integer and $\Lambda_{i,i}, i \neq 1$ are even integers, where l is the level of the hierarchical state and N_i is the number of particles in level i (N_1 is the number of electrons, N_2 is the number of condensed quasiparticles (or holes) of the first level (Laughlin) state, etc.). On the torus [33, 30, 31], we have a relation, $\sum_j \Lambda_{i,j} N_j = \delta_{i,1}\Phi$, and on the sphere [12, 13, 26], the relation is $\sum_j \Lambda_{i,j} N_j - \Lambda_{i,i} = \delta_{i,1}\Phi$. Following the discussion about the Laughlin state on Riemann surfaces, we expect that the relation is $\sum_j \Lambda_{i,j} N_j + (g-1)\Lambda_{i,i} = \delta_{i,1}\Phi$ for the hierarchical state on Riemann surfaces. We define an l dimensional integer lattice with basisvectors E_i and the inner products $E_i \cdot E_j = \Lambda_{i,j}$ (see Ref. [30, 31]). The above equation can be rewritten as

$$\sum_{i=1}^{l} N_i E_i + (g-1)(E_i \cdot E_i) E_i^\star = E_1^\star \Phi, \tag{11}$$

where $E_i^\star$ are the basisvectors of the dual lattice, defined by $E_i^\star \cdot E_j = \delta_{i,j}$. We can show that $E_i^\star \cdot E_j^\star = \Lambda_{i,j}^{-1}$. The quasiparticle is described by a vector $\mathcal{Q}_k = k_i E_i^\star$ (k_i is an integer) on the lattice spanned by $E_i^\star$ (see Refs. [27, 28]). The statistics parameter of this quasiparticle is $\theta_k = \mathcal{Q}_k \cdot \mathcal{Q}_k \pi = \sum_{i,j} k_i \Lambda_{i,j}^{-1} k_j$ and the charge is $Q_k = \mathcal{Q}_k \cdot E_1^\star = \sum_i k_i \Lambda_{i,1}^{-1}$. The mutual statistics between the quasiparticles $\mathcal{Q}_k$ and $\mathcal{Q}_{k'}$ is $\theta_{k,k'} = \mathcal{Q}_k \cdot \mathcal{Q}_{k'} \pi = \sum_{i,j} k_i \Lambda_{i,j}^{-1} k'_j$. If N quasiparticles, denoted by the vector $\mathcal{Q}_k$, are created, one has $\sum_{i=1}^{l} N_i E_i + (g-1)(E_i \cdot E_i) E_i^\star + N\mathcal{Q}_k = E_1^\star \Phi$. Taking the inner product of the two sides of this equation with $\mathcal{Q}_k$, one finds

$$\frac{N\theta_k}{\pi} + (g-1)\Delta \cdot \mathcal{Q}_k - Q_k \Phi = integer, \tag{12}$$

where $\Delta = \sum_{i=1}^{l}(E_i \cdot E_i)E_i^*$. Applying Eq. (8) to these quasiparticles and comparing it with Eq. (12), one gets

$$S_k = \frac{\theta_k}{2\pi} - \frac{\Delta \cdot \mathcal{Q}_k}{2} + \frac{n}{2(g-1)}, \quad n = integer, \ g \neq 1. \tag{13}$$

By using the argument in Ref. [17] (which we did for the quasiparticle of the Laughlin state), we can fix n in some cases. The charge of the quasiparticle E_i (denoted by the vector E_i) is $\delta_{i,1}$ and the statistics is $\theta = \delta_{i,1}\pi$. Thus the spin of this quasiparticle is $\frac{\delta_{i,1}}{2} + integer$. Because $E_i = \sum_j \Lambda_{i,j} E_j^*$, the quasiparticle E_i is a cluster which contains $\Lambda_{i,j}$ quasiparticles with vectors E_j^*. If $\det \Lambda$ and $g-1$ are coprime to each other, by using Eq. (10) for the cluster, we find that $S_i = \frac{E_i \cdot E_i}{2} - \frac{\Delta \cdot E_i}{2} - \frac{1}{2}$ for the quasiparticle E_j^* when $l-i$ is an even integer and $S_i = \frac{E_i \cdot E_i}{2} - \frac{\Delta \cdot E_i}{2}$ when $l-i$ is an odd integer. We will use the notation $i \in i_e$ if $l-i$ is an even integer. Generally, for the quasiparticle $\mathcal{Q}$, we find that

$$S_\mathcal{Q} = \frac{\mathcal{Q} \cdot \mathcal{Q}}{2} - \frac{\Delta \cdot \mathcal{Q}}{2} - \frac{1}{2}\Delta' \cdot \mathcal{Q}, \tag{14}$$

where $\Delta' = \sum_i E_i, i \in i_e$. If the quasiparticles have the standard spin-statistics relation, it is required that $\frac{\Delta \cdot \mathcal{Q}}{2} + \frac{1}{2}\Delta' \cdot \mathcal{Q}$ is an integer. Indeed, this number is always an integer for the Laughlin state. However, in the hierarchical state, this number may not be an integer. Thus the quasiparticles in the hierarchical state usually do not have the standard spin-statistics relation.

If $\det \Lambda$ and $g-1$ are not coprime to each other, there exist other solutions, not only Eq. (14). As we argued in the case of the Laughlin state, it is **unlikely** that these **other** solutions do give **true** QPS. Thus we suppose that Eq. (14) is the spin for quasiparticles and it does not depend on the topology of the underlying Riemann surface. The above method does not give any information about the QPS in the FQHE on the torus ($g = 1$). However, the above discussion strongly suggests that the QPS in the FQHE on the torus is also given by Eq. (14) which is thus supposed to give the QPS on any Riemann surface.

The Lagrangian for the long-distance physics of the Hall fluid on Riemann surfaces is [14],

$$\mathcal{L} = \frac{1}{4\pi}(\alpha_{\mu,i}\Lambda_{i,j}\epsilon^{\mu\nu\lambda}\partial_\nu\alpha_{\lambda,j} + 2A_\mu t_i \epsilon^{\mu\nu\lambda}\partial_\nu\alpha_{\lambda,i} + 2\omega s_i \epsilon^{0\nu\lambda}\partial_\nu\alpha_{\lambda,i}), \tag{15}$$

where ω is the connection one form (the curvature is given by $R = d\omega$). In the case of the hierarchical state, t_i is equal to $\delta_{i,1}$ and the matrix Λ is the one we gave above. By using Eq. (11), we can show that $s_i = \Lambda_{i,i}$. So s_i is a topologically independent constant. It is reasonable to believe that, for any kind of quantum Hall fluids, s_i is independent of the topology of the underlying Riemann surface. Due to the presence of the third term in the Lagrangian, the spin-statistics relation usually is not a generalized one [14].

Acknowledgments

I would like to thank T. Einarsson, Professor B. Dubrovin, Professor S. Cecotti, Professor R. Iengo and Professor J.M. Leinaas for enlightening discussions. The work is partially supported by EEC, Science Project SC1*-CT92-0789.

REFERENCES

1. J.M. Leinaas and J. Myrheim, *Nuovo Cimento B*, **37**, 1 (1977).
2. G.A. Goldin, R. Menikoff and D.H. Sharp, *J. Math. Phys.*, **22**, 1664 (1981).
3. F. Wilczek, *Phys. Rev. Lett.*, **49**, 957 (1982).
4. R.B. Laughlin, *Phys. Rev. Lett.*, **60**, 2677 (1988);
   V. Kalmeyer and R.B. Laughlin, *Phys. Rev. Lett.*, **59**, 2095 (1988);
   C.B. Hanna, R.B. Laughlin and A.L. Fetter, *Phys. Rev.*, **B40**, 8745 (1989);
   C.B. Hanna, R.B. Laughlin and A.L. Fetter, *Phys. Rev.*, **B39**, 9679 (1989);
   X.G. Wen, F. Wilczek, and A. Zee, *Phys. Rev. B*, **39**, 11413 (1989);
   Y.H. Chen, F. Wilczek, E. Witten, and B. Halperin, *Int. J. Mod. Phys. B*, **3**, 1001 (1989).
5. R. Prange and S. Girvin, "The Quantum Hall Effect", Springer-Verlag, New York, Heidelberg, 1990, 2nd ed. and references therein.
6. D. Arovas, J.R. Schriffer and F. Wilczek, *Phys. Rev. Lett.*, **53**, 722 (1984).
7. B.I. Halperin, *Phys. Rev. Lett.*, **52**, 1583 (1984); **52**, 2390(E) (1984).
8. F. Wilczek, "Fractional Statistics and Anyon Superconductivity", World Scientific, Singapore, 1990 and references therein.
9. R. Iengo and K. Lechner, *Phys. Rep. C*, **213**, 179 (1992) and references therein.
10. R.D. Tscheuschner, *Int. J. Theoretical Phys.*, **28**, 1269 (1989) and references therein.
11. J. Fröhlich and P.A. Marchetti, *Nucl. Phys. B*, **356**, 533 (1991).
12. D. Li, *Phys. Lett. A*, **169**, 82 (1992).
13. D. Li, *Phys. Rev. B*, **47**, 13370 (1993).
14. X.G. Wen and A. Zee, *Phys. Rev. Lett*, **69**, 953 (1992); Erratum: **69**, 3000 (1992).
15. S.L. Sondhi and S.A. Kivelson, *Phys. Rev. B*, **46**, 13319 (1992).
16. Y.-S. Wu, *Phys. Rev. Lett.*, **52**, 2103 (1984).
17. D. Thouless and Y.-S. Wu, *Phys. Rev. B*, **31**, 1191 (1985).
18. T. Einarsson, *Phys. Rev. Lett.*, **64**, 1995 (1990).
19. X.G. Wen, E. Daggoto and E. Fradkin, *Phys. Rev. B*, **42**, 6110 (1990).
20. T. Einarsson, *Mod. Phys. Lett. B*, **5**, 675 (91) and references therein.
21. T. Imbo and J. March-Russell, *Phys. Lett. B*, **252**, 84 (1990).
22. A.P. Balachandran, T. Einarsson, T.R. Govindarajan and R. Ramachandran, *Mod. Phys. Lett. A*, **6**, 2801 (1990).
23. F. Wilczek, *Phys. Rev. Lett.*, **69**, 132 (1992) and references therein.
24. R. Iengo and D. Li, Quantum mechanics and quantum Hall effect on Riemann surfaces, preprint, SISSA/ISAS/100/93/EP, to appear in *Nucl. Phys. B.*.
25. J. Fröhlich and A. Zee, *Nucl. Phys. B*, **364**, 517 (1991).
26. F.D.M. Haldane, *Phys. Rev. Lett.*, **51**, 605 (1983).
27. B. Blok and X.G. Wen, *Phys. Rev. B*, **42**, 8145 (1990); ibid. **43**, 8337 (1991).
28. N. Read, *Phys. Rev. Lett.*, **65**, 1502 (1990).
29. G. Moore and N. Read, *Nucl. Phys. B*, **360**, 362 (1991).
30. D. Li, *Int. J. Mod. Phys. B*, **7**, 2655 (1993).
31. D. Li, *Int. J. Mod. Phys. B*, **7**, 2779 (1993).
32. J.K. Jain, *Phys. Rev. Lett.*, **63**, 199 (1989); *Phys. Rev. B*, **40**, 8079 (1989); **41**, 7653 (1990); *Adv. in Phys.* **41**, 105 (1992).
33. F.D.M. Haldane and E.H. Rezayi, *Phys. Rev. B*, **31**, 2529 (1985).

PARTICIPANTS

D.B. Abraham	Oxford	United Kingdom
F. Acerbi	Trieste	Italy
M. Aizenman	Princeton	USA
R. Alicki	Gdańsk	Poland
G. Alli	London	United Kingdom
D. Alonso	Brussels	Belgium
R.F. Alvarez–Estrada	Madrid	Spain
N. Angelescu	Bucharest	Romania
B. Baumgartner	Wien	Austria
U. Behn	Leipzig	Germany
V.P. Belavkin	Nottingham	United Kingdom
K. Berndl	München	Germany
L. Bertini	Rome	Italy
L.V. Bogachev	Moscow	Russia
D. Bollé	Leuven	Belgium
S. Boraç	Berlin	Germany
C. Borgs	Berlin	Germany
K. Broderix	Göttingen	Germany
M. Broidioi	Leuven	Belgium
G. Casati	Como	Italy
F. Cerulus	Leuven	Belgium
J.T. Chayes	Los Angeles	USA
P. Collet	Paris	France
M. Courbage	Paris	France
M. Daumer	München	Germany
H. de Jong	Groningen	The Netherlands
A. DeMasi	L'Aquila	Italy
B. Derrida	Saclay	France
T.C. Dorlas	Swansea	United Kingdom
F. Dunlop	Paris	France
D.E. Evans	Swansea	United Kingdom
A. Evans	Cambridge	United Kingdom
M. Fannes	Leuven	Belgium
P.A. Ferrari	São Paulo	Brazil
A. Fledderjohann	Wuppertal	Germany
J. Fritz	Budapest	Hungary
Y.V. Fyodorov	Rehovot	Israel
G. Gallavotti	Rome	Italy

W. Gans	Berlin	Germany
G. Giacomin	Rutgers	USA
R. Gielerak	Wrocław	Poland
G. Gielis	Leuven	Belgium
M. Goldstein	Beer-Sheva	Israel
V. Gorunovich	Kiev	Ukraine
J. Gough	Maynooth	Ireland
J. Groeneveld	Utrecht	The Netherlands
J. Gruneberg	Dortmund	Germany
W. Haidegger	Wien	Austria
Z. Hasiewicz	Wrocław	Poland
P. Holický	Prague	Czech Republic
M. Hübner	München	Germany
J.O. Indekeu	Leuven	Belgium
V.I. Inozemtsev	Dubna	Russia
J. Jacobs	Leuven	Belgium
T. Kennedy	Tucson	USA
A. Klein	Irvine	USA
O. Knill	Zürich	Switzerland
T. Koma	Tokyo	Japan
B. Kümmerer	Tübingen	Germany
P. Kurasov	Stockholm	Sweden
V. Kurasov	Berlin	Germany
L.J. Landau	London	United Kingdom
J.T. Lewis	Dublin	Ireland
D. Li	Trieste	Italy
J. Lőrinczi	Groningen	The Netherlands
S.W. Lovesey	Oxon	United Kingdom
J.D.M. Maassen	Nijmegen	The Netherlands
J. Mackowiak	Toruń	Poland
C. Maes	Leuven	Belgium
W.A. Majewski	Gdańsk	Poland
D. Makowiec	Gdańsk	Poland
P.V. Malyshev	Kiev	Ukraine
P.A. Martin	Lausanne	Switzerland
T. Matsui	Tokyo	Japan
R. Mc Cann	Princeton	USA
A. Messager	Marseille	France
J. Miękisz	Leuven	Belgium
B. Momont	Leuven	Belgium
B. Nachtergaele	Princeton	USA
H. Narnhofer	Wien	Austria
J. Naudts	Antwerp	Belgium
A. Patrick	Leuven	Belgium
O. Penrose	Edingburgh	United Kingdom
C.E. Pfister	Lausanne	Switzerland
N. Poschadel	Jena	Germany
E. Presutti	Rome	Italy
J.V. Pulé	Dublin	Ireland
E. Rajczyk	Antwerp	Belgium

E. Raschhofer	Wien	Austria
A.L. Rebenko	Kiev	Ukraine
F. Redig	Antwerp	Belgium
S. Richter	Osnabrück	Germany
J. Rogiers	Leuven	Belgium
H. Roos	Göttingen	Germany
B. Rüdiger	Rome	Italy
J. Ruiz	Marseille	France
R.N. Sen	Beer–Sheva	Israel
G.L. Sewell	London	United Kingdom
S.B. Shlosman	Moscow	Russia
P. Siemion	Wrocław	Poland
W.I. Skrypnik	Kiev	Ukraine
J.P. Solovej	Princeton	USA
E.R. Speer	Rutgers	USA
H. Spohn	München	Germany
U.M. Studer	Leuven	Belgium
H. Tasaki	Tokyo	Japan
W. Troost	Leuven	Belgium
P. Tuyls	Leuven	Belgium
A.G. Ushveridze	Clausthal	Germany
M. Van Canneyt	Leuven	Belgium
F. Vanderseypen	Leuven	Belgium
K. Vande Velde	Leuven	Belgium
A. van Elst	Bonn	Germany
A.C.D. van Enter	Groningen	The Netherlands
S. Van Gulck	Antwerp	Belgium
J.H.M. van Leeuwen	Nijmegen	The Netherlands
A. Verbeure	Leuven	Belgium
R.F. Werner	Osnabrück	Germany
S.L. Woronowicz	Warsaw	Poland
H.T. Yau	New-York	USA
V.A. Zagrebnov	Marseille	France
M. Zahradnic	Prague	Czech Republic
A. Ziermann	Mons	Belgium

INDEX

Adiabatic theorem, 139–148, 151–154
Anharmonic oscillators, 203, 210
ANNI model, 459
Anomaly, 165
    chiral gauge, 226
Approximation
    adiabatic, 141
    Hartree, 393–394
    Hartree-Fock, 24
    Thomas-Fermi, 1

Band structure, 259–264
BBGKY hierarchy, 8, 19, 315–316
Beltrami-Laplace operator, 216, 246
Bethe Ansatz, 245–248, 385–392, 409–414
    algebraic, 387–389, 411–414
Bloch electrons, 420
Bohmian mechanics, 331–338, 429–434
Boltzmann
    equation, 5–8, 53, 64–65
    H-Theorem, 6–7
Boltzmann-Grad limit, 5
Boson field, 176, 361, 417–421
Brownian motion, 105, 109, 111, 441–444
Burgers equation, 44, 99, 265–269

Canonical commutation relations, 176, 361, 417–421
Cellular automata, 212–214
    probabilistic, 435–437
Central limit, 9, 176, 456
Chaos
    classical, 281
    quantum, 281–286, 289
Chern-Simons model, 405–407
Cluster integral, 425–428
Coarse graining, 241
Coherent state, 215–218, 221, 283
Cole-Hopf transformation, 266
Completely positive map, 115
Contact process, 435–440
Continuous media, 445–449
Coulomb system, 393–395
Critical exponent, 176
Critical fluctuations, 271

Crystal, 259–264
Current algebra, 226–227

Detailed balance condition, 120
Differential geometry, non-commutative, 343
Diffusion constant, 43, 101, 282
Diffusive limit, 43–51, 454–456
Dimerization, 83, 375
Dirac
    states, 419
    operator, discrete random, 323
Disordered media, 441
Dissipation, 7, 118
Dobrushin's uniqueness theorem, 122
Dykhne formula, 146, 150
Dynamics
    Glauber, 271–274, 279
    Langevin, 304
    Monte-Carlo, 303, 307
    stochastic, 11–12, 105–113, 115–123, 155, 435–440
Dyson
    interaction, 157
    series, 55

Effective action, 226
Effective potential, 170
Electron filling factor, 26
Empirical average, 45, 278
Entrance time, 73–74, 76
Entropy, 191, 204–213, 355
    relative, 47, 355, 378
Equation
    Boltzmann, 8, 53, 64–65
    Burgers, 44, 99, 265–269
    Euler-Maxwell, 17–20
    kinetic, 1–10
    Lifshitz, 305–306
    Liouville, 12, 20, 203
    Navier-Stokes, 45
    stochastic differential, 106, 111, 451–456
    Toda, 324
    transport, 1
    Ühling-Uhlenbeck, 6–8
    Vlasov, 7–9, 15–18

Equation (cont'd)
   Yang-Baxter, 250
   Young, 306
Equilibrium
   extremal state, 177
   local, 44–46, 304, 318–319, 445–449
   quantum, 7
   state, 188
Equivalence of ensembles, 94, 183, 192, 235, 295
Ergodicity, 73–75, 117–118, 203, 436
   constructive criteria, 435–440
Escape time, 331–338
Escape position, 331–338
Euclidian Green's functions, 339–345
Euler scale, 44–46
Euler-Maxwell equations, 17–20
Eulerian hydrodynamics, 12, 315–316
Exotic statistics, 363
Expanding maps, 73, 76
Expansion
   high temperature, 254–257
   low activity, 254–257

Falicov-Kimball model, 35, 458–459
Fermi
   bubble, 56–57, 63–64, 68
   gas, 3, 53, 56, 60, 65, 165
   lines, 55–56
   liquid, 166
   surface, 165
Fermion density fluctuation, 180
Ferromagnetism, itinerant electron, 23
Feynman diagram, 53–56, 62–64
Feynman-Kac formula, 115, 267, 460
Fluctuation, 4, 9, 176, 265, 272–274, 292, 299, 305, 455, 459
   critical, 271
   displacement operator, 177
   momentum operator, 177
   observables, 178, 309
   quantum, 175–176, 180, 309–310, 459
Formula
   Dykhne, 146, 150
   Feynman-Kac, 115, 267, 460
   Green-Kubo, 45, 49
   Landau-Zener, 145–146, 149–153
   Levy-Kintchin, 111
Fortuin-Kasteleyn representation, 86
Fourier's law, 11
Fractal, 286, 403
   dimension, 402
   multi-fractal, 286, 399–404
Functional integral, 217, 339–345

Geometrical phase, 153
Gibbs
   Ising-Gibbs measure, 157, 373–380
   measures, 156, 210, 295, 304, 307, 345, 373, 405
   non-Gibbs measures, 91, 96, 155–158,

Gibbs (cont'd)
   373–380
   representation of kernel, 81
Glauber evolution, 271–274, 279
Green-Kubo formula, 45, 49
Griffiths' singularity, 253
Ground state, 115, 161–162, 195, 241, 461

Hamiltonian systems, 203
Hard spheres model, 381–384
Harper model, 286
Hartree approximation, 393–394
Hartree-Fock approximation, 24
Heat bath, 103, 108
Heisenberg model, 240, 458–459
Hubbard model, 23, 35, 457–465
Hydrodynamic limit, 2, 11–21, 275–279, 315–316

Incompressible quantum fluid, 225–232
Index theorem, 217, 220
Infrared bounds, 239–242
Integrable models, 248–250, 259–264, 385–392, 423
Integral lattices, 227–231
Interfaces, 275–279
Irreversibility, thermodynamic, 446
Ising-Gibbs measure, 157
Ising model, 253, 347, 373, 399
   random field Ising model, 399
Isospectral deformations, 321–330

Kac potentials, 271–274
Kicked rotator, 282
Kinetic
   equations, 1–10
   limit, 315
KMS condition, 122, 211
Kohn anomaly, 176, 179–180
Kähler manifolds, 216

Landau level, 218, 221
Landau-Zener formula, 145–146, 149–153
Langevin dynamics, 304
Laplacian, discrete random, 321–330
Large deviations, 183, 188, 299, 347–354, 410
Lattice gas model, 161, 185, 190–191
Law of large numbers, 4
Lebowitz-Penrose limit, 271
Levy-Khinchin formula, 111
Lifshitz equation, 305–306
Limit law, 73
Limit
   Boltzmann-Grad, 5
   central, 176, 456
   diffusive, 43, 454–456
   high/low temperature, 423–428
   hydrodynamic, 2, 11–21, 275–279, 315–316
   kinetic, 315
   Lebowitz-Penrose, 271

Limit (cont'd)
  low density, 2–3, 5
  macroscopic, 213, 454–456
  mean field, 405, 393–397
  quantum central, 9, 455
  quantum large number, 454–455
  scaling, 2, 226, 401
  semi-classical, 147, 149, 151
Liouville equation, 12, 20, 203
Local equilibrium, 44–46, 304, 316–318, 445–449
Log Sobolev inequality, 46–47
Long range order, 239–242
Laughlin functions, 364
Low density limit, 2–3, 5
Luttinger model, 167, 172–173
Lyapunov exponent, 282

Macroscopic limit, 213, 454–456
Markov
  almost, 374
  global property, 158
  process, 115, 207–210, 438
Mean field limit, 393–397, 405
Mean field model, 2–3, 7, 309
Measure
  empirical, 45
  Gibbs, 156, 210, 295, 304, 307, 373, 405
  Ising-Gibbs, 157, 373–380
  non-Gibbs, 91, 96, ,155–158, 373–380
  operator-valued, positive, 335–337
  Wiener, 216–217, 221, 395, 441–444
Measurement problem, quantum, 331–338, 430
Mesoscopic scale, 99, 155, 225, 334
Mixing property, 74
Model
  ANNI, 459
  Chern-Simons, 405–407
  Falicov-Kimball, 35, 458–459
  hard spheres, 381–384
  Harper, 286
  Heisenberg, 240, 458–459
  Hubbard, 23, 35, 457–465
  integrable, 248–250, 259–264, 385–392, 423
  Ising, 253, 347, 373, 399
  lattice gas, 161, 185, 190–191
  Luttinger, 167, 172–173
  mean field, 2–3, 7
  non-linear Schrödinger, 409–415
  Overhauser, 309–313
  Potts, 84
  Schrödinger-Poisson, 393
  Sherrington-Kirkpatrick, 128–129
  solid on solid, 295, 299, 303–306
  spherical, 347–354
  Spin-Boson, 367–370
  velocity, 316
  voter, 274
  Yang-Yang, 410–411

Modular operator, 358
Momentum fluctuation operator, 177
Monte-Carlo dynamics, 303, 307
Monodromy matrix, 260, 387
Morse index, generalized, 327
Mott insulator, 29
Mourre's virial theorem, 369

N-level atom, 9
Navier-Stokes equation, 45
Néel order, 82, 239, 462
Neural networks, 399
Newtonian mean field, 15–16
Noise
  coloured, 109
  white, 105, 109
Nuclearity, 356–358

Occurrence time, 75
Order parameter, 239–242
Ornstein-Uhlenbeck process, 9
Overhauser model, 309–313

Parisi broken symmetry, 129
Path integral, 215
Peierls-Fröhlich instability, 179, 464
Phase transition, 11, 17, 19, 305
  hydrodynamical, 20
  long range order, 239–242
  order parameter, 239–242
  structural, 176–178
  symmetry breaking, 239–242, 312
Phonon, 176
Plasma, 11–14, 21
Poincaré recurrence, 73
Poisson process, 4, 76, 81, 85, 111, 217, 453–456
Potts model, 84
Principle of the largest term, 186
Process
  contact, 435–440
  with independent increments, 216
  Markov, 115, 209, 438
  Ornstein-Uhlenbeck, 9
  Poisson, 4, 76, 81, 85, 111, 217, 453–456
  reaction-diffusion, quantum, 451–456
  quantum stochastic, 104–105
  simple asymmetric exclusion, 43
  simple exclusion, 43
  two species simple exclusion, 91–92, 94
  Wiener, 456

Q-Independence, 361–362
Quantum anharmonic crystal, 176
Quantum central limit theorem, 9, 455
Quantum chaos, 281–286, 289
Quantum effects, 177
Quantum equilibrium, 7
Quantum fluctuation, 175–176, 180, 239
Quantum frustration, 193
Quantum group symmetry, 201

Quantum Hall effect, 215–217, 221, 225–232, 471
Quantum large number, 454–455
Quantum open system, 103–106
Quasi particle, 169–171

Random field, 65
　Gaussian, 277
Random operators, discrete
　Jacobi, 322
　Harper, 322–329
　Laplace, 321–330
Random map, 125, 131–133
Random matrices, 289
　Gaussian ensemble, 289
Random walk, 125, 134, 233–237
　intersection set, 233–237
　loop condensation, 233–237
Rate function, 206–209, 273, 350
Reaction-diffusion process
　Kawasaki-Glauber, 275–279
　quantum, 451–456
Recurrence times, 73
Reflection positivity, 341
Relative entropy, 47, 378
Renormalization group, 155, 169, 375
　block-spin renormalization, 156
　running couplings, 172
Replica theory, 125, 129
Reproducing kernel, 216, 221
Reversible state, 119, 209
Robinson tiles, 161

Scaling limit, 2, 226, 401
Scattering, 259–264, 331–338, 423–428, 451–456
　inelastic, 149
　matrix, 260, 425
Schrödinger model, non-linear, 409–415
Schrödinger-Poisson model, 393
Segregation principle, 36
Self-averaging property, non, 125–127, 132
Semi-classical limit, 147–151
Sensitive dependence, 75
Sherrington-Kirkpatrick model, 128–129
Shock, 91, 99
Simple exclusion process, 43
　asymmetric, 43
　two species, 91–92, 94
Solid on solid model, 295, 299, 303–306
Spectral gap, 200, 225, 259
Spherical model, 347–354
Spin chains, 245
Spin glass, 125–128, 254
Spin density wave, 309–313
Spin-Boson model, 367–370
Split property, 355–358
Squeezing, 176, 178
Stability, 162

State
　coherent, 215–216, 218, 221, 283
　Dirac, 419
　equilibrium, 188
　exposed, 194–195
　extremal equilibrium, 177
　finitely correlated, 198
　ground, 115, 161–162, 195, 239, 461
　invariant, 93, 117, 400
　non-periodic ground, 161–163
　pure, 199
　quantum equilibrium, 7
　reversible, 119, 209
　stationary, 110, 203
　valence bond solid, 196
Stochastic differential equation, 106, 111, 265–269, 271, 451–456
Stochastic dynamics, 11–12, 105–113, 115–123, 155, 435–440
Stochastic process
　with independent increments, 216
　quantum, 104–105
Symmetry breaking, 239–242, 312
Symplectic structure, 217, 220

Theorem
　adiabatic, 139–141, 151–153
　Boltzmann H-theorem, 6–7
　central limit, 176
　Dobrushin's uniqueness, 122
　index, 217, 220
　Mourre's virial theorem, 369
　quantum central limit, 9
Thermodynamic function, 183, 192
Thomas-Fermi theory, 1
Time operators, 331–338
Toda equation, 324
Topological electrodynamics, 405–407
Tomita structure, 376
Transfer matrix, 197, 291, 385
Transport equations, 1–8, 265–269
Two-level system, 108, 144, 150–151

Ühling-Uhlenbeck equation, 6–8
Uncertainty relation, 178, 283
Universality, 155

Velocity model, 316
Virial coefficient, 381–384, 423–428
Vlasov equation, 7–9, 15–18
Voter model, 274

Wannier wavefunction, 421
Weak coupling limit, 2–6, 53–54, 62–65
Weyl algebra, 176, 361, 417–421
　non regular representation, 417–421
Wiener
　measure, 216–217, 221, 395, 441–444
　process, 456

Wigner distribution, 3, 7–8
Wulff construction, 295–299

Yang-Baxter equation, 250, 385–392

Yang-Yang model, 410–411
Young's equation, 306

Zero-range potential, 423–427